X-RAY STRUCTURE DETERMINATION

X-RAY STRUCTURE DETERMINATION

A PRACTICAL GUIDE

Second Edition

GEORGE H. STOUT

LYLE H. JENSEN

Department of Biological Structure
University of Washington
Seattle, Washington

A Wiley-Interscience Publication

JOHN WILEY & SONS

New York • Chichester • Brisbane • Toronto • Singapore

Library of Congress Cataloging in Publication Data:

Stout, George H., 1932–
X-ray structure determination.

"A Wiley-Interscience publication."
Includes bibliographical references and index.
1. X-ray crystallography. I. Jensen, Lyle H.
II. Title.
QD945.S8 1989 548'.83 88-27931
ISBN 0-471-60711-8

Printed in the United States of America

10 9 8 7 6 5 4 3 2

To Kit and Mildred

PREFACE

The first edition of this book arose from our belief that X-ray crystallography was a powerful tool which could be applied by scientists whose main interests lay in other fields. This view has been entirely supported by the events since the late 1960s, and has been so widely accepted that there is now a risk that the field of X-ray crystallography of small molecules will be or has been relegated to a minor, service role. It is often felt, or so it seems, that all of the necessary knowledge is contained in various computer programs, and that these will serve by themselves to carry out an analysis to a uniquely correct result. We do not hold to this view and the main thrust of this edition is not merely that one can do the work oneself, but that it is necessary to understand what one is doing in order to be able to have confidence in the results.

The contents and organization of this book come very much out of our own experience and reflect our views of what is important and useful to the practical worker. Much of the material has been rearranged and rewritten from the first edition to reflect changes in the relative importance of various topics and the innovations in the field over the last twenty years. We believe, perhaps more strongly than ever, that "there appears to exist in crystallography, perhaps more than in most fields, a body of practical knowledge widely disseminated among members of the group but nowhere available in print to outsiders." We have continued to try to include as much of this material as possible.

We continue to offer our thanks to our colleagues who helped us with the first edition, and add our gratitude to all those, colleagues and students, who have contributed to our thoughts about crystallography in the intervening years. As before, our thanks go to our wives for continued understanding and support in what has turned out to be a *very* much longer undertaking than we originally imagined.

Seattle, Washington
March 1989

GEORGE H. STOUT
LYLE H. JENSEN

CONTENTS

X-RAY STRUCTURE DETERMINATION

INTRODUCTION

> The recent explosive growth in the development and distribution of electronic computers has reduced the labor associated with the determination of crystal structures by X-ray diffraction analysis immensely. As a result such analysis may now be used routinely to supply answers to problems arising in many unrelated fields. In the past, the common mode of attack has been for a worker, faced with a problem for which X-ray methods seemed appropriate, to attempt to find a crystallographer who would be willing to undertake the determination. As a result of a growing awareness of the capabilities of the method, however, the number of problems has been increasing at a greater rate than the number of crystallographers.
>
> It is the thesis of this book that the average chemist who is willing to devote some time to study, and who can obtain a suitable set of programs, can learn without great difficulty to perform crystallographic structural analyses for himself. The mathematical requirements are much less formidable than are widely supposed, the techniques of data collection are no more complicated than many other physico-chemical measurements, and the tedium of computation has been very largely removed by computers. Furthermore, no errors, other than the loss of a unique crystal, are irredeemable; one can always return to the point of error and find a new path.[1]

This introduction to the first edition of this book has been entirely justified by the results of the last twenty years. So much so, in fact, that we now feel the need to counterbalance the swing and to provide throughout this volume cautionary statements and evidence that the process of crystal

[1] G. H. Stout and L. H. Jensen, *X-Ray Structure Determination*, Macmillan, New York, 1968, p. 1.

structure determination is *not* as simple as is often supposed. The great advances in computer technology, the advent of automatic data collection systems, and the progress in dealing with the phase problem have all combined to relegate the crystallography of small molecules to a service function, almost comparable to infrared or NMR spectroscopy as yet another tool in the chemist's workbox. To some extent this view is justified, and in many cases structures can, indeed, be solved in a quick and routine fashion. Nevertheless, this process is also being accompanied by an increasing number of published errors and an unknown but significant number of structures which fail to solve by the routine methods but which prove amenable to more thoughtful attack. We hope to provide the readers of this much revised edition with the tools and insight that will enable them to make use of the modern techniques with understanding and judgment.

As before, the pattern of this book follows closely the development of an actual structural determination. After some introductory material on the nature of X-rays, the diffraction process, and the internal geometry of crystals, the selection and preparation of a crystal are considered. The techniques involved in the measurement of raw X-ray data are next taken up, followed by a discussion of the reduction of these data into the form in which they are usually used.

The second part of the book is devoted largely to a discussion of various methods of solving the "phase problem," the principal difficulty on the path from the raw data to the final answer. The past two decades have seen great advances in dealing with this problem, but it is still not a trivial barrier to the successful completion of a structural study. A number of methods are considered to provide the reader with insight into both common and uncommon approaches. Chapter 15 describes the process of completing the structure once the phase problem has been overcome.

The third part considers the processes of refinement by which an attempt is made to obtain the maximum information from the observed data and to evaluate its reliability. Finally, there is a discussion of some of the sources of errors in practice and interpretation, and of further knowledge which can be derived from the basic structure once it is known.

It is assumed without further comment that the reader has access to a modern computer.[2] More than any other factor, it has been the growth and widespread distribution of computers that have turned X-ray crystallography into a commonplace structural tool. Computers require programs, however, and obtaining, understanding, and using these programs often pose a serious problem. The material given here will help supply the background that is needed, but while the general characteristics of the more

[2]For discussion of hand calculations, which form a powerful educational tool, see S. C. Nyburg, *X-Ray Analysis of Organic Structure*, Academic Press, New York, 1961, pp. 77–84, 96–111; M. J. Buerger, *Crystal–Structure Analysis*, Wiley, New York, 1960, pp. 274–279 and references cited there; and H. Lipson and W. Cochran, *The Determination of Crystal Structures*, Bell, London, 1957, pp. 54–65, 76–97.

common sorts of programs are discussed, the wide variation in individual specimens prevents any detailed consideration.

No attempt is made here to cover the general problem of computer use. It is not necessary that the user of a computer know what is happening inside the machine, but some understanding of scientific programming and one of the languages used (commonly FORTRAN, much less often BASIC or C) is very valuable. Operating instructions are often written in the jargon of the programming languages, and unfamiliarity can lead to total frustration. Furthermore, it is often necessary to make changes in preexisting programs or to write small ones for special purposes, especially if one has any wish to progress beyond the "black-box" stage of operation.

The problem of obtaining programs is a real but not insuperable one. Most of the common operations have been programmed, generally repeatedly, for all of the ordinary varieties of computers. Crystallographers are usually very generous in regard to supplying copies within the limitations of their time, so the major difficulty is tracking down what is available. Probably the best single source for new material is the descriptions and abstracts of programs that appear in *Journal of Applied Crystallography*.

When selecting programs, the value of obtaining ones with good "write-ups" can hardly be overstressed. Furthermore, as is discussed in Chapter 7, it is advantageous to obtain the standard core of programs from a single source, so that they are matched and the output of one can serve directly as the input for another. A number of crystallographic "systems," collections of programs linked into a common unit, are available, some provided with diffractometers and others as stand-alone packages for computers of all sizes including PC-class machines.[3]

Since the structural interests of the authors of this work lie in the field of organic compounds, this volume shows a decided bias toward the problems likely to be encountered in the analysis of organic and organometallic structures. This is not to say that much of the material is not more generally applicable, but where there is a choice the discussion is restricted to crystals of low symmetry. This approach is not as unfair as might at first seem, since there are already available a number of works by crystallographers of more classical or mineralogical bent, which deal with the problems more characteristic of purely inorganic systems.

The orientation of this book is heavily toward the practical aspects of structure determination. Some theoretical material has been included where it is needed for intelligent application of the techniques described, but there has been no effort to justify all the statements made. Interested readers can find fuller details in the references given in the bibliographies following the various chapters.

[3]For an extensive survey and listing, see K. Huml in *Crystallographic Computing 3: Data Collection, Structure Determination, Proteins, and Data Bases*, G. H. Sheldrick, C. Krüger, and R. Goddard, Eds., Clarendon Press, Oxford, 1985, pp. 131–145. For a detailed description of one such system see S. R. Hall, J. M. Stewart, and R. J. Munn, *Acta Cryst.*, **A36**, 979 (1980).

Similarly, the choice of literature to be cited has been selective rather than exhaustive. An attempt has been made to provide references that would be illuminating to students with relatively limited crystallographic backgrounds. Little attention has been paid to questions of priority, and much use has been made of the secondary literature.[4] Where a secondary reference has been used repeatedly in a chapter, it appears in the footnotes as a short form and the full citation is given at the end of the chapter.

Many new references have been supplied for this revision, but many old ones have been retained as being more suited for introductory purposes. Unfortunately, many of the classic texts are out of print and have not been adequately replaced, so they need to be searched out in libraries.

Although crystal structures now appear in many journals, particularly those dealing with particular classes of compounds, the bulk of important papers on techniques appear in *Acta Crystallographia* (*Acta Cryst.*), published by the International Union of Crystallography (IUCr). The IUCr has also published four volumes that contain an immense amount of practical information used constantly by crystallographers. We shall cite these works in the shortened form ***International Tables**, Vol. I–IV, A* corresponding to the full references given in Table A.

We suggest that the novice intending to carry out a structure determination read Chapters 1, 2, 10, and 19 to get a brief introduction to the problems ahead. The actual analysis can be begun with Chapter 4, and carried on in sequence, a chapter or two at a time. Chapters 2 and 3 are vital and will probably require several rereadings at various stages.

TABLE A Full Citations for *International Tables* Volumes

Vol. I	*International Tables for X-Ray Crystallography*, Vol. I, N. F. M. Henry and K. Lonsdale, Eds., Kynoch Press, Birmingham, England, 1952.
Vol. II	*International Tables for X-Ray Crystallography*, Vol. II, J. S. Kaspar and K. Lonsdale, Eds., Kynoch Press, Birmingham, England, 1959.
Vol. III	*International Tables for X-Ray Crystallography*, Vol. III, C. H. Macgillavry and G. D. Rieck, Eds., Kynoch Press, Birmingham, England, 1962.
Vol. IV	*International Tables for X-Ray Crystallography*, Vol. IV, J. A. Ibers and W. C. Hamilton, Eds., Kynoch Press, Birmingham, England, 1974.
Vol. A	*International Tables for Crystallography*, Vol. A, T. Hahn, Ed., D. Riedel, Dordrecht, Netherlands, 1983.

[4]Excellent lists of literature references prior to 1959 can be found for many of the topics covered here in M. J. Buerger, *Crystal–Structure Analysis*, Wiley, New York, 1960.

PART I

PRELIMINARY STAGES

CHAPTER 1

X-RAYS

Although most crystallographers are only indirectly concerned with X-rays themselves, an understanding of the origins and properties of X-rays is necessary for their proper use in crystallographic studies. The information contained in this chapter provides a background for the application of X-ray methods to structural problems.

1.1. ORIGIN[1]

X-rays lie in the electromagnetic spectrum between ultraviolet light and gamma radiation and have an approximate range of wavelengths of 0.1–100 Å.[2] They are usually produced by rapidly decelerating fast-moving electrons and converting their energy of motion into a quantum of radiation. The wavelengths produced will depend on the energy of the electrons; we shall be concerned with X-rays having a wavelength of about 1 Å. X-rays are also emitted by certain radioactive isotopes, for example, ^{55}Fe. Although such sources are convenient for testing and calibration, they have not been used for diffraction purposes.

To generate X-rays, electrons are accelerated by an electric field and directed against a metal target, which slows them rapidly by multiple

[1] *International Tables*, Vol. III, pp. 71–78.

[2] Although the *Ångstrom unit* (1 Å $= 10^{-8}$ cm) is not a standard SI unit, angstroms are extremely useful for work in the range of molecular dimensions since bond lengths then have single-digit values. They are still more commonly used by crystallographers than the nearest SI equivalent, the nanometer (1 Å = 0.1 nm).

collisions. Under the usual conditions most of the electrons are not brought to a full stop by a single collision, and a continuum of radiation is formed (Fig. 1.1). The minimum wavelength of this *white radiation* is determined by the accelerating voltage and can be calculated from

$$\lambda_{\min} = \frac{12{,}398}{V_{\text{acc}}} \text{ Å} \qquad (1.1)$$

The greatest intensity occurs at a somewhat longer wavelength. As the voltage is increased, not only are the cutoff and peak intensity moved to shorter, more penetrating wavelengths, but also the total intensity increases even though the electron current remains the same.

While the distribution of intensity in the white radiation depends primarily on the accelerating voltage and only to a small extent on the nature of the target material, X-ray spectra show in addition a number of sharp spikes of high intensity whose positions change from one material to another (Fig. 1.2). These peaks are the ***characteristic lines*** for the element of which the target is made.[3] When the electrons bombarding the target reach certain critical energies (***threshold potentials***) they are capable of knocking electrons out of their atomic orbitals. In particular, at energies of about 10,000 eV (for elements with atomic number ~30) they can remove electrons from the innermost (K) shell. The vacancy in the K shell is then filled by the descent of an electron from the next higher shell (L) or the one above that (M). The decrease in potential energy in going from the higher level to the lower appears as radiation, and as the energies of the shells are

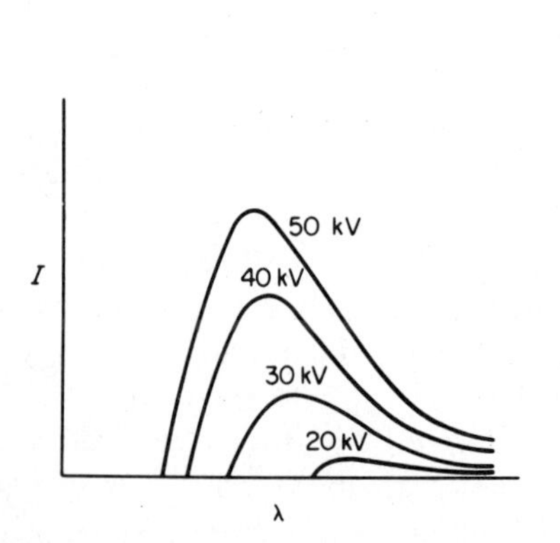

Figure 1.1. Continuous X-ray spectra as a function of accelerating voltage.

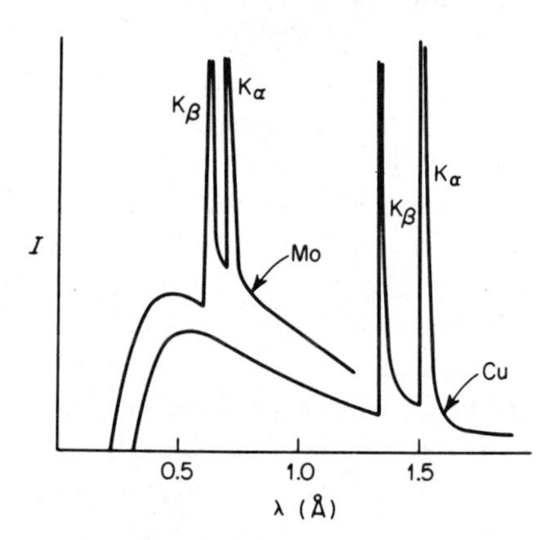

Figure 1.2. X-ray spectra with characteristic peaks: Mo K_α, 50 kV; Cu K_α, 35 kV.

[3]For a complete list of wavelengths, see *International Tables*, Vol. IV, pp. 5–43.

well defined, each transition gives a nearly monochromatic line. The principal peaks are

$$K_{\alpha1}, K_{\alpha2} \qquad L \rightarrow K$$

$$K_{\beta1}, K_{\beta2} \qquad M \rightarrow K$$

Because the difference in energy between L and K is less than that between M and K, K_α is always at a longer wavelength than K_β. The lines are close doublets because transitions can occur from two possible electronic configurations, which differ slightly in energy. $K_{\alpha1}$ is twice as intense as $K_{\alpha2}$ and about three to six times as strong as $K_{\beta1}$. $K_{\beta2}$ is usually so weak that it is ignored.

As the atomic number (Z) of the target element increases, the characteristic lines shift to shorter wavelengths, and one can, in principle, select a target to give almost any desired value for the K_α line. In practice, however, one is limited to materials that are conductive, solid, dense, and high-melting, that is, to metals. Fortunately, the transition elements of the first and second long periods ($Z = 21$–30 and 39–48) meet these requirements and have characteristic radiation in the region that is most useful for crystal structure analysis.

1.2. ABSORPTION AND FILTERING[4]

For reasons that will become clear later, the radiation used for most diffraction work should be as nearly monochromatic as possible. The K_α lines fulfill this requirement, but the presence of the accompanying K_β is a nuisance. Fortunately, selective filters may be found that will remove the K_β to any desired extent, with a relatively much smaller loss of K_α.

The absorption of X-rays by a solid follows, like so many other absorption phenomena, the equation

$$I/I_0 = e^{-\mu\tau} \tag{1.2}$$

where μ is the linear absorption coefficient and τ is the path length through the solid. If one plots μ against λ for a given element, one finds a set of curves of the shape $\mu = k\lambda^3$ that are connected by sharp jumps (Fig. 1.3). These discontinuities are *absorption edges* and occur at the wavelength at which the incident X-ray quantum is just energetic enough to knock an electron out of an atomic orbital. In particular, the K absorption edge of an element lies slightly to the short-wavelength side of the K_β lines for that element. Like the characteristic lines, the absorption edge shifts to longer wavelengths with decreasing Z, and the magnitude of this shift is such that

[4] *International Tables*, Vol. III, pp. 71–78.

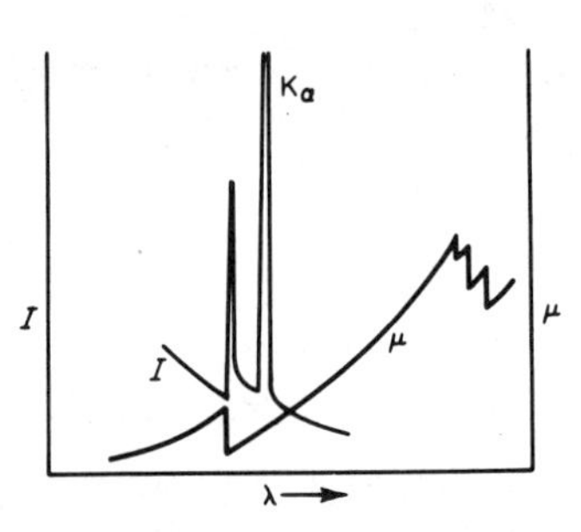

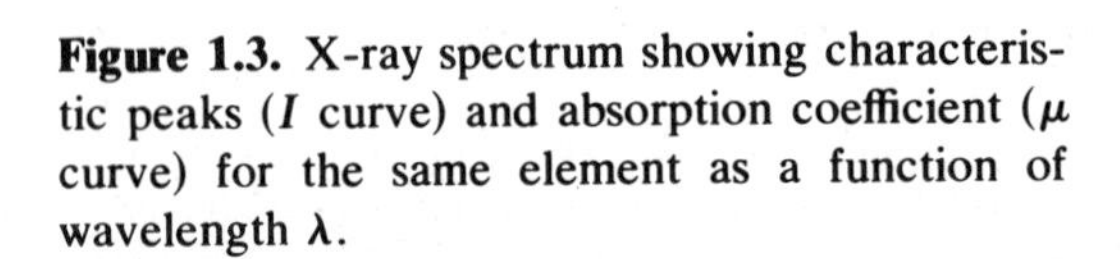

Figure 1.3. X-ray spectrum showing characteristic peaks (I curve) and absorption coefficient (μ curve) for the same element as a function of wavelength λ.

the edge for an element of atomic number $Z-1$ falls between the K_α and K_β lines of the element with an atomic number one higher, Z.[5] Consequently an element can be used as a selective filter to remove the K_β line from the spectrum of the next higher element. Figure 1.4 shows the effect of niobium ($Z=41$) as a filter for molybdenum ($Z=42$) radiation.

For targets chosen from the second long period, $Z-1$ and $Z-2$ are both possible K_β filters. Either is acceptable, and in practice considerations of availability and convenient physical form will decide. By going one or two atomic numbers lower, it is possible to find filters that will remove the K_α radiation as well, a technique (the ***balanced filter method***[6]) that is occasionally used for obtaining accurate measurements of the effects of background. Table 1.1 gives filters suitable for both purposes as well as the thickness of filter needed to reduce K_β to $0.01K_\alpha$.

An alternative method of obtaining monochromatic X-rays is by the use of a crystal monochromator.[7] In this case the direct beam from the X-ray tube is reflected (diffracted) from a large single crystal of a suitable material before being used. Because of the nature of X-ray diffraction (see Chapter

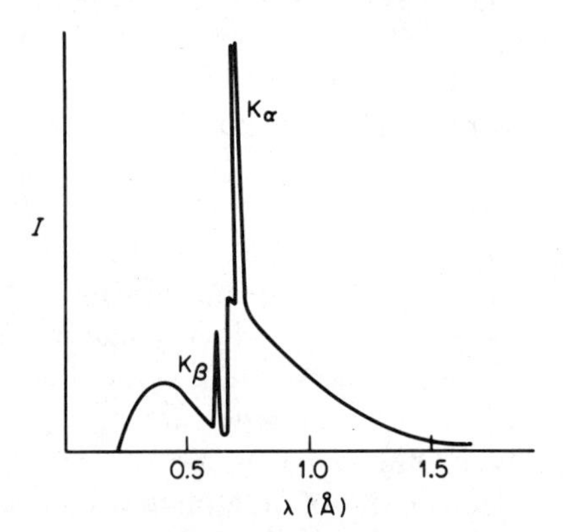

Figure 1.4. X-ray spectra showing effects of a β filter. Molybdenum radiation (50 kV) with 0.001-in. niobium β filter.

[5] This is actually true only for elements with $Z \leq 70$, but these include all the target materials whose characteristic radiations are commonly used.

[6] *International Tables*, Vol. III, pp. 78–79.

[7] *International Tables*, Vol. III, pp. 79–86.

2) only a very narrow band of wavelengths appears at a given crystal setting, and so the resulting beam is very nearly monochromatic. Although elegant in principle, this technique suffers from a number of disadvantages, among them the loss of intensity occurring during the process and certain other problems. Nevertheless crystal monochromators are now widely used in conjunction with diffractometers (see Section 5.10).

1.3. SELECTION OF RADIATION

The choice of radiation to be used for collecting diffraction data depends on a number of factors. Copper K_α was for many years the conventional choice for use with organic crystals and is still the best if the diffraction data are to be recorded photographically (see Chapters 5 and 6). It is sufficiently penetrating that it does not suffer too badly from absorption in the crystal or while passing through air, and at the same time it is recorded with reasonable efficiency on film. When used with a diffraction apparatus that can record all of the reflections theoretically accessible, it can provide enough data to more than adequately determine a structure.

The second common choice is Mo K_α. This is a more penetrating radiation and is less well adapted to use with film because of the low efficiency of recording. Nevertheless it is used to some extent in cases in which the apparatus is limited in its ability to record the available data (see the discussion of the precession camera, Section 5.9). It is generally the radiation of choice for use with diffractometers since these instruments (Section 5.10) use scintillation counters, which have nearly perfect counting efficiency for Mo K_α. Under these conditions, it is possible to use the harder radiation effectively and gain the advantage of its lesser absorption.

Wavelengths longer than that of Cu K_α do not usually offer any advantages except in the special case discussed below. If the crystal cell is very large, as in proteins, long-wavelength radiation will give better separation between the reflections but at the cost of losing some data. Absorption corrections are also increasingly a problem.

The elements present in the crystal may play a role in the choice of radiation. If one of them lies only a few atomic numbers below the target material, the X-rays will fall close to its K absorption edge. The consequences will be high specific absorption, causing a decrease in the accuracy of intensity measurements, and a large amount of diffusely scattered radiation (*X-ray fluorescence*) at the characteristic wavelength of the absorbing atom. This last is produced as the vacancy in the K shell of the absorber is filled by electrons from higher shells. It has the effect of increasing the background against which reflections must be measured. X-ray fluorescence will occur any time the incident radiation has a wavelength less than that of an absorption edge of the elements in the crystal, but it is less severe when the two are well separated. Elements

TABLE 1.1 Target Materials and Associated Constants

	Cr	Fe	Cu	Mo
Z	24	26	29	42
α_1 (Å)	2.2896	1.9360	1.5405	0.70926
α_2 (Å)	2.2935	1.9399	1.5443	0.71354
$\langle\alpha\rangle^a$ (Å)	2.2909	1.9373	1.5418	0.71069
β_1 (Å)	2.0848	1.7565	1.3922	0.63225
β filt	V, 0.4 mil[b]	Mn, 0.4 mil	Ni, 0.6 mil	Nb, 3 mils
α filt	Ti	Cr	Co	Y
Resolution (Å)	1.15	0.95	0.75	0.35
Critical potential (kV)	5.99	7.11	8.98	20.0

[a] $\langle\alpha\rangle$ is the intensity-weighted average of α_1 and α_2 and is the figure usually used for the wavelength when the two lines are not resolved.
[b] 1 mil = 0.001 in. = 0.025 mm.

below calcium cause little difficulty because the fluorescent radiation is so soft that it is absorbed by air.

Although specific absorption has the disadvantages described, it also produces some useful effects that may be lumped together as ***dispersion*** or ***anomalous scattering***. These can sometimes provide helpful information as part of a structural analysis, and the radiation may then be chosen to maximize rather than minimize them.

Table 1.1 lists some of the commonly used target materials and gives various associated constants. The resolution value is an indication of the minimum resolvable separation of two carbon atoms, for example, when viewed with the radiation in question. It is only a rough approximation, however, as the actual value depends on the properties of the crystal studied.

1.4. X-RAY TUBES: CONSTRUCTION AND GEOMETRY[8]

The basic parts of an X-ray tube are a source of electrons and a metal anode that emits the X-rays. Tubes can be classified according to the nature of these parts.

The common ***hot cathode tube*** is very much like a television tube in construction. Electrons are liberated from a heated filament and accelerated by a high voltage toward the target anode (Fig. 1.5*a*). The tube is constructed as a permanently sealed unit, costs about $3500 (1988), and has a working life about 5000–10,000 hours. The drawbacks are its tendency to become "dirty" and emit spurious radiation as it ages, and the restriction to a single anode material. The first phenomenon is due to the slow deposition

[8] Guinier, *X-Ray Crystallographic Technology*, Chapter 2.

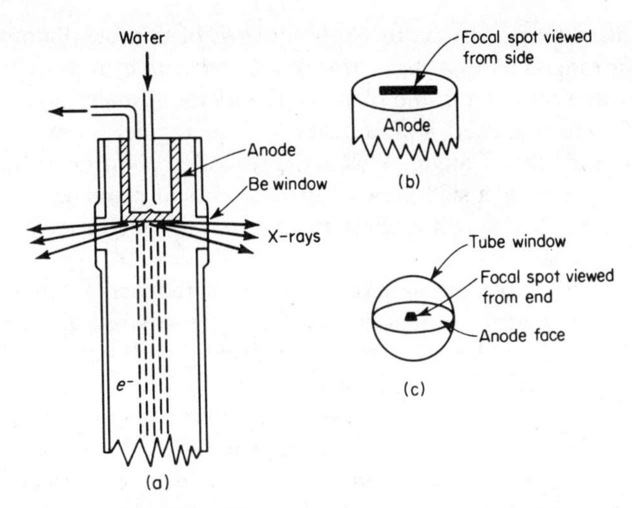

Figure 1.5. (*a*) Section along the axis of an X-ray tube. (*b*) Anode with focal spot viewed from side. (*c*) Focal spot viewed through tube window.

onto the anode and windows of tungsten and other elements (especially iron) evaporated from the filament assembly. In well-designed commercial tubes the problem is usually not a serious one. In attempts to escape both of these disadvantages, tubes have been built that could be demounted for cleaning or change of anode, but they are now rarely used.

Because of the high voltage through which the electrons are accelerated, the power dissipated at the anode is quite large (500–1500 W). In order to keep the anode from melting, it is made hollow and is cooled internally with circulating water. All commercial X-ray power supplies are provided with pressure- or flow-sensitive controls to shut down the electron beam if the cooling water flow falls below safe limits. Despite the cooling, however, the heat resistance of the anode still determines the maximum power of the tube. To increase the possible output and thus lessen the time required for obtaining data, ***rotary anode tubes*** are sometimes used.[9] Here the electron beam is directed onto the edge or face of a rotating wheel of the target material. With proper design, the origin of the X-rays remains in a fixed location but the anode moves under it and the heat generated is spread over a much larger area. Such tubes can handle at least ten times the power of those with stationary anodes. Unfortunately, they are expensive and complicated, requiring continuous evacuation and some extra maintenance, but they have now become relatively trouble-free and are widely used.

In a hot cathode tube, the beam of electrons arrives at the anode as an image of the filament, that is, a line $\sim$10 mm $\times$ 1 mm (Fig. 1.5*b*). The

[9] See, for example, A. Taylor, *J. Sci. Instr.*, **26**, 225 (1949); *Rev. Sci. Instr.*, **27**, 757 (1956); D. A. Davies, A. McL. Mathieson, and G. M. Stiff, *Rev. Sci. Instr.*, **30**, 488 (1959).

X-rays are taken from the tube through *windows* of thin beryllium foil. Two of these are arranged so that they view the X-ray source from the end at an angle only a little away from the plane of the anode (usually 1–6°); thus the line image is foreshortened and appears as a rectangle, 1 mm × 0.2–1 mm depending on the takeoff angle. This is the so-called *point focus* (Fig. 1.5*c*). A third window affords a side view of the source and provides a *line focus*, which is used for X-ray diffraction from powder specimens but not for single-crystal studies.

Microfocus tubes have the apparent focal spot reduced to about 0.1 mm on a side. These require particular care in alignment but are now widely used. They have particular advantages for dealing with very small crystals where most of the usual beam is wasted.

The operating conditions for most X-ray tubes lie in the ranges 30–60 kV and 10–35 mA of tube current, depending on the anode material. The working voltage is several times the threshold potential because both the relative and absolute intensity of the characteristic radiation increase with voltage.

1.5. SYNCHROTRON RADIATION[10]

A recent addition to the list of X-ray sources are the giant particle accelerators known as storage rings. In these, electrons or positrons are injected into a very large evacuated loop and kept circulating at relativistic velocities by energy pumped in from powerful radio-frequency sources. To constrain the circulating particles to the chamber, external magnets, often superconducting, provide an inward acceleration. A consequence of this acceleration is the emission of *synchrotron radiation*. This radiation is a broad-spectrum (white) emission that resembles the white radiation of an X-ray tube but is immensely more intense (100 to 10^4 times the intensity of a conventional characteristic line). It is obtained from the storage ring through *beam lines* that are tangent to the circulating particle current. Because the particles are tightly bunched in the chamber, the emission is not continuous but occurs as a series of extremely short (~1 ns) pulses separated by longer (but only relatively, a few hundred nanoseconds) gaps. For most purposes, this pulsation is not a problem.

The radiation wavelength is a function of the size (radius) of the storage ring and the energy at which it is operated. It is described in terms of the *critical wavelength* λ_c given by

$$\lambda_c = 5.59 R/E^3 \tag{1.3}$$

[10]R. Fourme and R. Kahn, in *Crystallography in Molecular Biology*, D. Moras, J. Drenth, B. Strandberg, D. Suck, and K. Wilson, Eds., Plenum, New York, 1987, pp. 27–44.

where R is the radius of curvature (in meters) in the magnetic field and E is the particle energy in gigaelectronvolts. This wavelength is approximately the most intense emission and is the midpoint of the distribution of radiated energy. The large rings have λ_c in the region of a few angstroms and contain in the mix of radiation a large amount in the 0.8–2.0 Å wavelengths of interest to crystallographers.

Rings particularly intended for radiation production contain additional devices (*wigglers* and *undulators*) to cause additional bending of the particle beam at locations that optimize the radiation generation. Wigglers cause sharper than usual bends in the particle current (and then a reverse bend so that there is no net deviation), leading to a shift of the radiation to shorter wavelengths and increasing the intensity in the range of interest to the crystallographer. Undulators create a series of small deviations such that the emissions produced at each turn are subject to interference with radiation from all the other turns. This results in an approximately monochromatic radiation source (or one in which the emission is concentrated in a few harmonically related lines) with vastly enhanced intensity (as much as 10^4) at the selected wavelengths.

The desired wavelength is usually selected by crystal monochromators (see above), often in connection with complicated focusing and reflecting arrangements to concentrate more of the available intensity on the sample. By suitable adjustments of the operating conditions almost any wavelength can be selected, and this ability to tune the radiation is potentially almost as important as the intensity gain.

The drawback to synchrotron sources is that they are extremely large and expensive installations, normally located at some distant place and much in demand, so the scheduling of data collection is complicated. Nonetheless there are cases for which synchrotrons are the only satisfactory tool, and their use is increasing, especially for the study of macromolecules.

1.6. SAFETY[11]

Like most scientific tools, X-ray equipment can be dangerous if mishandled. Although the radiation produced by the sources used is comparatively nonpenetrating, it presents a serious health hazard if proper precautions are not taken. The very fact of its "softness" is significant, since it means that the ionizing effect of the radiation is concentrated in a shallow region near

[11] *International Tables*, Vol. III, pp. 333–338; International Union of Crystallography, Commission on Crystallographic Apparatus, *Acta Cryst.*, **16**, 324, (1963); Radiation Safety for X-Ray Diffraction and Fluorescence Analysis Equipment, in *NBS Handbook*, Vol. 111, National Bureau of Standards, Boulder, CO, 1971; E. B. M. Martin, *A Guide to the Safe Use of X-Ray Diffraction and Spectrometry Equipment*, Science Reviews Ltd., Leeds, England, 1983; C. S. Barrett, J. V. Gilfrich, R. Jenkins, D. E. Leyden, J. C. Ross, and P. K. Predecki, Eds., *Advances in X-Ray Analysis*, Vol. 30, Plenum, New York, 1987.

the surface of the body. The direct beam is particularly dangerous, since it represents a concentrated radiation source (several hundred roentgens per second) and brief exposure can cause serious injuries. Any diffraction equipment must have a beam stop to receive the direct beam that passes through the sample. Whenever it is necessary to work without a beam stop, for example, when aligning cameras, the beam intensity should be reduced to the lowest possible value and the experimenter should take particular care to keep his or her hands clear of it. Shutters should always be provided for covering the exit windows of the tube and should be kept closed whenever diffraction is not actually being recorded.

Scattered radiation is another problem, especially around the junction between the tube and the collimator. This junction should be designed with a labyrinth type of coupling so that no radiation can escape. The mere fact that the window cannot be seen is not enough, since double scattering can occur. An excellent study of this problem and of shutter designs has been made by the Apparatus and Standards Committee of the American Crystallographic Association,[12] and it should be consulted by anyone setting up a crystallographic laboratory. Scattering can also occur from the edges of the beam stop, from the crystal and its mounting, and from any other objects that intrude into the direct beam. In order to eliminate any slight risk from scattered radiation, modern diffractometers are generally furnished with surrounding shields. (Because of the softness of the radiation, Plexiglas is quite adequate, so it is possible to watch the operations easily.) Camera installations often do not have these, and although it should not be a problem it is advisable to check occasionally with a survey counter (be sure that it will detect soft radiation!) or by placing film badges near the cameras for several weeks.

The question of legal responsibility in X-ray diffraction laboratories is uncertain but increasingly serious. Licensing and inspection requirements vary greatly from state to state and from country to country. Liability in case of any mishaps is potentially a major problem despite the excellent safety record of such laboratories (X-ray analysis laboratories have not always been so fortunate).

If it is necessary to open the power supply, the operator should remember that it operates at very high voltages and great care must be taken. All power should be turned off, and any large condensers (which may be very large indeed) should be shorted out, allowed to stand 10–15 min, and shorted out again before any work is done.

[12]K. E. Beu, *American Crystallographic Association Apparatus and Standards Committee Report*, No. 1, 1962, Goodyear Atomic Corporation, P.O. Box 628, Piketon, OH 45661.

BIBLIOGRAPHY

Cullity, B. D., *Elements of X-Ray Diffraction*, Addison-Wesley, Reading, MA, 1956, Chapter 1.

Guinier, A., *X-Ray Crystallographic Technology*, Hilger and Watts, London, 1952, Chapters 1 and 2.

Henry, N. F. M., H. Lipson, and W. A. Wooster, *The Interpretation of X-Ray Diffraction Photographs*, Macmillan, London, 1960, pp. 25–33.

CHAPTER 2

DIFFRACTION OF X-RAYS

In this chapter, the groundwork is laid for an understanding of crystals and their interaction with X-rays. A development of the nature of lattices is followed by the derivation of Bragg's law. The concept of the reciprocal lattice is then introduced, and finally, the diffraction of X-rays is interpreted and discussed in terms of it.

2.1. LATTICES, PLANES, AND INDICES

The constancy of the external forms of well-developed crystals early led to the idea that they were built from blocks of a unit structure regularly repeated in space. Quantitative studies of the interfacial angles confirmed this idea and established a crystal as being constituted of these units stacked side to side in three dimensions.

As an introduction to crystal structures, it is convenient to consider as an analogy a two-dimensional array. The nature of such an array is illustrated in Fig. 2.1 by the regular repetition of a simple motif, a comma in this case, by translation along two directions in the plane. It is convenient to refer the array of commas to a lattice or grid system. Such a grid can be chosen in different ways. For an unlimited lattice there are, in fact, an infinite number of ways of drawing the grid lines, and in principle all are equally valid, although in practice matters of convenience and convention dictate the choice. Figure 2.2 shows three different ways of drawing the grid lines for the array in Fig. 2.1. It is important to note that the grid lines are drawn at equal intervals corresponding to the repeat distances of the array and that as a consequence *the surroundings of each grid line intersection, or lattice*

Figure 2.1. Regular two-dimensional array.

point, are identical. In finite lattices this is, of course, not true for points at the edges, but we will be dealing with lattices of such size that for practical purposes they can be considered infinite.

That area in Fig. 2.2 set off by successive grid lines in two grid-line directions is termed a *unit cell*. In a three-dimensional lattice, the unit cell is a volume element and can be defined as the *parallelepiped whose edges are successive grid lines* (Fig. 2.3). The unit cell is the "unit structure" referred to above that produces the macroscopic crystal when stacked side to side in three dimensions.

Note that a unit cell may contain a whole unit of the repeating motif as in Fig. 2.2b, or it may contain parts of several as in Figs. 2.2*a* and *c*. Although the unit cells can differ in both shape and position, nevertheless they have the same area (volume) as long as there is the equivalent of one unit motif per unit cell.

It must be emphasized that the lattice is a purely imaginary construct but a very necessary and useful one. It functions as a coordinate system to which the actual structure is referred. The three grid line directions or coordinate axes are termed x, y, and z and are ordinarily chosen to form a right-handed system, that is, a system in which the digits of the right hand point respectively along $+x$, $+y$, and $+z$ as indicated in Fig. 2.4. An equivalent view is that of a right-handed screw, which advances along $+z$ when rotated from $+x$ to $+y$, along $+x$ when rotated from $+y$ to $+z$, or along $+y$ when rotated from $+z$ to $+x$. The lengths of the unit cell edges along the x, y, and z axes, respectively, are labeled a, b, and c. These same symbols are often used for the axes as well. The angles between the axes are α, β, and γ, with α between b and c, β between a and c, and γ between a and b (see Fig. 2.3).

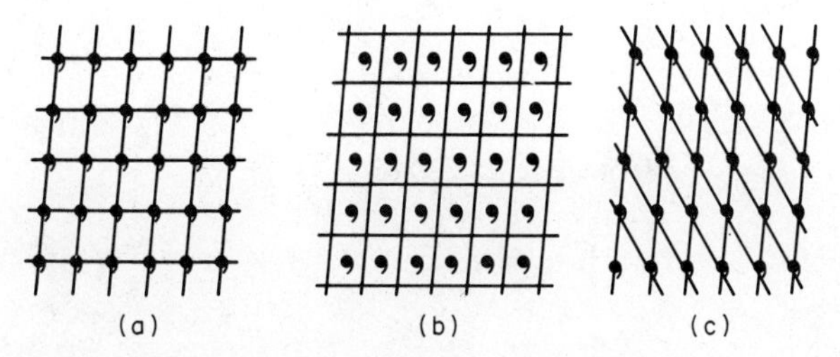

Figure 2.2. Three different grid systems referred to the array of Fig. 2.1.

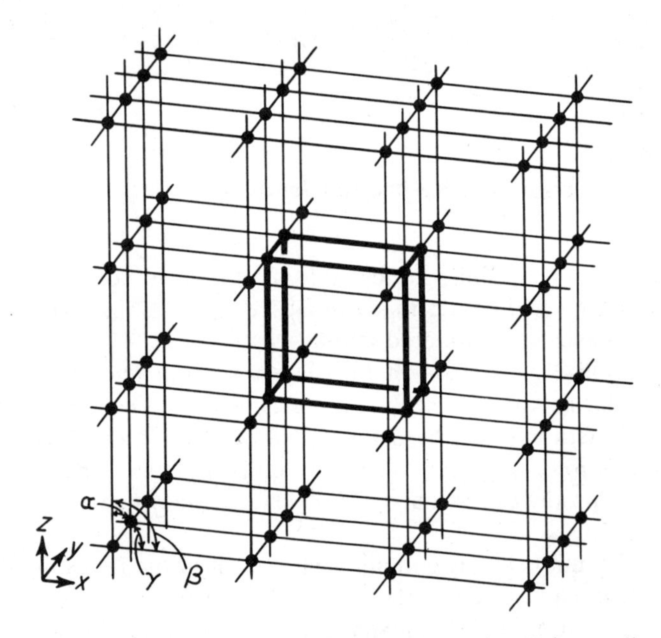

Figure 2.3. Three-dimensional lattice, showing unit cell (heavy lines).

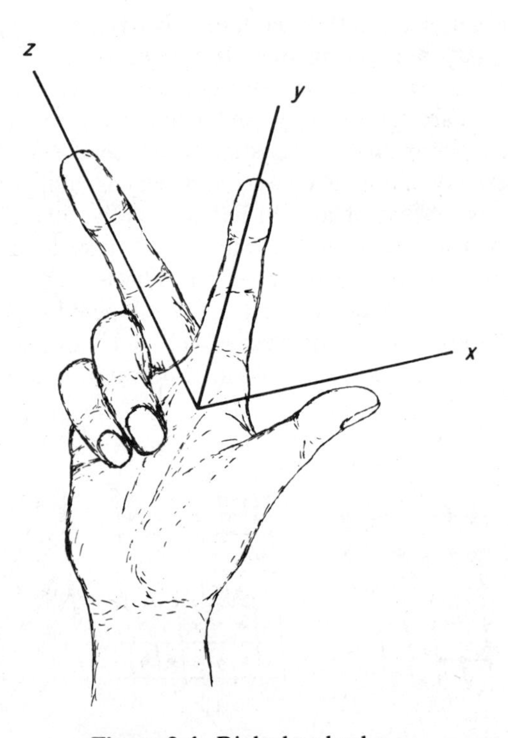

Figure 2.4. Right-hand rule.

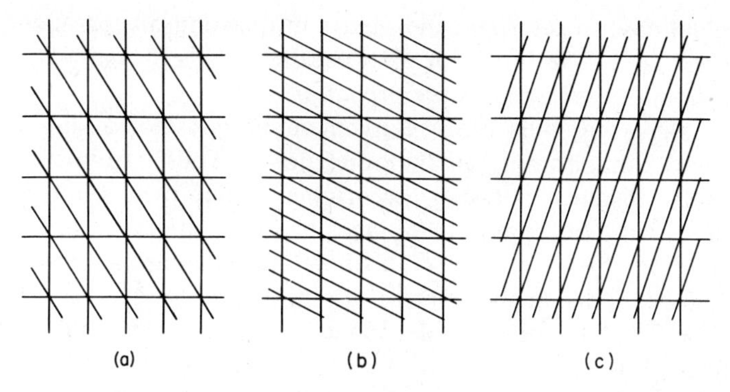

Figure 2.5 Three families of lattice "planes" in a two-dimensional lattice.

For use in the subsequent sections, it is necessary to consider sets of parallel lattice planes constructed so that for any given set every lattice point lies on some member of it. Figures 2.5*a*, *b*, and *c* show the same two-dimensional lattice with different sets of "planes" constructed in this way. These sets of planes are identified by three numbers in the three-dimensional case, one corresponding to each axis. It follows from the repetitive nature of the lattice that when such planes cut an edge of the unit cell the edge is always divided into an integral number of equal parts. These are, of course, common fractions of the unit translation; $1/1, 1/2, \ldots, 1/n$. The fractional intercepts on the three unit cell axes are used as a basis for a triple of numbers, the indices that characterize uniquely each possible set of planes. These indices are obtained by considering some lattice point as the origin and proceeding from it along the axes until the first member of a set of planes is reached. When the intercepts of the plane on the axes are expressed as fractions of the unit cell edge, their reciprocals are just the desired indices. Thus, in Fig. 2.6, plane 1 has intercepts $\frac{1}{2}, \frac{1}{2}, 1$ on the x, y, and z axes, respectively, and has indices (2, 2, 1).[1]

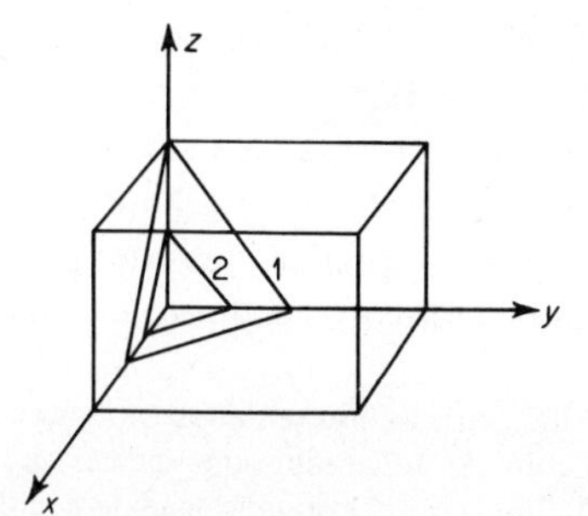

Figure 2.6. Example of a lattice plane and an interleaving plane.

[1]Alternatively, these indices can be regarded as showing the number of cuts made by the set of planes in one unit translation along each axial direction.

A special case arises when the planes in question are parallel to one or more of the axes. The intercept on a parallel axis is formally at infinity, so the corresponding index is its reciprocal, 0.

As we shall see, it is often convenient to consider additional sets of parallel planes that do not contain lattice points. These are interleaved so as to divide the distance between the original lattice planes into equal fractions. Thus, if the original lattice planes had the indices (h, k, l),[2] resulting from intercepts $1/h$, $1/k$, and $1/l$, a single interleaving plane would cut the axes at half these distances from the origin, that is, $\frac{1}{2}h$, $\frac{1}{2}k$, and $\frac{1}{2}l$, and have indices $(2h, 2k, 2l)$. For example, plane 2 in Fig. 2.6 has intercepts $\frac{1}{4}, \frac{1}{4}, \frac{1}{2}$ and indices $(4, 4, 2)$.[3] In general, the addition of $n-1$ interleaving planes between each pair of the set (h, k, l) will give the family of planes with indices (nh, nk, nl).

Classical crystallography was concerned with lattice planes primarily as they were exhibited in the exterior faces of the crystal. In this sense, the planes $(2, 2, 1)$ and $(4, 4, 2)$ are indistinguishable, and so the use of so-called Miller indices, that is, those having no common factor,[4] became customary. In dealing with X-ray diffraction, however, we shall find the general indices useful. Thus when indices are referred to subsequently without qualification, they will be general rather than Miller indices.

Miller indices can be used in any of three ways: to designate a set of lattice planes, a particular member of the set, or the face of a macroscopic crystal parallel to the set. Thus (010) is used to indicate the set of planes parallel to the xz plane and at $y = 0, b, 2b, 3b, \ldots$; the particular plane at $y = 0$; or the crystal face parallel to these. In contrast, $[x, y, z]$ indicates a direction in the lattice, that of a line segment from the origin to the point x, y, z. Frequently x, y, z are integers and may be confused with Miller indices. This occurs because locations in and around the unit cell are often specified by using the lengths of the cell edges as units. On this basis the lattice point at a distance a from the origin along the x axis has the coordinates $1, 0, 0$, and the line segment connecting it to the origin is [100]. Thus the three cell axes are [100], [010], and [001], while (100), (010), and (001) are the yz, xz, and xy planes, respectively (Fig. 2.7).[5] Note that the plane (100) intersects the axis [100], although the two are not necessarily perpendicular, and similarly for the other planes and axes.

A third, related notation is {100}, and so on, which refers to a set of face planes that are equivalent by the symmetry of the crystal (see Chapter 3). Such a set is called a crystal *form*. The number of planes included will depend on the particular crystal and its symmetry; thus for a cubic crystal,

[2]These symbols are the ones that are universally used for the general representation of indices.
[3]Where there is no danger of misunderstanding, the commas separating the indices are customarily omitted. Thus, (4, 4, 2) is usually seen as (442), but (12, 4, 2) would remain as such.
[4]It can be shown that one of the characteristics of the indices of planes of lattice points is that they are relatively prime, that is, they have no common factor.
[5]These are more commonly referred to as the a, b, and c axes and the bc, ac, and ab planes.

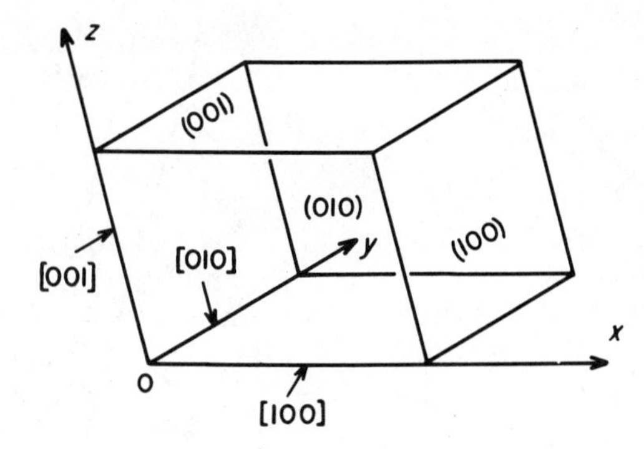

Figure 2.7. Unit cell showing bounding planes and edges.

{100} includes (100), (010), (001), ($\bar{1}$00), (0$\bar{1}$0), and (00$\bar{1}$),[6] while for a triclinic crystal it may include only (100).

The indices for the various planes in a lattice will depend on just how the unit cell is chosen. Transformation of the indices in changing from one unit cell to another follows certain well-established rules, which are treated in more detail in specialized texts.[7,8]

2.2. X-RAY DIFFRACTION

The diffraction of X-rays by crystals was discovered by Max von Laue in 1912, and the sequence of events that led to the discovery is one of the most fascinating chapters in the history of science.[9] Although X-rays had been discovered in 1895 by Roentgen, their nature was not known. During the years following their discovery, a number of determined efforts were made to prove them particles or waves. It was not, in fact, until diffraction by crystals was observed that their wave character was proved.

Following the experimental observation of X-ray diffraction early in 1912, von Laue showed that the phenomenon could be described in terms of diffraction from a three-dimensional grating.[10] In the same year, however, while engaged in experimental studies, W. L. Bragg[11] noticed the similarity of diffraction to ordinary reflection and deduced a simple equation treating diffraction as "reflection" from planes in the lattice. In order to

[6]It is standard crystallographic practice to place minus signs *over* indices. The plane ($\bar{1}$00) is one that cuts the $-x$ axis at $-a$ and lies parallel to the y and z axes.

[7]*International Tables*, Vol. I, pp. 15–21.

[8]Buerger, *X-Ray Crystallography*, pp. 10–28.

[9]See P. P. Ewald in *Fifty Years of X-Ray Diffraction*, P. P. Ewald, Ed., International Union of Crystallography, Utrecht, 1962, pp. 6–75.

[10]M. von Laue, *Sitz. math. phys. Klasse Bayer. Akad. Wiss.*, 303 (1912).

[11]W. L. Bragg, *Proc. Camb. Phil. Soc.*, **17**, 43 (1913).

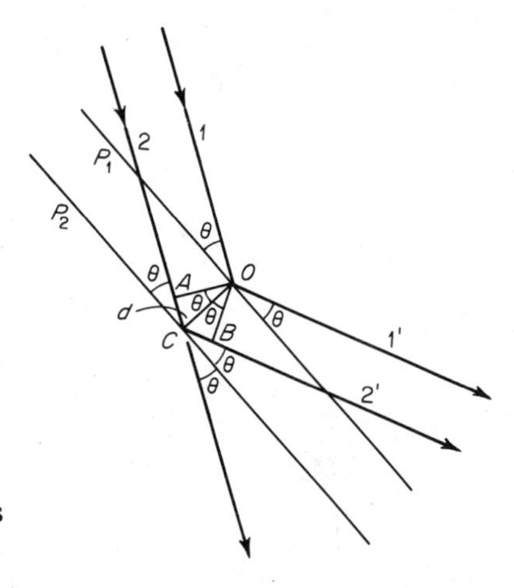

Figure 2.8. Construction showing conditions for diffraction.

derive the equation, we consider an X-ray beam incident on a pair of parallel planes P_1 and P_2 with interplanar spacing d (Fig. 2.8). The parallel incident rays 1 and 2 make an angle θ with these planes. Electrons assumed at O and C will be forced to vibrate by the oscillating field of the incident beam and, as vibrating charges, will radiate in all directions. For that particular direction[12] where the parallel secondary rays 1′ and 2′ emerge at angle θ as if reflected from the planes, a diffracted beam of maximum intensity will result if the waves represented by these rays are in phase. Dropping perpendiculars from O to A and B, respectively, it becomes evident that $\angle AOC = \angle BOC = \theta$. Hence $AC = BC$, and waves in ray 2′ will be in phase, that is, crest to crest, with those in 1′ if $AC + CB$ $(= 2AC)$ is an integral number of wavelengths λ. This is expressed by the equality

$$2AC = n\lambda \tag{2.1}$$

where n is an integer. By definition, $AC/d \equiv \sin\theta$, and by substitution in Eq. (2.1),

$$2d \sin\theta = n\lambda \tag{2.2}$$

This is *Bragg's law*.

The process of reflection is described above in terms of incident and reflected rays each making an angle θ with a fixed crystal plane. It can also be viewed as involving a fixed incident beam, in which case reflection occurs from planes set at the angle θ with respect to the beam and generates a reflected ray deviating through 2θ. This description resembles

[12]It can be shown (see Appendix A) that if one considers the simultaneous reflection of a large number of such rays, interference prevents the appearance of radiation in any other direction.

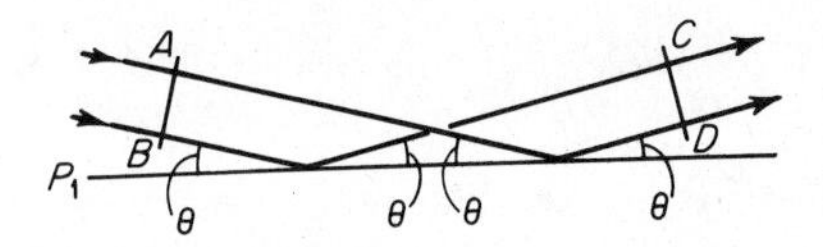

Figure 2.9. Reflection from two points in a plane.

more closely the actual experimental arrangements and is the more commonly used.

Though Fig. 2.8, on which the derivation is based, represents a special case, the result is, in fact, general. What holds for points O and C can be shown to hold for any pair of points in planes 1 and 2. This can be seen by recalling the elementary treatment of reflection. Figure 2.9 shows two parallel incident and reflected rays, where the angles θ between the rays and the plane are all equal. The path length by way of these rays (or any parallel rays) from the incident wave front AB to the emergent wave front CD is always the same provided the incident rays, the normals to the plane P at the points of incidence, and the reflected rays all lie in a plane. Thus the emergent rays shown, or any others similarly constructed, are in phase. It can now be seen by reference to Fig. 2.8 that any ray parallel to 1′ but from another point in plane P_1 will differ in phase by just $n\lambda$ not only from ray 2′ emergent from point C but also from any other parallel ray from any other point in plane P_2.

In the derivation of Bragg's law, electron density was assumed to be in the planes of Fig. 2.8. In actual structures, however, the electron density is distributed throughout the unit cell and does not lie in special planes. Nevertheless, the derivation is valid, since it can be shown that waves scattered from electron density not lying in the plane P (Fig. 2.10) can be added to give a resultant as if reflected from the plane. It is the variation in these resultants that accounts for the differing intensities of reflection observed for various planes (see Chapter 8).

In the derivation, reflections from only two planes were considered. In such a case the diffraction maxima would be broad and the various diffracted rays would virtually merge. In crystals, however, many hundreds or thousands of planes make up each of the mosaic blocks (see Chapter 4) that constitute the macroscopic crystal. Under these conditions, the diffraction maxima will be sharp and will occur only at clearly defined values of θ. Indeed, for real crystals the breadth of a reflection as measured in terms of the range of θ over which it can be observed is usually a small fraction of a degree and due mostly to the slight misalignment of the mosaic blocks.

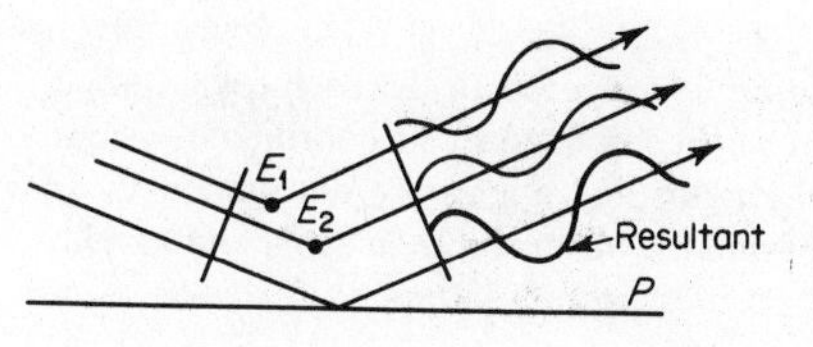

Figure 2.10. Diffraction from E_1 and E_2 as if reflected from plane P.

2.3. THE RECIPROCAL LATTICE

From a consideration of Bragg's law in the form

$$\sin\theta = \frac{n\lambda}{2}\left(\frac{1}{d}\right) \tag{2.3}$$

it is seen that $\sin\theta$ is inversely proportional to d, the interplanar spacing in the crystal lattice. Since $\sin\theta$ is a measure of the deviation of the diffracted beam from the direct beam, it is evident that structures with large d will exhibit compressed diffraction patterns, and conversely for small d. Interpretation of X-ray diffraction patterns would be facilitated if the inverse relation between $\sin\theta$ and d could be replaced by a direct one. What amounts to this can be achieved by constructing a reciprocal lattice based on $1/d$, a quantity that varies directly as $\sin\theta$.

The reciprocal lattice can be defined as follows: Consider normals to all possible direct lattice planes (h, k, l) to radiate from some lattice point taken as the origin. Terminate each normal at a point at a distance $1/d_{hkl}$ from this origin, where d_{hkl} is the perpendicular distance between planes of the set (hkl). The set of points so determined constitutes the *reciprocal lattice*. A two-dimensional example of the process is shown in Fig. 2.11, in which reciprocal lattice (r.l.) points corresponding to the direct lattice planes (0, 1), (1, 1), (2, 1), and (3, 1) are designated by asterisks and assigned the indices of the planes they represent.

That a set of points as defined above is in fact a lattice can be seen by referring to Fig. 2.12. Take the direct[13] lattice point O as the origin and let

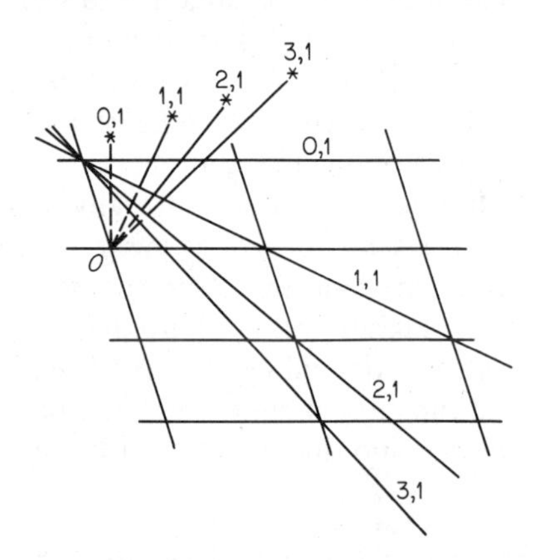

Figure 2.11. "Planes" in two-dimensional direct space represented by points in reciprocal space.

[13]The crystal lattice is commonly called the *direct lattice* as opposed to the reciprocal lattice. Each of these lattices can be thought of as existing in a space defined by its coordinate system, that is, *direct space* and *reciprocal space*. Reciprocal lengths and angles are designated by an asterisk appended to the corresponding direct space symbols, for example, a^*, b^*, and c^* play the same role in the reciprocal lattice as a, b, and c do in the direct lattice.

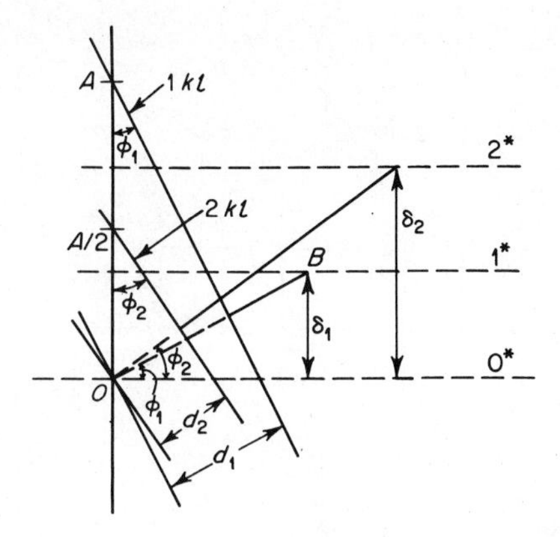

Figure 2.12. Construction showing that reciprocal points form a lattice (see text).

OA be a unit translation along some axial direction (say x) in direct space. All planes having their h index 0 will be parallel to OA, and their normals will lie in the plane perpendicular to OA through O, the trace of which is labeled 0*. By definition, the corresponding r.l. points will also lie in 0*. We can further show that all direct lattice planes with x intercepts at A, that is, with their h index 1, will have r.l. points lying in a plane 1*; those with x intercepts at $OA/2$, that is, with h index 2, will have their r.l. points lying in a plane 2*; and so on. Thus for a plane $(1, k, l)$ in Fig. 2.12,

$$d_{1kl} = OA \sin \phi_{1kl} \tag{2.4}$$

$$OB = 1/d_{1kl} = 1/(OA \sin \phi_{1kl}) \tag{2.5}$$

$$\delta_{1kl} = OB \sin \phi_{1kl} = 1/OA \tag{2.6}$$

In the same way it can be shown that

$$\delta_{2kl} = 1/\tfrac{1}{2}OA = 2(1/OA) \tag{2.7}$$

and in general

$$\delta_{nkl} = n(1/OA) \tag{2.8}$$

Thus the various r.l. points are segregated according to their h indices into equally spaced parallel layers. That these layers are perpendicular to the x axis follows from the fact that the first, 0*, must be. In the same way it can be shown that perpendicular to the other two direct lattice axial directions there are also equally spaced planes of r.l. points. Therefore the points are arranged periodically in three independent directions and so constitute a lattice. The appearance of the reciprocal lattice is as shown in Fig. 2.13. Every point lies at the intersection of three planes of points parallel to the sides of some reciprocal unit cell. Each of these planes is numbered as in

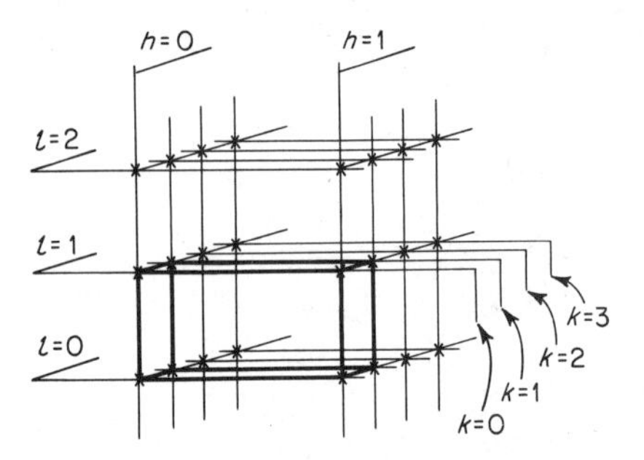

Figure 2.13. Indexing the reciprocal lattice.

Fig. 2.12, beginning with 0 for the one through the origin, and each contains those r.l. points having one index constant. The three numbers determined for a point by the intersection of three such planes are just the indices of the set of planes in direct space represented by that point.

The r.l. points shown as asterisks in Fig. 2.11 derive from planes of direct lattice points, so their indices correspond to Miller indices. The reciprocal lattice also includes, however, many points whose indices have common factors, such as 224 and 336. These may be considered as representing parallel families of planes not all of which contain lattice points, or, more simply, different orders of diffraction from the fundamental sets of planes. Thus in Bragg's law [Eq. (2.2)] the integer n is the order of a particular reflection. It is simply the number of wavelengths difference in path length for rays reflected from successive planes. But n need not be included explicitly when diffraction is considered in terms of the reciprocal lattice. This can be seen by rewriting Eq. (2.2) in the two forms

$$\frac{2 \sin \theta}{\lambda} = \frac{n}{d} \tag{2.9}$$

and

$$\frac{2 \sin \theta}{\lambda} = \frac{1}{d/n} \tag{2.10}$$

It is evident that $\sin \theta$ for the nth-order reflection from planes with spacing d [Eq. (2.9)] is the same as that for the first-order reflection from planes with spacing d/n [Eq. (2.10)].

The n orders of reflection possible from a particular set of planes thus correspond in the reciprocal lattice to n points at distances $i(1/d) = 1/(d/i)$, $i = 1, 2, \ldots, n$, from the origin. The order of a reflection is implicit in the indices and is, in fact, just the common factor by which general indices and the corresponding Miller indices are related. We shall follow the convention that when indices or Miller indices refer to planes in the lattice or faces of crystals, they will be surrounded by parentheses—thus, (hkl)—as in de-

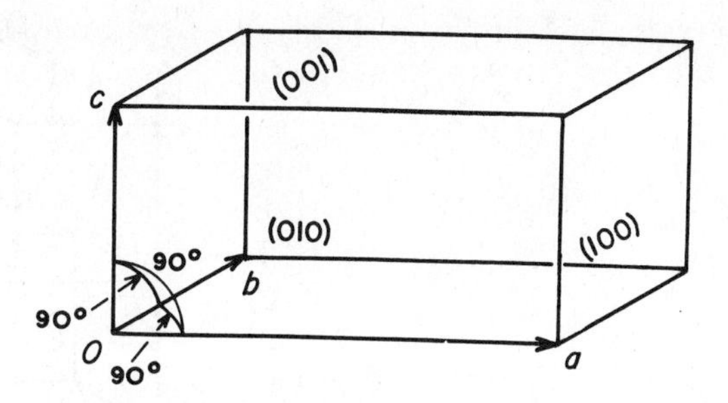

Figure 2.14. Orthorhombic direct cell.

scriptive crystallography; when they refer to an X-ray reflection or a point in reciprocal space, they will be written simply as *hkl*.

The relationships between the direct and reciprocal lattices in three dimensions depend on the angles between the axes in the direct lattice. Consider the cases shown in Figs 2.14, 2.16, and 2.19. In Fig. 2.14 the three direct axes are mutually perpendicular but the axial lengths are not the same (an orthorhombic cell, see Chapter 3). The planes (100), (010), and (001) are perpendicular to the *a*, *b*, and *c* axes, respectively, so their normals are along these axes. Thus the r.l. point 100 can be considered to be on the *a* axis at a distance $1/a$ from the origin. The line connecting this point to the origin is one edge of the unit cell in reciprocal space and is thus a^*. In the same way the reciprocal axes b^* and c^* can be seen to coincide with the corresponding direct axes *b* and *c*, and all three reciprocal axes are mutually perpendicular. The result is a reciprocal unit cell with the relationships summarized in Table 2.1 and pictured in Fig. 2.15.

Matters become somewhat more complex when the direct axes are not orthogonal. Figure 2.16 shows the case of a monoclinic cell in which two axes (by convention *a* and *c*) meet in an obtuse angle while the third (*b*) is perpendicular to both. The normal to (100) still lies in the *ac* plane and is still perpendicular to *b* but no longer coincides with *a*. Similarly, the normal to (001) also lies in *ac* but does not coincide with *c*. A plane view of these relationships as seen down the *b* axis is shown in Fig. 2.17. The r.l. point 100 will lie as usual on the normal to (100) at a distance $1/d_{100}$ from the origin. Note, however, that d_{100} is the interplanar separation measured

TABLE 2.1 Orthorhombic Direct/Reciprocal Relationships

$a^* = 1/a$	$a = 1/a^*$	$\alpha = \beta = \gamma = \alpha^* = \beta^* = \gamma^* = 90°$
$b^* = 1/b$	$b = 1/b^*$	$V^* = 1/V = a^*b^*c^*$
$c^* = 1/c$	$c = 1/c^*$	$V = 1/V^* = abc$

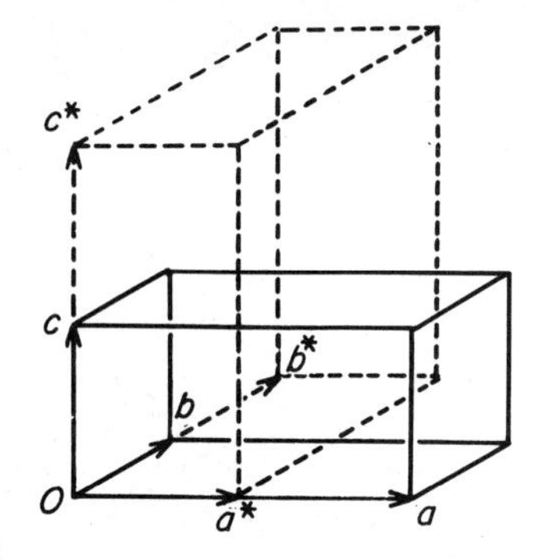

Figure 2.15. Orthorhombic direct and reciprocal cells.

along the (100) normal and is not, in this case, equal to a. If the ac angle is β,

$$d_{100} = a \sin(180° - \beta) \quad (2.11)$$

or

$$d_{100} = a \sin \beta \quad (2.12)$$

Thus the distance to the r.l. point 100, that is, a^*, is given by

$$a^* = 1/(a \sin \beta) \quad (2.13)$$

Similarly,

$$c^* = 1/(c \sin \beta) \quad (2.14)$$

On the other hand, since b is perpendicular to the ac plane (010), b^* still lies along it and

$$b^* = 1/b \quad (2.15)$$

The angular relationships between the two cells can be deduced from Figs. 2.16 and 2.17 and are given in Table 2.2, while Fig. 2.18 shows the shapes of a monoclinic cell and its reciprocal.

In the general, triclinic, case none of the axes are perpendicular, and as a

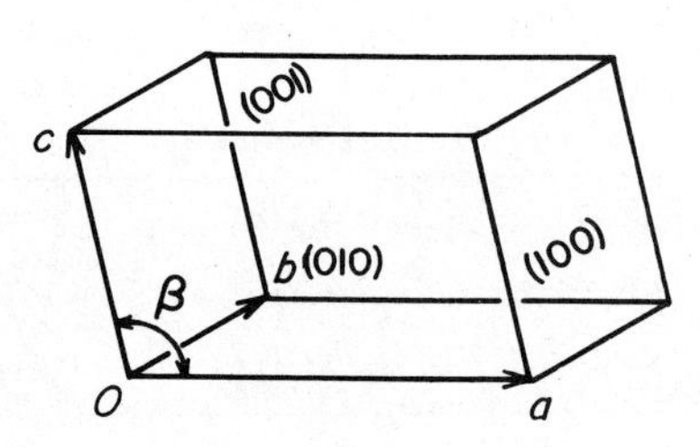

Figure 2.16. Monoclinic direct cell.

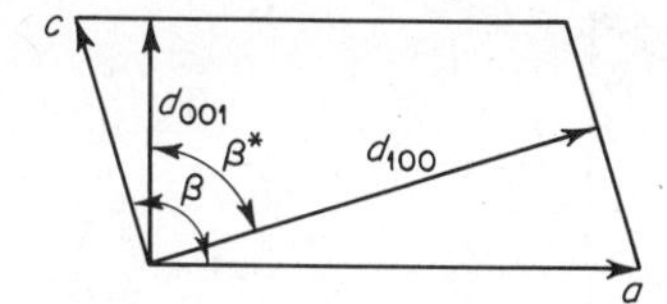

Figure 2.17. The *ac* face of the cell in Fig. 2.16.

result none of the plane normals that define the reciprocal axes coincide with their direct counterparts. Figure 2.19 shows this arrangement. The relationships between the direct and reciprocal axes are quite complex and will not be derived here. One form in which they can be expressed is given in Table 2.3, although others are possible and may be advantageous in particular cases.[14] A comparison of Tables 2.1–2.3 shows that the equalities in the first two can be easily obtained from those in the third by setting the proper angles to 90°. Special relationships for the other crystal classes (Chapter 3) can be similarly derived from the general expressions as needed.

It is important to note that the relationship between the direct and reciprocal axes is strictly reciprocal. That is, whatever is true in going from one to the other is equally true if the process is reversed. Put another way, any true statement about the two lattices remains true if all the unstarred symbols are replaced by starred ones and the starred by unstarred. An examination of Table 2.3 will reveal examples.

One vital conclusion that may be reached either from this principle or from an analysis similar to that at the beginning of this section relates the planes of r.l. points to the direct lattice axes. By construction, an r.l. axis (e.g., a^*) must be perpendicular to one face of the unit cell [e.g., (100), the *bc* plane]. By changing stars, it can be seen that similarly the b^*c^* plane of the reciprocal lattice must be perpendicular to the *a* direct axis. This plane contains the r.l. points $0kl$ and is parallel to those containing $1kl$, $2kl$, and so on. This result is general, and thus any direct axis has a family of r.l. planes perpendicular to it, each plane containing points of constant value for the index associated with that axis (Fig. 2.20).

TABLE 2.2 Monoclinic Direct/Reciprocal Relationships

$a^* = 1/(a \sin \beta)$	$a = 1/(a^* \sin \beta^*)$	$\alpha = \gamma = \alpha^* = \gamma^* = 90°$
$b^* = 1/b$	$b = 1/b^*$	$\beta^* = 180° - \beta$
$c^* = 1/(c \sin \beta)$	$c = 1/(c^* \sin b^*)$	$V^* = 1/V = a^*b^*c^* \sin \beta^*$
	$V = 1/V^* = abc \sin \beta$	

[14]Buerger, *X-Ray Crystallography*, pp. 347–362.

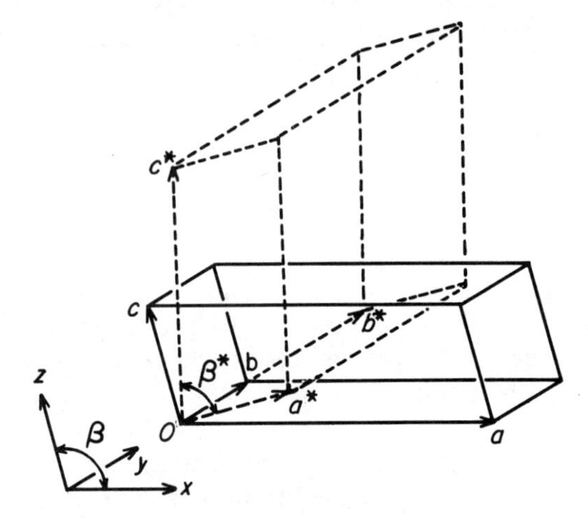

Figure 2.18. Monoclinic direct and reciprocal cells.

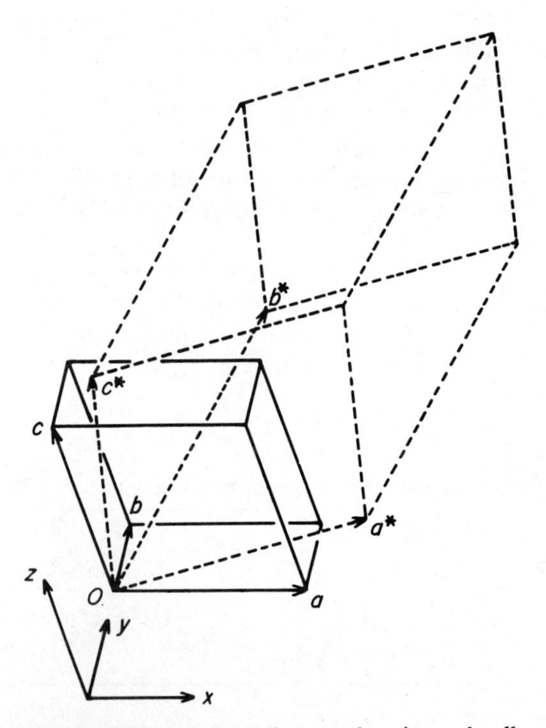

Figure 2.19. Triclinic direct and reciprocal cells.

TABLE 2.3 Triclinic Direct and Reciprocal Relationships

$$a^* = \frac{bc \sin \alpha}{V} \qquad a = \frac{b^* c^* \sin \alpha^*}{V^*}$$

$$b^* = \frac{ac \sin \beta}{V} \qquad b = \frac{a^* c^* \sin \beta^*}{V^*}$$

$$c^* = \frac{ab \sin \gamma}{V} \qquad c = \frac{a^* b^* \sin \gamma^*}{V^*}$$

$$V = \frac{1}{V^*} = abc\sqrt{1 - \cos^2 \alpha - \cos^2 \beta - \cos^2 \gamma + 2 \cos \alpha \cos \beta \cos \gamma}$$

$$V^* = \frac{1}{V} = a^* b^* c^* \sqrt{1 - \cos^2 \alpha^* - \cos^2 \beta^* - \cos^2 \gamma^* + 2 \cos \alpha^* \cos \beta^* \cos \gamma^*}$$

$$\cos \alpha^* = \frac{\cos \beta \cos \gamma - \cos \alpha}{\sin \beta \sin \gamma} \qquad \cos \alpha = \frac{\cos \beta^* \cos \gamma^* - \cos \alpha^*}{\sin \beta^* \sin \gamma^*}$$

$$\cos \beta^* = \frac{\cos \alpha \cos \gamma - \cos \beta}{\sin \alpha \sin \gamma} \qquad \cos \beta = \frac{\cos \alpha^* \cos \gamma^* - \cos \beta^*}{\sin \alpha^* \sin \gamma^*}$$

$$\cos \gamma^* = \frac{\cos \alpha \cos \beta - \cos \gamma}{\sin \alpha \sin \beta} \qquad \cos \gamma = \frac{\cos \alpha^* \cos \beta^* - \cos \gamma^*}{\sin \alpha^* \sin \beta^*}$$

2.4. BRAGG'S LAW IN RECIPROCAL SPACE

In X-ray crystallography the most important property of the reciprocal lattice is that it allows a simple visualization of Bragg's law that is much more convenient in practice than the one used in the derivation above. Imagine a crystal in a beam of X-rays of wavelength λ, and consider the a^*c^* section of its reciprocal lattice (Fig. 2.21*a*). Assuming that the crystal is oriented so that the X-ray beam is parallel to this a^*c^* plane, draw a line XO in the direction of the beam and passing through the r.l. origin O. Finally, describe a circle of radius $1/\lambda$ with its center C on XO and located so that O falls on its circumference.

Now consider the properties of an r.l. point P lying on this circle. The

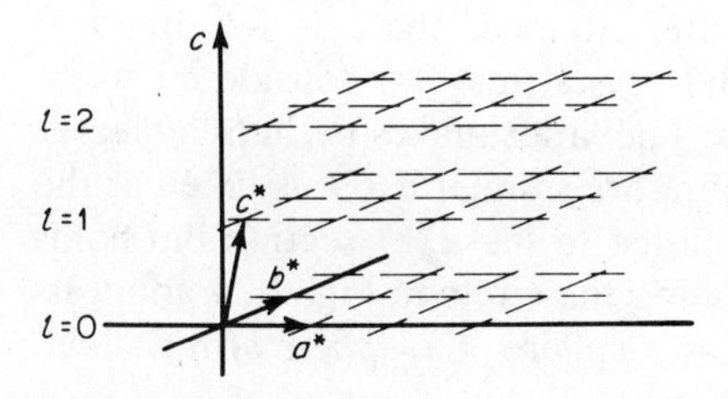

Figure 2.20. Reciprocal lattice levels a^*b^* perpendicular to a direct axis c.

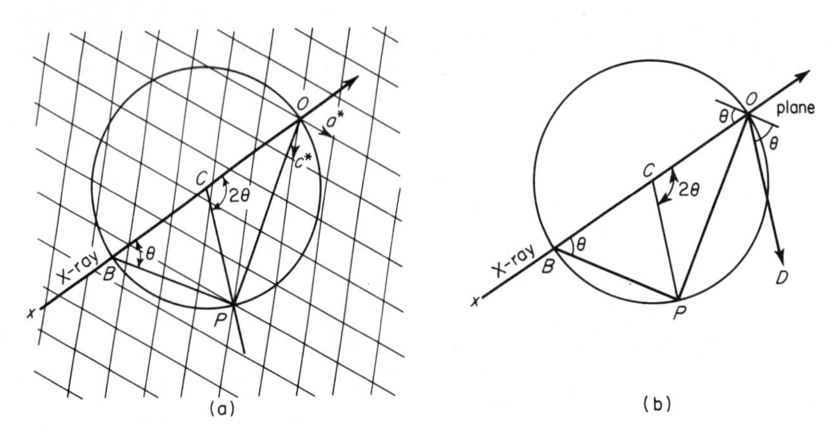

Figure 2.21. Diffraction in terms of the reciprocal lattice. (*a*) The reciprocal lattice and the sphere of reflection; (*b*) the direct plane and the reflected ray.

angle *OPB* is inscribed in a semicircle and thus is a right angle. Therefore,

$$\sin OBP = \sin \theta = \frac{OP}{OB} = \frac{OP}{2/\lambda} \tag{2.16}$$

$$\sin \theta = (OP/2)\lambda \tag{2.17}$$

But since *P* is a reciprocal lattice point, the length of *OP* is by definition equal to $1/d_{hkl}$. Substituting gives

$$\sin \theta = 1\lambda/2d_{hkl} \tag{2.18}$$

or

$$1\lambda = 2d \sin \theta \tag{2.19}$$

which is just Bragg's law with $n = 1$.

This derivation is completely general and implies that whenever an r.l. point coincides with a circle constructed as described, Bragg's law is satisfied and reflection occurs. The reflecting plane is perpendicular to *OP* (Fig. 2.21*b*), hence parallel to *BP*, and makes an angle θ with *BO*. Thus the direction of the diffracted beam *OD* is parallel to *CP*, making, as required, an angle of 2θ with the incident beam. Since, experimentally, the diffraction is viewed from an approximately infinite distance, there is no practical distinction between *OD* and *CP*, so the crystal may be considered to be centered at either *C* or *O* as one desires. The latter makes perhaps a clearer picture for a beginner, but later derivations are simpler if *CP* is taken as the diffracted ray. The construction is not limited to the a^*c^* section but holds for all points on circles produced by rotating the circle in Fig. 2.21 about its diameter *OB*. The figure so generated is a sphere, the *sphere of reflection*.

Thus for any point on the surface of this sphere, the same conditions hold as for point P in Fig. 2.21, and the conditions for Bragg reflection are fulfilled for the related direct space plane. By rotating the lattice about its origin, various r.l. points can be brought into coincidence with the surface of the sphere of reflection and the corresponding reflection can be observed.

When the construction of the reciprocal lattice is carried out as described here, its size depends solely on the dimensions of the direct unit cell. The various repeat distances are expressed in units of reciprocal angstroms (Å^{-1}). Changing the wavelength of the incident radiation has the effect of enlarging (shorter λ) or shrinking (longer λ) the size of the sphere of reflection (Fig. 2.22). Since the diameter of this sphere is $2/\lambda$, every r.l. point within that distance of the origin can be brought into coincidence with its surface, that is, into position to reflect. Thus every r.l. point within a sphere of radius $2/\lambda$, the *limiting sphere* (Fig. 2.23*a*), is a potential reflection, and the total number of reflections possible for any substance is limited to the number of r.l. points within this sphere. As the sphere grows, so will the number of reflections that can be observed. If there is only one point per unit reciprocal cell, the total number of possible reflections, N, is approximately

$$N = \frac{\frac{4}{3}\pi(2/\lambda)^3}{\text{vol. reciprocal cell}} \tag{2.20}$$

$$N = \frac{33.5}{\text{vol. reciprocal cell} \times \lambda^3} \tag{2.21}$$

or

$$N = 33.5\,\frac{\text{vol. direct cell}}{\lambda^3} \tag{2.22}$$

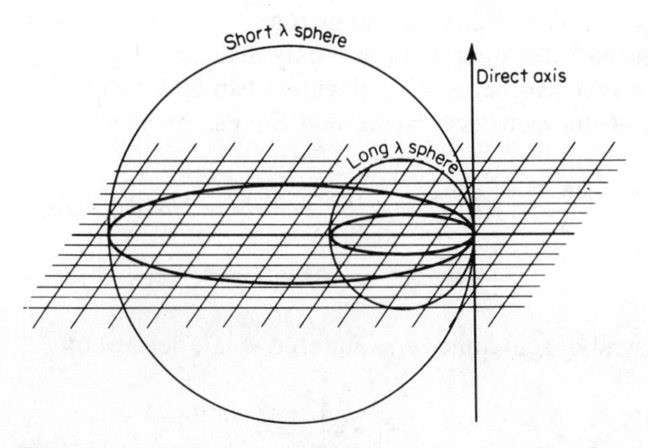

Figure 2.22. Spheres of reflection for two wavelengths.

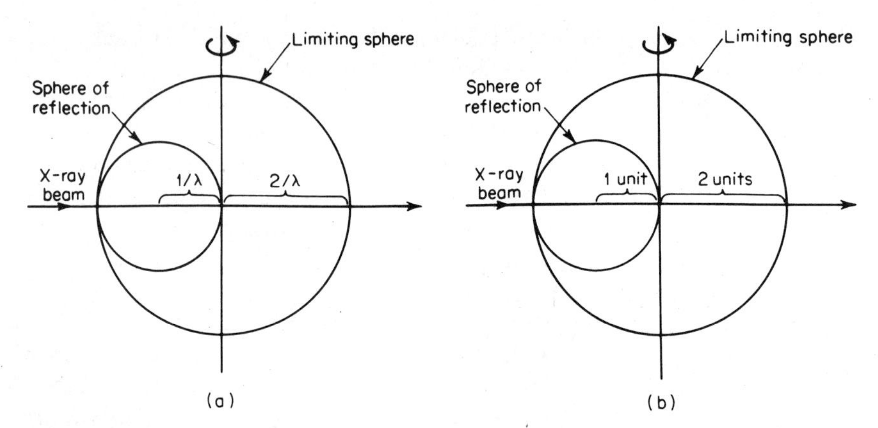

Figure 2.23. Sections through the sphere of reflection and the limiting sphere (*a*) for reciprocal lattice defined as $1/d$, (*b*) for reciprocal lattice defined as λ/d.

For example, a rectangular cell of dimensions $8 \times 10 \times 20$ Å has a volume of 1600 $Å^3$. For Cu K_α radiation ($\lambda = 1.5418$ Å) there are 14,600 reflections accessible, while for Mo K_α ($\lambda = 0.7107$ Å) the number climbs to a staggering 149,000. Fortunately not all these reflections are unique, and only a fraction of the total need be measured. In the following chapters we shall see how this reduction arises as a consequence of the symmetry of the atomic arrangement in crystals.

Quantities other than $1/d_{hkl}$ are often used to describe the distances of the r.l. points from the origin. Of these the most common is $(\sin\theta)/\lambda$, which by Bragg's law is

$$\frac{\sin\theta}{\lambda} = \frac{1}{2d_{hkl}} \tag{2.23}$$

The units are still reciprocal angstroms, although they are usually not expressed, but the quantities are only half as large as $1/d$. A general expression that can be used to calculate $(\sin\theta)/\lambda$ can be derived from the geometry of the reciprocal lattice and Eq. (2.23); this is

$$\frac{\sin\theta}{\lambda} = \tfrac{1}{2}(h^2a^{*2} + k^2b^{*2} + l^2c^{*2} + 2hka^*b^*\cos\gamma^* + 2hla^*c^*\cos\beta^* + 2klb^*c^*\cos\alpha^*)^{1/2} \tag{2.24}$$

Another symbol sometimes encountered is $|S|$, defined by

$$|S| = \frac{1}{d_{hkl}} = \frac{2\sin\theta}{\lambda} \tag{2.25}$$

TABLE 2.4 Limiting Values of Reciprocal Lattice Measures and Resolution

	Cu K_α	Mo K_α
λ	1.5418 Å	0.7107 Å
$[(\sin\theta)/\lambda]_{max}$	0.648 Å^{-1}	1.407 Å^{-1}
$\lvert S\rvert = (1/d_{hkl})_{max}$	1.296 Å	2.814 Å^{-1}
$(d_{hkl})_{min} = \lambda/2$	0.7709 Å	0.3554 Å
Resolution	0.71 Å	0.33 Å

This arises from a vector derivation of Bragg reflection and is the magnitude of the scattering vector S normal to the reflecting plane. A little care should be exercised, however, since $|S|$ is occasionally equated with $(4\pi \sin\theta)/\lambda$ as a result of the use of different units in the derivation.

Since the maximum possible value of $\sin\theta$ is 1, Eqs. (2.23) and (2.25) can be used to define limits to these various quantities for each radiation. Examples are given in Table 2.4.

In can be shown[15] that ideally the limit of resolution—the minimum separation at which two atoms can be distinguished—is given by

$$\mathrm{LR} = 0.92 d_{min} \tag{2.26}$$

In practice the limit can conservatively be taken as d_{min}. Thus the resolution attainable with data to the limit of the Cu K_α sphere is less than 0.8 Å, far more than is needed to resolve atoms, while cutting the data halfway to the limit will give a resolution of 1.5 Å, just showing separation of atomic peaks (but with much overlap).

An alternative but entirely equivalent construction defines the reciprocal lattice in dimensionless units of λ/d_{hkl}. The sphere of reflection then has a radius of one r.l. unit (1 r.l.u.) and the limiting sphere, 2 r.l.u. (Fig. 2.23*b*). Under these conditions, a change in λ has the effect of shrinking (shorter λ) or expanding (larger λ) the reciprocal lattice, while the sphere of reflection remains always the same size (Fig. 2.24). The two views lead invariably to the same predictions with regard to the process of diffraction, but one may be more convenient than the other for particular purposes. Thus the variable reciprocal lattice is often used for describing the mechanics of recording diffraction effects (Chapters 5 and 6) while the constant reciprocal lattice is better for discussing resolution and related phenomena (Chapters 9 and 15).

Thorough familiarity with the reciprocal lattice is essential in applying it to the interpretation of diffraction effects, and practice constructions similar to that in Fig. 2.21 are an excellent aid in visualizing and understanding just

[15] R. E. Stenkamp and L. H. Jensen, *Acta Cryst*, **A40**, 251 (1984).

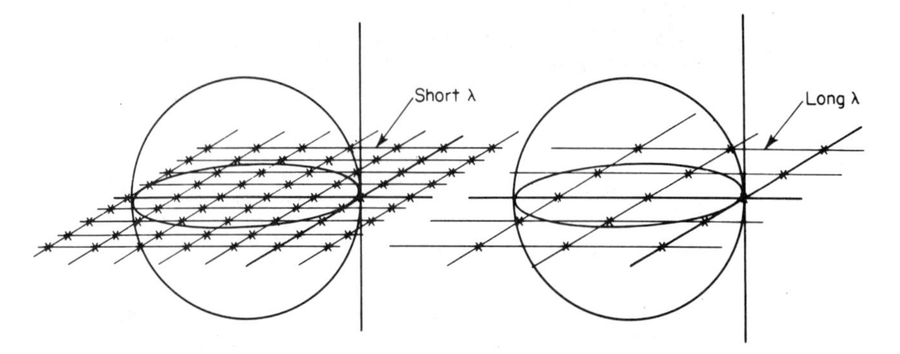

Figure 2.24. Spheres of reflection and reciprocal lattices for two wavelengths.

how diffraction occurs. Sections of the reciprocal lattice should be drawn to appropriate scale on tracing paper (1 Å^{-1} = 10 cm is convenient). These are then laid over a sheet on which is drawn the outline of the sphere of reflection and pinned at the position corresponding to the origin. The r.l. sections may then be rotated, and the conditions for Bragg reflection are achieved for any plane (***hkl***) when the corresponding r.l. point intersects the trace of the sphere of reflection.

An example of the use of the reciprocal lattice in interpreting diffraction is shown by Fig. 2.25, which illustrates the relationship between it and the sphere of reflection in one of the common modes for recording diffraction. A direct lattice axis is taken to be vertical, and the X-ray beam is assumed to be perpendicular to it. Since levels of r.l. points are normal to the direct axis (Fig. 2.20) it is evident that rotation about the axis will cause each level of points to intersect the sphere in a circle.

The diffracted rays will pass from the center of the sphere through these circles, forming cones, the zero level being a flat cone. The cone axes are coincident and are parallel to the rotation axis.

The direction of the diffracted beams as well as the order of their

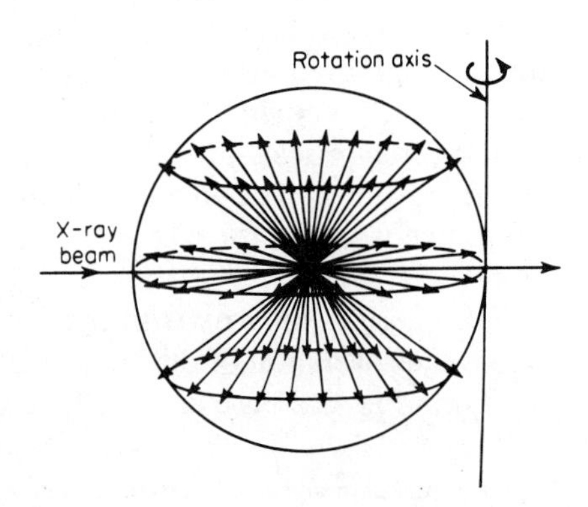

Figure 2.25. Cones of diffracted rays for rotation about a direct axis.

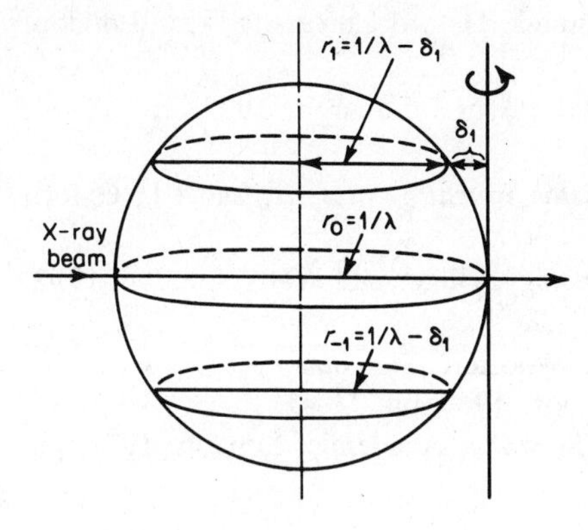

Figure 2.26. Traces of intersection of upper-level r.l. nets with the sphere of reflection.

appearance can be predicted for the upper-level reflections by using the same construction described above (Fig. 2.21). It should be noted, however, that for upper levels the r.l. planes cut the sphere of reflection in circles whose radius is less than $1/\lambda$, but that the point about which rotation occurs is still $1/\lambda$ from the center of the circle (Figs. 2.25 and 2.26).

KEY FORMULAS

Bragg's Law

$$2d \sin \theta = n\lambda \tag{2.2}$$

Reciprocal Axes

$$a^* = 1/d_{100} \qquad b^* = 1/d_{010} \qquad c^* = 1/d_{001}$$

Orthorhombic direct/reciprocal relationships: See Table 2.1.
Monoclinic direct/reciprocal relationships: See Table 2.2.
Triclinic direct/reciprocal relationships: See Table 2.3.

$$\frac{\sin \theta}{\lambda} = \tfrac{1}{2}(h^2 a^{*2} + k^2 b^{*2} + l^2 c^{*2} + 2hka^* b^* \cos \gamma^* + 2hla^* c^* \cos \beta^* + 2klb^* c^* \cos \alpha^*)^{1/2} \tag{2.24}$$

BIBLIOGRAPHY

Buerger, M. J., *X-Ray Crystallography*, Wiley, New York, 1942, pp. 4–47, 347–362.

Bunn, C. W., *Chemical Crystallography*, 2nd ed., Oxford University Press, London, 1961, pp. 120–125.

Gay, P., *The Crystalline State*, Oliver and Boyd, Edinburgh, 1972, pp. 15–33, 56–140.

Glusker, J. P., and K. N. Trueblood, *Crystal Structure Analysis*, 2nd ed., Oxford University Press, New York, 1985, pp. 11–41.

Guinier, A., *X-Ray Crystallographic Technology*, Hilger and Watts, London, 1952, Chapter III.

Henry, N. F. M., H. Lipson, and W. A. Wooster, *The Interpretation of X-Ray Diffraction Photographs*, Macmillan, London, 1960, pp. 34–41.

Jeffery, J. W., *Methods in X-Ray Crystallography*, Academic, London, 1971, pp. 18–31.

CHAPTER 3

CRYSTALS, SYMMETRY, AND SPACE GROUPS

In this chapter we shall discuss some aspects of crystal symmetry that are important for the interpretation of reciprocal lattice images recorded in various ways. Moreover, knowledge of the symmetry relationships among the atoms, ions, or molecules in a structure simplifies the practical operations of determining crystal structures since it reduces the number of unique parameters that must be determined.

3.1. CRYSTAL SYSTEMS AND SYMMETRY[1]

There are seven three-dimensional coordinate systems that are useful in describing crystals and are the basis for their classification. In general, the unit cell is characterized by six parameters, three axial lengths and three interaxial angles (Fig. 3.1). The lengths of the unit cell edges are designated a, b, c, and the interaxial angles α, β, γ. The angle α is between b and c, β is between a and c, while γ is between a and b.

In Table 3.1 the crystal systems are listed with the unit cell parameters that characterize each. In the most general system, triclinic, all six lattice parameters can assume any value. In the other crystal systems, symmetry decreases the number of independent parameters to the values given in column 2. Certain simple conventions have been followed in tabulating parameters. In the monoclinic system, one of the axes is unique in the sense that it is perpendicular to the other two. This axis is conventionally taken as b so that $\beta \neq 90°$. A further convention requires that the a and c axes be

[1]See the bibliography at the end of the chapter for references for further reading.

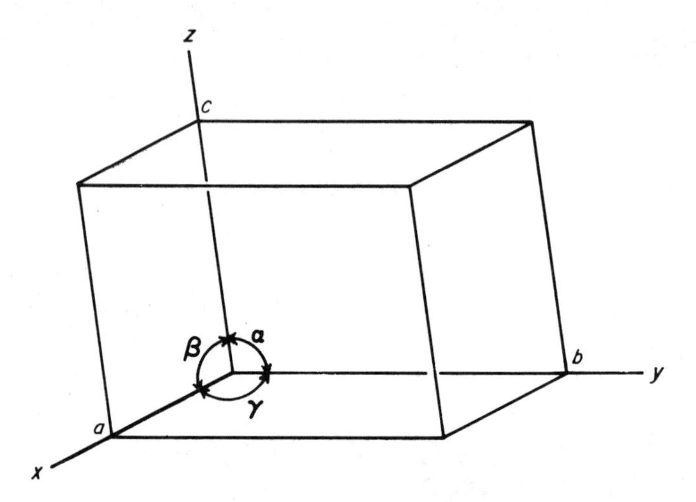

Figure 3.1. Unit cell.

chosen so that $\beta > 90°$. In the hexagonal and tetragonal systems, one axis is also differentiated from the others by its symmetry properties. In these systems the unique axis is conventionally c.

The equalities in Table 3.1 are, in fact, identities. For example, the equality of the a and b axes in the tetragonal system is a consequence of the identity of structure in tetragonal crystals along these two mutually perpendicular axial directions. On the other hand, it is possible for a crystal of one system to have parameters that mimic those of another without actually belonging to it. Thus crystals have been found that have α, β, and γ apparently equal to 90° yet are triclinic because of their lack of internal symmetry. It is, in fact, the symmetry that is fundamental, and the characteristic unit cell parameters are dependent quantities. Because of its fundamental importance in crystal structure analysis, symmetry will be dis-

TABLE 3.1 The Seven Crystal Systems

Crystal System	Number of Independent Parameters	Parameters	Lattice Symmetry
Triclinic	6	$a \neq b \neq c$; $\alpha \neq \beta \neq \gamma$	$\bar{1}$
Monoclinic	4	$a \neq b \neq c$; $\alpha = \gamma = 90°$; $\beta > 90°$	$2/m$
Orthorhombic	3	$a \neq b \neq c$; $\alpha = \beta = \gamma = 90°$	mmm
Tetragonal	2	$a = b \neq c$; $\alpha = \beta = \gamma = 90°$	$4/mmm$
Trigonal			
rhombohedral lattice	2	$a = b = c$; $\alpha = \beta = \gamma \neq 90°$	$\bar{3}m$
hexagonal lattice	2	$a = b = c$; $\alpha = \beta = 90°$; $\gamma = 120°$	$6/mmm$
Hexagonal	2	$a = b = c$; $\alpha = \beta = 90°$; $\gamma = 120°$	$6/mmm$
Cubic	1	$a = b = c$; $\alpha = \beta = \gamma = 90°$	$m3m$

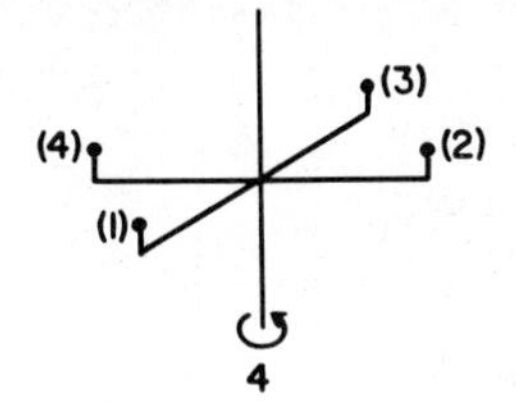

Figure 3.2. A 4-fold rotation axis.

cussed in some detail and illustrated in the lattices of the seven crystal systems. Our description will be based on two simple types of symmetry: (1) rotation and (2) reflection.

Rotation is necessarily about an axis, and it is designated as n-fold if on rotating $360°/n$ the resulting arrangement is equivalent to the one initially present. For example, the vertical axis in Fig. 3.2 is a 4-fold axis because rotation of $360°/4 = 90°$ about this axis leaves the set of four points in a position indistinguishable from the initial position. Rotation axes are designated by an integer, and only 1-, 2-, 3-, 4-, and 6-fold axes can occur in normal crystals.

If a plane exists in a structure such that every part on one side of the plane is related to a part on the other side as if reflected, the structure is said to possess a *plane of symmetry* or *mirror plane*. Such a plane is designated by the letter m, and two points thus related are shown in Fig. 3.3.

These simple symmetry elements can be combined to produce more complex symmetry. The combination of a 2-fold axis with a mirror plane perpendicular to it is an example that results in a very commonly occurring symmetry. In Fig. 3.4 an asymmetric unit represented by a comma (1) is rotated 180° about the z axis to the intermediate position (2). On reflection in the xy plane, (2) is transformed into (3). Two units such as (1) and (3) are said to be *centrosymmetrically related*. Either unit can be derived from the other by a single operation called *inversion*, amounting to reflection in a point midway between them. This point is called the *inversion center* or

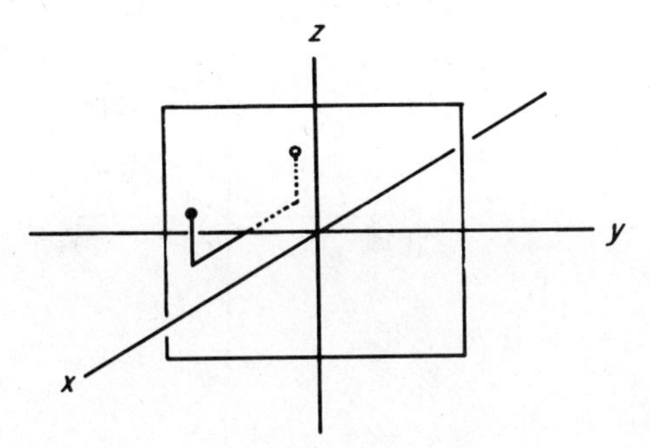

Figure 3.3. A mirror plane.

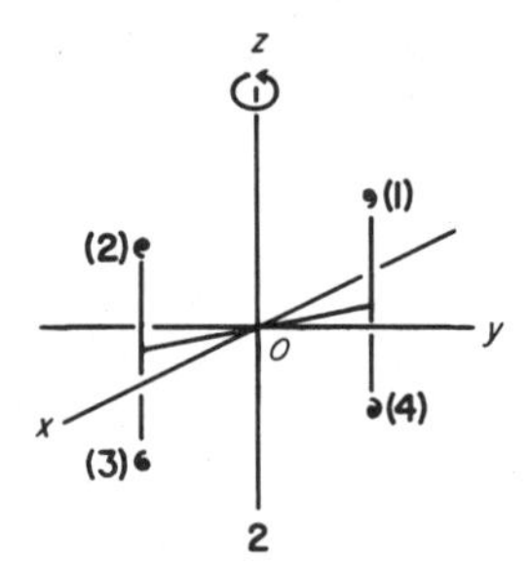

Figure 3.4. Center of symmetry produced by a 2-fold axis combined with reflection.

center of symmetry. In Fig. 3.4 the point O where the xy plane intersects the axis of rotation is the center of symmetry.

In Fig. 3.4 the unit (3) can equally well be derived from (1) by the application of the symmetry elements in the inverse order. Reflections of (1) in the xy plane transforms it into (4), which when rotated 180° gives (3). If the coordinates of unit (1) are x, y, z, then unit (3) has coordinates $\bar{x}$, $\bar{y}$, $\bar{z}$. This is the analytical relation between centrosymmetrically related points if the center of symmetry and the origin of the coordinate system coincide. The result follows from the application of the individual symmetry operations. Note that rotation of 180° about z transforms x, y, z for (1) to $\bar{x}$, $\bar{y}$, z for (2). Reflection across the plane $z = 0$ alters the sign of z and results in coordinates $\bar{x}$, $\bar{y}$, $\bar{z}$ for (3). In exactly the same way that the combination of a 2-fold axis and a mirror perpendicular to it results in a center of symmetry, so a 2-fold axis through a center of symmetry results in a mirror, and a mirror plane through a center of symmetry results in a 2-fold axis. Thus the three symmetry elements are intimately related; the presence of any two is invariably accompanied by the third.

Another set of symmetry elements that occurs in crystals is a rotation axis combined with an inversion center to produce a *rotary inversion axis*. A 4-fold rotary inversion axis is illustrated in Fig. 3.5. The unit (1) is rotated 360°/4 about the vertical axis in the direction indicated by the arrow and inverted through O, giving (2). Rotating another 360°/4 and inverting through O gives (3); another 360°/4 rotation and inversion gives (4); and

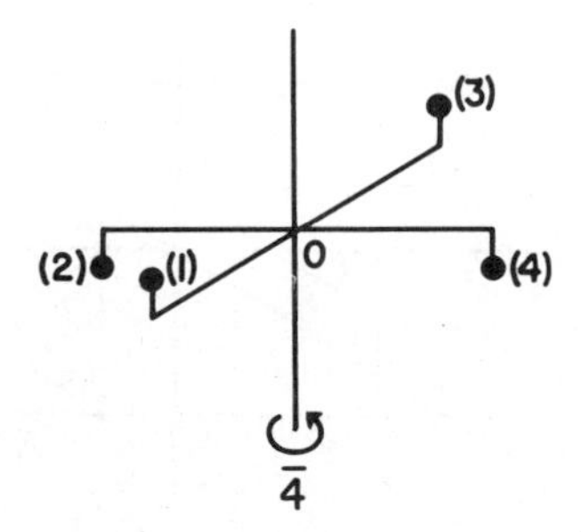

Figure 3.5. A 4-fold rotary inversion axis.

finally another 360°/4 rotation and inversion gives a position coincident with (1). Rotary inversion axes are designated by a bar over an integer. Thus the 4-fold rotary inversion axis is designated $\bar{4}$ (read: bar four).

If $\bar{4}$ is taken as the c axis of a tetragonal lattice with the inversion point at $z = 0$, then the equivalent positions (1)–(4) in Fig. 3.5 can be written down in analytical form at once. Point (1) is taken as x, y, z and the other three are then (2) $\bar{y}$, x, $\bar{z}$; (3) $\bar{x}$, $\bar{y}$, z; and (4) y, $\bar{x}$, $\bar{z}$.

We can now explain and justify the symbols used in Table 3.1 to indicate the lattice symmetry for each crystal system. Every lattice is inherently centrosymmetric since the lattice points and points midway between them are centers of symmetry. This is evident for the lattice points since their distributions in opposite directions about any one lattice point are identical. It is also true for the points midway between the lattice points, as can be seen by reference to Fig. 3.6, where each such point in *one* unit cell has been marked by an asterisk. An inversion center is the only symmetry possessed by the triclinic lattice, and it is indicated as a 1-fold inversion axis, $\bar{1}$; that is, rotation of 360°/1 followed by inversion through the origin. All other lattices display additional symmetry.

The symbols $2/m$ (read: 2 over m) for the monoclinic lattice indicate a 2-fold axis with a mirror plane perpendicular to it. In this system convention dictates the use of the b axis, the axis perpendicular to the ac plane (Fig. 3.7), as unique. Consideration of an extended monoclinic lattice shows that a rotation of 360°/2 about b brings the lattice points into coincidence, establishing b as a 2-fold axis. Similarly, 2-fold axes parallel to b occur at the midpoints of the a and c cell edges and the center of the ac face.

In view of the fact that b is perpendicular to ac, planes of the set (010) are mirror planes, reflecting lattice points on one side into those on the

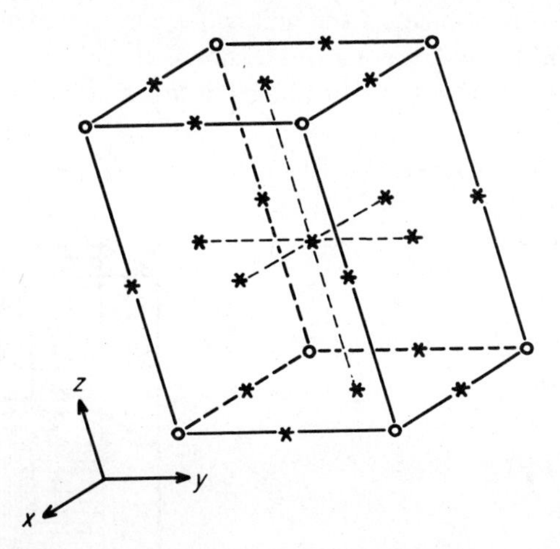

Figure 3.6. Triclinic unit cell showing centers of symmetry at corners of the unit cell (○) and between them (*).

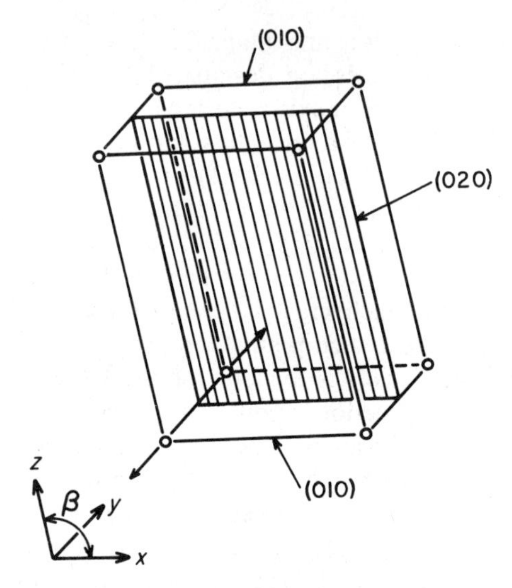

Figure 3.7. Monoclinic unit cell showing two planes of the set (010) and one axis of the set [010]. The interleaving mirror plane (020) is also shown.

other side. In addition, it will be evident that parallel interleaving planes occur, for example, the shaded plane at $b/2$ in Fig. 3.7. It is a frequent feature in lattices that primary symmetry elements give rise to interleaving elements of the same kind.

The symmetry of the orthorhombic lattice is designated by *mmm*, indicating three mutually perpendicular mirror planes. It should be evident at this point that as a consequence of the mutually orthogonal axes in this system the three pairs of faces of the unit cell are mirror planes. These three sets of planes with the interleaving ones, shown at $a/2$, $b/2$, and $c/2$ in Fig. 3.8, constitute the full set of mirror planes in this system. These are not the

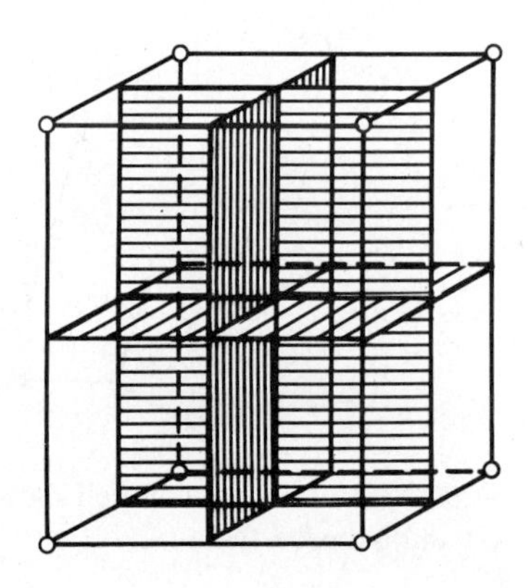

Figure 3.8. Unique mirror planes in an orthorhombic unit cell.

only symmetry elements in the orthorhombic lattice, however. In fact, the a, b, and c axes are 2-fold with interleaving 2-fold axes at the midpoints of the cell edges and through the centers of each face. Thus the full symmetry of the orthorhombic lattice is $2/m$, $2/m$, $2/m$, but the designation *mmm* used here is sufficient and implies the 2-fold axes actually present.

The symbols $4/m$ *mm* designate the symmetry of the tetragonal lattice. The first pair of symbols represent the 4-fold c axis and the (001) planes perpendicular to it (Fig. 3.9). The second pair of symbols, *mm* in the secondary and tertiary positions, represent two mirror planes parallel to the c axis and intersecting at 45° in a line coincident with this axis. The secondary m represents the plane perpendicular to the x axis and repeated at 90° by the 4-fold axis. The tertiary m represents the planes at 45° to these. Interleaving planes occur parallel to (100), (010), and (001) at $a/2$, $b/2$, and $c/2$, respectively. Only the last at $c/2$ is shown in Fig. 3.9. In addition, an extra 4-fold axis is parallel to c and passes through the center of the *ab* face of the unit cell. Other symmetry elements occur in the tetragonal lattice, but the designation $4/m$ *mm* (usually written $4/mmm$) used in Table 3.1 is sufficient and implies the full symmetry of the system.

The trigonal system is unique in the sense of involving two lattices: the rhombohedral and the hexagonal. Depending on the symmetry characterizing the structures, some trigonal crystals are naturally referred to a rhombohedral framework, others more easily to a hexagonal one.

The unit cell in a rhombohedral lattice may be thought of as a cube that has been compressed or stretched along a body diagonal. This diagonal is a

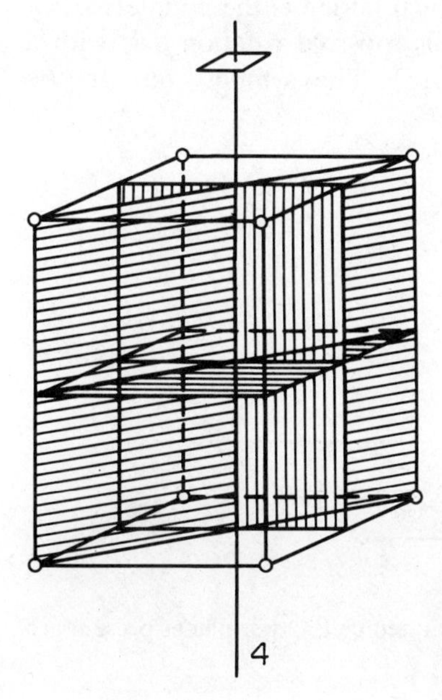

Figure 3.9. Unique mirror planes in a tetragonal unit cell. Other planes present are generated from these by symmetry.

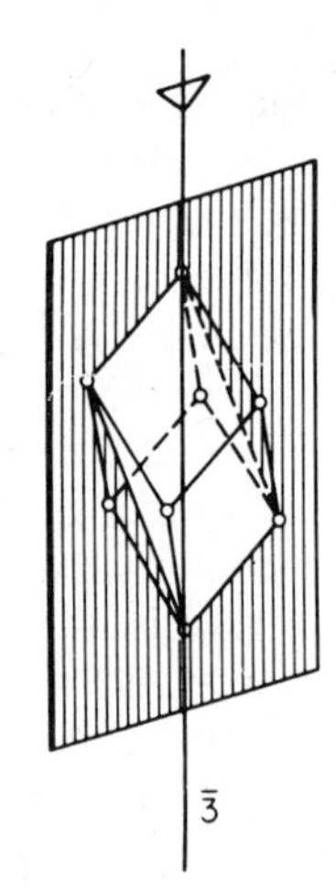

Figure 3.10. Unique mirror plane in a rhombohedral unit cell. Other planes present are generated from this one by symmetry.

3-fold rotary inversion axis and lies in the intersection of three mirror planes. Only $\bar{3}$ and one of the mirror planes are shown in Fig. 3.10, but these are sufficient since $\bar{3}$ generates the additional mirrors at 60° to that shown. Thus, $\bar{3}m$ indicates the essential symmetry of the rhombohedral lattice.

It may come as a surprise that any trigonal structure should be referred to a hexagonal lattice having as one symmetry element a 6-fold axis. As we shall see, however, it is not necessary for the internal structure to use all of the lattice symmetry, and the 3-fold axis of the trigonal system is an allowed subelement of the 6-fold axis of the lattice.

The designation $6/m\ mm$ for the hexagonal lattice is the counterpart of that for the tetragonal lattice. The c axis is a 6-fold rotation axis with a mirror plane perpendicular to it (Fig. 3.11). The symbols mm in the

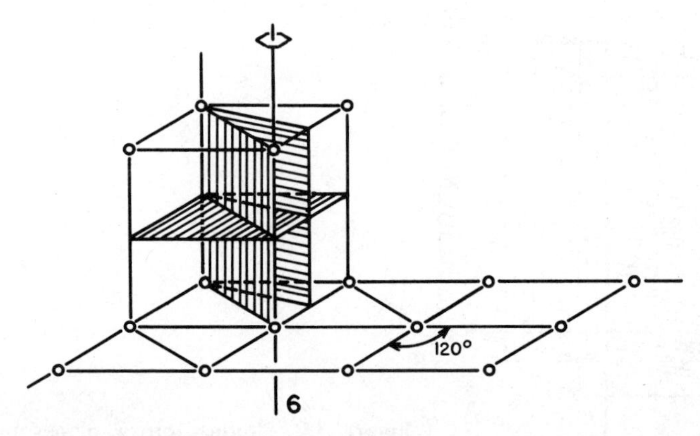

Figure 3.11. Unique mirror planes in a hexagonal unit cell. Other planes present are generated from these by symmetry.

secondary and tertiary positions represent two sets of mirror planes parallel to the c axis and intersecting at 30° in a line coincident with it. The secondary m represents the plane perpendicular to the x axis and repeated at 60° intervals. The tertiary m designates planes at 30° to these (parallel to the x axis and repeated at 60° intervals).[2]

The symbols in $m3m$ designate the cubic lattice. We note that since the number 3 does not appear first it does not represent a principal 3-fold axis. Instead, it refers to the set of four 3-fold body diagonals of a cubic unit cell, and the following m refers to the set of six diagonal planes in which these axes lie. One of the 3-fold body diagonals is shown in Fig. 3.12, along with one of the three diagonal planes in which it lies (shaded horizontally). The first m in the symmetry designation refers to the set of three planes parallel to the faces of the unit cell. One plane of the set is shown as the diagonally shaded plane in Fig. 3.12. The three sets of symmetry elements designated by $m3m$ define the cubic lattice, and if present in the lattice, they ensure that it is cubic. All the additional symmetry elements, the six 2-fold and three 4-fold axes, follow from those designated.

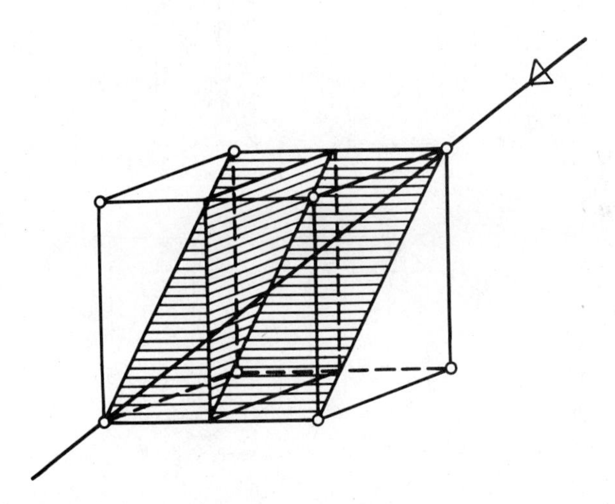

Figure 3.12. Unique mirror planes in a cubic unit cell. Other planes present are generated from these by symmetry.

[2]Hexagonal lattices are often distinguished by the use of four indices (h, k, i, l) to characterize planes. One of these indices $[i = -(h + k)]$ is redundant, but its use helps preserve symmetry. For further discussion, see Buerger, *Elementary Crystallography*, pp. 103–106, and N. F. M. Henry, H. Lipson, and W. A. Wooster, *The Interpretation of X-Ray Diffraction Photographs*, Macmillan, London, 1960, pp. 12–15, 19–23.

3.2. NONPRIMITIVE LATTICES

The simple lattices in each of the crystal systems in Table 3.1 have lattice points only at the corners of the unit cell; that is, there is the equivalent of one lattice point per unit cell. Such lattices are termed primitive and are designated by the letter *P* preceding the symmetry symbols (except in the rhombohedral lattice, where *R* is used). Thus the seven primitive lattices are denoted as $P\bar{1}$, $P2/m$, $Pmmm$, $P4/mmm$, $R\bar{3}m$, $P6/mmm$, and $Pm3m$.

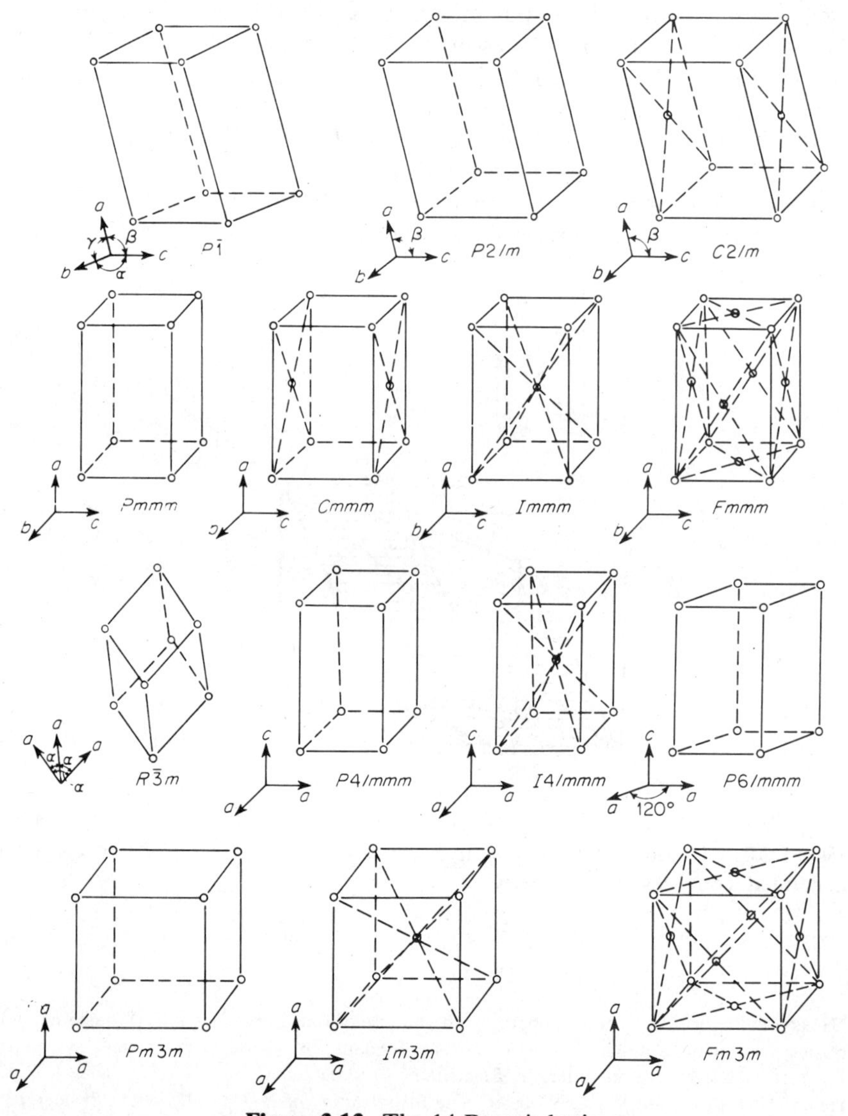

Figure 3.13. The 14 Bravais lattices.

In his study of lattices, Bravais discovered some that were more complex but still conformed to the symmetry of one of the seven crystal systems. These *nonprimitive* lattices contain two or more lattice points (*doubly primitive*, *triply primitive*, etc.) per unit cell and can be most simply viewed as the combination of a primitive lattice with one or more offset identical copies of itself. Thus the lattice points and every symmetry element in the primitive lattice are reproduced with a corresponding offset. The 14 Bravais lattices (Fig. 3.13) consist of seven primitive and seven nonprimitive lattices.

A nonprimitive lattice with a pair of lattice points centered on opposite faces of the unit cell is designated A, B, or C depending on whether the bc, ac, or ab faces are centered. If there is a lattice point at the body center of a unit cell, it is designated by I (inner). If all faces have lattice points at their centers, the designation is F.

In any array of lattice points, it is always possible to choose a primitive triclinic cell regardless of the symmetry present. But to disregard that symmetry would be to neglect the simplification it provides and lose the advantages that result from classification of crystals according to Table 3.1. In selecting the unit cell, therefore, a cardinal rule is to choose in such a way that it conforms to the symmetry actually present. In addition, there are conventions that bring a degree of standardization to the choice of a cell, although there is often latitude that may lead to different lattice designations. For example, in Fig. 3.14 the unit cells outlined by the light grid lines have their ab faces centered and the lattice would be designated C. An alternative and equally good set of unit cells would correspond to the cell outlined by the heavy lines. Those points at the center of the ab faces of the former cells are at the body centers of the latter, and the designation would be I. Both have the symmetry of the monoclinic lattice, but conventionally the C-centered cell is chosen. Occasionally, however, convenience or structural considerations may favor the use of a nonstandard cell.

Note that in the monoclinic system a unit cell is never chosen in such a way that the lattice designation is B. In this case a primitive unit cell of half the size, still monoclinic, can always be found (Fig. 3.15). For the same

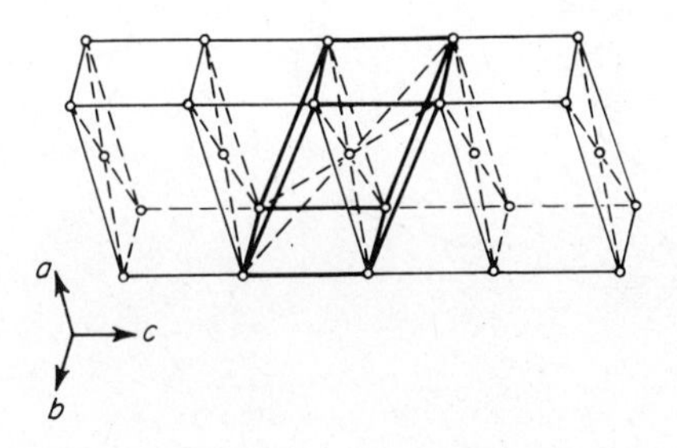

Figure 3.14. Choice of C or I unit cells in monoclinic lattices.

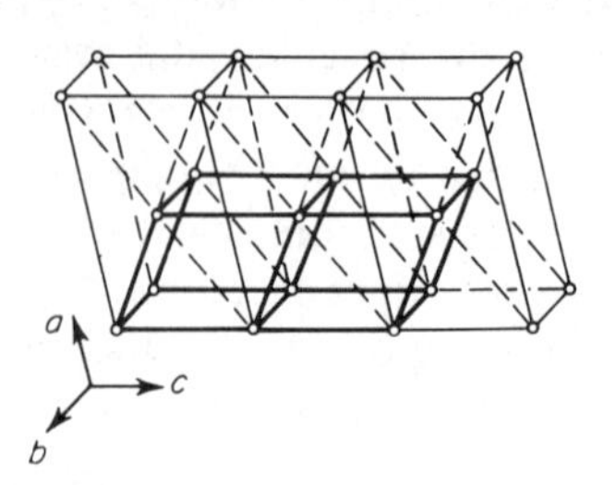

Figure 3.15. Relationship between monoclinic B and P lattices.

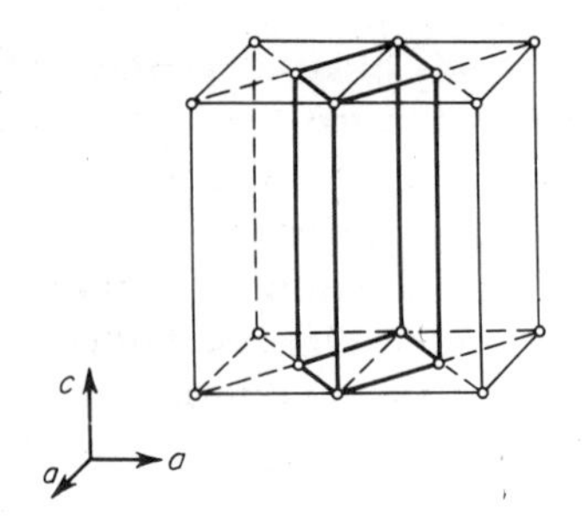

Figure 3.16. Relationship between tetragonal C and P lattices.

reason, no C tetragonal lattice is listed in Fig. 3.13. Such a lattice is, in fact, nothing but a simple tetragonal lattice in a different orientation. Figure 3.16 shows the relationship between a C-centered tetragonal unit cell and the primitive unit cell of half the size.

In the hexagonal system it is possible, and sometimes convenient, to choose an orthogonal set of axes by choosing a doubly primitive unit cell. How this can be done is illustrated in Fig. 3.17, where it is evident that one of the equal axes and c remain unchanged, while the third axis, b, is perpendicular to them. The resulting cell is C-centered and has twice the volume of the original hexagonal cell. It is clear that although the lattice is now orthorhombic, there are still only two independent lattice parameters, since $b = 2a \cos 30°$.

As noted earlier, the rhombohedral and hexagonal lattices are related and their unit cells interconvertible. Thus the primitive unit cell of a hexagonal lattice can be transformed to a triply primitive rhombohedral cell with three times the volume, the c axis of the hexagonal cell becoming the body diagonal of the rhombohedral cell. Conversely, the primitive unit cell of a rhombohedral lattice can be transformed to a triply primitive hexagonal cell with three times the volume, the $\bar{3}$ axis of the rhombohedral cell becoming the c axis of the hexagonal cell. Although the hexagonal cell is

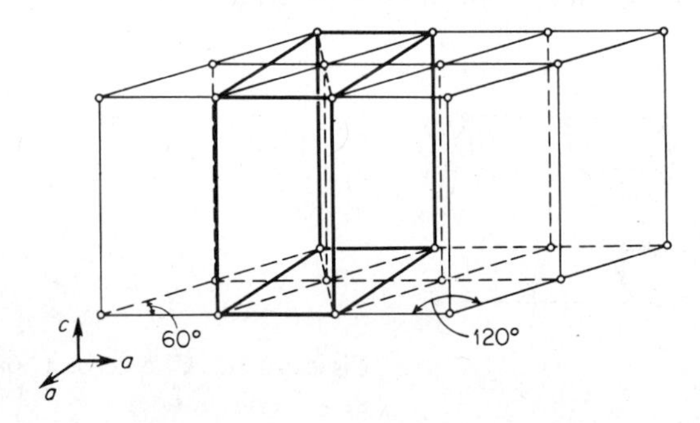

Figure 3.17. Possible choice of an orthorhombic unit cell in the hexagonal system.

larger, the transformation is usually advantageous for illustrative and computational convenience for structures initially referred to a rhombohedral lattice.[3]

The ability to consider alternative unit cells of quite different shapes points up the important fact that the native crystal lattice is presented to us only as a set of equivalent points. How these are connected is a matter of convention and judgment, made more difficult in the presence of experimental error. It is sometimes neither simple nor obvious, and is discussed in practical terms in Chapter 5.

It is left as an exercise to show that the symmetry of each nonprimitive lattice in Fig. 3.13 conforms to that of the crystal system with which it is associated. Furthermore, readers should convince themselves that in each of the nonprimitive Bravais lattices the environments of all lattice points are, in fact, identical. It is an instructive exercise to extend these considerations to other lattices, postulating additional possible systems and showing that they are equivalent to those listed in Fig. 3.13.

3.3. POINT GROUPS[4–6]

If the faces of crystals are considered as planes defining an idealized solid figure, it is found that they are related by just those types of symmetry operations already found in lattices, namely, rotary axes, mirrors, centers of symmetry, and rotary inversion axes. Therefore, crystals can be classified in terms of the group[7] of symmetry operations relating their faces. Each of these groups, known as point groups, represents one of the possible unique combinations of crystallographic symmetry elements. These operations may be visualized in terms of the relationships among the faces of the crystal (Fig. 3.18*a*) or as the symmetry operations relating the points of an assemblage (Fig. 3.18*b*). In Fig. 3.18*a* an idealized crystal is shown referred to coordinate axes x, y, z. By inspection, it is evident that the faces are reflected in the xy, xz, and yz planes, that is, there are three mutually perpendicular mirror planes. Figure 3.18*b* shows an assemblage of points related by the same group of symmetry operations as in Fig. 3.18*a*, while

[3] *International Tables*, Vol. I, pp. 15–21.

[4] *International Tables*, Vol. I, pp. 25–34.

[5] McWeeny, *Symmetry*, pp. 54–85.

[6] For excellent descriptions of point group symmetry in terms of the external forms of crystals, see Phillips, *An Introduction to Crystallography*, pp. 103–151, and Buerger, *Elementary Crystallography*, pp. 112–170.

[7] The term "group" is used here and in connection with space groups in a strict mathematical sense. It is not necessary, however, to understand group theory in order to follow the subsequent discussion, since the individual groups may be considered merely as collections of symmetry elements. For good introductions to the mathematical details, see McWeeny, *Symmetry*, and Prince, *Mathematical Techniques in Crystallography and Materials Science*, Chapter 2.

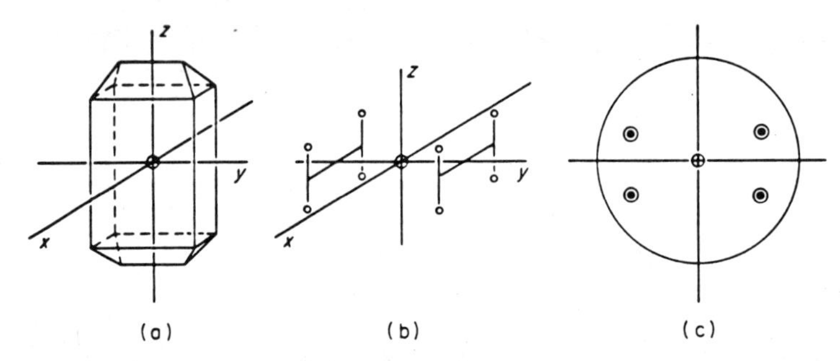

Figure 3.18. (*a*) Crystal with symmetry *mmm*. (*b*) Set of points related by symmetry *mmm*. (*c*) Plane representation of symmetry *mmm*.

Fig. 3.18*c* shows a convenient plane representation of these operations. Solid circles represent points above the plane of the paper, while open circles represent those below. Solid and open circles superimposed represent two points related by reflection in the plane of the paper.

It is characteristic of point groups that there is a point in space about which or through which the symmetry elements may be considered to operate and which remains unmoved by the operations. This point is taken as the origin when the group is referred to coordinate axes and is indicated by ⊕ in Fig. 3.18.

The symmetries of the 32 point groups are illustrated by the diagrams in Fig. 3.19. They are tabulated in columns according to the principal axis, which is assumed to be normal to the plane of the paper. In the first row of the figure, only the principal axis of symmetry is present. In succeeding rows, additional symmetry is added as indicated at the left. Of the 42 combinations thus obtained, only 32 are unique point groups. The remaining 10 are alternative representations of members of the unique set. These have been designated by indicating the equivalent symmetry within their circles.

The symbols used for the essential symmetry of each point group have been chosen to emphasize the principal axis. When a different designation is commonly used, it is also included. That there may be additional symmetry beyond the essential is seen by considering the point group in the orthorhombic system designated $2/mm$. These symmetry elements are inevitably associated with two additional 2-fold axes and an additional mirror plane, as can be seen from the diagram for this point group in Fig. 3.19. The full symmetry is designated by $2/m2/m2/m$. The symbols $2/mm$ are sufficient, however, to specify the essential symmetry, and this form is used here to emphasize the presence of a 2-fold axis. A more conventional designation for this point group is *mmm*, which emphasizes the equivalence of the three axes.

The 32 point groups are divided among the seven crystal systems. The

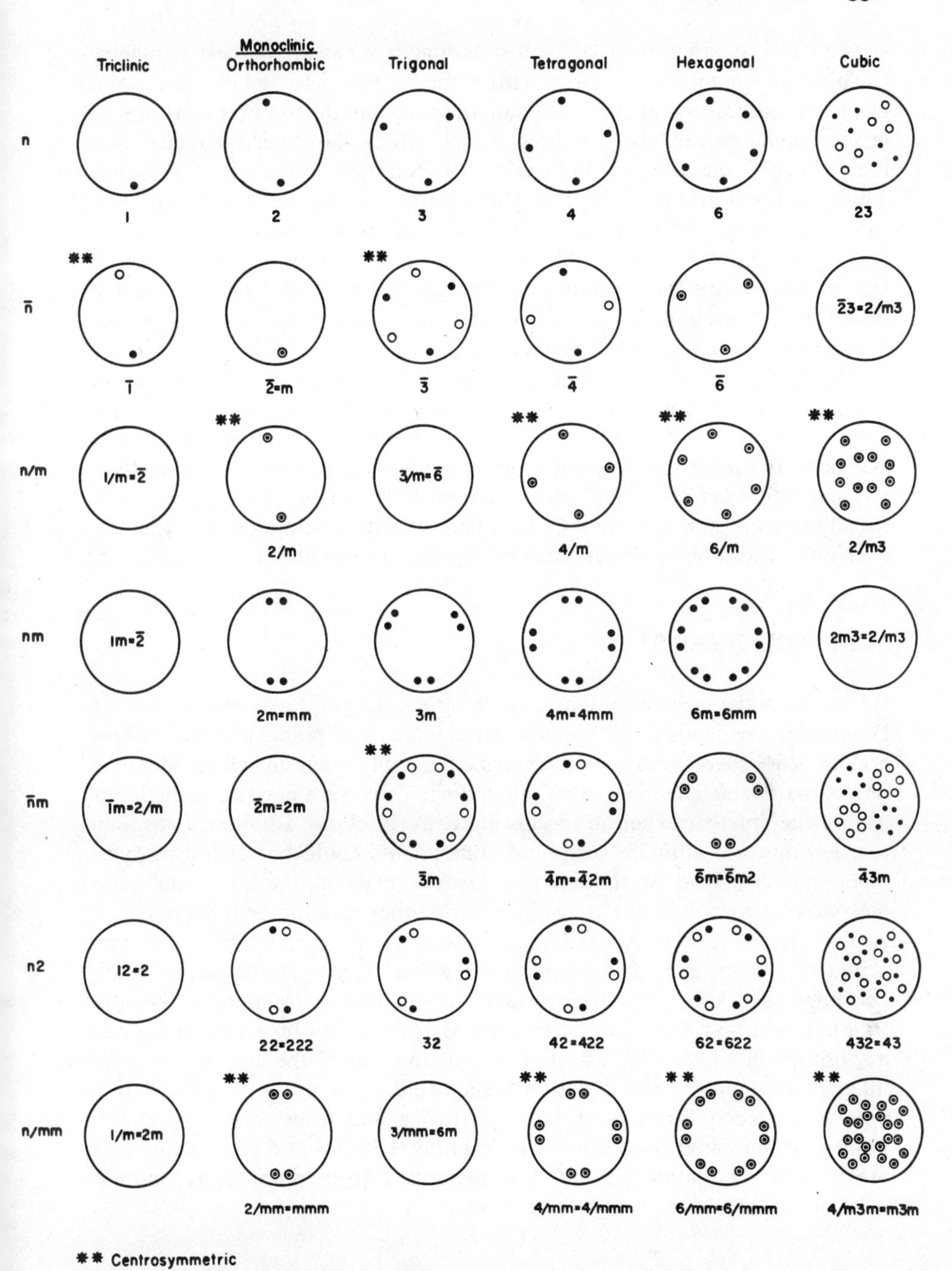

Figure 3.19. Plane representations of the 32 point groups.

lattice that is characteristic of each system always has the highest symmetry (*holohedral* symmetry) possible within the system and defines the point group of highest symmetry that can be accommodated. Point groups of lesser symmetry can also be fitted to the lattice, however, provided that their effect on the axes and the angles between them agree with the lattice. Thus, the point groups 2, *m*, and 2/*m* require one axis to be perpendicular to the plane of the other two, and they fall in the monoclinic system. Likewise the point groups 222, 2*mm* (more commonly written as *mm*2), and *mmm* require three mutually perpendicular axes, and they fall in the orthorhombic system. In a similar way, the other point groups can be assigned to the appropriate crystal system.

The faces that develop when crystals grow are a direct consequence of the internal arrangements of atoms, ions, or molecules, and the relationships among the faces of ideally developed crystals always conform to the symmetry of one of the 32 point groups. In practice, however, the growth of any individual crystal is sufficiently nonuniform that the ideal symmetry is usually obscure and may only be revealed by extensive optical examination of many crystals or by the application of other techniques.

3.4. SPACE GROUPS[8]

It can be shown rigorously that combining the 32 point groups with the 14 Bravais lattices leads to 230 unique arrangements of points in space. These are the 230 *space groups* that describe the only ways in which identical objects can be arranged in an infinite lattice. They were derived in the latter part of the nineteenth century following Bravais' classic studies. It was then realized that the ultimate structural components could be related by symmetry not displayed by the external form of crystals, the additional symmetry elements combining translation with either rotation or reflection.

The combination of a rotation axis and a translation parallel to the axis produces a *screw axis*. The direction of such an axis is usually along a unit cell edge, and the translation must be a subintegral fraction of the unit translation in that direction. Screw axes are designated by an integer n and a subscript m where $n = 1, 2, 3, 4$, or 6 is the fold of the axis and m is an integer less than n. Thus 3_1 designates a 3-fold screw axis with a translation between successive points of 1/3 (m/n) of a unit translation (Fig. 3.20). Point 2 is generated from point 1 by rotating $+360°/3$ and advancing $+1/3$ of the unit translation.[9] Point 3 is generated from point 2 by another rotation of 360°/3 and an advance of 1/3.

[8]For an extensive discussion of space groups, involving, however, some nonstandard choices of origin, see M. J. Buerger, *Elementary Crystallography*, Wiley, New York, 1963, pp. 235–273.

[9]The positive sense of rotation and the advance in screw axes are always related as in a right-hand screw.

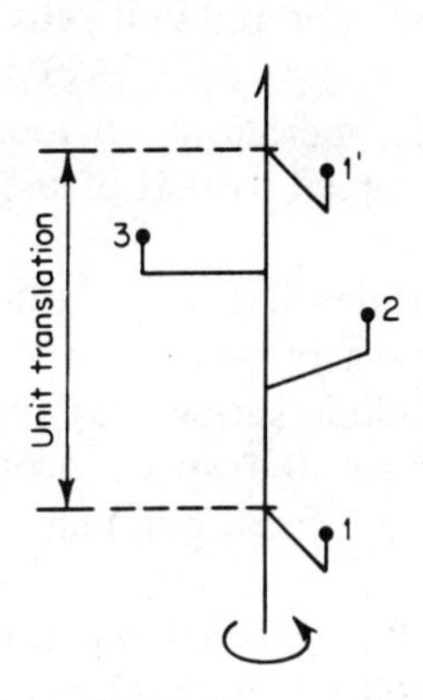

Figure 3.20. Screw axis 3_1.

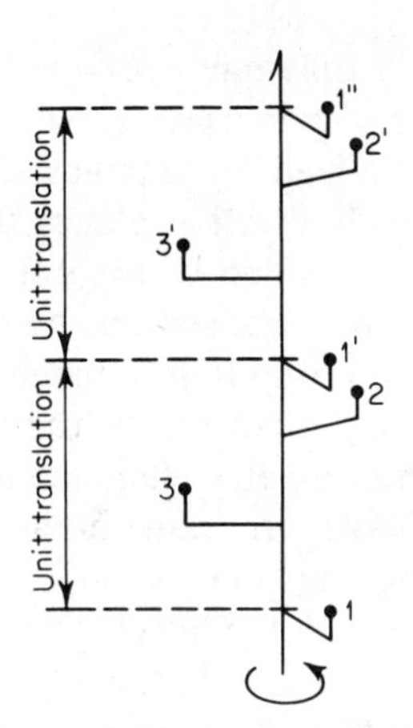

Figure 3.21. Screw axis 3_2.

In a similar way, 3_2 indicates a 3-fold screw axis with a translation of 2/3 of the unit translation. Figure 3.21 shows the relationship between successive points. Point 2 is generated from point 1 by rotating +360°/3 and advancing 2/3. Point 3′ arises from an additional rotation of +360°/3 and an additional translation of +2/3. The rest of the points in Fig. 3.21, 1′, 2′, and 3, are obtained from points 1, 2, and 3′ by a unit translation. Comparison of Figs. 3.20 and 3.21 will indicate that 3_1 and 3_2 are related in the same way as right- and left-handed screws; that is, they are enantiomorphs.

It is characteristic of an n-fold screw axis that the position of the nth point laid down differs from the initial point by an integral number of unit translations; that is, the positions of these points within their respective cells are identical.

The combination of a mirror plane and a translation parallel to the reflecting plane produces a *glide plane*. The translation in such a plane is along an edge or face diagonal of the unit cell, and in most cases its magnitude is half the axial or diagonal length. A glide plane is designated by a, b, or c if the translation is $a/2$, $b/2$, or $c/2$ and by n if the translation is $(a+b)/2$, $(a+c)/2$, or $(b+c)/2$. In Fig. 3.22 an a glide reflecting in (010) is illustrated.

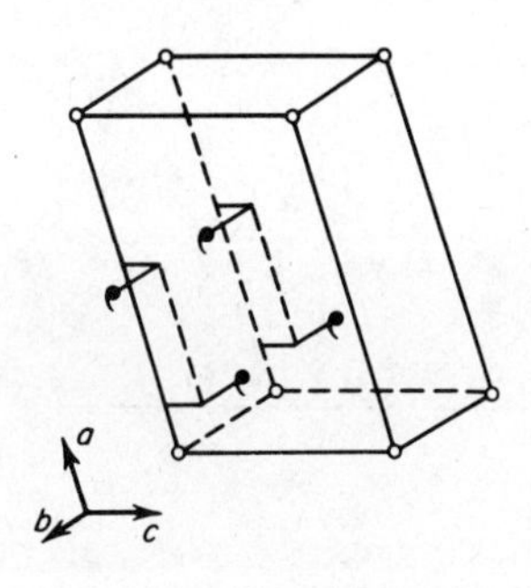

Figure 3.22. Glide plane a.

There is only one additional type of glide plane, the diamond glide, *d*. It can occur only in space groups with face- or body-centered unit cells, and it is characterized by a translation $(a+b)/4$, $(a+c)/4$, or $(b+c)/4$. It is characteristic of glide planes that after two glide operations (or four for *d* glides) the position of the point laid down is identical to that of the initial point plus unit translations on one or two axes.

A space group is designated by a capital letter identifying the lattice type (*P*, *C*, etc.) followed by the point group symbols in which rotation and reflection symmetry elements are extended to include screw axes and glide planes.[10] We are now in a position to illustrate the space groups by combining the point groups one by one with the corresponding lattices, beginning with the lowest symmetry.

In order to discuss the effect of symmetry on the cell contents, it is usual to consider its effect on some arbitrary general point *x*, *y*, *z*. The symmetry operations act on *x*, *y*, *z* to produce new coordinates for *equivalent positions*, that is, sites identical in all respects as seen by the molecule (see Tables 3.2 and 3.3).

TABLE 3.2 Some Symmetry Elements and Their Equivalent Positions

		Equivalent Positions	
Axis 2	Parallel to *a*	x, y, z	$x, \bar{y}, \bar{z}$
2	*b*	x, y, z	$\bar{x}, y, \bar{z}$
2	*c*	x, y, z	$\bar{x}, \bar{y}, z$
2_1	*a*	x, y, z	$x+\frac{1}{2}, \bar{y}, \bar{z}$
2_1	*b*	x, y, z	$\bar{x}, y+\frac{1}{2}, \bar{z}$
2_1	*c*	x, y, z	$\bar{x}, \bar{y}, z+\frac{1}{2}$
Plane *m*	Perpendicular to *a*	x, y, z	$\bar{x}, y, z$
m	*b*	x, y, z	$x, \bar{y}, z$
m	*c*	x, y, z	$x, y, \bar{z}$
a	*b*	x, y, z	$x+\frac{1}{2}, \bar{y}, z$
a	*c*	x, y, z	$x+\frac{1}{2}, y, \bar{z}$
b	*a*	x, y, z	$\bar{x}, y+\frac{1}{2}, z$
b	*c*	x, y, z	$x, y+\frac{1}{2}, \bar{z}$
c	*a*	x, y, z	$\bar{x}, y, z+\frac{1}{2}$
c	*b*	x, y, z	$x, \bar{y}, z+\frac{1}{2}$
n	*a*	x, y, z	$\bar{x}, y+\frac{1}{2}, z+\frac{1}{2}$
n	*b*	x, y, z	$x+\frac{1}{2}, \bar{y}, z+\frac{1}{2}$
n	*c*	x, y, z	$x+\frac{1}{2}, y+\frac{1}{2}, \bar{z}$
d	*a*	x, y, z	$\bar{x}, y+\frac{1}{4}, z+\frac{1}{4}$
d	*b*	x, y, z	$x+\frac{1}{4}, \bar{y}, z+\frac{1}{4}$
d	*c*	x, y, z	$x+\frac{1}{4}, y+\frac{1}{4}, \bar{z}$

[10] *International Tables*, Vol. I, pp. 45–56, 74–346.

TABLE 3.3 Symbols for Symmetry Elements

Symmetry	Symbol	Designation If Parallel to Plane of Projection	Designation If Perpendicular to Plane of Projection
Center	$\bar{1}$	○	○
2-Fold axis	2	⟷	● (lens-shaped)
3-Fold axis	3	—	▲
4-Fold axis	4	—	■
6-Fold axis	6	—	● (hexagon)
2-Fold screw axis	2_1	half-arrow line (⟵⟶ with half arrowheads)	[2-fold screw symbol]
3-Fold screw axis	3_1	—	[3-fold screw symbol]
3-Fold screw axis	3_2	—	[3-fold screw symbol]
4-Fold screw axis	4_1	—	[4-fold screw symbol]
4-Fold screw axis	4_2	—	[4-fold screw symbol]
4-Fold screw axis	4_3	—	[4-fold screw symbol]
6-Fold screw axis	6_1	—	[6-fold screw symbol]
6-Fold screw axis	6_2	—	[6-fold screw symbol]
6-Fold screw axis	6_3	—	[6-fold screw symbol]
6-Fold screw axis	6_4	—	[6-fold screw symbol]
6-Fold screw axis	6_5	—	[6-fold screw symbol]
Mirror	*m*	┐	——
a Glide plane	*a*	┐ with downward arrow	----
b Glide plane	*b*	┐ with leftward arrow	----
c Glide plane	*c*	—	......
n Glide plane	*n*	corner with diagonal arrow	-·-·-
d Glide plane	*d*	corner symbol with arrows marked $\frac{1}{8}$ and $\frac{3}{8}$	-→·-·-

Point Group 1

Combining point group 1 with the triclinic lattice leads to space group *P*1, which is illustrated in Fig. 3.23 by the pattern of commas associated with the lattice points projected parallel to the *c* axis and onto the *C* face. No symmetry operation is involved except for the identity axis. Only a single comma occurs in each unit cell, and it is at an arbitrary position *x*, *y*, *z*, the + sign indicating the location above the base of the unit cell.

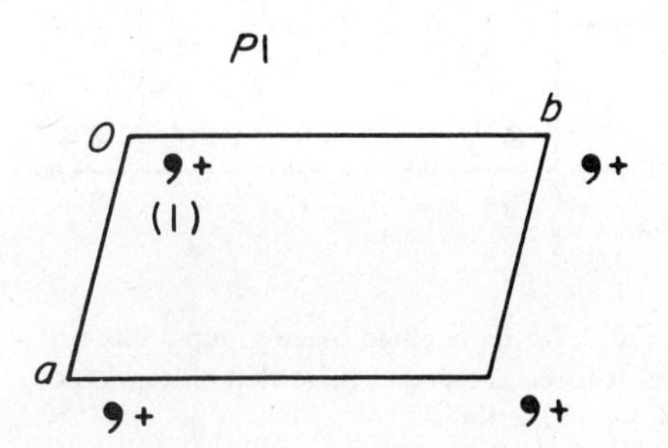

Figure 3.23. *P*1, equivalent positions *x*, *y*, *z*.

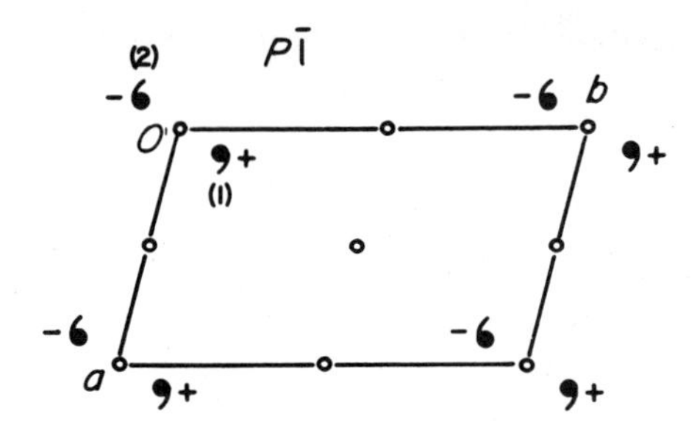

Figure 3.24. $P\bar{1}$, equivalent positions (1) x, y, z; (2) $\bar{x}, \bar{y}, \bar{z}$.

Point Group $\bar{1}$

The next point group, $\bar{1}$, is centrosymmetric and is consistent with the full symmetry of the triclinic lattice. Combining the two leads to the centrosymmetric space group $P\bar{1}$ (Fig. 3.24), two commas being associated with each lattice point and with each unit cell. The two associated with a lattice point are related by the center of symmetry at the point; the two within a unit cell are related by the center of symmetry at $\frac{1}{2}, \frac{1}{2}, \frac{1}{2}$. Centrosymmetric point groups always lead to centrosymmetric space groups, and, as we shall see in Chapter 8, considerable simplification results by having the unit cell origin coincide with a center of symmetry.

Point Group 2

The next three point groups (see Fig. 3.19) have either a 2-fold axis, a mirror plane, or both. These symmetry elements are inconsistent with the triclinic lattice but consistent with the monoclinic.[11] Here we have two lattices, P and C (or A, which merely exchanges labels on a and c; or I, which is a different choice of unit cell). Combining point group 2 with the P lattice leads to space groups $P2$ and $P2_1$, since both the 2-fold rotation and

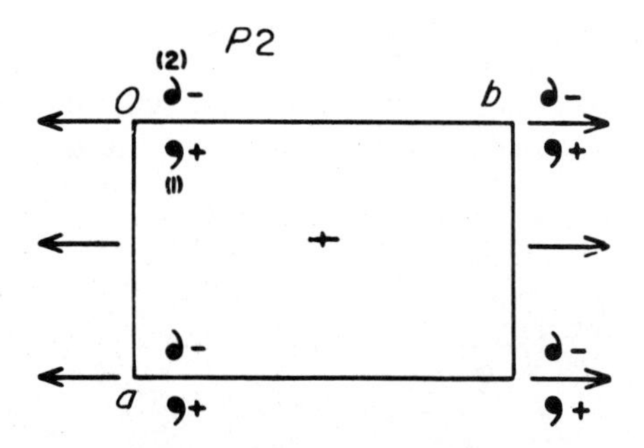

Figure 3.25. $P2$, equivalent positions (1) x, y, z; (2) $\bar{x}, y, \bar{z}$.

[11]Two settings are listed in *International Tables*, Vol. I, for monoclinic space groups. The first setting has c unique, the second setting, b unique. It is recommended there that all reports of monoclinic structures follow the historic b unique second setting.

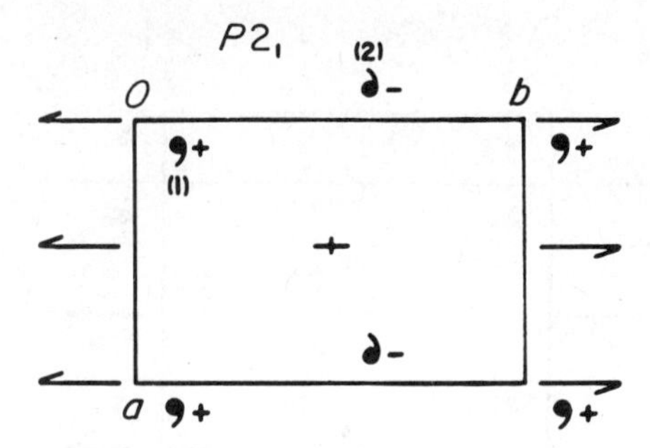

Figure 3.26. $P2_1$, equivalent positions (1) x, y, z; (2) $\bar{x}, y+\frac{1}{2}, \bar{z}$.

screw axes must be considered. The two space groups are illustrated by projections on the C faces in Figs. 3.25 and 3.26, where the 2 and 2_1 axes are taken to coincide with the b axes, giving rise to accompanying 2 and 2_1 axes at $a/2$.

Combining point group 2 with the monoclinic C lattice leads to space group $C2$ and provisionally to $C2_1$. The two are illustrated by the C face projections in Figs. 3.27 and 3.28. In Fig. 3.27 the 2-fold rotation axes are taken coincident with the unit cell b axes and necessarily recur at $a/2$ because the lattice is C-centered. Here we see a new feature, accompanying axes of a different kind, the 2_1 axes at $a/4$ and $3a/4$, midway between the 2-fold rotation axes. In considering $C2_1$ we see in Fig. 3.28 that taking the 2_1 axes coincident with b (and repeated at $a/2$) leads to accompanying 2-fold rotary axes at $a/4$ and $3a/4$. At first sight the patterns of commas in Figs. 3.27 and 3.28 appear to differ. Shifting the origin by $a/4$, however, moving from O to O' in Fig. 3.28, causes it to fall on a 2-fold rotation axis as it does in Fig. 3.27. The *pattern* of commas in the new unit is clearly the same as that in the unit cell of Fig. 3.27, except for orientation and elevation, but these are arbitrary. Thus $C2_1$ is not a new space group.

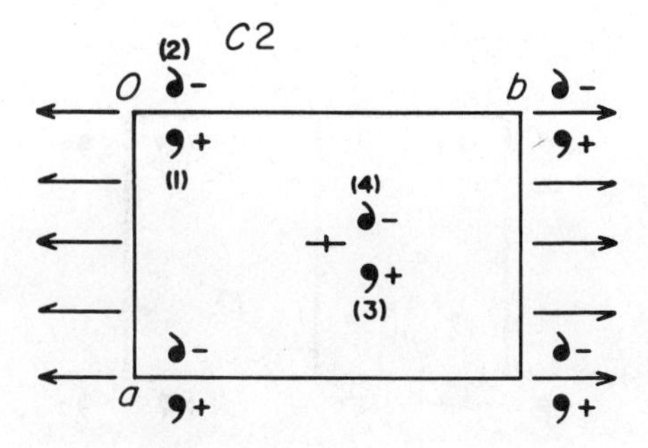

Figure 3.27. $C2$, equivalent positions (1) x, y, z; (2) $\bar{x}, y, \bar{z}$; (3) $x+\frac{1}{2}, y+\frac{1}{2}, z$; (4) $\bar{x}+\frac{1}{2}, y+\frac{1}{2}, \bar{z}$.

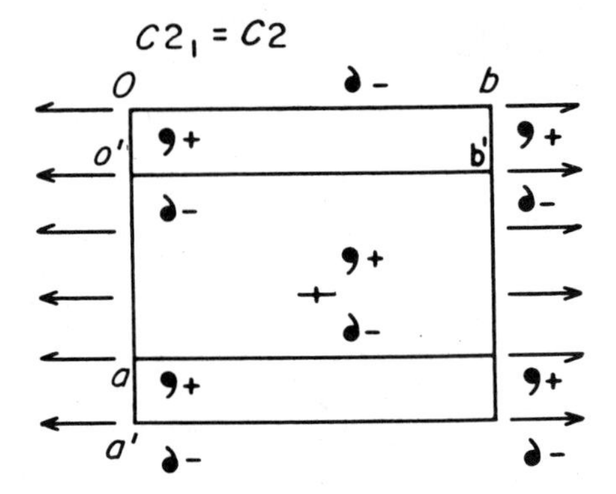

Figure 3.28. $C2_1$, to be compared with Fig. 3.27. See text.

Point Group *m*

Turning to point group *m* and considering both mirror planes and glide planes, we expect four space groups, the planes being necessarily perpendicular to the *b* axis of the unit cell. In the case of glide planes, the direction of the glide may be *a*, *c*, or *n*. Convention dictates *c*. Accordingly, the possible space groups are *Pm*, *Pc*, *Cm*, and *Cc*. These are illustrated in Figs. 3.29–3.32. In Fig. 3.29 the mirror planes in *Pm*, seen edge on and shown as heavy lines, coincide with the *B* faces of the unit cell, leading, as we have seen before, to interleaving planes occurring at $b/2$. The space group *Pc* in Fig. 3.30 is similar, except that the glide planes, shown as dotted lines, replace the mirror planes.

Figure 3.31 illustrates space group *Cm*. The mirror planes, coincident with the *B* faces of the unit cell, necessarily repeat at $b/2$ because of the *C* centering. Again we have interleaving planes, but here they are of a different type, glide planes at $b/4$ and $3b/4$ represented by the dashed lines. That these are *c* glides is readily confirmed by noting the relationship between positions such as (1) and (4). The pair of mirror-related commas associated with the origin is repeated by the *C* centering at the same position relative to the point $\frac{1}{2}, \frac{1}{2}, 0$.

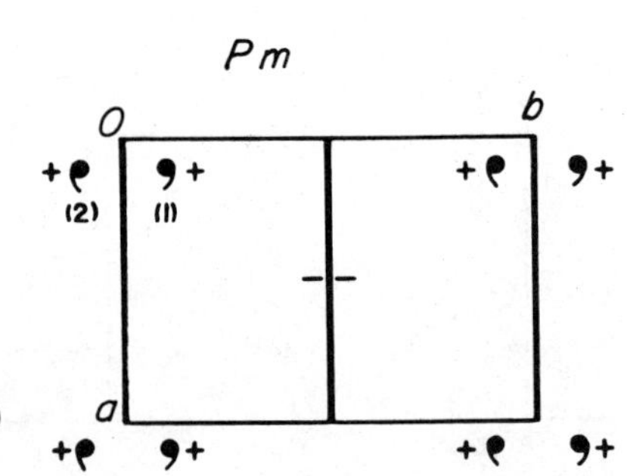

Figure 3.29. *Pm*, equivalent positions (1) x, y, z; (2) $x, \bar{y}, z$.

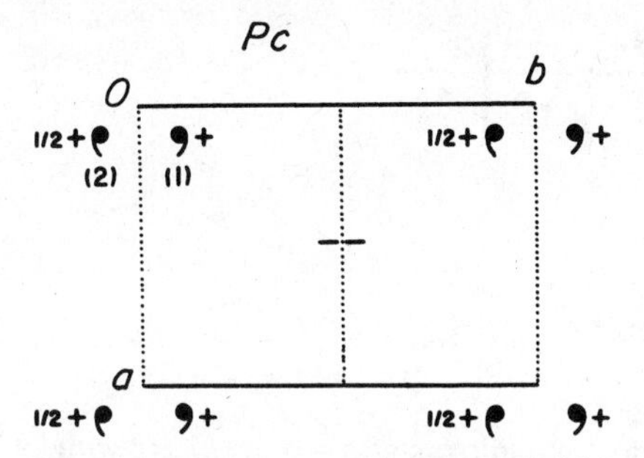

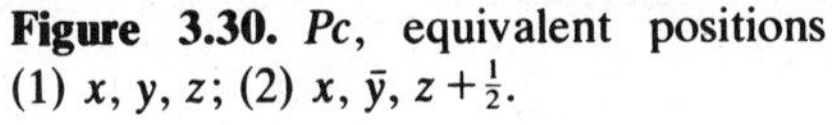

Figure 3.30. *Pc*, equivalent positions (1) x, y, z; (2) $x, \bar{y}, z+\frac{1}{2}$.

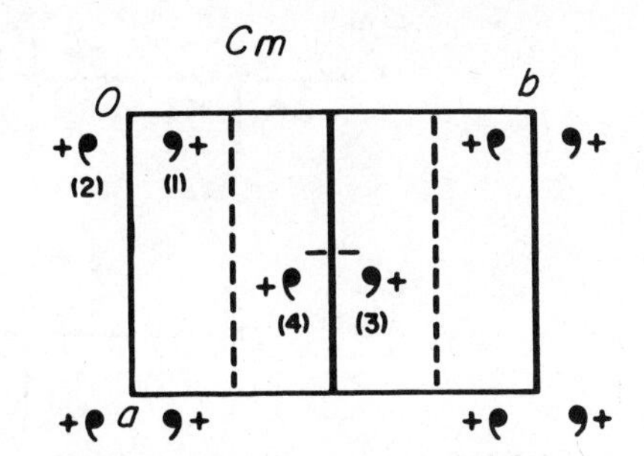

Figure 3.31. *Cm*, equivalent positions (1) x, y, z; (2) $x, \bar{y}, z$; (3) $x+\frac{1}{2}, y+\frac{1}{2}, z$; (4) $x+\frac{1}{2}, \bar{y}+\frac{1}{2}, z$.

An understanding of space group *Cc* (Fig. 3.32) follows directly from the preceding. The interleaving planes at $b/4$ and $3b/4$ are again glide planes but are now *n* glides represented by the dot-dashed lines.

Point Group 2/*m*

Combining point group $2/m$ with the monoclinic *P* lattice leads to the space groups $P2/m$, $P2_1/m$, $P2/c$, and $P2_1/c$. In $P2/m$ the center of symmetry lies at the intersection of the 2-fold rotary axis and the mirror plane. Thus choosing the 2-fold axis coincident with *b* and the mirror planes coincident with the *B* faces of the unit cell places the center of symmetry at the origin. Figure 3.33 shows the *C* projection of space group $P2/m$.

For the space groups $P2_1/m$, $P2/c$, and $P2_1/c$ the centers of symmetry do not lie at the intersections of the axes and planes because 2_1 and *c*

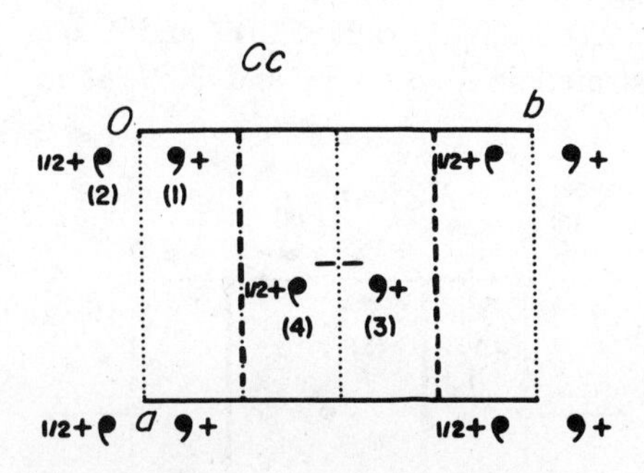

Figure 3.32. *Cc*, equivalent positions (1) x, y, z; (2) $x, \bar{y}, z+\frac{1}{2}$; (3) $x+\frac{1}{2}, y+\frac{1}{2}, z$; (4) $x+\frac{1}{2}, \bar{y}+\frac{1}{2}, z+\frac{1}{2}$.

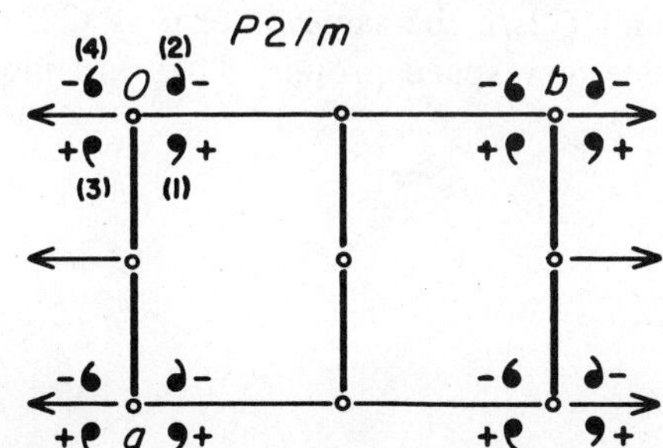

Figure 3.33. $P2/m$, equivalent positions (1) x, y, z; (2) $\bar{x}, y, \bar{z}$; (3) $x, \bar{y}, z$; (4) $\bar{x}, \bar{y}, \bar{z}$.

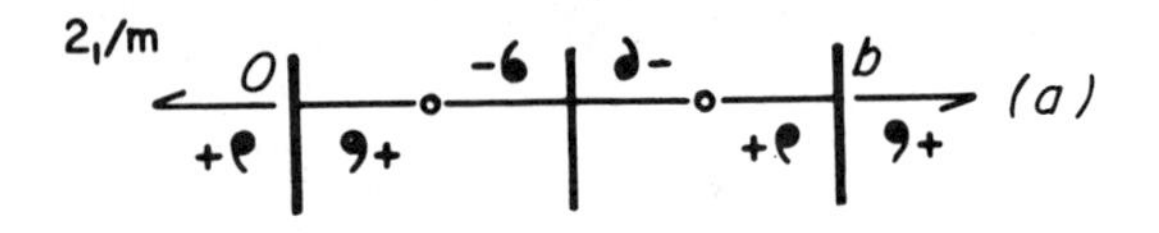

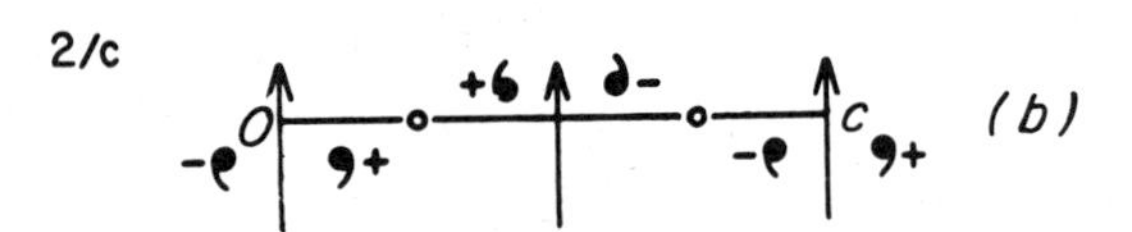

Figure 3.34. (*a*) $2_1/m$ with screw in *b* direction and mirrors at $b = 0, \frac{1}{2}$. (*b*) $2/c$ with *c*-glide perpendicular to *b* and axes at $c = 0, \frac{1}{2}$.

involve translations. Where centers of symmetry occur can be determined from the partial views of the symmetry elements. For symmetry $2_1/m$, with 2_1 coincident with *b*, Fig. 3.34*a* shows centers at $b/4$ and $3b/4$ and mirror planes at $b = 0$ and $b = 1/2$. Shifting the center to $b = 0$ places the mirror planes at $b/4$ and $-b/4 = 3b/4$. In the same way for symmetry $2/c$, Fig. 3.34*b* shows a center at $c/4$ with 2-fold axes at $c = 0$ and $c = \frac{1}{2}$. Shifting the center to $c = 0$ places the 2-fold axes at $c/4$ and $-c/4 = 3c/4$. In a similar way for symmetry $2_1/c$ it can be shown that with the center of symmetry at the origin, 2_1 axes occur at $c/4$ and $3c/4$ with glide planes at $b/4$ and $3b/4$.

Figure 3.35 shows the symmetry elements and equivalent positions for space group $P2_1/m$ with a center of symmetry at the origin. Figures 3.36 and 3.37 are similar views for space groups $P2/c$ and $P2_1/c$. Note that in the latter two figures the 2 and 2_1 axes are at $c/4$ (and $3c/4$). Thus, for example, the *z* coordinates for equivalent positions (1) in both figures generate $\bar{z} + \frac{1}{2}$ for equivalent positions (3).

In combining point group $2/m$ with the monoclinic *C* lattice we have $C2/m$, $C2_1/m$, $C2/c$, and $C2_1/c$. In Figs. 3.27 and 3.28 we saw that the 2 and 2_1 axes in the *C* lattice generated identical equivalent positions. In the same way it can be shown that $C2_1/m$ generates the same pattern as $C2/m$, and $C2_1/c$ the same pattern as $C2/c$. Accordingly, only $C2/m$ and $C2/c$ are new space groups. They are illustrated in Figs. 3.38 and 3.39. It is

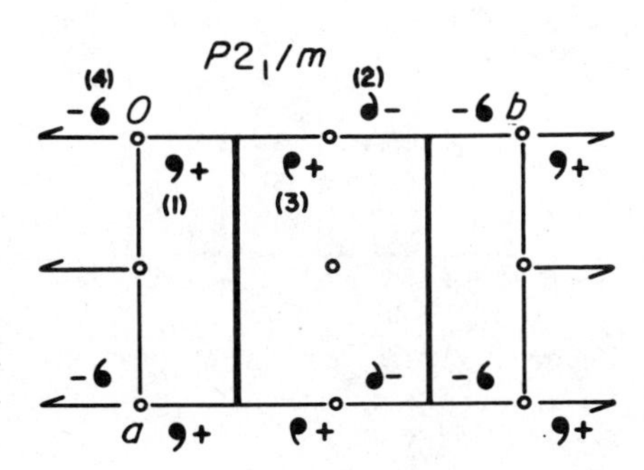

Figure 3.35. $P2_1/m$, equivalent positions (1) x, y, z; (2) $\bar{x}, y + \frac{1}{2}, \bar{z}$; (3) $x, \bar{y} + \frac{1}{2}, z$; (4) $\bar{x}, \bar{y}, \bar{z}$.

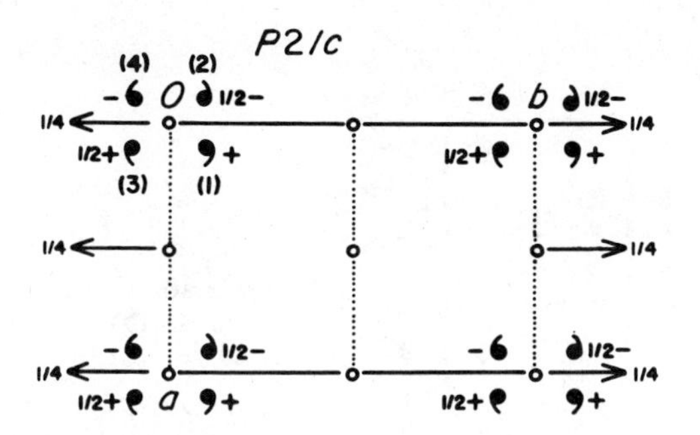

Figure 3.36. $P2/c$, equivalent positions (1) x, y, z; (2) $\bar{x}, y, \bar{z}+\frac{1}{2}$; (3) $x, \bar{y}, z+\frac{1}{2}$; (4) $\bar{x}, \bar{y}, \bar{z}$.

instructive to compare these figures for $C2/m$ and $C2/c$ with Figs. 3.33 and 3.36 for $P2/m$ and $P2/c$ and with Figs. 3.29 and 3.30 for Pm and Pc.

This completes the triclinic and monoclinic space groups and illustrates how the simpler symmetry elements are arranged singly or in combination and how they relate to the various patterns of equivalent positions in these two crystal systems. The same approach used for the lower symmetry space groups can be applied to those with higher symmetry, but their increasing complexity and sheer number preclude further development here.

A complete list of the 230 space groups in standard orientation is given in Table 3.4. They are tabulated with alternative orientations in *International Tables*.[12]

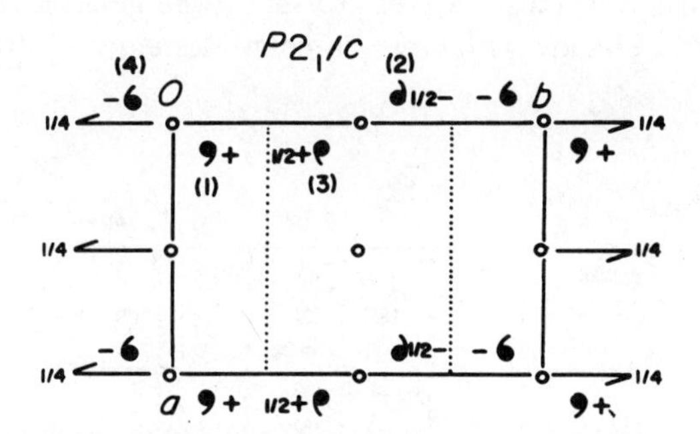

Figure 3.37. $P2_1/c$, equivalent positions (1) x, y, z; (2) $\bar{x}, y+\frac{1}{2}, \bar{z}+\frac{1}{2}$; (3) $x, \bar{y}+\frac{1}{2}, z+\frac{1}{2}$; (4) $\bar{x}, \bar{y}, \bar{z}$.

[12] *International Tables for Crystallography*, Vol. A, T. Hahn, Ed., Reidel, Dordrecht, 1983, pp. 102–707. A special situation exists for the higher symmetry trigonal space groups referred to hexagonal axes and where the identity symbol 1 appears (see Table 3.4). Fourteen such groups occur and are paired, with the identity symbol in either the secondary or tertiary position. For example, the symbols $1m$ in $P31m$ represent the x and y (and i) axes as 1-fold (identity) with mirror planes perpendicular to the tertiary directions 30° from the axes. In other words, the axes lie in the mirrors. The symbols $m1$ simply exchange the positions of the symmetry elements.

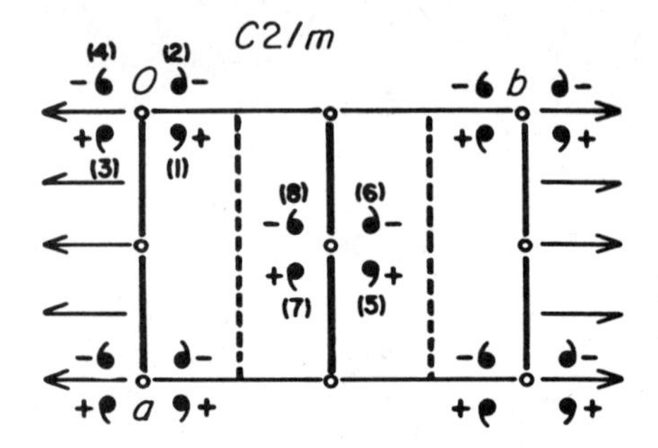

Figure 3.38. $C2/m$, equivalent positions (1) x, y, z; (2) $\bar{x}, y, \bar{z}$; (3) $x, \bar{y}, z$; (4) $\bar{x}, \bar{y}, \bar{z}$; (5) $x+\frac{1}{2}, y+\frac{1}{2}, z$; (6) $\bar{x}+\frac{1}{2}, y+\frac{1}{2}, \bar{z}$; (7) $x+\frac{1}{2}, \bar{y}+\frac{1}{2}, z$; (8) $\bar{x}+\frac{1}{2}, \bar{y}+\frac{1}{2}, \bar{z}$.

It is evident from the foregoing discussion that any space group is related to the point group resulting when the lattice symbol is deleted and remaining symbols with translation are replaced by the corresponding symbol without translation. Thus both $I4/mcd$ and $P4_2/nmc$ belong to point group $4/mmm$. A necessary consequence is that all those space groups, and only those, that reduce to centrosymmetric point groups are centrosymmetric. Consequently one can say on inspection that space group $Cmc2_1$, which reduces to $mm2$, is noncentrosymmetric, while $Fddd$, which reduces to mmm, is centrosymmetric.

Computational work in crystal structure determination requires the equivalent positions for the proper orientation of the space group to which the structure conforms.[13,14] These were developed for the triclinic and monoclinic space groups and are included in Figs. 3.23–3.33 and 3.35–3.39. The positions listed in those figures are said to be general because they are not restricted to coincide with any symmetry element. The related positions

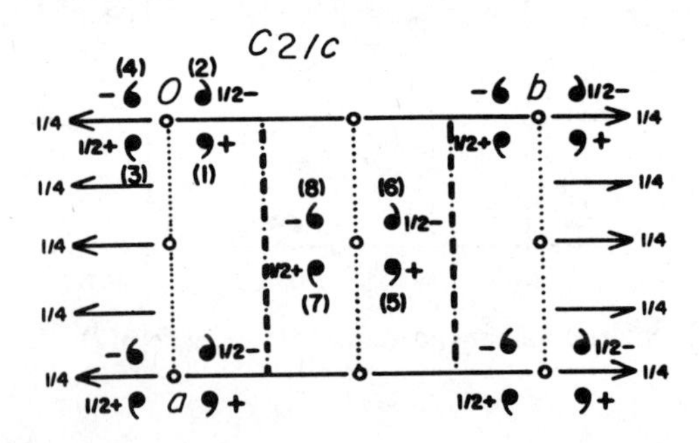

Figure 3.39. $C2/c$, equivalent positions (1) x, y, z; (2) $\bar{x}, y, \bar{z}+\frac{1}{2}$; (3) $x, \bar{y}, z+\frac{1}{2}$; (4) $\bar{x}, \bar{y}, \bar{z}$; (5) $x+\frac{1}{2}, y+\frac{1}{2}, z$; (6) $\bar{x}+\frac{1}{2}, y+\frac{1}{2}, \bar{z}+\frac{1}{2}$; (7) $x+\frac{1}{2}, \bar{y}+\frac{1}{2}, z+\frac{1}{2}$; (8) $\bar{x}+\frac{1}{2}, \bar{y}+\frac{1}{2}, \bar{z}$.

[13] Buerger, *Elementary Crystallography*, pp. 460–470.

[14] H. Lipson and W. Cochran, *The Determination of Crystal Structures*, G. Bell and Sons, London, 1957, pp. 21–27.

for the lower-symmetry space groups can be written down by inspection, noting the kind and location of the symmetry elements. For any space group, the general positions can be derived by elementary mathematical methods and are listed in standard orientation for all space groups in *International Tables*.[12]

Also listed for each space group are the special positions where two or more general positions coalesce on one or more symmetry elements of the space group. An atom, ion, or molecule can occupy such a position only if it possesses symmetry conforming to that of the position. The number of positions in a set of special positions is always less than the number of general positions in the cell. Thus in $P2_1/c$, for example, the site $0, 0, 0$ gives rise only to the equivalent position $0, \frac{1}{2}, \frac{1}{2}$ under the influence of the symmetry elements. As a result, two centrosymmetric objects with their centers placed at these points fulfill the symmetry requirements of the space group and may constitute the entire contents of the unit cell.[15]

Space groups containing translation elements, that is, nonprimitive lattices, screw axes, and glide planes, yield diffraction patterns in which certain classes of reflections are absent. Such systematic absences are termed *space group extinctions* or simply *extinctions*. We shall reserve a general analytical derivation of these extinctions until Chapter 8, but their use in space group determination is discussed in Chapter 5.

3.5. MATRIX REPRESENTATION OF SYMMETRY

Although the algebraic representation of equivalent positions, as was done above and in *International Tables*, is convenient and useful for calculation in specific space groups, it lacks generality. This is particularly unfortunate for machine computation, where it is necessary to provide separate code for each case. A more general representation arises by realizing that the symmetry operations of the 32 crystallographic point groups can all be regarded as generalized rotations (which include reflection in a plane and inversion through a point). Such rotations can be represented as a 3×3 matrix,[16] which operates on the vector representing the coordinates of a point, $[x, y, z]$ in the general case. The symmetry elements of space groups are given by exactly the same rotation matrices plus an additive vector that gives the translations ($0, \frac{1}{2}, \frac{1}{3}, \frac{1}{4}$, or $\frac{1}{6}$) along the three axes. Thus the equivalent positions in space groups are given by the matrix equation

$$\mathbf{R}\mathbf{x} + \mathbf{t} = \mathbf{x}' \tag{3.1}$$

[15]For further discussion, see H. Lipson and W. Cochran, *The Determination of Crystal Structures*, G. Bell and Sons, London, 1957, pp. 45–47; M. J. Buerger, *Crystal-Structure Analysis*, Wiley, New York, 1960, pp. 249–258.

[16]See Appendix C for an introduction and references to matrix methods.

TABLE 3.4 Space Groups in Standard Orientations[a]

System	Point Group	Space Group						Fraction
Triclinic	1	$P1$						1/2
	$\bar{1}$	$P\bar{1}$						
Monoclinic	2	$P2$	$P2_1$	$C2$				1/4
	m	Pm	Pc	Cm	Cc			
	$2/m$	$P2/m$	$P2_1/m$	$C2/m$	$P2/c$	$P2_1/c$	$C2/c$	
Orthorhombic	222	$P222$	$P222_1$	$P2_12_12$	$P2_12_12_1$	$C222_1$	$C222$	1/8
		$F222$	$I222$	$I2_12_12_1$				
	$mm2$	$Pmm2$	$Pmc2_1$	$Pcc2$	$Pma2$	$Pca2_1$	$Pnc2$	
		$Pmn2_1$	$Pba2$	$Pna2_1$	$Pnn2$	$Cmm2$	$Cmc2_1$	
		$Ccc2$	$Amm2$	$Abm2$	$Ama2$	$Aba2$	$Fmm2$	
		$Fdd2$	$Imm2$	$Iba2$	$Ima2$			
	mmm	$Pmmm$	$Pnnn$	$Pccm$	$Pban$	$Pmma$	$Pnna$	
		$Pmna$	$Pcca$	$Pbam$	$Pccn$	$Pbcm$	$Pnnm$	
		$Pmmn$	$Pbcn$	$Pbca$	$Pnma$	$Cmcm$	$Cmca$	
		$Cmmm$	$Cccm$	$Cmma$	$Ccca$	$Fmmm$	$Fddd$	
		$Immm$	$Ibam$	$Ibca$	$Imma$			
Tetragonal	4	$P4$	$P4_1$	$P4_2$	$P4_3$	$I4$	$I4_1$	1/8
	$\bar{4}$	$P\bar{4}$	$I\bar{4}$					
	$4/m$	$P4/m$	$P4_2/m$	$P4/n$	$P4_2/n$	$I4/m$	$I4_1/a$	
	422	$P422$	$P42_12$	$P4_122$	$P4_12_12$	$P4_222$	$P4_22_12$	1/16
		$P4_322$	$P4_32_12$	$I422$	$I4_122$			
	$4mm$	$P4mm$	$P4bm$	$P4_2cm$	$P4_2nm$	$P4cc$	$P4nc$	
		$P4_2mc$	$P4_2bc$	$I4mm$	$I4cm$	$I4_1md$	$I4_1cd$	
	$\bar{4}2m$	$P\bar{4}2m$	$P\bar{4}2c$	$P\bar{4}2_1m$	$P\bar{4}2_1c$	$P\bar{4}m2$	$P\bar{4}c2$	
		$P\bar{4}b2$	$P\bar{4}n2$	$I\bar{4}m2$	$I\bar{4}c2$	$I\bar{4}2m$	$I\bar{4}2d$	

	$4/mmm$	$P4/mmm$	$P4/mcc$	$P4/nbm$	$P4/nnc$	$P4/mbm$	$P4/mnc$	
		$P4/nmm$	$P4/ncc$	$P4_2/mmc$	$P4_2/mcm$	$P4_2/nbc$	$P4_2/nnm$	
		$P4_2/mbc$	$P4_2/mnm$	$P4_2/nmc$	$P4_2/ncm$	$I4/mmm$	$I4/mcm$	
		$I4_1/amd$	$I4_1/acd$					
Trigonal/rhombohedral	3	$P3$	$P3_1$	$P3_2$	$R3$			1/6
	$\bar{3}$	$P\bar{3}$	$R\bar{3}$					
	32	$P312$	$P321$	$P3_112$	$P3_121$	$P3_212$	$P3_221$	1/12
		$R32$						
	$3m$	$P3m1$	$P31m$	$P3c1$	$P31c$	$R3m$	$R3c$	
	$\bar{3}m$	$P\bar{3}1m$	$P\bar{3}1c$	$P\bar{3}m1$	$P\bar{3}c1$	$R\bar{3}m$	$R\bar{3}c$	
Hexagonal	6	$P6$	$P6_1$	$P6_5$	$P6_2$	$P6_4$	$P6_3$	1/12
	$\bar{6}$	$P\bar{6}$						
	$6/m$	$P6/m$	$P6_3/m$					
	622	$P622$	$P6_122$	$P6_522$	$P6_222$	$P6_422$	$P6_322$	1/24
	$6mm$	$P6mm$	$P6cc$	$P6_3cm$	$P6_3mc$			
	$\bar{6}m\bar{2}$	$P\bar{6}m2$	$P\bar{6}c2$	$P\bar{6}2m$	$P\bar{6}2c$			
	$6/mmm$	$P6/mmm$	$P6/mcc$	$P6_3/mcm$	$P6_3/mmc$			
Cubic	23	$P23$	$F23$	$I23$	$P2_13$	$I2_13$		1/24
	$m3$	$Pm3$	$Pn3$	$Fm3$	$Fd3$	$Im3$	$Pa3$	
		$Ia3$						
	432	$P432$	$P4_232$	$F432$	$F4_132$	$I432$	$P4_332$	1/48
		$P4_132$	$I4_132$					
	$\bar{4}3m$	$P\bar{4}3m$	$F\bar{4}3m$	$I\bar{4}3m$	$P\bar{4}3n$	$F\bar{4}3c$	$I\bar{4}3d$	
	$m3m$	$Pm3m$	$Pn3n$	$Pm3n$	$Pn3m$	$Fm3m$	$Fm3c$	
		$Fd3m$	$Fd3c$	$Im3m$	$Ia3d$			

[a]The 11 Laue symmetries are separated by horizontal lines.

TABLE 3.5 Symmetry Operations and Matrices

Symmetry Operation	Orientation	Equivalent Positions	Rotation Matrix	Translation Vector
1	Any	x, y, z	$\begin{bmatrix}1&0&0\\0&1&0\\0&0&1\end{bmatrix}$	$\begin{bmatrix}0\\0\\0\end{bmatrix}$
$\bar{1}$	Any	(1) x, y, z (2) $\bar{x}, \bar{y}, \bar{z}$	$\begin{bmatrix}\bar{1}&0&0\\0&\bar{1}&0\\0&0&\bar{1}\end{bmatrix}$	$\begin{bmatrix}0\\0\\0\end{bmatrix}$
2	[010]	(1) x, y, z (2) $\bar{x}, y, \bar{z}$	$\begin{bmatrix}\bar{1}&0&0\\0&1&0\\0&0&\bar{1}\end{bmatrix}$	$\begin{bmatrix}0\\0\\0\end{bmatrix}$
2_1	[010]	(1) x, y, z (2) $\bar{x}, y+\frac{1}{2}, \bar{z}$	$\begin{bmatrix}\bar{1}&0&0\\0&1&0\\0&0&\bar{1}\end{bmatrix}$	$\begin{bmatrix}0\\\frac{1}{2}\\0\end{bmatrix}$
m	⊥ to [010]	(1) x, y, z (2) $x, \bar{y}, z$	$\begin{bmatrix}1&0&0\\0&\bar{1}&0\\0&0&1\end{bmatrix}$	$\begin{bmatrix}0\\0\\0\end{bmatrix}$
c	⊥ to [010]	(1) x, y, z (2) $x, \bar{y}, z+\frac{1}{2}$	$\begin{bmatrix}1&0&0\\0&\bar{1}&0\\0&0&1\end{bmatrix}$	$\begin{bmatrix}0\\0\\\frac{1}{2}\end{bmatrix}$

where **R** is the rotation matrix (0s and 1s), **x** is a column matrix of the coordinates of a point, **t** is a column matrix of the components of the translation vector, and **x**′ is a column matrix of the transformed coordinates.

Table 3.5 lists the rotation and translation matrices for the symmetry operations we have been using above. In general, the appropriate matrices can be obtained by considering the *International Tables* listing for any space group. The first row of the matrix converts x into the variable part of the new x' and can be written down by inspection; the second takes y into y', and similarly for the third. The corresponding translation vector consists of the constant fractional quantities that are part of the new coordinates. Thus for a 2-fold axis for which the equivalent positions are x, y, z and $\bar{x}, y, \bar{z}$, the first line of the matrix is $[\bar{1}\ \ 0\ \ 0]$ to take x to $\bar{x}$. The second is $[0\ \ 1\ \ 0]$ to leave y unchanged, and the third is $[0\ \ 0\ \ \bar{1}]$ to take z to $\bar{z}$. This matrix applies to both 2 and 2_1, the difference between the two being that 2 has the translation vector $[0\ \ 0\ \ 0]$, that is, no translation, while 2_1 has $[0\ \ \frac{1}{2}\ \ 0]$ corresponding to the constant term in $\bar{x}, y+\frac{1}{2}, \bar{z}$.

To illustrate the use of matrix methods, consider the derivation of the equivalent positions for space group $P2_1/c$. First, apply the 2_1 rotation matrix and translation to x, y, z [position (1)] to generate position (2):

$$\begin{bmatrix} \bar{1} & 0 & 0 \\ 0 & 1 & 0 \\ 0 & 0 & \bar{1} \end{bmatrix} \begin{bmatrix} x \\ y \\ z \end{bmatrix} + \begin{bmatrix} 0 \\ \frac{1}{2} \\ 0 \end{bmatrix} = \begin{bmatrix} -x \\ y+\frac{1}{2} \\ -z \end{bmatrix} \tag{3.2}$$

Next, apply the transformation for the c glide to positions (1) and (2) to generate (3) and (4):

$$\begin{bmatrix} 1 & 0 & 0 \\ 0 & \bar{1} & 0 \\ 0 & 0 & 1 \end{bmatrix} \begin{bmatrix} x \\ y \\ z \end{bmatrix} + \begin{bmatrix} 0 \\ 0 \\ \frac{1}{2} \end{bmatrix} = \begin{bmatrix} x \\ -y \\ z+\frac{1}{2} \end{bmatrix} \tag{3.3}$$

$$\begin{bmatrix} 1 & 0 & 0 \\ 0 & \bar{1} & 0 \\ 0 & 0 & 1 \end{bmatrix} \begin{bmatrix} -x \\ y+\frac{1}{2} \\ -z \end{bmatrix} + \begin{bmatrix} 0 \\ 0 \\ \frac{1}{2} \end{bmatrix} = \begin{bmatrix} -x \\ -y-\frac{1}{2} \\ -z+\frac{1}{2} \end{bmatrix} \tag{3.4}$$

Rearranging and remembering that translations of $+\frac{1}{2}$ and $-\frac{1}{2}$ are effectively identical, we have for the equivalent positions

(1) x, y, z; (2) $\bar{x}, \frac{1}{2}+y, \bar{z}$; (3) $x, \bar{y}, \frac{1}{2}+z$; (4) $\bar{x}, \frac{1}{2}-y, \frac{1}{2}-x$

Note that point (4) is related to (1) by the $\bar{1}$ matrix and the translation vector $\{0, \frac{1}{2}, \frac{1}{2}\}$, implying that (1) and (4) are related by a center of symmetry at $0, \frac{1}{4}, \frac{1}{4}$. Shifting the origin to this point transforms the equivalent positions to the standard form found in *International Tables* and given in Fig. 3.37.

3.6. INTENSITY-WEIGHTED RECIPROCAL LATTICE

Each ray reflected from a given set of planes in a crystal has an inherent intensity that depends on the electron distribution. Since each reciprocal lattice point corresponds to one such ray, each has associated with it an intensity. A reciprocal lattice in which each point has been assigned a weight equal to the intensity of the corresponding diffracted ray is termed the *intensity-weighted reciprocal lattice*. The possible symmetries of such lattices, the *Laue symmetries*, are just those of the 11 centrosymmetric point groups that result when a center is added to each of the 32 point groups. These are indicated in Fig. 3.19 and are set off by horizontal lines in Table 3.4. The reason for the added center in the diffraction pattern stems from the fact that diffraction effects are inherently centrosymmetric. That this is so may be seen by noting that r.l. point $\overline{hkl}$, as well as hkl, is defined in terms of the set of direct lattice planes hkl. Both r.l. points hkl and $\overline{hkl}$ lie on the normal to the plane through the origin but in opposite directions (Fig. 3.40). In Fig. 3.41 the single plane through the origin of the set hkl is shown with the associated r.l. points hkl and $\overline{hkl}$. Figure 3.41*a* shows

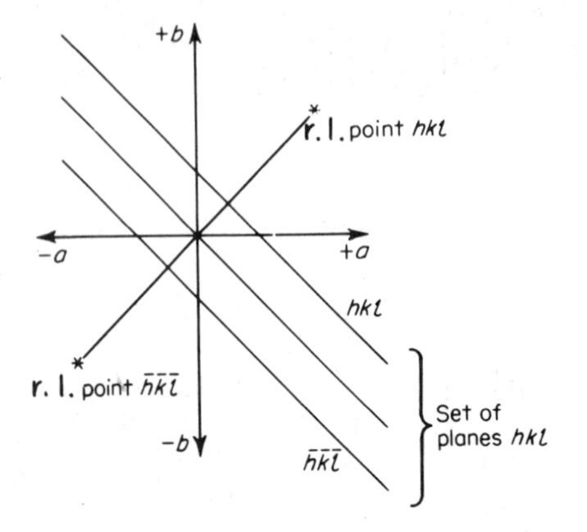

Figure 3.40. Reciprocal lattice points *hkl* and $\overline{hkl}$.

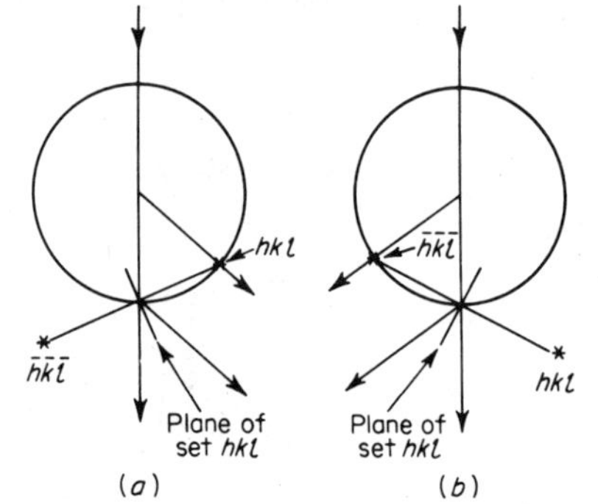

Figure 3.41. (*a*) Reciprocal lattice point *hkl* in position to reflect. (*b*) Reciprocal lattice point $\overline{hkl}$ in position to reflect.

reflection *hkl* in position to reflect; Fig. 3.41*b* shows $\overline{hkl}$ in a similar position. It is evident that these two positions correspond to reflection from opposite sides of the set of planes *hkl* so that $I_{hkl} = I_{\overline{hkl}}$. This is *Friedel's law*.[17]

At the end of Chapter 2, it was shown that the number of possible reflections for any substance is equal to the number of r.l. points within the limiting sphere. It was indicated there that only a fraction of these, however, were unique. This fraction depends on the Laue symmetry and is listed in Table 3.4 for each Laue class. How this arises as a consequence of symmetry will be dealt with in Chapter 6.

In this and the preceding chapter, we have treated lattices, diffraction of X-rays, and symmetry. These are basic to the field of X-ray crystallography, and the concepts developed in these chapters will be applied in Chapter 5 and subsequently to the more practical matters of recording the reciprocal lattice and interpreting the results, collecting intensity data, and deriving structures from it. Before turning to these topics, however, Chapter 4 will provide an introduction to some of the practical details of preparing crystal specimens.

BIBLIOGRAPHY

Buerger, M. J., *Elementary Crystallography*, Wiley, New York, 1956.

Bunn, C. W., *Chemical Crystallography*, 2nd ed., Oxford University Press, London, 1961, pp. 10–63, 231–258.

[17]Friedel's law does not hold for diffraction from substances in which one or more of the elements present have an absorption edge near that of the wavelength being used (see Section 8.13). Even for such cases, however, deviations are usually not large.

Dunitz, J. D., *X-Ray Analysis and the Structure of Organic Molecules*, Cornell University Press, Ithaca, NY, 1979, pp. 73–96.

Gay, P., *The Crystalline State*, Oliver and Boyd, Edinburgh, 1972, pp. 141–178.

Glusker, J. P., and K. N. Trueblood, *Crystal Structure Analysis*, 2nd ed., Oxford University Press, New York, 1985, Chapter 7.

Luger, P., *Modern X-Ray Analysis on Single Crystals*, Walter de Gruyter, Berlin, 1980, pp. 111–147.

McWeeny, R., *Symmetry: An Introduction to Group Theory and Its Applications*, Macmillan, New York, 1963.

Phillips, F. C., *An Introduction to Crystallography*, 3rd ed., Wiley, New York, 1963.

Prince, E., *Mathematical Techniques in Crystallography and Materials Science*, Springer-Verlag, New York, 1982, pp. 20–48.

Sands, D. E., *Vectors and Tensors in Crystallography*, Addison-Wesley, Reading, MA, 1982.

CHAPTER 4

CRYSTALS AND THEIR PROPERTIES

We now turn to some practical considerations of crystals and their handling that are preliminary to recording the X-ray diffraction pattern. In particular, it is necessary to have methods for obtaining satisfactory single crystals and criteria as a guide in choosing a suitable specimen. Once chosen, the crystal must be mounted and properly aligned in preparation for further experimental work.

4.1. CRYSTALLIZATION[1]

The process of crystallization is one of ordering, wherein randomly arranged ions, atoms, or molecules in the gas or liquid phase take up regular positions in the solid state. Indeed, it is the order in the structure of solids that is responsible for such distinctive properties as sharp melting points and Bragg reflection.

It is generally agreed that the initial stage in the process of crystallization involves the phenomenon of nucleation, the precise mechanism of which need not concern us here. Once formed, however, the nuclei grow by deposition on the crystallite faces. The process may be considered to be a dynamic equilibrium between particles in the fluid phase and those in the solid. The crystal grows when the forward rate in the equilibrium

$$\text{fluid} \rightleftharpoons \text{solid}$$

[1]For an introduction to the theory and practice of crystallization, see G. F. Reynolds in *Physics and Chemistry of the Organic Solid State*, Vol. I, D. Fox et al., Eds., pp. 223–286.

exceeds the reverse rate. These rates will depend on the nature of the forces at the crystal surface, on the concentration of the crystallizing substance, and on the nature of the medium in the vicinity of the growing crystal. In general, the faces of the crystal that grow most rapidly are those to which the crystallizing particles are bound most securely.

It is a general rule that the most rapidly growing faces of crystals are those that are the smallest and least well developed. Thus the external form of a crystal, its ***habit***, should be noted, since it sometimes hints at the internal structure.[2] Furthermore, the external form of a crystal is useful in the initial selection of axes, and once these have been correlated with the form, additional specimens are easily mounted about any axis.

4.2. GROWING CRYSTALS

Although crystals can be grown by deposition from any sort of fluid phase, for most organic and many inorganic compounds the most practical method involves crystallization from a solution in a suitable solvent. Crystals can be grown from a saturated solution in any of several ways, all of which serve to raise the solute concentration above that which can be supported by the solution: slow evaporation, slow cooling, diffusion. One of the simplest methods is to allow a saturated or nearly saturated solution to stand under conditions where the solvent evaporates slowly. The best crystals are usually produced when the solution is free from mechanical vibration and allowed to evaporate without disturbance. Some solutions, however, show a marked tendency to become supersaturated. When crystallization is then finally induced, it is usually so rapid that only microscopic crystals result. Gross supersaturation can often be prevented by mild mechanical shock or by seeding the solution with a few crystal fragments of the solute.

Slow evaporation often deposits crystals as a microcrystalline crust on the walls of the container just at the surface of the solution. As the solvent evaporates, the solution recedes, leaving the crust in a position where it is not effective in inducing good crystal growth. Furthermore, it is so difficult to control evaporation from small volumes of solution that appreciable amounts of solvent, a few milliliters or more, are required.

The second method commonly employed is slow cooling. Cooling rates from elevated temperatures for solutions contained in laboratory vessels of ordinary size are often too rapid to produce anything but microscopic crystals. The rate of cooling can be slowed somewhat by placing the vessel containing the solution in a large container of hot water or by surrounding it with sufficient insulating material. The whole process can also be carried out to advantage in a Dewar vessel. Extremely slow rates of cooling can be

[2]For a discussion, see P. Hartman in *Physics and Chemistry of the Organic Solid State*, Vol. I, D. Fox et al., Eds., pp. 369–409.

obtained in a well-insulated and thermostated oven by reducing the thermostat settings gradually.

An alternative to slow cooling is to establish a thermal gradient across the vessel containing the solution. The magnitude of the gradient is gradually increased by lowering the temperature of the heat sink until crystallization occurs. With a horizontal gradient, convection currents are set up that continuously supply the colder surface with fresh solution so that growth on any nuclei formed is sustained. If the gradient is too large, however, convection may be so rapid as to impair proper crystal growth. Often only a small gradient is needed, and it can be obtained by the simple expedient of placing the vessel containing the solvent against a cold outside window. Slow evaporation may be allowed to occur simultaneously to bring the solution to saturation. If a vertical thermal gradient is used, the cold side will usually be on the bottom, thus preventing convection. The heat sink can be a cold plate maintained at low temperature by circulating cold water. An elegantly simple device for producing thermal and concentration gradients has been described by Hope.[3]

If difficulty is experienced in growing satisfactory crystals from single solvents by either of the above methods, mixtures of two or more solvents should be tried. Changing the nature of the solvent often has a pronounced effect on the habit and size of crystals grown, since properties that may influence crystal growth, such as density, dielectric constant, viscosity, and solvation as well as solubility, can be varied in a controlled way over wide ranges and adjusted to fit the particular circumstances.

A method that can often give good single crystals with milligram amounts of solute involves altering the solvent by vapor diffusion as in Fig. 4.1. A solution of the substance in solvent S_1 is contained in tube T. A second solvent, S_2, placed in the closed beaker B, is chosen that produces with S_1 a mixed solvent in which the solute is less soluble than in S_1 alone. Slow diffusion of S_2 into T (and S_1 out) will cause crystallization, which under favorable conditions can result in the growth of relatively large single

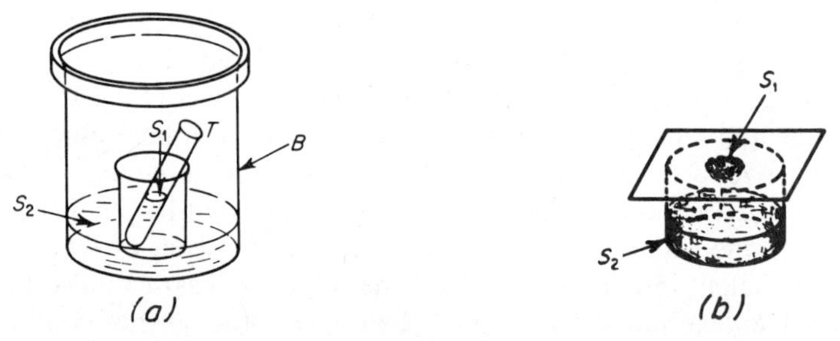

Figure 4.1. Crystallization by vapor diffusion.

[3]H. Hope, *J. Appl. Cryst.*, **4**, 333 (1971).

crystals. If S_2 is more volatile than S_1, it is possible for the volume of solution to increase during crystallization and thus avoid the troublesome crusts that often form when solutions evaporate. In order to obtain good mixing of S_1 and S_2 as they diffuse together, it is helpful that their densities be either relatively similar, or, better, that S_2 be denser than S_1.

A microscopic version of the same process involves putting the solute in a drop of solvent S_1 hanging from a cover glass over a quantity of S_2 in a microbeaker (Fig. 4.1*b*). Because of the small volumes involved, it is important to have a good seal, impervious to the solvents used, between the cover glass and the beaker.

Diffusion in the liquid phase can also be used to grow crystals. Figure 4.2 shows one arrangement that depends on the differences in density to maintain the initial separation between solvents S_1 and S_2. The solute must always be dissolved in the solvent in which it is more soluble. Crystals appear at the interface between S_1 and S_2 and grow as the solvents mix by diffusion.

Diffusion can sometimes be used to advantage with a single solvent where the substance to be crystallized is formed in solution. For example, reactants A and B may be allowed to diffuse together and form the product C (Fig. 4.3). If C is insoluble, it will crystallize in the zone where the reactants mix.[4]

Satisfactory crystals can sometimes be grown from the melt by slow cooling. This method is not amenable to the wide variations in conditions attainable with the methods already outlined, and it has only limited application in growing crystals of organic compounds. For certain inorganic substances and metals, however, melting and crystallizing in a zone growth technique are capable of producing very large and pure single crystals.

Sublimation is another method of limited applicability, but one that sometimes works well. If a solid vaporizes readily without decomposition, sublimation can be considered as an alternative to other methods. If a gas condenses rapidly as a solid on a relatively cold surface, crystal growth is impaired and only a microcrystalline mass results. However, if sublimation

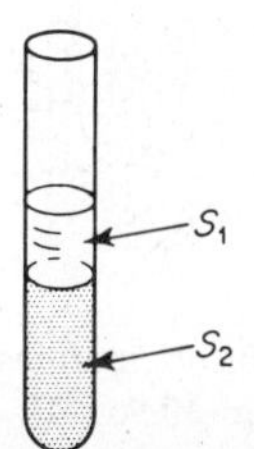

Figure 4.2. Crystallization by solute diffusion.

[4]See S. A. Martin and H. M. Haendler, *J. Appl. Cryst.*, **11**, 62 (1978). Similar methods involving diffusion in silicate gels have also been used; see C. Bridle and T. R. Lower, *Acta Cryst.*, **19**, 483 (1965).

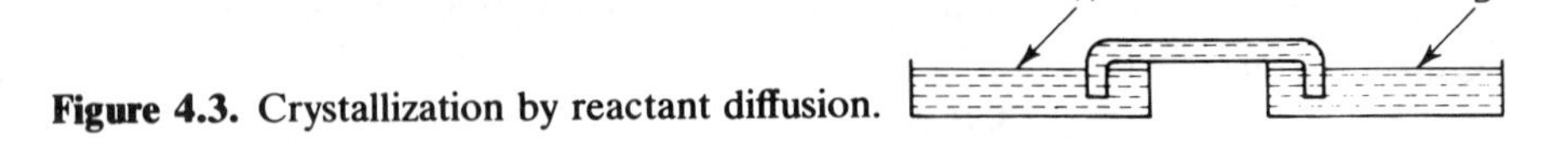

Figure 4.3. Crystallization by reactant diffusion.

is carried out *slowly* at room or elevated temperatures so that larger crystals grow at the expense of smaller ones, very well formed specimens can sometimes be obtained.

This discussion of crystal growth has covered in general terms the most common methods of producing suitable single crystals for X-ray diffraction measurements. It has been kept general because growing crystals is very much an art, and for any particular substance the method to be chosen and the variations in its use must usually be decided on the basis of exploratory experiments. While initial attempts to grow crystals may be disappointing, it has been our experience that a surprising number of compounds that at first seemed incapable of producing satisfactory crystals did yield them when the proper conditions were found.

4.3. CHOOSING A CRYSTAL

If a crystal is to be satisfactory for collecting X-ray diffraction data, two main requirements must be met: (1) It must possess uniform internal structure, and (2) it must be of proper size and shape.

To fulfill the first requirement, a crystal must be pure at the molecular, ionic, or atomic level. It must be a single crystal in the usual sense, that is, it should not be twinned[5] or composed of microscopic subcrystals. It should not be grossly fractured, bent, or otherwise physically distorted. It need not, however, have particularly uniform or well-formed external faces.

Almost all real single crystals are, in a sense, imperfect, being composed of slightly misaligned minute regions (*mosaic blocks*, Fig. 4.4). The misalignment is small, however—not more than 0.1–0.2° for most crystals. This

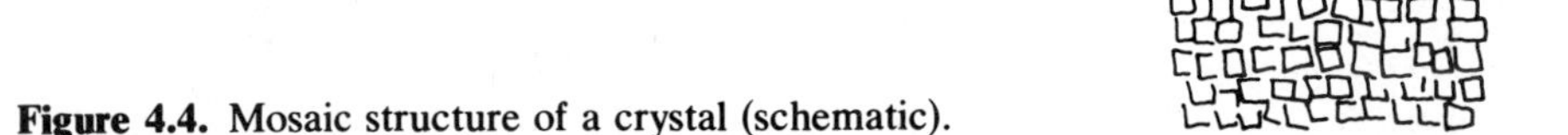

Figure 4.4. Mosaic structure of a crystal (schematic).

[5]A full discussion of the phenomenon of twinning—the existence of two different orientations of a lattice in what is often apparently one crystal—would take us too far afield. The phenomenon, although not common in organic crystals, does occur, however. It is somewhat more common in minerals, which tend to crystallize in space groups of higher symmetry. For discussion, examples, and leading references, see Buerger, *Crystal–Structure Analysis*, pp. 53–68; J. D. Dunitz, *Acta Cryst.*, **17**, 1299 (1964); F. H. Herbstein, *Acta Cryst.*, **17**, 1094 (1964); **18**, 996 (1965); **19**, 590 (1965); A. Santoro, *Acta Cryst.*, **A30**, 224 (1974).

imperfection is, in fact, desirable, since the diffracted intensities are much greater from crystals with such a *mosaic structure* than from the unusual perfect crystals in which the lattice planes traverse the whole crystal without appreciable distortion or discontinuity (see extinction in Section 17.6). Sometimes crystals are deliberately subjected to thermal shock, such as that produced by repeated dipping in liquid nitrogen, to ensure a more pronounced mosaic structure.

Crystals can be screened rapidly be examination between "crossed" Polaroid sheets or more conveniently with a polarizing microscope. If rotated about an axis normal to the polarizing material, the crystals should either appear uniformly dark in all positions or be bright and extinguish, that is, appear uniformly dark, once every 90°. Crystals that are made up of two or more fragments with different orientations will often reveal themselves by displaying both dark and light regions at one time. The ultimate evidence, however, of the internal structure of a crystal is furnished by the diffraction pattern itself. The reflections that appear should be single spots, without "tails" or streaks connecting them (other than those due to white radiation, see Section 6.1), and the pattern should be indexable in terms of a single lattice.

The choice of a suitable size for a crystal specimen depends on a balance of a number of contradictory factors. In the usual diffraction equipment the primary beam has a plateau of uniform intensity with dimensions 0.5 mm × 0.5 mm or a bit larger. In single-crystal work, the specimen should not exceed this size, that is, all parts of the crystal should be exposed to the same radiation intensity. The preferred crystal size is somewhat smaller, however, generally 0.1–0.3 mm, partly because of the difficulties in aligning the crystal precisely with a plateau of the same size.

Another effect important in determining a suitable crystal size is absorption of X-rays by the crystal. As is common in absorption processes, the intensity I of a beam after passing through a thickness τ of absorber is given by

$$I = I_0 e^{-\mu\tau} \tag{4.1}$$

where I_0 is the intensity of the incident beam and μ is the *linear absorption coefficient*. A different absorption coefficient can be defined by rewriting Eq. (4.1) in the form

$$I = I_0 e^{-\tau\rho(\mu/\rho)_\lambda} \tag{4.2}$$

where $(\mu/\rho)_\lambda$ is defined as the *mass absorption coefficient* for the wavelength used and ρ is the density.[6] For a compound made up of P_1% by

[6]The mass absorption coefficients of most elements for common characteristic radiations are tabulated in *International Tables*, Vol. III, pp. 161–165.

mass of element E_1, P_2% of E_2, and so on, and having a density ρ, the linear absorption coefficient for radiation of specified wavelength λ is given by the expression

$$\mu_\lambda = \rho \sum_n \frac{P_n}{100} \left(\frac{\mu}{\rho}\right)_\lambda, E_n \tag{4.3}$$

Since the intensities of the diffracted rays from a given crystal are proportional to the amount of material present in the specimen, there is an advantage in selecting as large a crystal as possible. Because of absorption, however, there is an optimum thickness. For any greater thickness, diffracted rays that have passed through the crystal will show a decrease in intensity. This optimum thickness is a function of the linear absorption coefficient, and thus of the wavelength, and is given by

$$t_{\text{opt}} = 2/\mu \tag{4.4}$$

For most organic compounds lacking heavy atoms, $\mu < 10\ \text{cm}^{-1}$ for Cu K_α radiation. The optimum thickness for organic crystals is, therefore, greater than 0.2 cm and so is rarely approached in practice. The absorption coefficient increases rapidly with atomic number (except at an absorption edge) and the optimum crystal thickness is proportionally less. Thus for crystalline iodine, $\mu = 1450\ \text{cm}^{-1}$ and the optimum thickness is a mere 0.0014 cm. On the other hand, the absorption coefficient of a given substance decreases with decreasing wavelength, except for local increases at wavelengths just less than those corresponding to absorption edges of its components. Thus for Mo K_α radiation the absorption coefficients for organic substances are less than $1\ \text{cm}^{-1}$, and for crystalline iodine $\mu = 183\ \text{cm}^{-1}$.

A serious problem associated with absorption, even for crystals of less than optimum size, arises from the fact that the incident and diffracted rays may have different average path lengths in the crystal for different reflections. As a result these reflections will suffer from absorption to varying extents, and systematic errors will be introduced into the observed intensities. It is possible to make a calculated correction for this effect if the shape of the crystal is known exactly (Section 7.2), but the computations are time-consuming and often not very accurate. Thus it is best to avoid the problem as much as possible by proper[7] choice of the crystal and the experimental conditions.

Although the effects of absorption were often overlooked in the past, and may still be ignored in many cases if the highest precision is not needed, it is well to calculate the possible errors for a specific case before assuming that

[7]For a more detailed experimental discussion of the intensity variations arising from absorption, see J. W. Jeffery and K. M. Bose, *Acta Cryst.*, **17**, 343 (1964).

they can be tolerated. Clearly the worst case is that of a flat plate, for which the paths of rays passing through the longest and shortest dimensions may differ by a factor of 10 or more. The amounts of absorption in these two directions can be calculated from Eq. (4.1) or (4.2) and represent approximately the extreme intensity variations that can be expected. The results are often disconcerting. For example, an organic compound containing one bromine for approximately 40 carbons and oxygens has $\mu \approx 20\ \text{cm}^{-1}$ for Cu K_α radiation. If a flat needle with a cross section of $0.005\ \text{cm} \times 0.025\ \text{cm}$ is mounted along the needle axis, a zero-level reflection with theoretical intensity 1.0 will have actual values of approximately 0.90 to 0.61, depending on the crystal orientation. These errors are generally expressed in terms of the correction A^* required to convert an observed intensity to its true value. Thus for this case A^* would range from $1/0.9 = 1.11$ to $1/0.61 = 1.64$. This range corresponds to ~20% variation about a mean intensity value, an effect too large to be passed off without some thought, although it will not in most cases prevent solution of the structure.

Two paths are open for reducing these errors. The first involves using a more penetrating radiation to diminish the amount of absorption that occurs in the crystal. Thus if Mo K_α is used in place of Cu K_α, μ for the case given above falls to $\sim 12\ \text{cm}^{-1}$,[8] the corrections will range between 1.06 and 1.35, and the error will be reduced to ~±12%. For this reason, molybdenum

TABLE 4.1 Mass and Linear Absorption Coefficients for Three Assumed Compositions

	I C only; Assumed Density 1.0 g/cm³	II 20C + 1Cl; Assumed Density 1.2 g/cm³	III 20C + 1I; Assumed Density 1.5 g/cm³
Mass absorption coefficient μ/ρ:			
Ag	0.40	1.09	6.84
Mo	0.62	2.01	13.24
Cu	4.60	17.64	104.70
Linear absorption coefficient μ:			
Ag	0.40	1.31	10.26
Mo	0.62	2.41	19.86
Cu	4.60	21.17	157.05

[8]This example fails to point up fully the usual advantages of Mo K_α. Since the excitation energy of bromine lies closely below the energy of Mo K_α radiation, its absorption coefficient is abnormally large and the improvement over Cu K_α is relatively small. For most elements, the ratio $\mu(\text{Mo})/\mu(\text{Cu}) \approx \frac{1}{9}$ and the improvement is much greater.

TABLE 4.2 Factors by which Diffracted Intensities from Cylindrical Crystals of Different Radii Must be Multiplied to Compensate for Absorption[a]

			(I) 2θ				(II) 2θ				(III) 2θ		
R (cm)		μR	0°	90°	180°	μR	0°	90°	180°	μR	0°	90°	180°
0.005	Ag	0.0020	1.00	1.00	1.00	0.0066	1.01	1.01	1.01	0.0513	1.09	1.09	1.09
	Mo	0.0031	1.01	1.01	1.01	0.0120	1.02	1.02	1.02	0.0993	1.18	1.18	1.18
	Cu	0.0230	1.04	1.04	1.04	0.1058	1.19	1.19	1.19	0.7852	3.62	3.18	2.79
0.010	Ag	0.0040	1.01	1.01	1.01	0.0131	1.02	1.02	1.02	0.1026	1.18	1.18	1.18
	Mo	0.0062	1.01	1.01	1.01	0.0241	1.04	1.04	1.04	0.1986	1.40	1.39	1.37
	Cu	0.0460	1.08	1.08	1.08	0.2117	1.43	1.42	1.40	1.5705	11.8	7.16	5.08
0.020	Ag	0.0080	1.01	1.01	1.01	0.0262	1.05	1.05	1.05	0.2052	1.41	1.40	1.38
	Mo	0.0125	1.02	1.02	1.02	0.0482	1.09	1.09	1.09	0.3972	1.94	1.88	1.80
	Cu	0.0920	1.17	1.17	1.17	0.4234	2.03	1.96	1.87	3.140	80.6	18.2	10.0

[a]Factors are tabulated for each of three λ and three 2θ values for compositions I, II, and III in Table 4.1.

radiation is generally used with diffractometers, where copper has no compensating advantage in detection efficiency.

The second alternative is to shape the crystal so that the effects of absorption are the same for all reflections or, at worst, depend only on θ. One obvious form is a sphere, and methods have been devised for generating spheres from crystals.[9] Fortunately, however, the geometry of the common methods of collecting intensity data is such that a cylinder, mounted along its axis, is nearly as good. Since many crystals grow naturally as needles, it is often much easier to shape these into cylinders than into spheres.[10] Mounting is also simplified since the unique axial direction is conserved.

How much shaping is to be done will depend on the seriousness of the absorption error and the amount of the crystal that can be discarded. In the example above, the largest true cylinder that could be generated would have a diameter of 0.05 mm, and over 80% of the original mass would be lost. The result would be a large net decrease in intensity and an increase in the *random* errors associated with each measurement, although the *systematic* errors due to absorption would be reduced. A practical compromise would be to remove about half the crystal width to reduce the maximum path length and round the edges so that the final cross section was elliptical rather than round. This process would reduce the maximum errors for Cu K_α to ±3–5% at a cost of about half the intensity.

For the most accurate work, or if μ is very large, a final correction is needed even for cylindrical or spherical crystals. The corrections for both spheres and cylinders are tabulated in *International Tables*[11] as a function of θ and μR (R is the sphere or cylinder radius in centimeters). For most organic crystals these secondary effects are small, and their neglect causes only a slight increase in the error of the parameters describing the thermal motion of the atoms in the crystal. If either μ or R is large, the effects can become quite significant (Tables 4.1 and 4.2), but in most cases still larger errors will be present as a result of failure to achieve a perfectly regular form by shaping.

4.4. CRYSTAL MOUNTING[12]

For single-crystal diffractometry it is convenient to have the crystal mounted so that it can be moved for proper alignment and centering in the camera. The customary device used for this purpose is a *goniometer head* (Fig. 4.5), which has two arcs (A and B), which allow the crystal to be

[9] W. L. Bond, *Rev. Sci. Instr.*, **22**, 344 (1951).

[10] J. Peterson, L. K. Steinrauf, and L. H. Jensen, *Acta Cryst.*, **13**, 104 (1960); G. H. Stout and L. H. Jensen, *X-Ray Structure Determination*, Macmillan, New York, 1968, pp. 71–73.

[11] *International Tables*, Vol. II, pp. 291–305.

[12] *International Tables*, Vol. III, pp. 21–34.

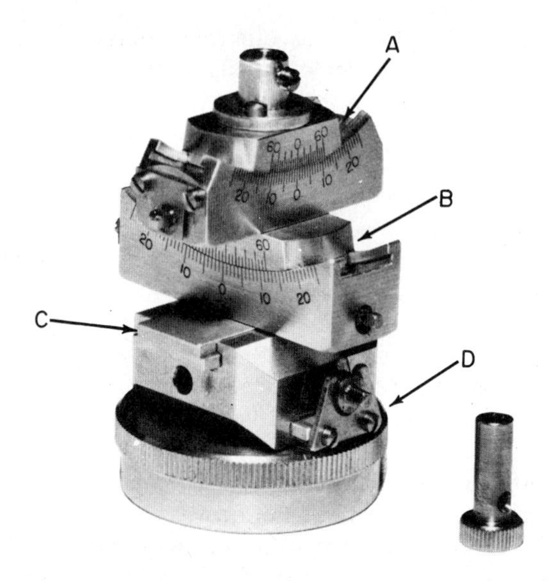

Figure 4.5. A standard goniometer head. (Courtesy of Charles Supper Co.)

tipped ±20° in each of two perpendicular planes, and two sledges (C and D), which allow it to be shifted until it is on the central axis of the head. The head is designed to be screwed onto a rotating spindle in the camera, and the crystal can be adjusted (*aligned*) so that either a direct or a reciprocal crystal axis is on the axis of rotation.

Since the portion of the mounting immediately adjacent to the crystal often protrudes into the X-ray beam, it must be amorphous if it is not to give rise to spurious diffraction spots. It should also be small and composed of elements of low atomic weight so as to keep the amount of diffusely scattered radiation to a minimum. The usual support is a very fine glass fiber about 20–70 μm in diameter to which the crystal is attached by an adhesive. The glass fiber in turn is fastened with sealing wax (before mounting the crystal!) to a metal pin suitable for insertion into the goniometer head to be used (Fig. 4.6). Most U.S. heads are adapted to a

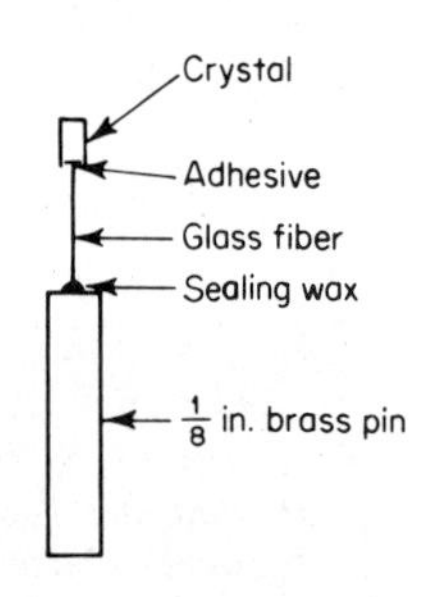

Figure 4.6. Mounted crystal.

1/8-in. pin, which is usually made of brass with a small axial hole into which the glass fiber can be inserted for better support. A small grooved collar on the pin makes it easier to grasp while adjusting it in the head.

Crystals that are unstable in air are mounted in capillaries (quartz or Lindemann glass), sealed with sealing wax, and mounted in turn in brass pins. Such mountings have the disadvantages of causing additional diffuse scattering and of potentially extending far enough to interfere with parts of the camera or diffractometer. In addition, the crystal is not supported as firmly, risking shifting during the ensuing measurements.

Many materials have been used to attach crystals to the fiber, for example, gum arabic in water, shellac in alcohol, library glue, sealing wax (for minerals), and thinned Duco or other similar cements, but epoxy glues seem superior in strength and convenience. Their solvent resistance makes them particularly suitable for use with crystals that are to be shaped by solvent saw methods. If the crystal is somewhat sensitive to the atmosphere, it can even be covered with a drop of epoxy with very little impairment of the subsequent photographs. Crystals that are to be used for data collection at low temperature can even be frozen to the fiber.[13]

The length of the mounting pin and fiber are critical only if they are to be used with a eucentric head on a diffractometer (see Section 5.10). Under these conditions the correct position of the crystal is determined by the geometry of the system, and the range of adjustment available is small. For such an apparatus, the combined length of the pin and fiber should be between 13 and 14 mm. Photographic methods are less demanding since the head position can be adjusted to accommodate a much larger variation. In this case, a fiber length of ~1 cm coupled with a pin of 10–15 mm is often convenient. Such a mounting will usually put the crystal at some distance from the center(s) of curvature of the adjusting arcs; so changes in the arcs will require translational corrections to bring the crystal back to the central axis of the head. Nevertheless, having the crystal well away from the head has its compensating advantages.

Small crystals are best mounted under a binocular, wide-field dissecting microscope of 10–20× power, although with practice an ordinary monocular instrument set to its lowest power may be used. Many organic compounds crystallize either as needles or thin plates. In the absence of contrary indications, the needles are best mounted to rotate around their needle axis, whereas the thin plates should be placed with the axis of rotation in the plane of the plate and parallel to a well-defined edge if possible. These choices may have to be changed as more information becomes available, but they are convenient and very often the best. For these orientations, the crystal is placed on a microscope slide and pushed gently with a needle or a glass fiber until the desired end or edge projects beyond the edge of the slide. The slide is placed on the microscope stage on

[13]H. Hope, *Acta Cryst.*, **B44**, 22 (1988).

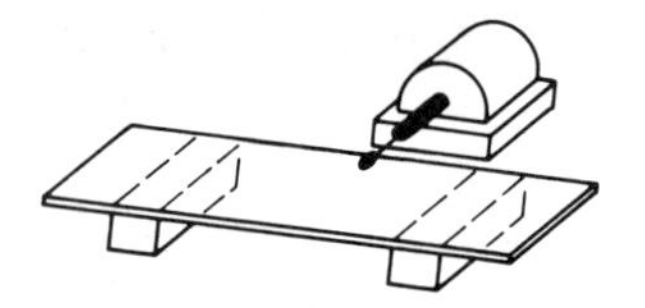

Figure 4.7. Simple method of mounting crystals.

a pair of blocks of suitable height (about $\frac{1}{2}$–$\frac{3}{4}$ in., but best determined by actual trial) and the fiber, bearing on its tip a drop of cement, is moved parallel to the stage until it touches the crystal. During this last operation a micromanipulator can be used, or the pin can be simply inserted in a hole in a third block (a cork with one edge cut off is excellent) that is cut to such a thickness that the elevation of the fiber is equal to the height of the spacing blocks plus the thickness of the microscope slide (Fig. 4.7). Alternatively, one can use a mound of modeling clay on a microscope slide to support the pin.

Epoxy glues harden relatively slowly, and it may be necessary to leave the block in place for several hours in order to obtain a firm set. Various brands have markedly different characters, some being thick and some thin when mixed, and a few trials are useful to find one with the right balance of properties. The slow setting can be turned to advantage with a fairly thick cement, which will hold the crystal balanced on the fiber and yet give enough that it can be teased into a desired orientation with a fine needle. When the cement has hardened, the pin and its crystal are transferred from the block to the goniometer head with the aid of tweezers and locked into place by a setscrew.

If the pin can be rotated in the head, or if one is suitably foresighted during the mounting operation, it is possible to have a pair of major faces in the plane of one of the arcs of the head and perpendicular to the other (Fig. 4.8). This arrangement is highly desirable for photographic alignment. The crystal can be roughly aligned by setting the arcs so that the assumed axis appears parallel to the axis of rotation, but a more precise method is usually necessary, as the crystal should be aligned to within at least 30 minutes of arc. Several methods are available. If the crystal has several well-formed faces and the proper equipment is at hand, it can be aligned optically (see below). If one of these conditions is not met, the whole operation can be

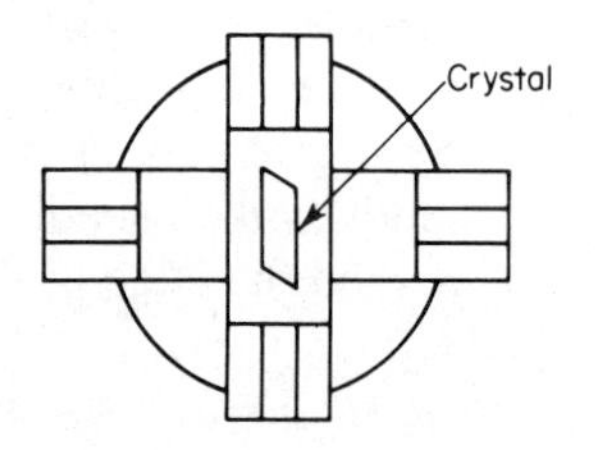

Figure 4.8. Preferred orientation, with prominent crystal faces parallel to an arc of the goniometer head.

carried out on the X-ray camera (see Chapter 5). Even ill-formed fragments with no visible edges or faces can be aligned in this way, but the process is tedious, and a preliminary visual alignment, even if very rough, simplifies matters greatly. Today, many crystals are "aligned" automatically by diffractometer-setting programs that search out a few reflections and *determine* the crystal orientation rather than *adjusting* it to a preferred arrangement. This approach, although certainly convenient, has the disadvantages of leaving one at the mercy of the machine's limited intelligence and making it difficult to obtain photographs to have a more thorough view of the reciprocal lattice. Thus we strongly recommend taking the trouble to actually align a crystal, at least if it is being examined for the first time.

4.5. OPTICAL ALIGNMENT[14]

The optical examination of crystals and their faces is a field of study in its own right, and any thorough discussion of it is impossible here. It is possible, however, to make use of some of its simpler techniques in order to align crystals rapidly and easily. The principal instrument used in the optical study of crystal forms is the two-circle goniometer (Fig. 4.9). This consists of a light source, a crystal support, and a combination microscope/telescope for observing either the crystal or the image of the light source that is reflected from its faces. The crystal is mounted in a goniometer head, which in turn is attached to a graduated circle (A, Fig. 4.9) capable of being rotated through 360°. This circle is carried on an arm that can be rotated about a vertical axis and whose position can be read off on a second circle (B, Fig. 4.9) on the base of the instrument. By the use of these two circles the crystal can be rotated about both its mounting axis and another axis perpendicular to it. Consequently, all the crystal faces can be set to a position in which their normals bisect the angle between the light source and the telescope; that is, the signal from the former is reflected into the latter.

Since optical methods of crystal alignment involve observing reflections of the source from the crystal, they can be applied only to those crystals that have reasonably well developed faces. As a rule it is more important that the faces be smooth and clean than that they be large; crystals of a size suitable for X-ray studies can give perfectly usable images. On the other hand, adequate reflections can often be observed from rather unpromising crystals; so a trial alignment is usually worth the effort. If even an ap-

[14]C. W. Wolfe, *Manual for Geometrical Crystallography*, Edwards Brothers, Ann Arbor, MI 1953, pp. 125–148; P. Terpstra and L. W. Codd, *Crystallometry*, Academic, New York, 1961, pp. 335–386. The optical goniometer also serves as an excellent model and training tool for the X-ray diffractometer since its geometry is very similar and it is possible to develop a feel for the effects of changing the various angles.

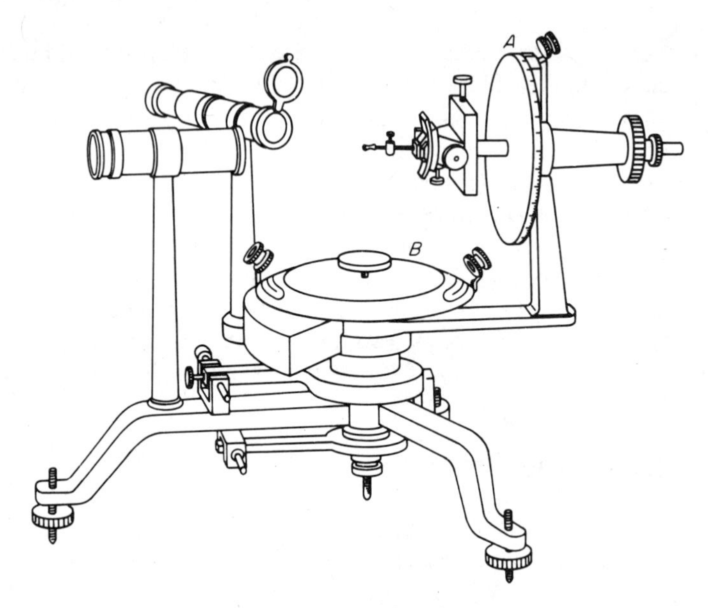

Figure 4.9. Optical goniometer.

proximate alignment can be achieved, it will reduce the time needed for refinement by X-ray methods.

The preliminary adjustments are made with the microscope set to show the crystal itself. These consist merely of adjusting the head arcs and sledges until the desired axis appears roughly coincident with the rotation axis of the head. For the finer adjustments, one lens is removed from the microscope to convert it to a telescope, and this is then focused on a pattern at the light source. Usually several possible patterns are provided, the most common being a graduated series of crosses. The size of the source image is chosen to minimize the amount of extraneous scatter from objects other than the crystal faces.

The face reflections can be used in a number of ways to achieve alignment. We shall consider only one, however. This is applicable to the common case of a prism or needle crystal in which there is a set of faces developed parallel to the needle axis and in which the mounting is along this axis. It is best if there are several pairs of faces available (the two members of each pair being opposite), but a minimum of two, nonopposing faces can serve. The desired condition of alignment is that shown in Fig. 4.10*a*. The crystal axis coincides with the rotation axis of the mount, and the normals for the parallel belt of faces (*zone*) are all perpendicular to it. Thus as the crystal is rotated about its axis the successive reflections will all appear at the same position in the telescope. If the crystal is misaligned, the condition of Fig. 4.10*b* will appear for one face of a pair, while that of Fig. 4.10*c* will be true after a rotation of 180°. Thus the reflection in Fig. 4.10*b*

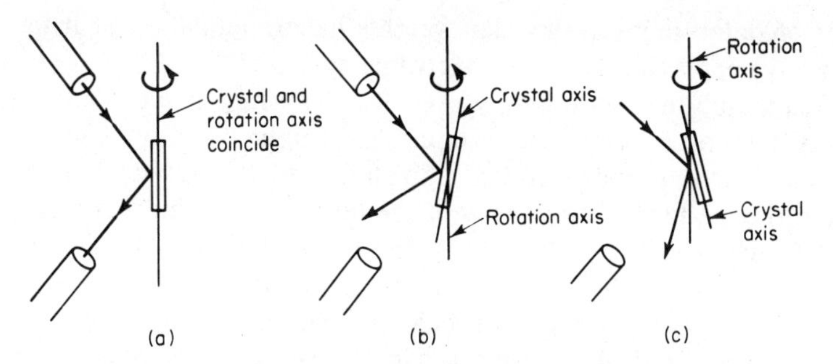

Figure 4.10. Alignment of a crystal with the normal to the rotation axis bisecting the angle between the collimator and the telescope. (*a*) Correct alignment. (*b*) Misset crystal. (*c*) Crystal as in (*b*) after rotation of 180°.

will appear at one side of the telescope field and that of Fig. 4.10*c* at the other. The correction is made by moving the arcs of the goniometer head to bring one reflection halfway toward the other. It is obviously easier to do this if the entire correction is controlled by one arc, and for this reason it was recommended that the crystal be mounted so that at least one major face is perpendicular to an arc. If this has been done, and if it is the reflection from this face that is being observed, that arc is now correctly set, and a repetition of the process using another pair of faces and making the correction only on the second arc should complete the alignment. In practice a process of successive approximations may be needed, but since it is always possible to see what one is doing, the method becomes very rapid after a little practice. The results, even with a poor crystal, are usually precise to ±0.1–0.2°, quite adequate for most X-ray studies.

It has been suggested[15] that, in the absence of an optical goniometer, a low-powered laser beam reflects from small crystal faces well enough to serve for alignment directly on the camera or diffractometer.

4.6. MEASUREMENT OF CRYSTAL PROPERTIES

There are several physical properties of crystals that can be determined early in a crystallographic investigation and can prove useful. One of these is the density of the crystal.[16] This can be combined with the measured dimensions of the unit cell to give an accurate value for the total molecular weight of its contents. From this it is usually possible to deduce the number of molecules in the cell and their individual weights and be warned of possible solvent content. In the past it was also one of the most accurate

[15]G. Moss and and Z. Bernea, *J. Appl. Cryst.*, **9**, 510 (1976).
[16]*International Tables*, Vol. III, pp. 21–34.

methods of determining molecular weights experimentally, but it has now been largely supplanted by mass spectrometry.

The technique most commonly used for determining the density (ρ) of crystals is flotation in a mixture of liquids, one lighter and one heavier than the crystal, whose proportions are adjusted until the crystal remains suspended in the medium. The density of the liquid, determined by weighing a sample of known volume,[17] is that of the crystal. A pair of solvents often used are heptane ($d = 0.684$) and carbon tetrachloride ($d = 1.589$), whose mixtures have the advantages of being comparatively poor solvents for many compounds and of covering a wide range of densities. For nonpolar organic compounds, which are likely to be soluble in most dense organic solvents, concentrated solutions of inorganic salts such as KI may be used. These pose problems of evaporation and high surface tension, however, although the latter effect may be reduced by the judicious addition of wetting agents.

In practice, a sample of several crystals is placed in one of the solvents in a centrifuge tube, and the other solvent is added in small portions until the crystals are suspended without rising or falling. The method is quite sensitive, and a small drop of pure solvent added to a mixture may carry it from one side of equilibrium to the other. If further sensitivity is desired, the equilibrium mixture can be checked by centrifuging the sample briefly and observing whether the crystal remains suspended under the added force. The crystals used do not have to be large, but it will be found that if they are smaller than ~0.1 mm on a side they will be slow to move and will be particularly sensitive to convection currents, so centrifuging will probably be necessary.

Since crystals often contain occluded air or solvent as regions of lower density, it is advisable to use more than one crystal and to regard as the equilibrium mixture the one of highest density that still suspends one or more crystals. With reasonable care, particularly with respect to having all the solutions and vessels at one temperature, the accuracy of such a determination is usually better than ±1%. This method is capable of high precision but depends for its success on the careful elimination of all sources of error from the crystals themselves as well as from the measuring techniques.

The densities of most organic solids fall within the range 0.9–1.7 g/mL, and certain generalizations may be made. Compounds with a large proportion of hydrogen tend to be less dense than those without. Planar molecules, which can pack together well, have fairly high densities, but these are lowered by loose, irregular side chains. As would be expected, the addition of heavy atoms increases the density.

Once the volume of the unit cell (in cubic angstroms) has been determined (see Chapter 5) the mass of material in it (in daltons) is obtained by

[17]See also S. Zamil, R. Pludow, and A. F. Fucaloro, *J. Appl. Cryst.*, **11**, 163 (1978).

dividing the mass in grams ($V\rho \times 10^{-24}$) by the mass of a hypothetical atom of atomic weight 1.0. Thus

$$M = \frac{V\rho \times 10^{-24}}{1.6604 \times 10^{-24}} = 0.60226\, V\rho \qquad (4.5)$$

The mass per asymmetric unit can be found by dividing M by the number of asymmetric units appropriate to the space group (see Section 5.12), and the quotient will, in most cases, be the molecular weight. It is possible, however, for it to be either a fraction of the formula weight (usually one-half) if molecular symmetry allows the molecule to occupy two or more asymmetric units, or a multiple, if there is more than one molecule per asymmetric unit. Usually one can expect agreement of 1–1.5% between observed and calculated values. For inorganic compounds, Eq. (4.5) is generally used in a rearranged form to calculate a theoretical density that is reported for comparison with the observed value.

Other physical methods have sometimes been applied to crystals in an effort to determine whether they contain an inversion center among their symmetry elements. Older techniques included tests for piezoelectricity and pyroelectricity,[18] but these are difficult or impossible to perform on small organic crystals. A very effective modern method involves observing the generation of second harmonics when a laser beam is shone on powdered crystals lacking a center.[19] The technique is easy and sensitive but unfortunately requires specialized equipment that is not readily available.

KEY FORMULAS

Absorption

Linear absorption coefficient μ

$$\mu_\lambda = \rho \sum_n \frac{P_n}{100} \left(\frac{\mu}{\rho}\right)_\lambda, \mathrm{E}_n \qquad (4.3)$$

for the n elements E, each constituting P_n% of the crystal.

$$I = I_0 e^{-\mu\tau} \qquad (4.1)$$

Unit Cell Mass

$$M = 0.60226\, V\rho \qquad (4.5)$$

[18] *International Tables*, Vol. I, pp. 41–43.
[19] J. P. Dougherty and S. K. Kurtz, *J. Appl. Cryst.*, **9**, 145 (1976).

BIBLIOGRAPHY

Buckley, H. E., *Crystal Growth*, Wiley, New York, 1951.

Buerger, M. J., *Crystal-Structure Analysis*, Wiley, New York, 1960, pp. 53–76.

Fox, D., M. M. Labes, and A. Weissberger, Eds., *Physics and Chemistry of the Organic Solid State*, Vol. I, Interscience, New York, 1963.

Jeffery, J. W., *Methods in X-Ray Crystallography*, Academic, London, 1971, pp. 143–148.

CHAPTER 5

GEOMETRIC DATA COLLECTION

The collection of crystallographic diffraction data poses two quite different problems. The first is the determination of the geometry of diffraction, from which the size, shape, and symmetry of the reciprocal and direct lattices may be calculated. The second is the assignment of an observed intensity to every point in the reciprocal lattice (or to every reflecting plane in real space). This intensity may ultimately be related to the distribution of diffracting electrons in the unit cell. This chapter deals with the first problem, and Chapter 6 with the second.

For routine structural studies the only practical means of collecting the desired data require that diffraction patterns be obtained from a single crystal mounted so that the various lattice planes may be brought successively into reflecting positions. This process must be combined with some means of recording both the locations and intensities of the diffracted X-rays. Historically, photographic film served both ends, but at present radiation counters mounted on automatic diffractometers are used almost exclusively for small molecules. The diffracted intensities are counted directly, and the angular settings at which they occur serve to define the cell geometry. *Nevertheless, nothing can supplant photographic methods for the purpose of obtaining a general view of the geometry and symmetry of the reciprocal lattice*. A disturbing number of errors have arisen recently in space group and lattice assignments. Many of these could have been avoided if photographs of the crystal lattice had been taken rather than depending solely on individual reflections found by an automated diffractometer.

A large number of cameras have been devised for taking reciprocal lattice photographs, but in practice only two basic types need be considered. These are the rotation/oscillation/Weissenberg geometry and the Buerger precession camera.

5.1. ROTATION AND OSCILLATION THEORY

The rotation/Weissenberg camera served for many years as the standard data-collecting device in X-ray crystallography. Photographs taken with it contained all the information needed for a structure solution. It has been largely supplanted by diffractometers in modern practice but is still an excellent teaching tool. Furthermore, an understanding of its principles provides a needed familiarity with the practical application of reciprocal lattice concepts.

The basis of rotation photography is simple. If a crystal is mounted so that it can be rotated about a direct crystal axis (assume *a*) (Fig. 5.1), by the definition of the reciprocal lattice (Chapter 2) the b^*c^* reciprocal plane is perpendicular to the direct *a* axis and therefore perpendicular to the axis of rotation. This plane contains all the reflections for which the *h* index is zero, and the planes for which $h = 1$, 2, and so on are parallel to it and also perpendicular to the axis of rotation. As the crystal is turned, these reciprocal planes may be regarded as turning with it and cutting the sphere of reflection (Fig. 5.2) (see Section 2.4). The intersection of each of the planes with the sphere is a circle, and as each reciprocal lattice (r.l.) point touches the surface of the sphere the conditions for reflection are met, and a diffracted ray passes from the center of the sphere of reflection through this point of contact (Fig. 5.3). As the crystal rotates, successive points along a reciprocal lattice line will be brought into reflecting position (Fig. 5.4). The origin of the reciprocal lattice, point 000, is always in contact with the reflecting circle for the zero level, but as this is also in line with the direct X-ray beam, its reflection cannot be measured directly.

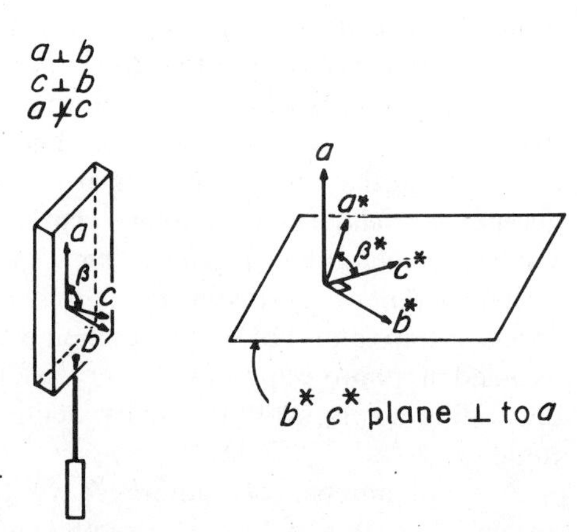

Figure 5.1. Monoclinic crystal mounted for rotation about *a*.

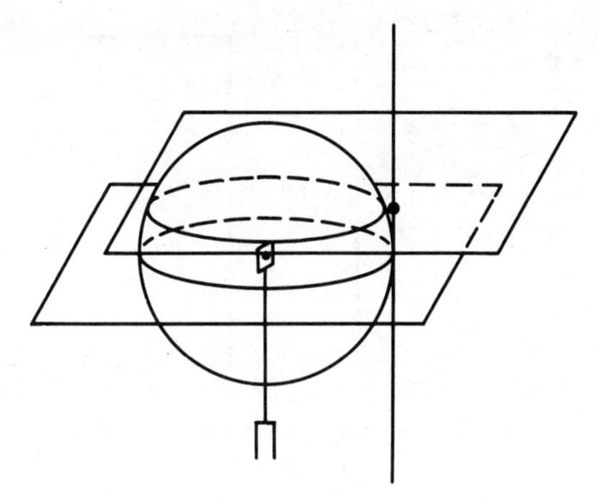

Figure 5.2. Rotation of a crystal about a direct axis and the intersection of the r.l. planes with the sphere of reflection.

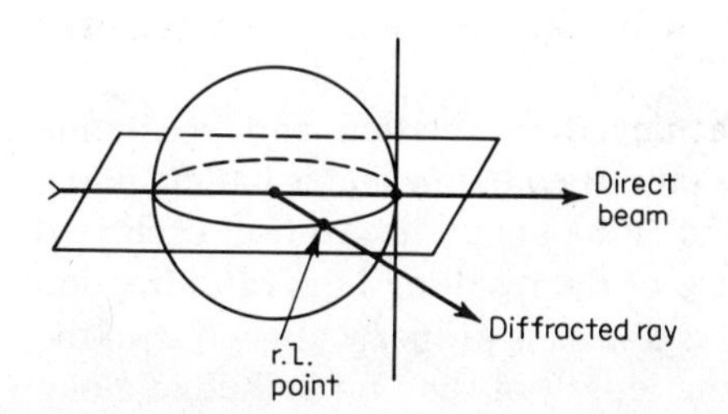

Figure 5.3. Reciprocal lattice condition for diffraction.

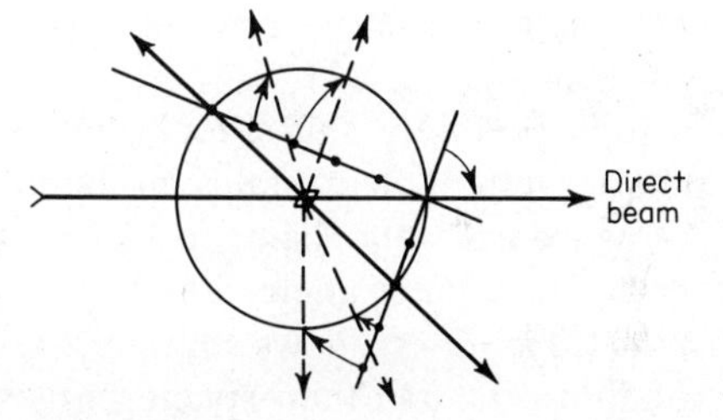

Figure 5.4. Successive diffraction by rows of r.l. points. View down rotation axis.

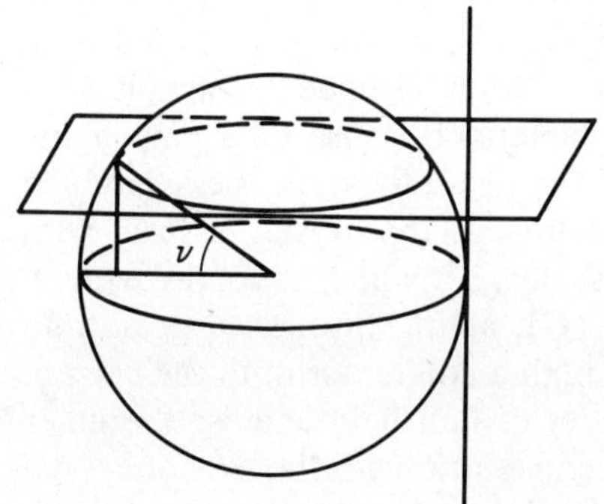

Figure 5.5. Diffraction from an upper level of a reciprocal lattice.

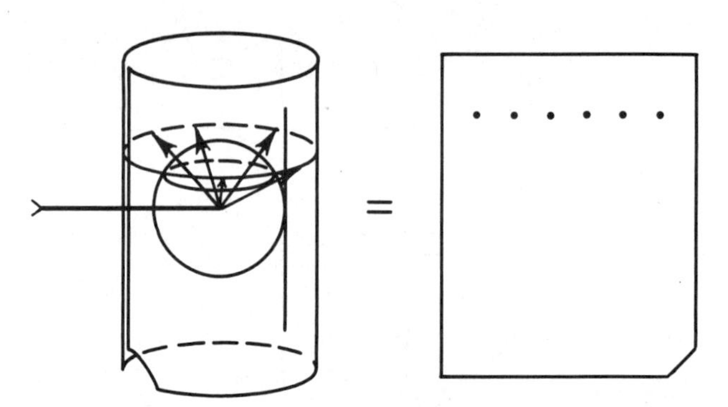

Figure 5.6. Recording a single layer line on cylindrical film.

For levels above zero, several arrangements are possible. The most important fact at the moment, however, is that any ray passing through a point on the circle cut by one of the upper-level planes has the same vertical angle ν (Fig. 5.5). Thus, if the crystal (which may be regarded as being at the center of the sphere of reflection) is surrounded by a cylinder of film as shown in Fig. 5.6, the extended rays will all intersect the film at the same height and the resulting spots will lie on a straight line when it is unrolled.

It is precisely this effect that is achieved in rotation and oscillation photographs. The crystal is mounted as described below and is bathed in the X-ray beam while being either rotated 360° about its axis or oscillated through a smaller angle. The appearance of the resulting lines of diffraction spots (*layer lines*) shows whether the crystal axis is properly aligned with the rotation axis, and their spacing gives the length of the unit cell edge along this axis.

5.2. ROTATION AND OSCILLATION PRACTICE

The Weissenberg camera, which is also used for rotation and oscillation photographs, is shown in Figs. 5.7 and 5.11. The crystal is mounted on a goniometer head (Section 4.4), which is in turn fastened by means of a threaded ring to the spindle (A) of the camera. Because of a pin on the spindle and a matching slot in the base of the head, they may be assembled in only one position, and thus the orientation of the crystal is fixed with respect to the spindle and the *azimuth* scale (B), which revolves with it. The spindle is rotated by a small motor (C), acting through a gear train (D), while the X-ray beam is directed through a collimator onto the crystal. The diffracted rays are received by a sheet of film held against the inner surface of a fixed, cylindrical shell that is concentric with the axis of crystal rotation. After exposure the film is developed, fixed, and washed, revealing

Figure 5.7. Weissenberg camera. (Courtesy of Charles Supper Co.)

the diffraction pattern in the form of characteristic layer lines (Figs. 5.6 and 5.8). Figure 5.9 shows a drawing of the camera with the reciprocal lattice as it should be visualized.

The standard Weissenberg film holder has a nominal diameter of

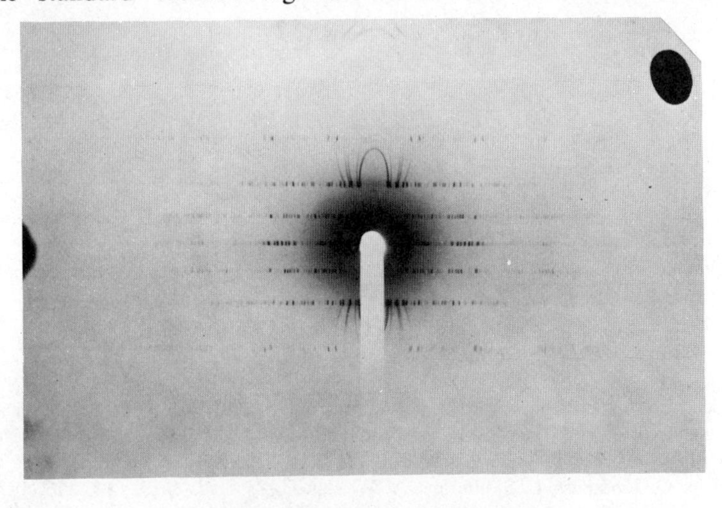

Figure 5.8. Rotation photograph.

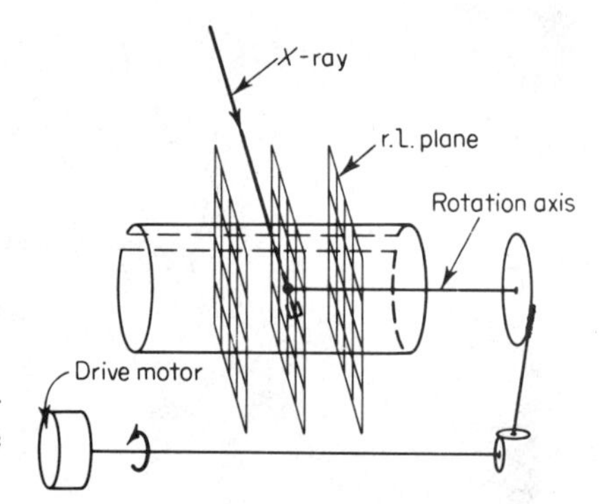

Figure 5.9. Schematic drawing of a Weissenberg camera and three reciprocal lattice levels.

57.3 mm. This value is chosen because 1 mm across the unrolled film is then equal to 2° around the circumference of the camera. Because of the thickness of the film and irregularities in the way in which it is placed in the camera, this value is rarely exact, and for careful measurements of cell constants each camera should be calibrated with a crystal of known spacings.

The usual film is ~5 in. × $6\frac{3}{8}$ in., but the long dimension will vary slightly for different makes of cameras. It should be cut as accurately as possible from a sheet of 5 × 7 X-ray film,[1] since a film that is too long is difficult to load and one that is too short is not held firmly against the camera. One corner is clipped for orientation. The film is placed in a black paper envelope, and it is usually necessary to protect the side that will be exposed to light by an additional sheet of black paper to avoid dark spots from pinhole leaks. The paper used to separate the sheets of film in the box is usually suitable.

Figure 5.10 shows the important features of the cassette. As it is loaded

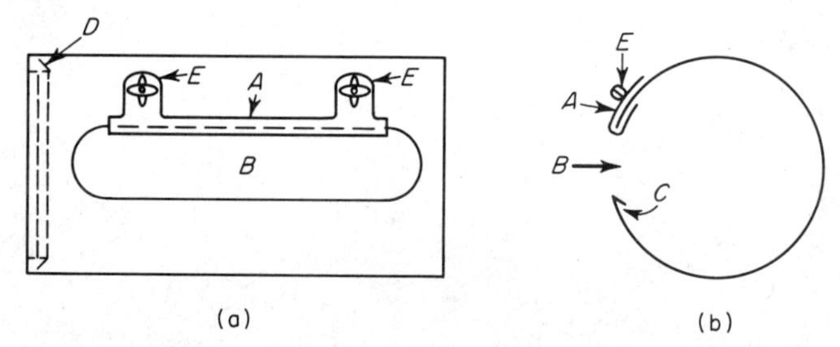

Figure 5.10. Weissenberg cassette. (*a*) Side view; (*b*) end view.

[1]The choice of film is unimportant for survey and geometrical photographs but significant for intensity measurements. Numerous detailed studies have been made of the properties of particular films, but these become obsolete as the films described go off the market. See S. Abrahamson, O. Lindqvist, L. Sjölin, and A. Wlodawer, *J. Appl. Cryst.*, **14**, 256 (1981) for references and an outline of the important questions to be answered.

in a darkroom, a first trial with a dummy piece of film will provide practice and simplify the actual operation. The spring clip (*A*) is removed from the cassette, and the film package, rolled into a cylinder, is inserted from one end or the other, depending on the design. The clipped corner should always be placed in one specific orientation, usually the upper right-hand corner as the film is viewed through the collimator slot (*B*) in the cassette. The far edge of the film is placed firmly in the groove (*D*) at the end of the cassette. A fingernail run around the groove after the film is in place will ensure that it is correctly located and that the envelope is not caught on the edge. The spring clip (*A*) is replaced, making sure that its lip is hooked over the upper end of the film, and is fastened down with its quick-release catches (*E*). The last edge is flattened against the cassette walls either by a split ring that serves as the end of the cassette body or by a cylindrical flat spring that fits inside the cassette. The whole assembly may be completed by the addition of a cylindrical cap to one end of the cassette. This last item is not required, but it reduces the possibility of scattered X-rays in the room in which the camera is located.

The main body of the camera (Fig. 5.11) consists of four parts: (*A*) a fixed base, which rests on three legs whose heights are adjustable; (*B*) a movable base, which may be translated to one side by means of a crank and screw; (*C*) the motor, gear train, cassette guides, and crystal mount, which may be rotated about an axis approximately perpendicular to the table; and (*D*) the film holder base, which rests on tracks and may be slid to and fro parallel to the axis of crystal rotation. When the camera is used for Weissenberg photographs, a split nut is engaged, and this base is driven

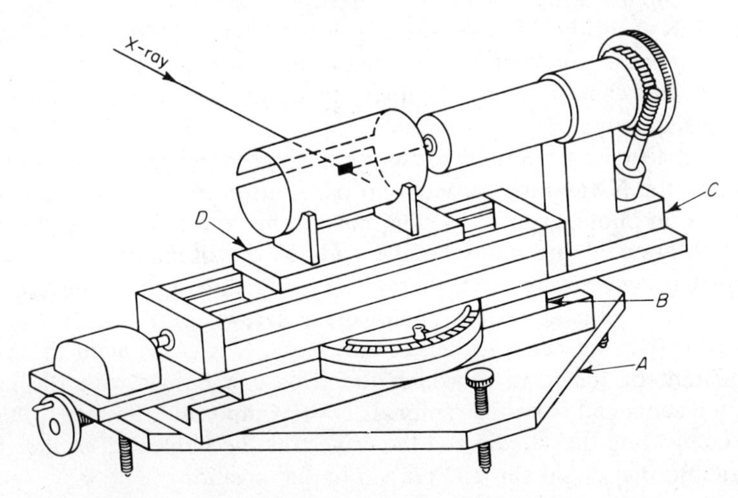

Figure 5.11. Schematic view of a Weissenberg camera.

back and forth by a thread on the shaft linking the motor and the crystal mounting. For rotation photographs, however, the nut is disengaged and the film holder is locked in one position.

The collimator, which is used to provide a nearly parallel beam of X-rays, is held by an arm attached to the back of the movable base (*B*). While the camera is being set up, the base (*B*) is usually shifted to the end of its translation so that the collimator is free of the X-ray tube and can be removed and replaced. It is essential that the crystal be centered on the axis of rotation and positioned so that it is in line with the opening of the collimator. Otherwise it will spend part or all of its time outside of the X-ray beam, and part of the diffraction pattern will be lost. For testing the centering, the crystal spindle may be released from the drive mechanism by loosening a screw or pin that is mounted on the azimuth ring; the crystal can then be rotated freely by hand. The spindle can be adjusted back and forth by means of a screw at the outer end of its shaft, and the centering is provided by the sledges on the head. In some cameras the rotating crystal and the collimator can be observed simultaneously with a small demountable telescope attached to the camera base; in others it is necessary to observe the crystal through the collimator tube with the aid of a low-power magnifying glass.

When the crystal is properly mounted, the direct beam stop is passed over it and slid into place on the fixed sleeve surrounding the rotation spindle. This stop serves to prevent the direct beam from striking the cassette and giving rise to large amounts of scattered radiation. The film holder is then attached to its base, which is shifted until the crystal is located at about the middle of the film and is then locked in place. The collimator, which must be removed in order to mount the film holder, is replaced, and the movable base is translated until the opening of the collimator aligns with the beam from the tube. The scale setting that corresponds to this position should, in principle, have to be determined only once, but it is well to check it from time to time, especially if several people are using the camera or if the intensity of the photographs appears to decrease suddenly.

In order to take a rotation photograph, the camera motor need only be turned on, the X-ray tube started, and the shutter on the tube opened. For an oscillation photograph, provision must be made for reversing the direction of rotation of the motor at the ends of the oscillation. A reversing microswitch is generally used, the switch being thrown by adjustable trips driven by some linkage to the main drive screw, for example, by being fastened to the ring bearing the azimuth scale. It is convenient to have an arrangement of this kind in which the trips can be set with the crystal rotation disengaged from the motor. It is very important, however, that the screw connecting the spindle and the drive train be tightened, or the crystal and azimuth dial might slip and remain in one position.

5.3. ROTATION AND OSCILLATION USES

Rotation and oscillation photographs are generally used to align crystals and to measure the cell edge along the axis of rotation. In some cases preliminary symmetry information can be obtained, but this is more easily done with an additional photograph using a precession camera.

Alignment

Although diffractometers are ordinarily programmed to cope with the problem of determining the orientation of a crystal in a totally general position, it is an excessively tedious exercise using photographs. In general it is possible to align a crystal by eye (or, better, with an optical goniometer, using the classical methods of observing light reflecting from faces; see Section 4.5) so that a direct axis is approximately along the rotation axis. In this case the layer lines in oscillation photographs will not be straight or will be tilted with respect to the central axis of the film. In a rotation photograph of a misaligned crystal the individual layer lines fan out as they approach the edges of a film.

The effects of small misalignments can be considered as resulting from errors of orientation in two planes; both contain the rotation axis, but one (A) is parallel to and the other (B) perpendicular to the X-ray beam (Fig. 5.12). To simplify the calculation of corrections, the azimuth of the camera should be set so that the arcs of the goniometer head lie approximately in these planes.

Since the reciprocal lattice is always aligned with the crystal, any tilt in the crystal is reflected in a corresponding tilt in the reciprocal lattice. Thus if there is a setting error in plane A (e.g., the crystal is tipped toward the X-ray source), the reciprocal lattice is tipped as shown in Fig. 5.13*a*. The zero plane cuts the sphere of reflection in a circle as usual, but the X-ray beam intersects this circle only at the r.l. origin, that is, the point at which the beam leaves the sphere of reflection. The trace on the film made by the diffracted rays passing from the crystal through this circle is shown in Fig. 5.13*b*. The occurrence of such a curved layer line is here called *bow*.

If the error is in plane B, the r.l. planes are tipped as in Fig. 5.13*c* and

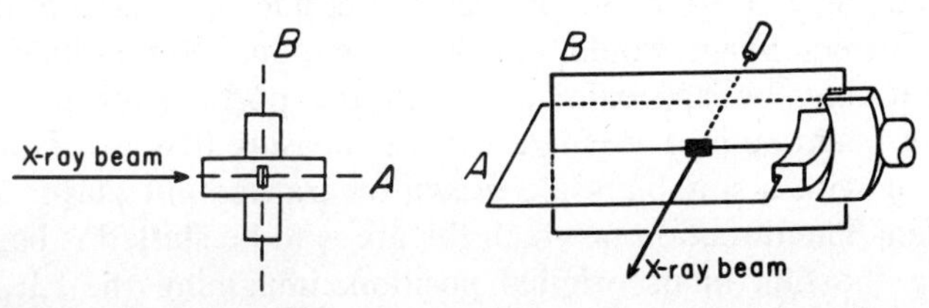

Figure 5.12. Schematic diagram of a goniometer head and its relationship in two perpendicular error planes.

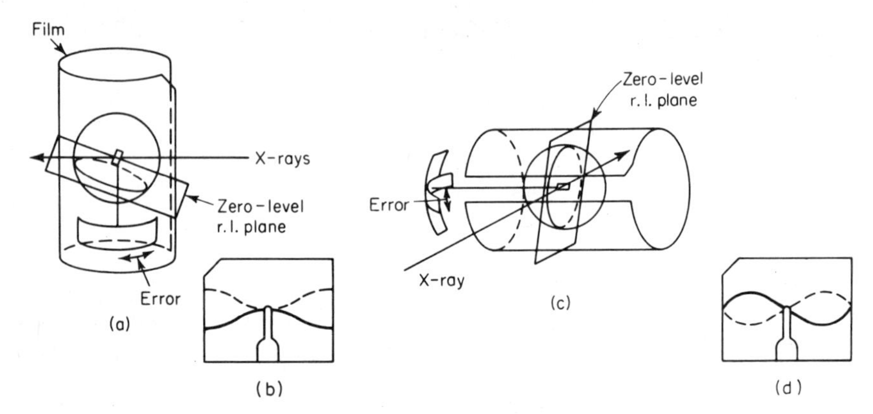

Figure 5.13. Alignment errors and their effects. (*a*) Error in horizontal plane (*A*, Fig. 5.12) only; seen from above. (*b*) Zero-layer trace on film resulting from setting as in (*a*); dashed lines indicate trace after azimuth is changed by 180°. (*c*) Error in vertical plane (*B*, Fig. 5.12) only. (*d*) Zero-layer trace resulting from (*c*); dashed lines indicate trace after azimuth is changed by 180°.

the circle cut on the sphere of reflection is rotated about the X-ray beams so that it is not perpendicular to the axis of crystal rotation. The trace (Fig. 5.13*d*) is then tilted with respect to the central axis of the film. This condition is called *tilt*. In both cases, if the crystal is rotated through 180°, a mirror image pattern is formed.

The usual misset crystal is in error on both arcs, and the pattern resulting from a short oscillation at one setting will show a combination of both tilt and bow. Two methods are useful for obtaining the corrections.

1. A photograph is taken at one setting using a short oscillation range (±5–15°) and unfiltered radiation. If layer lines cannot be recognized, shortening the oscillation range will often cause them to appear. Usually an exposure of 1 hr is enough; in many cases, a shorter exposure will serve. Without moving the film, the crystal is rotated 180° and a second similar exposure of one-half the duration is made. The resulting photograph will look more or less like Fig. 5.13*d*. The correction in tilt (error in plane *B*) is one-half the angle between the two zero-level lines measured at the point at which the direct beam would intersect the film. This point is not seen directly but may be approximated from the portions of the layer lines nearest the beam stop. The angle can be measured by a rotating hairline and scale or with a suitable scale drawn on transparent plastic with a fine drafting pen. The direction in which the arc is to be shifted is best obtained by holding the film in its original position, imagining the r.l. plane that produced one of the zero-level lines (the lighter one, unless the crystal has been returned to its original azimuth), and deciding whether the plane must turn clockwise or counterclockwise in order to be perpendicular to the

rotation axis. The arc is then corrected in this direction by the proper amount. The corrections for the second arc are obtained by rotating the crystal holder 90° and repeating the process.

A modification of this procedure consists of using the Weissenberg layer line screen (Section 5.4) and taking four oscillation photographs at 90° increments in azimuth, shifting the cassette laterally between exposures.[2] Angular measurements on these allow the calculation of corrections for both head arcs.

2. The second method is faster, in that only one film is needed, and more precise, but it places certain additional requirements on the crystal. A double oscillation photograph is taken as above, the only difference being that between the first and second exposures the film holder is shifted by a small amount. Two millimeters is usually a convenient shift, but whatever the value it should be measured as accurately as possible (±0.1 mm or less). Most cameras are provided with a suitable scale; if not, an accurate vernier caliper is quite effective for measuring the gap between the film holder and some fixed point on the base.

After the film is developed the distance between the two images of the zero-layer line is measured in millimeters at points 90° (45 mm for a camera with $D = 57.3$ mm) to the right ($\Delta r'$) and left ($\Delta l'$) of the direct beam point. These differences are positive if the lines have the same relative positions as at the center of the film, and negative if they have crossed (Fig. 5.14). The points chosen for measurement may vary by ±3–5° from the 90° position but should be as close as possible to this value. If no trace can be found in this region, the measurements can be made at other points, but the calculations are more involved.[3] It is better to try first increasing the exposure time, and if this fails, to use method 1 and refine from Weissenberg photographs (Section 5.7).

The true error is calculated by subtracting the known film shift (Δf) from

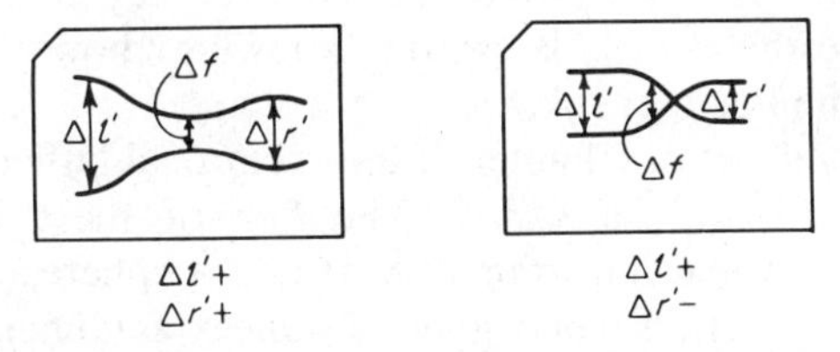

Figure 5.14. Appearance of zero-layer lines on double oscillation photographs of a misaligned crystal.

[2]I. R. Hanson, *J. Appl. Cryst.*, **14**, 353 (1981).

[3]I. Garaycochea and H. Cid-Dresdener, *Acta Cryst.*, **14**, 200 (1961) and references cited there. See also P. T. Davies, *Acta Cryst.*, **14**, 1295 (1961).

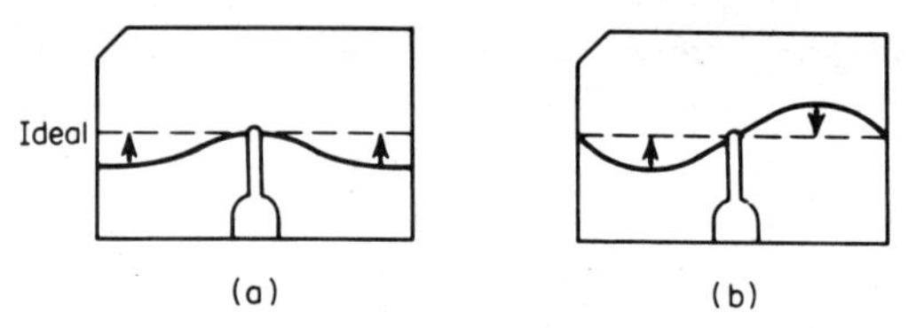

Figure 5.15. Appearance of zero-layer lines in oscillation photographs. (*a*) Correction needed for bow only; (*b*) correction needed for tilt only.

both quantities:

$$\Delta r = \Delta r' - \Delta f; \qquad \Delta l = \Delta l' - \Delta f \tag{5.1}$$

The corrections in degrees are then given by

$$\text{tilt error} = \left|\frac{\Delta r - \Delta l}{2}\right|; \qquad \text{bow error} = \left|\frac{\Delta r + \Delta l}{2}\right| \tag{5.2}$$

Conventions are often given by which the direction of the correction may be determined from the sign of the error, but direct comparison of the film and the crystal, as described under method 1, is probably more straightforward. Consider a single zero-layer trace, and remember that it is to be converted into a straight line perpendicular to the central axis of the film. The correction for tilt is obtained as described above; that for bow by determining which way the plane should be rotated to bring it perpendicular to the crystal rotation axis (see Fig. 5.13*a*). In general the correction should be such as to bring the points near the edges of the film into collinearity with the unmoving center of the zero-layer line (Fig. 5.15).

Measurement of Axial Length

Analysis of diffractometer setting angles gives cell parameters in a fraction of the time and with several times the accuracy of most photographic methods. Nevertheless it is worth knowing how to measure these parameters on film as a check.

As was pointed out in Chapter 2, the reciprocal lattice can be defined in terms of either $1/d_{hkl}$ or λ/d_{hkl}. The first of these leads to distances measured in reciprocal angstroms (Å^{-1}) and a sphere of reflection with a radius of $1/\lambda$ Å^{-1}. The second gives distances as dimensionless quantities generally called *reciprocal lattice units* (r.l.u.), and the radius of the sphere of reflection is 1 r.l.u. Of the two systems, the latter has been more commonly used in discussions of r.l. recordings since it relates the reciprocal lattice more directly to the observed diffraction pattern. We shall use this approach here and in Chapter 6.

With regard to interpreting rotation photographs, consideration of Figs.

5.5 and 5.6 shows that the distance between the zero- and nth-layer lines on a rotation photograph is proportional to the perpendicular distance d_n^* between the zero and nth levels of r.l. points. Figure 5.16 shows the arrangement, and it follows that d_n^* (in r.l.u.) is given by

$$\frac{d_n^*}{1} = \frac{y_n}{(R^2 + y_n^2)^{1/2}} = \sin \tan^{-1}\left(\frac{y_n}{R}\right) \tag{5.3}$$

$$d_1^* = d_n^*/n \tag{5.4}$$

R is the true film radius (note that the usual figure of 57.3 mm is a diameter), and y_n is the distance on the film, in the same units, from the zero-layer line to the nth-layer line. Because the direct axis, which is the axis of rotation, is perpendicular to the r.l. levels and thus parallel to d^*, the repeat distance r along this axis is given by

$$r = \frac{\lambda}{d_1^*} = \frac{n\lambda}{\sin \tan^{-1}(y_n/R)} \tag{5.5}$$

regardless of the crystal class involved. Note, however, that it is not necessarily correct to equate d_1^* with the reciprocal axis repeat (r^*) corresponding to r. If r, r^*, and d^* coincide, $r^* = d_1^*$; but if r is an inclined axis, r^* will not be the perpendicular distance between r.l. layers and will not coincide with d^* (Fig. 5.17).

The graph in Fig. 5.18 provides means of estimating axial lengths rapidly as a function of the distance between the zero- and first-layer lines.[4] For more accurate measurements the distance between corresponding layer

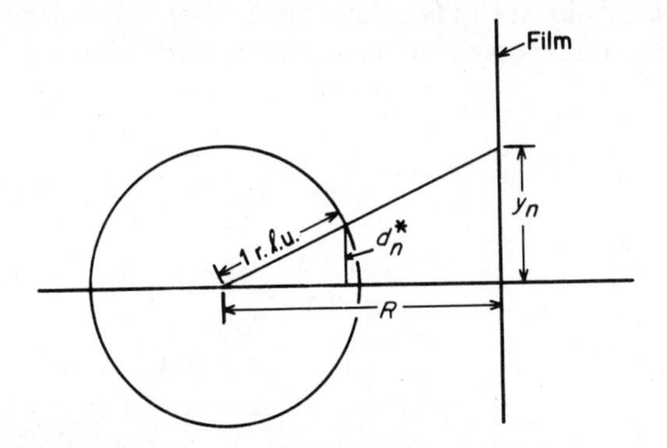

Figure 5.16. Relation between d_n^* in the reciprocal lattice and spacing between the zero- and nth-layer lines on a rotation photograph.

[4]For tabulated values of r as a function of y measured to the nearest 0.1 mm, see *International Tables*, Vol. III, pp. 101–121.

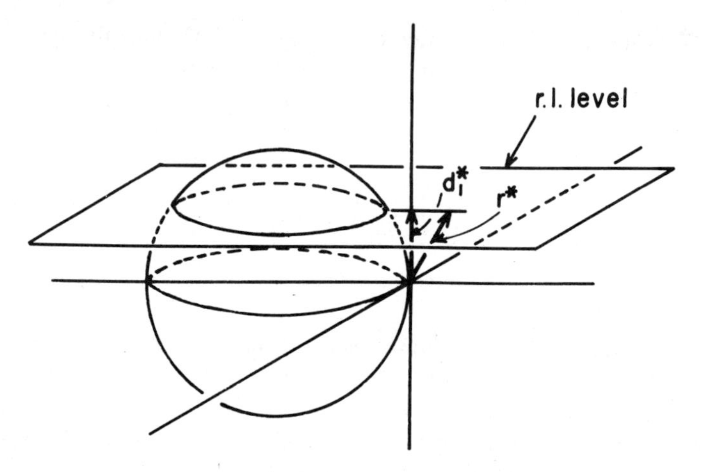

Figure 5.17. Relation between d_1^* and r^*, reciprocal lattice repeat, for an inclined axis.

lines above and below the center (i.e., between n and $-n$) is measured to ± 0.05 mm. The resulting value is divided by 2 and used for the distance y_n The calculation of r is carried out for each layer that is observed, and the results are then averaged. It will frequently occur that the value calculated from the first level varies markedly from the others because of the small separation measured on the film. If so, it may be omitted from the averaging.

The results obtained by this method are generally precise to ~0.2–0.5%, but the accuracy can be considerably poorer. Small errors in alignment of the crystal (<1°) are not serious, but changes in the effective radius of the camera can be significant. These are most often caused by errors in the camera body, the thickness of the film envelope and film, and film shrinkage

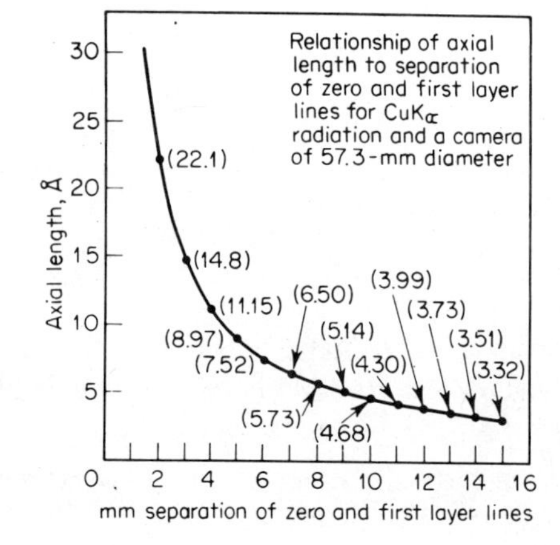

Figure 5.18. Axial lengths as a function of 0–1 layer line spacing on rotation or oscillation photographs. For Cu K_α radiation and a camera of 57.3 mm diameter.

after development. The usual effect is to decrease the apparent camera diameter by 0.5–1.0 mm and systematically alter the apparent cell constants by 1–2%.

5.4. ZERO-LEVEL WEISSENBERG THEORY[5]

The serious disadvantage of rotation and oscillation methods is that the information contained in an entire two-dimensional reciprocal lattice plane is condensed into a one-dimensional layer line. Indexing reflections becomes very tedious, and many films are necessary to avoid possible overlap of spots.[6] For these reasons, it is desirable to have a means of recording data so that a single r.l. plane is mapped onto an entire sheet of film. Indexing is simplified greatly if some straightforward geometric relationship can be established between the lattice and its photographic image.

The first widely used method that accomplished this was proposed by Weissenberg in 1924.[7] Mechanically, the technique is very simple. A single layer line is selected by a slotted screen that stops all other diffracted beams (Fig. 5.19); as the crystal is rotated, the film is moved past the slot

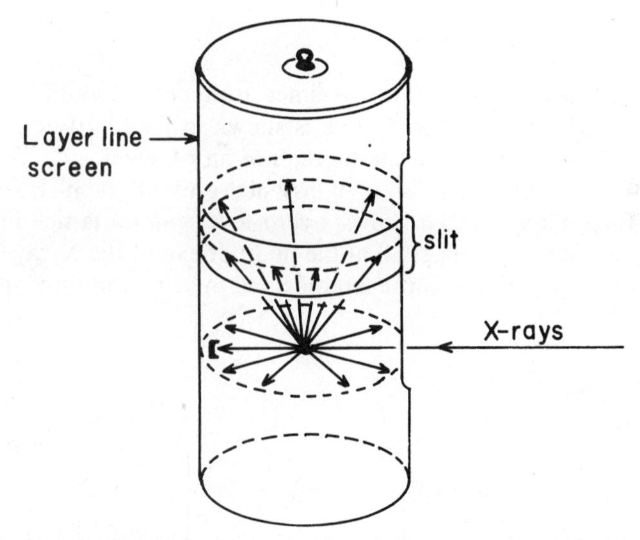

Figure 5.19. Weissenberg layer line screen and its function.

[5]Buerger, *X-Ray Crystallography*, Chapter 12; Jeffery, *Methods in X-Ray Crystallography*, pp. 188–198.

[6]Nevertheless, this method, with oscillation ranges as small as 1° and computerized indexing, has been widely used for collecting intensity data in protein crystallography. It has the advantage that all reflections are recorded as they occur, minimizing the total X-ray exposure required for the sensitive crystal. In the Weissenberg method many reflections occur that are lost to the screen.

[7]K. Weissenberg, *Z. Physik*, **23**, 229 (1924).

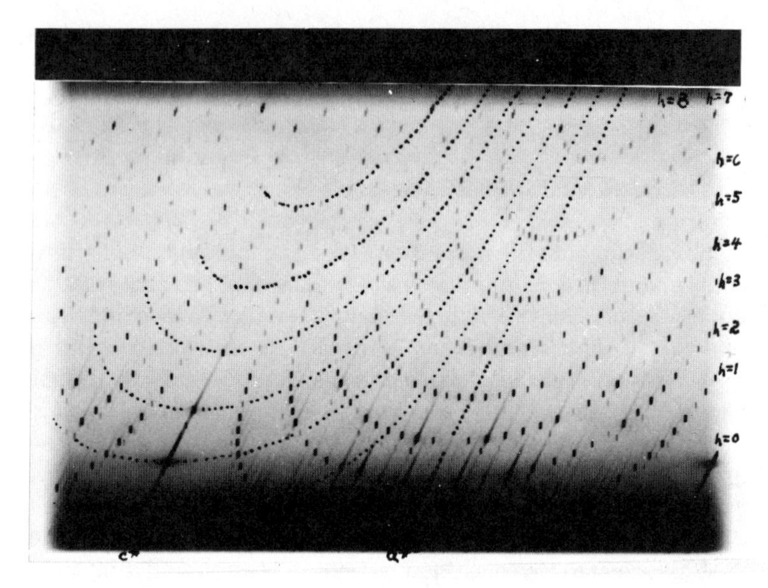

Figure 5.20. Half a Weissenberg photograph showing indexed r.l. lines. Dotted lines added for $l = 0, 4, 8, 12, 16, 20, 24$.

and reflections that occur at different times are recorded at different points on the film. Half a typical photograph is shown in Fig. 5.20, together with its interpretation at a distorted photograph of an r.l. level.

The theoretical interpretation of Weissenberg photographs is reasonably straightforward. Figure 5.21*a* shows a zero-level central lattice line that is tangent to the sphere of reflection at the point of exit of the X-ray beam. As the crystal is rotated, the reciprocal lattice and this line will turn with it. The

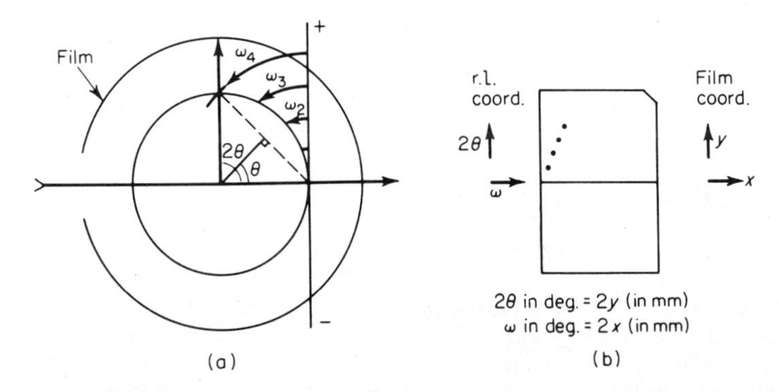

Figure 5.21. (*a*) Zero-level central lattice line. (*b*) Appearance of zero-level Weissenberg corresponding to (*a*).

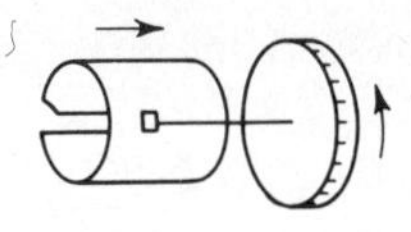

Figure 5.22. Weissenberg coupled as right-hand thread.

points equally spaced along the line will give rise to reflections (Fig. 5.21*b*) as they pass through the sphere of reflection. The angle through which the *n*th point must travel before it reaches the sphere is ω_n. The diffracted ray makes an angle $2\theta_n$ with the direct beam. If 2θ is bisected (Fig. 5.21*a*), θ can be seen to be equal to ω, since their sides are mutually perpendicular. Thus $2\theta_n = 2\omega_n$, or in general $2\theta = 2\omega$, which is the equation of a straight line in a coordinate system with 2θ along one axis and ω along the other.

Since the film is shaped into a cylinder around the sphere of reflection and parallel to the axis of crystal rotation, the distance of a spot from the center line is proportional to the angle 2θ of the diffracted beam. As was pointed out earlier, most cameras are designed so that 1 mm in the 2θ direction on the unrolled film is equal to 2° in 2θ.

If the film is moved past the slot in the layer screen at a constant rate while the crystal is being rotated uniformly, the position of the spots in the direction along the central line of the film will be proportional to ω, the crystal rotation. Thus the condition that we observe the Weissenberg diffraction pattern in a coordinate system based on 2θ and ω is satisfied, and we can expect to find central lattice lines appearing as straight lines on the photographs (Fig. 5.21*b*). For convenience, the gearing that couples the film and crystal drives is usually chosen so that 1 mm of film translation is equal to 2° of crystal rotation. Thus the scales in the 2θ and ω directions are the same, and the lines corresponding to central lattice lines have a slope of 2.[8]

Most Weissenberg cameras are constructed so that the film cassette moves to the right as the crystal rotates clockwise, that is, the coupling is that of a right-hand thread (Fig. 5.22). Under these conditions, if the unrolled film is viewed as though from the X-ray source, that is, with the

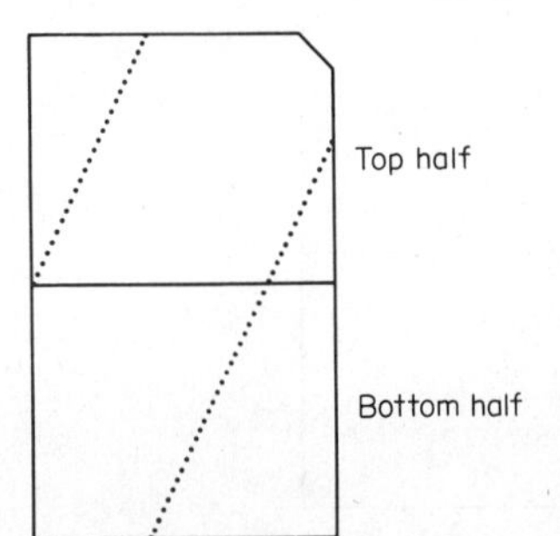

Figure 5.23. Appearance of a central lattice line on a zero-level Weissenberg film.

[8] A camera in which this is true is said to have "undistorted scale."

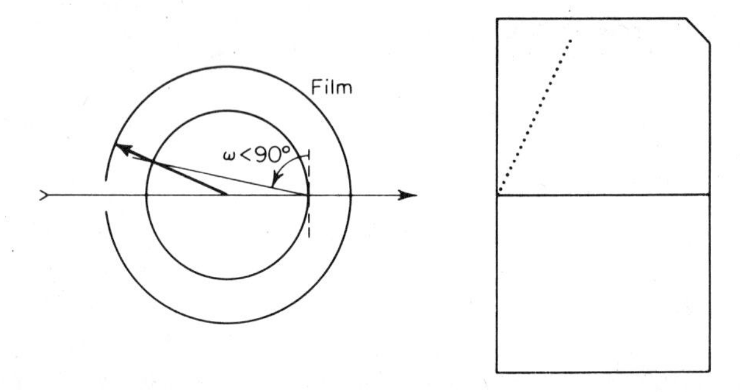

Figure 5.24. (*a*) Central lattice line for $\omega < 90°$. (*b*) Appearance on a zero-level Weissenberg.

clipped corner (Section 5.2) at the upper right, the axial lines will have positive slope (Fig. 5.23). We shall assume this point of view in all subsequent discussion.

As the angle of rotation for the lattice line of Fig. 5.21 approaches 90°, the diffracted beams approach more and more closely to the incident beam (Fig. 5.24) until finally they no longer intercept the film. When the rotation is slightly greater than 90°, the diffracted ray again strikes the film, but now on the lower half (Fig. 5.25). Thus the trace of a central lattice line runs off the top of the film and reappears on the bottom. As rotation continues, the trace approaches the central line of the film. At $\omega = 180°$ the condition shown in Fig. 5.26 is reached. The lattice line is again tangent to the sphere of reflection as in Fig. 5.21, but if the portion that caused reflections as ω changed from 0° to 180° is considered the positive portion of a coordinate

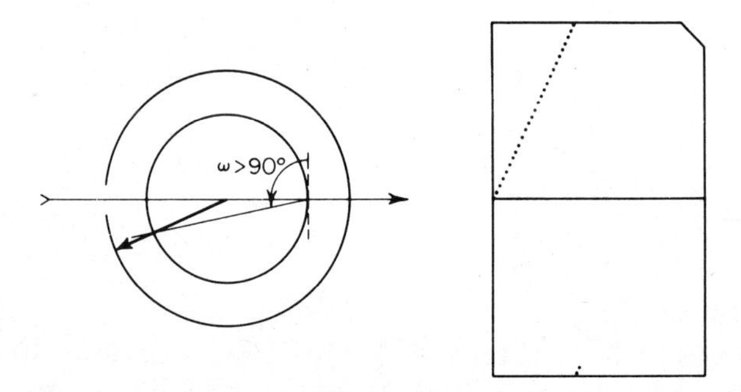

Figure 5.25. (*a*) Central lattice line for $\omega > 90°$. (*b*) Appearance on a zero-level Weissenberg.

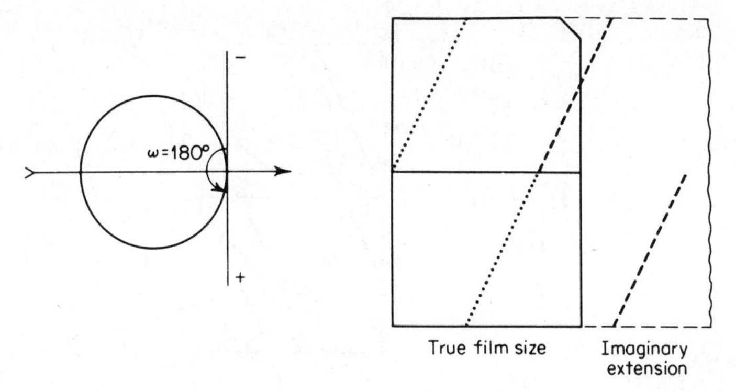

Figure 5.26. (*a*) Central lattice line for $\omega = 180°$. (*b*) Appearance on a zero-level Weissenberg. The trace produced as ω changes from 180° to 360° is shown by dashed lines.

axis, that which will pass through the sphere as ω changes from 180° to 360° is negative. Thus there is a change of sign as the trace of the lattice line crosses the center line of the film, and the reflections in the top half have indices opposite in sign from those in the bottom half. If the line that we have been considering is one of the r.l. axes (e.g., a^*), the indices in the positive direction will be $h00$ and those in the negative direction $\bar{h}00$. If it is a more general line, for example, the one passing through 101, 202, and so on, the negative will be $\bar{1}0\bar{1}$, $\bar{2}0\bar{2}$, and so on.

Consider a reciprocal axis a^* tangent to the sphere of reflection and the second axis c^* making an angle β^* with it (Fig. 5.27*a*). As the crystal is

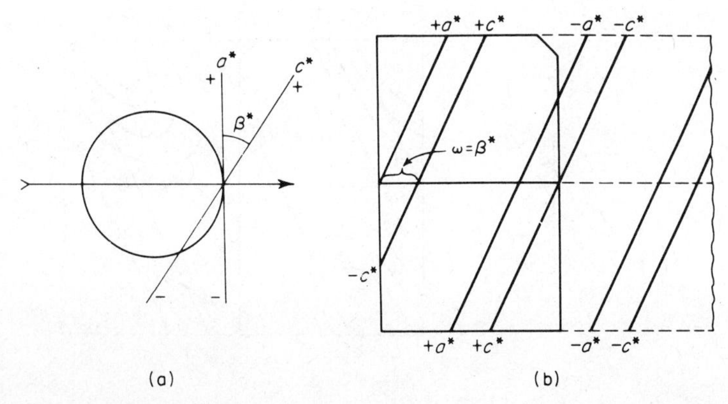

Figure 5.27. (*a*) Axial lines (central lattice lines) a^* and c^*. (*b*) Appearance of axial lines on a zero-level Weissenberg.

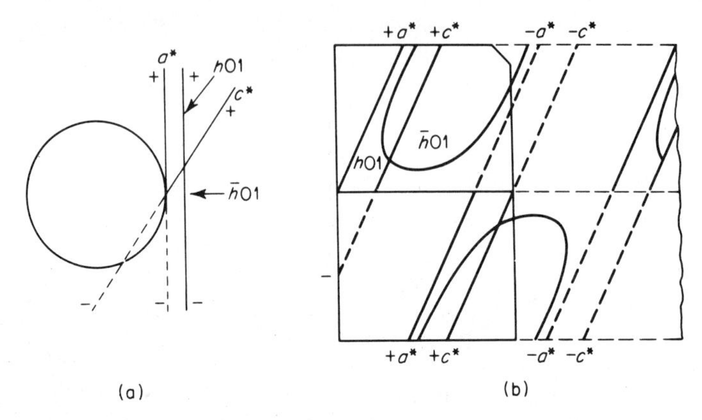

Figure 5.28. (*a*) The a^* and c^* axial lines and $h01$ noncentral lattice line parallel to a^*. (*b*) Appearance of zero-level Weissenberg.

rotated through 360°, the trace of a^* will be generated on the film as described above. When $\omega = \beta^*$, however, c^* will then be tangent to the sphere and on further rotation will generate the same kind of trace. Thus the behavior of the two axes will be the same and will lead to two families of straight lines offset from each other by $\omega = \beta^*$ degrees (Fig. 5.27*b*).

The behavior of noncentral lattice lines is slightly more complicated, although the nature of the trace on the film can be deduced by arguments similar to the above. We will content ourselves by looking at the results, however. Consider a lattice line parallel to one r.l. axis, for example, the line containing the points $h01$ (Fig. 5.28*a*). Note that the sign of the h index changes as this line crosses the c^* axis. As the lattice is rotated through 360°, the trace produced will be like that shown in Fig. 5.28*b*. Note

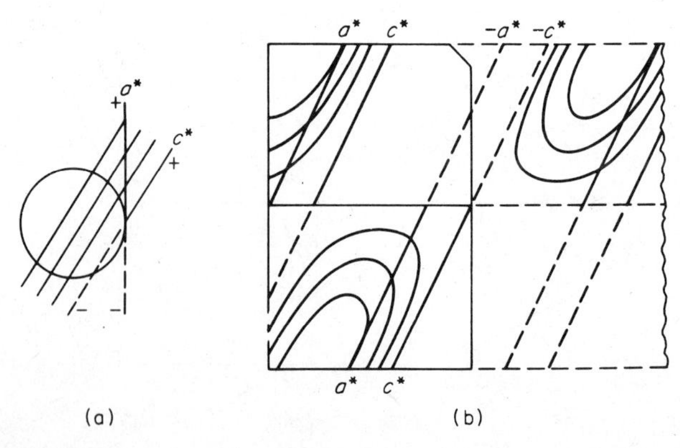

Figure 5.29. (*a*) The a^* and c^* axial lines and $10l$, $20l$, and $30l$ lattice lines. (*b*) Appearance of zero-level Weissenberg.

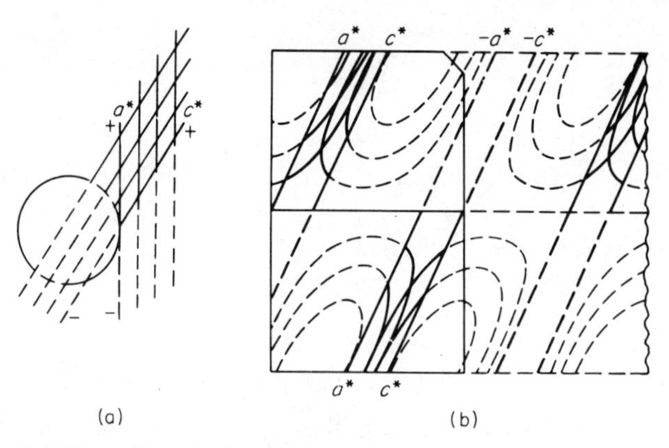

Figure 5.30. (*a*) The a^* and c^* axial lines and $h01$, $h02$, $h03$, $10l$, $20l$, and $30l$ lattice lines. (*b*) Appearance of zero-level Weissenberg.

that the trace of $h01$ always lies between the traces of the positive parts of a^* and c^* and that $\bar{h}01$ correspondingly always lies between $-a^*$ and c^*, reflecting the relationship in the r.l. net.

A family of lines (e.g., $10l$, $20l$, etc.) parallel to a reciprocal axis produces a set of similar nonintersecting curves (*festoons*) at increasing distances from the center of the film (Fig. 5.29). Corresponding lattice lines on the opposite side of the axis (i.e., $\bar{1}0l$, $\bar{2}0l$, etc.) produce similar curves that fill the remaining quadrants of the film. Note that $\bar{1}0\bar{l}$ lies between $-a^*$ and $-c^*$ while $\bar{1}0l$ is between $-a^*$ and c^*.

Festoons of similar shape will be produced by any set of parallel noncentral lattice lines, and in particular by those parallel to the second axis.

The image on the film of the r.l. net will be the superposition of these two families of festoons (Figs. 5.30 and 5.20). The festoons $h01$ and $10l$ have only 101 in common and so will cross at that point and only that point. Thus any reflection can be identified by determining the constant indices of the two festoons on which it lies. The points $h0l$ will always lie between $+a^*$ and $+c^*$, $\bar{h}0l$ between $-a^*$ and $+c^*$, and so on.

5.5. UPPER-LEVEL WEISSENBERG THEORY[9]

There are several ways in which Weissenberg photographs of upper levels can be obtained. One of these, the *equi-inclination Weissenberg*, is so advantageous that it is used almost exclusively, so we shall limit our discussion to it.

For rotation and zero-level Weissenberg photographs, Fig. 5.31 shows the relationships among the crystal, the levels of the reciprocal lattice, the

[9]Buerger, *X-Ray Crystallography*, Chapter 14.

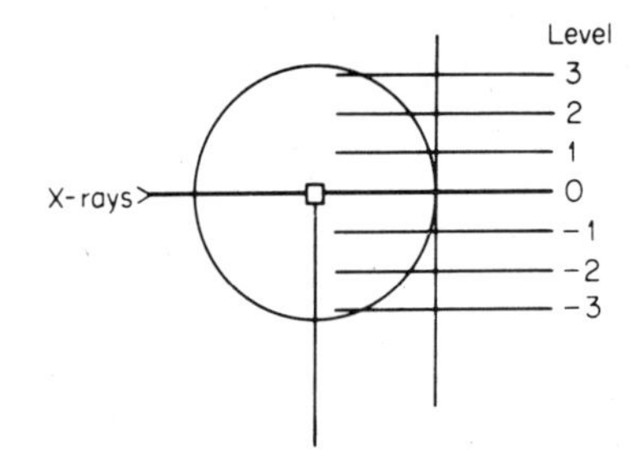

Figure 5.31. Sphere of reflection and reciprocal lattice levels for X-ray beam normal to direct axis.

X-ray beam, and the sphere of reflection as seen from the side. If the X-ray beam is tilted with respect to the crystal, the conditions of Bragg's law require that the origin of the reciprocal lattice remain at the point at which the beam exits from the sphere of reflection. At the same time the reciprocal lattice itself remains aligned with the crystal. Figure 5.32 shows the resulting arrangement. If the tilt μ of the X-ray beam is chosen properly, the point of intersection between the axis about which the reciprocal lattice rotates and the r.l. layer that is to be examined can be made to fall on the sphere of reflection. Since this point is on the rotation axis, it will remain fixed as the reciprocal lattice turns with the crystal and can be regarded as the origin about which the selected layer rotates.

Two cases need to be considered. In the first, the reciprocal axis corresponding to the direct axis of rotation is perpendicular to the layer nets. This is true for all orthorhombic axes and for monoclinic b^*. The reciprocal axis then coincides with the rotation axis and passes through the upper-level rotation origin as defined above. This rotation origin is then the coordinate origin for the layer net as well (Fig. 5.33). Under these conditions the net cuts the sphere of reflection in a circle as it is rotated about its origin, which lies on this circle. The conditions are exactly those of the zero-level Weissenberg photograph except that the radius of the circle is $\cos \mu$ r.l.u. and thus less than the unit radius of the sphere of reflection. Consequently the appearance of the upper-level Weissenberg photographs is the same as that of the zero-level photographs, and the methods of indexing are exactly the same. As μ increases and the intercepted circle grows smaller, the number of points that pass through it decrease, and the

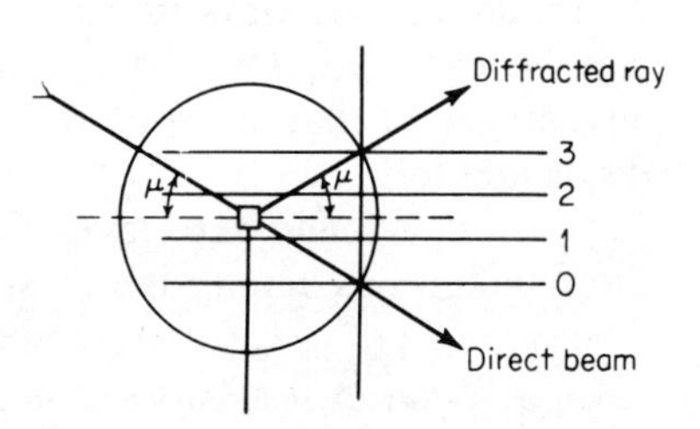

Figure 5.32. Equi-inclination geometry.

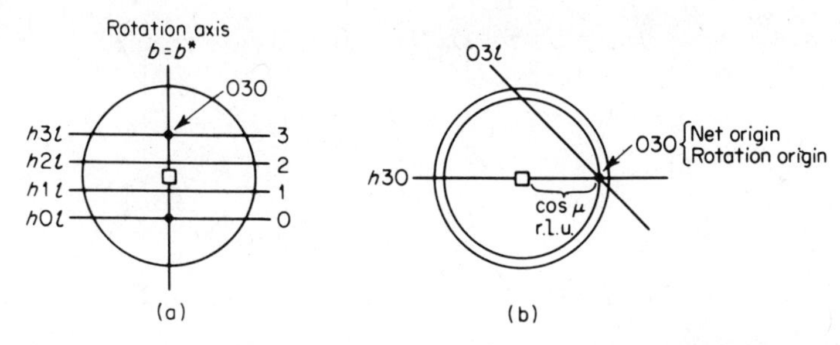

Figure 5.33. Equi-inclination geometry, rotation about an orthogonal axis. (*a*) Sphere of reflection and r.l. levels as viewed along the X-ray beam. (*b*) Sphere of reflection as seen along the rotation axis showing trace of intersection of third reciprocal level ($k = 3$) with the sphere.

festoons move farther apart and become fewer in number. This effect is small at first but increases fairly rapidly at high values of μ.

For crystals in which the axis of rotation and its corresponding reciprocal axis do not coincide, for example, monoclinic rotated about a or c and triclinic, the analysis is not quite so simple. In these cases (see Fig. 5.34) the coordinate origin of the net, which lies on the reciprocal axis, does not coincide with the rotation origin, which lies on the direct axis. In a monoclinic crystal the deviation is always in the a^*c^* plane, and the $h0n$ or $n0l$ lines, that is, the nth-level central lines corresponding to the a^* or c^* axes, will always pass through the rotation origin even though this does not occur at the point $00n$ or $n00$. Figure 5.34 shows the case where the rotation axis is c and $n = 3$. Since these "level axes" pass through the rotation origin, they will appear as straight lines on the photograph. They will also show right–left symmetry about their traces.

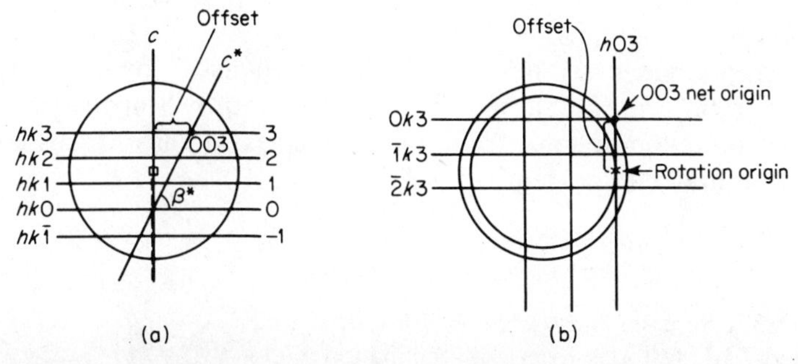

Figure 5.34. Equi-inclination geometry, rotation about monoclinic c axis. (*a*) Sphere of reflection and r.l. levels as viewed along the X-ray beam. (*b*) Sphere of reflection as seen along the rotation axis, showing some lattice lines in the $hk3$ section.

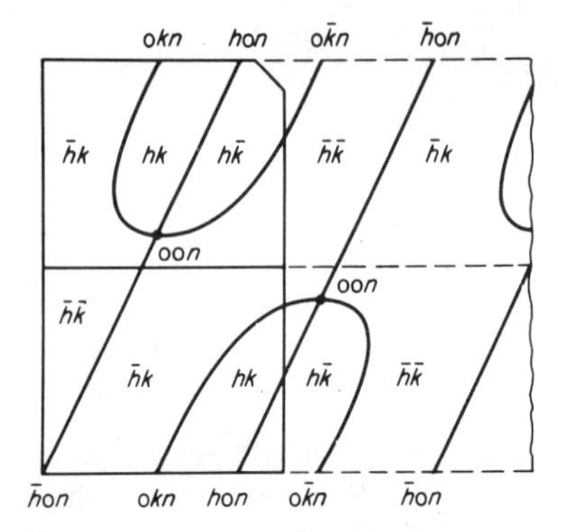

Figure 5.35. Traces ($0kn$ and $h0n$) on the nth-level equi-inclination Weissenberg of a monoclinic crystal rotated about c.

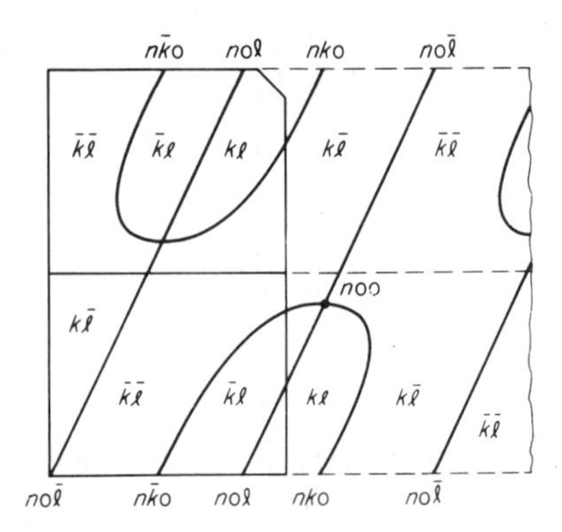

Figure 5.36. Traces ($nk0$ and $n0l$) on the nth-level equi-inclination Weissenberg of a monoclinic crystal rotated about a.

On the other hand, the "level axis" corresponding to b^* does not pass through the rotation origin and so will appear as a festoon rather than as a straight line. The traces of $0kn$ or $nk0$ with respect to $h0n$ or $n0l$ are shown in Figs. 5.35 and 5.36. It is perhaps illuminating to consider the zero-level trace of b^* not as a straight line but rather as a flat-bottomed U (Fig. 5.37), the limiting case of a festoon. As this is shifted in the direction of $+a^*$ (or c^*) it assumes the character of a normal festoon.

A consequence of this shift is that the line $\bar{1}kn$ (or $nk\bar{1}$) is moved toward the rotation axis (Fig. 5.34) and may pass more or less through the rotation origin. It will then appear as a straight line on the film, and this can lead to disastrous errors in indexing if the danger is not recognized. Another result is that lattice points of positive h or l index are moved away from the level rotation origin and out of the area that passes through the sphere of reflection, while points of negative index move into this area. Consequently there are more reflections of negative than of positive h or l indices on upper-level photographs.

In the triclinic case, neither of the "level axes" passes through the rotation origin (Fig. 5.38), and consequently both will appear as festoons. The interpretation follows from the principles already discussed and is considered in detail by Henry et al.[10]

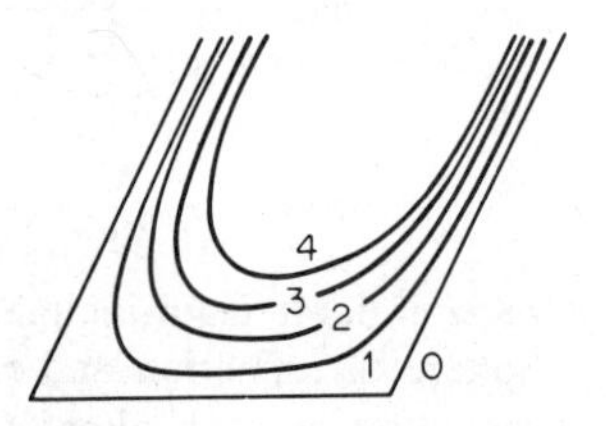

Figure 5.37. Superimposed traces of $0k0$, $0k1$, $0k2$, $0k3$, and $0k4$, festoons from equi-inclination Weissenberg photographs of monoclinic crystal rotated about c.

[10]Henry et al., *The Interpretation of X-Ray Diffraction Photographs*, pp. 142–158.

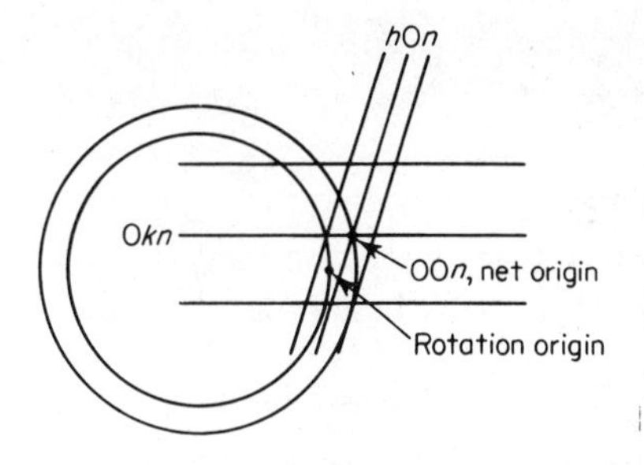

Figure 5.38. Equi-inclination geometry, triclinic crystal rotated about c. Sphere of reflection and some lattice lines on the nth level.

5.6. WEISSENBERG PRACTICE

The general design of a Weissenberg camera was described in Section 5.2. The film holder is engaged on the worm screw that leads from the motor to the crystal mounting and is moved in strict coincidence with the crystal rotation. At either end of its travel the film holder (or an adjustable stop) strikes a reversing switch, which changes the direction of the crystal rotation and sends the film back the other way. Thus a Weissenberg photograph is based on an oscillatory movement rather than a pure rotation. The usual free translation of the film carriage is 100–110 mm, corresponding to an oscillation range of 200–220°. In most cases this will allow all the unique reflections of a level to be recorded on one film, but in some cases two films will be required.

Weissenberg exposures may vary from less than an hour to 100 hr or more. The usual organic crystal will require an exposure of several hours to show a reasonable number of reflections, but this is very strongly dependent on the nature and size of the crystal. The exposure time needed can be reduced by decreasing the azimuth range through which the crystal oscillates, but only at the cost of omitting some reflections entirely.

The one extra piece of equipment used in the Weissenberg technique is a *layer line* screen (Fig. 5.19), which is fitted around the axis of crystal rotation and prevents diffracted beams not in the desired layer from reaching the film. This screen is only a little smaller than the inner diameter of the film holder and is normally supported by being a close sliding fit on the nonrotating sleeve through which the crystal spindle projects.

The inclination angle μ is adjusted by releasing a clamping screw and rotating the drive and crystal unit about a vertical axis perpendicular to the X-ray beam. The amount of rotation can be read to the nearest 0.1° by means of a vernier and scale.

The simplest method for determining the proper value of μ is to calculate d^* [Eq. (5.6)], the r.l. distance between adjacent layers. The distance to the nth layer is merely

$$d_n^* = nd^* \tag{5.6}$$

and

$$\mu = \sin^{-1}(d_n^*/2) \tag{5.7}$$

The derivation of this follows from Fig. 5.32.

Some cameras are built so that μ can be set in either direction. For the usual operation, the angle between the incident X-ray beam and the crystal tip should be acute.

In zero-level photographs the layer line screen is set by eye so that the crystal is centered in the slit opening. This position usually corresponds to the zero position marked off on a scale on its mount. Since the diffracted beams of upper levels come from the crystal in directions that are not perpendicular to the rotation axis, the layer line screen must be shifted to select the desired cone. The shift toward the crystal tip is measured on the scale and is

$$\text{shift} = r_s \tan \mu \tag{5.8}$$

where r_s is the radius of the screen. It is convenient to make a plot of shift versus μ for a given camera and post it for reference.

The width of the slit in the layer line screen is usually adjustable. A narrow slit reduces the amount of scattered radiation that strikes the film and gives a cleaner photograph. It also allows the study of layer lines that are so close that more than one will pass through the normal opening. On the other hand, a narrow slit places more stringent requirements on the accuracy of alignment since wavy layer lines may be partly cut off by the layer screen.

5.7. WEISSENBERG USES

Since Weissenberg photographs give a view of an entire r.l. level, they can provide almost any desired information about that level. In particular they can be used for improving alignment, measuring cell constants, determining the space group, and measuring intensities. We shall consider only the first two uses here.

Alignment

Zero-level Weissenberg photographs can serve as the basis for a very simple, fast, and accurate method for aligning crystals that allows one to make corrections in both arcs from a single photograph.

Figure 5.39 shows the appearance of Weissenberg traces of central lattice lines when the crystal is slightly misset. The solid lines represent the ideal positions of the traces, which make an angle τ with the central line,

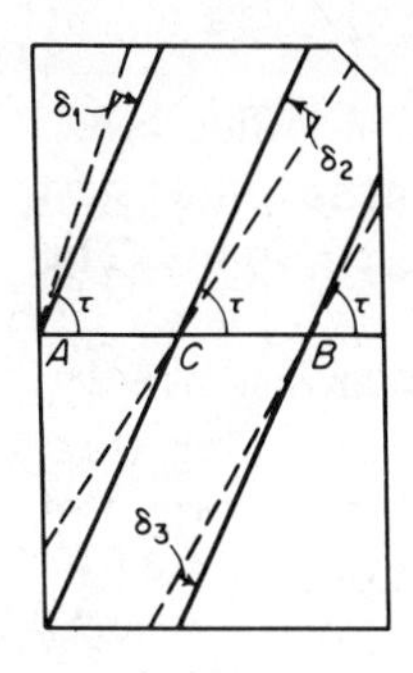

Figure 5.39. Appearance of central lattice lines on a zero-level Weissenberg film of a misset crystal.

while the dashed lines are the observed positions. These are in error by δ_1 and δ_2, respectively. The r.l. lines to be used are chosen so that they are approximately parallel to the adjusting arcs of the goniometer heads, that is, so that at the azimuths corresponding to the points A, B, and C one arc is parallel and one perpendicular to the X-ray beam. The lines do not need to be axes, and they do not even need to contain a large number of points.

In practice, the angles ($\angle A$, $\angle B$, and $\angle C$) between the lattice line traces and the central streak are measured. For small errors (1–2°) it is true that

$$\delta_1 = \left| \frac{\angle A - \angle B}{2} \right| \tag{5.9}$$

and

$$\tau = \frac{\angle A + \angle B}{2} \tag{5.10}$$

The correction to be applied is given by the expression

$$\Delta_i = \delta_i / \sin \tau \tag{5.11}$$

which, for a camera of undistorted scale, is

$$\Delta_i = \delta_i / 0.89 = 1.12\delta_i \tag{5.12}$$

The correction for the second arc is obtained by repeating the above process at point C. Then

$$\delta_2 = |\angle C - \tau| \tag{5.13}$$

and the correction is calculated from Eq. (5.11) or (5.12).

In order to apply the corrections, the azimuth is set to the value corresponding to point A, and the arc perpendicular to the X-ray beam is adjusted in the same sense (clockwise or counterclockwise) as is needed to bring the observed trace to the ideal position.

Cell Constants

Since the Weissenberg photograph is a map of the reciprocal lattice layer, the most straightforward method for obtaining cell constants is to measure the r.l. constants and then convert them into direct lattice constants. The measurements are generally made on the zero-level film.

In order to obtain accurate r.l. constants, the perpendicular distance $2y_n$ between the corresponding reflections on the axial lines is measured (Fig. 5.40). The distance ξ (xi) of any point on the zero level from the origin of the reciprocal lattice is given according to the construction of Fig. 5.41 by

$$\theta = \frac{y/2}{R} \text{ radians} \tag{5.14}$$

$$\xi/2 = \sin\theta = \sin(y/2R) \tag{5.15}$$

$$\xi = 2\sin(y/2R) \quad \text{(argument in radians)} \tag{5.16}$$

where R is the camera radius. For reflections measured as above, and since tables of sines in degree arguments are more common than those for radians, this becomes

$$\xi_n = 2\sin(57.30 \times 2y_n/4R) \quad \text{(argument in degrees)} \tag{5.17}$$

When a camera has a nominal diameter of 57.3 mm, this equation is approximated by

$$\xi_n = 2\sin y_n \quad (y \text{ in mm} = \text{angle in deg}) \tag{5.18}$$

For an r.l. axis the quantity ξ_1 corresponds to the axial repeat a^*, b^*, or c^*. Usually, however, this is the least accurately determined distance, and a

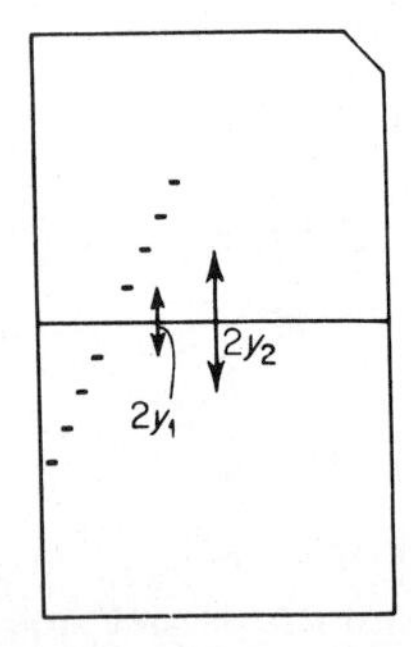

Figure 5.40. Distances measured on Weissenberg film and used to calculate interplanar spacings.

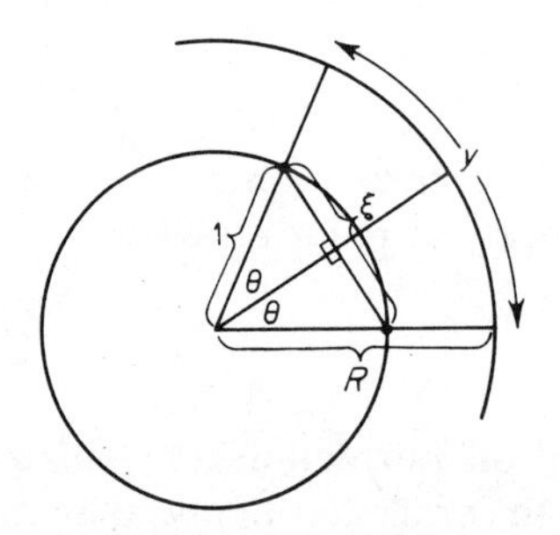

Figure 5.41. Geometry showing relationship among y, R, and θ.

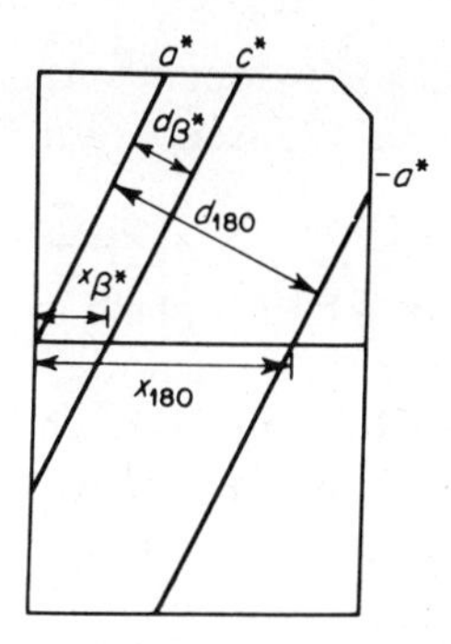

Figure 5.42. Diagram showing distances measured in determining reciprocal interaxial angles.

much better figure can be obtained by using the relationship

$$\xi_1 = \xi_n / n \tag{5.19}$$

and averaging the results.

By this method one can obtain the spacings along the two r.l. axes in the plane being studied. The angle between these axes is determined by the ratio of the distance between the two axes to the distance between the two traces of a single axis; in Fig. 5.42,

$$\beta^* = \frac{d_{\beta^*}}{d_{180}} \times 180° = \frac{x_{\beta^*}}{x_{180}} \times 180° \tag{5.20}$$

These distances can be measured either along the central line of the film or perpendicularly to the axial lines. The former has the advantage that most cameras have $x_{180} = 90$ mm and so

$$\beta^* = 2x_{\beta^*} \qquad (x_{\beta^*} \text{ in mm}) \tag{5.21}$$

Because of film shrinkage, Eq. (5.21) is only approximate, and Eq. (5.20) should be used to obtain accurate values.

If the rotation axis is both a direct and reciprocal axis, the combination of oscillation and zero-level Weissenberg photographs allows the measurement of all the unit cell parameters. If it is not (triclinic or monoclinic not mounted around b), upper-level photographs and additional measurements are required. Various methods have been proposed,[11] but it is difficult to get high accuracy.

5.8. PRECESSION THEORY[12]

The only other widely used single-crystal X-ray camera is the precession camera, designed by M. J. Buerger of M.I.T. in the early 1940s. Like the

[11]See H. Herbert, *Acta Cryst.*, **A34**, 946 (1978); P. de Meester, *Acta Cryst.*, **A36**, 732 (1980).
[12]M. J. Buerger, *The Precession Method in X-Ray Crystallography*, Wiley, New York, 1964; Henry et al., *The Interpretation of X-Ray Diffraction Photographs*, pp. 132–142; Jeffery, *Methods in X-Ray Crystallography*, pp. 218–236.

Weissenberg it is a moving crystal, moving film device for mapping one reciprocal lattice level onto one sheet of film, but unlike the Weissenberg it does this in such a way as to provide an ***undistorted*** record from which the angles and distances of the lattice can be read off more or less directly (Fig. 5.44). It is obvious that such an undistorted photograph of a lattice plane is easier to index, measure, and search for symmetry than the rather distorted view afforded by the Weissenberg technique. Although either method can be used alone, the most fruitful results are obtained when both are applied, taking advantage of the fact that for a single crystal mounting they provide complementary views of the reciprocal lattice.

Consider a direct lattice axis that is parallel to the incident beam. The zero-level r.l. net perpendicular to it is thus tangent to the sphere of reflection at the origin (Fig. 5.43*a*). If the crystal is turned through a small angle $\bar{\mu}$ about an axis (not necessarily a crystal axis) perpendicular to the beam, the situation of Fig. 5.43*b* is attained. The r.l. net now cuts the

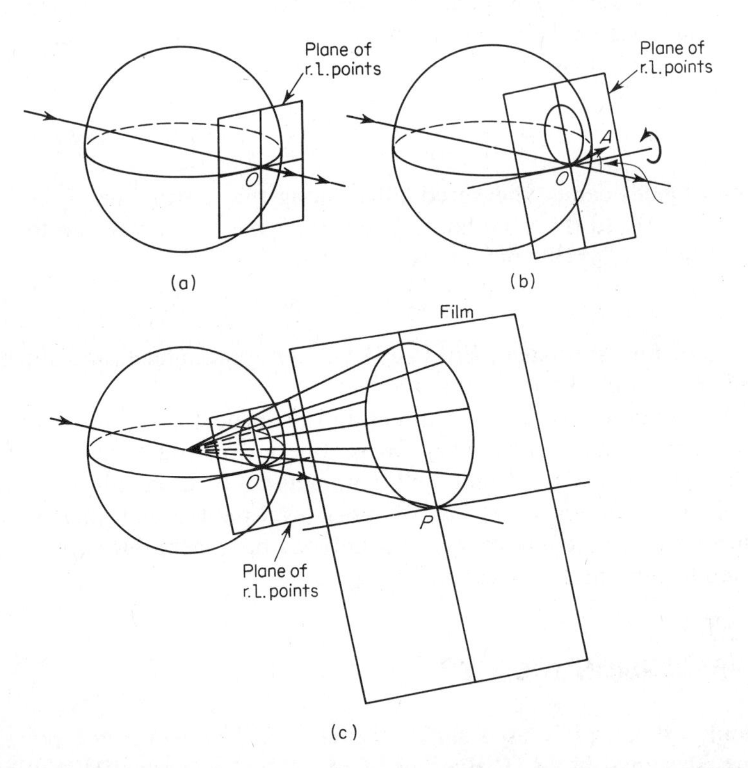

Figure 5.43. Precession geometry. (*a*) Reciprocal lattice plane perpendicular to X-ray beam (tangent to sphere of reflection). (*b*) Reciprocal lattice plane in (*a*) rotated slightly about horizontal axis showing circular intersection with sphere of reflection. (*c*) Film parallel to r.l. plane showing circular trace on film of circular intersection with sphere of reflection.

sphere that passes through the origin O. If the crystal is moved in such a way that the direct axis AO revolves about the beam, keeping the constant angular separation $\bar{\mu}$, the intercepted circle will revolve about the origin. At any instant (e.g., that corresponding to Fig. 5.43*c*) those r.l. points that are lying on the circle will give rise to reflections. If a film lies parallel to the r.l. plane, the diffracted rays passing from the center of the sphere through the points on the circle will project an undistorted image of their relationship onto the film. If, as the crystal axis is rotated about the beam, the film precesses about the point F (the theoretical intersection of the beam with the film) so as to remain always parallel to the r.l. plane, the same will be true for all orientations of the axis. Since P corresponds to the projection of the r.l. point O, all the other r.l. points will be projected onto the film in the proper relationship to it, and the net result will be a picture showing the r.l. net in undistorted form (Fig. 5.44).

Although we have been considering the zero-level r.l. net, this is not the only one that is passing through the sphere and causing reflections. Consequently it is usually necessary to use a suitable screen to isolate the

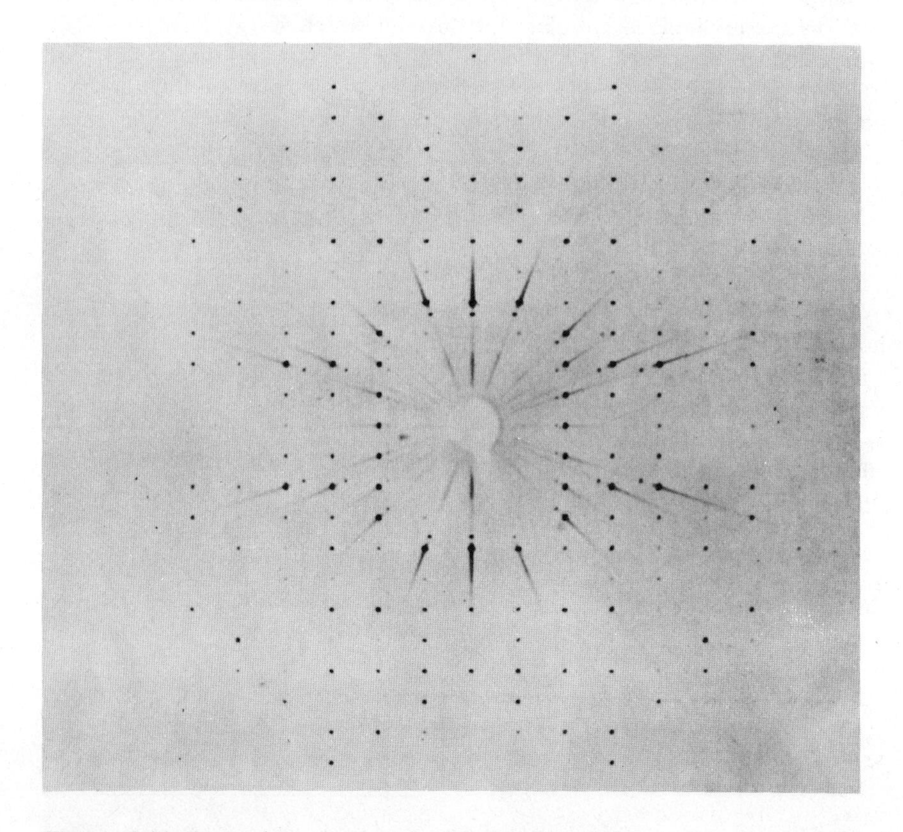

Figure 5.44. Precession photograph. Myoinositol, $0kl$ net. (Courtesy of Dr. L. Sieker.)

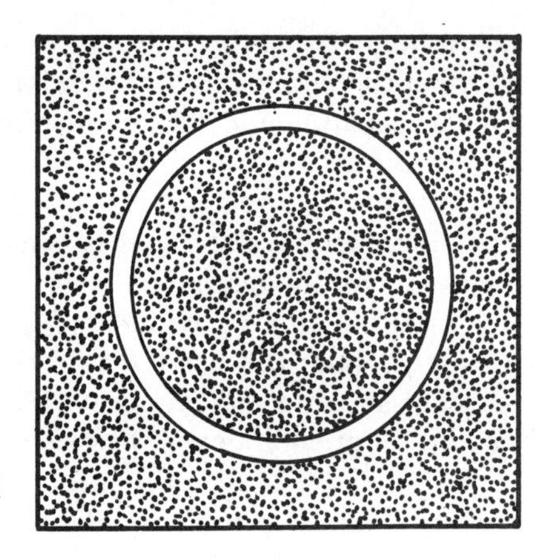

Figure 5.45. Precession camera level screen.

reflections corresponding to the net under study. Such a screen is a flat metal plate containing a circular slot through which X-rays can pass (Fig. 5.45). This is placed parallel to the film and between it and the crystal. The arrangement is such that the cone of diffracted rays from the plane of interest will pass through the clear ring on their way to the film (Fig. 5.46). As the crystal moves, the level screen precesses to keep the normal through its center coincident with that through the center of the reflecting circle. In this way the arrangement of Fig. 5.46 is maintained while the crystal axis revolves about the beam.

By suitable adjustment of the positions of the film and screen it is possible to exclude the zero-level reflections and to obtain upper-level photographs with a precession camera. The theory follows straightforwardly from that given above, and since it is discussed in detail in Buerger's monograph we shall not go into it here. Upper-level precession photographs are less commonly used than zero-level ones, partly because they suffer from a blank region in the center that makes them less suitable for obtaining intensity data.

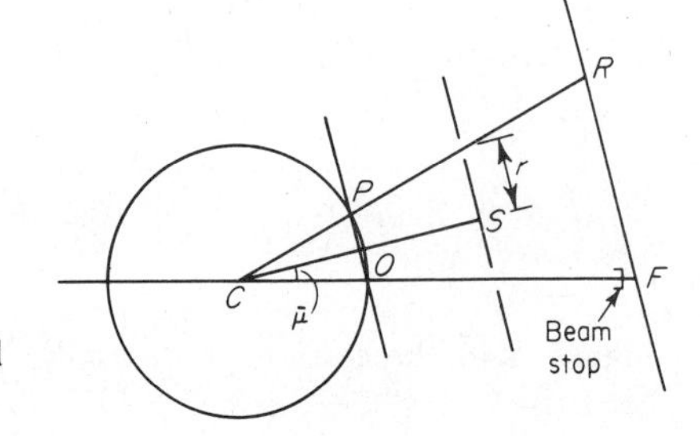

Figure 5.46. Precession geometry, viewed perpendicular to the X-ray beam.

Figure 5.47. Precession camera. (Courtesy Charles Supper Co.)

5.9. PRECESSION PRACTICE

The precession camera (Fig. 5.47) appears somewhat more complicated than the Weissenberg, principally because of the linkages required to cause the crystal, level screen, and film to perform their requisite motions in unison. There are also more instrumental parameters that can be varied.

Geometry

The first of these parameters is the film-to-crystal distance. The film is a flat sheet (usually 5×5 in.) mounted in a holder that may be set at various distances from the crystal. The greatest separation is usually[13] 60.0 mm and is the setting for zero-level photographs, the film being advanced toward the crystal for recording upper levels. It is important that the crystal-to-film distance be known accurately, since it is the scaling factor that is used to convert the separation of reflections on the film into corresponding separations in the reciprocal lattice. Figure 5.46 shows the relationship. By similar triangles,

$$\frac{RF}{CF} = \frac{PO}{CO} = \frac{PO}{1} \tag{5.22}$$

[13]Cameras with larger spacings (up to 100 mm) can be obtained and are convenient when dealing with crystals having very large unit cells, such as proteins, since they provide greater magnification of the reciprocal lattice.

that is, the distance in r.l. units of the r.l. point P from the origin O is just the ratio of the distance of the corresponding reflection R from the film center F to the crystal-to-film distance CF. Thus if the distance of a reflection to the film center is measured in millimeters, division by 60.0 (in the usual case) will give the distance of the corresponding r.l. point from the origin. If the point is an axial one, the r.l. distance can then be converted into the direct axis repeat by the formulas of Chapter 2.

Equation (5.22) can also be used to point up one of the limitations of the precession method. For a standard film, the edges are about 60 mm from the center. The maximum distance in the reciprocal lattice that can be recorded is therefore

$$\frac{60}{60} = 1 \text{ r.l.u.} \tag{5.23}$$

considerably less than the 2 r.l.u. available to the Weissenberg. For this reason precession cameras are often used with molybdenum rather than copper radiation. The shorter (0.7107 Å as compared to 1.542 Å) wavelength of the former compresses the reciprocal lattice by a factor of about 2.2 and allows precession photographs to show all the lattice points that can be observed with a Weissenberg camera and Cu K_α.

The second parameter is the precession angle $\bar{\mu}$, that is, the angle between the direct axis and the beam. It can be seen from Fig. 5.46 that

$$OP = 2 \sin \bar{\mu} \tag{5.24}$$

so, using the maximum value of 1 from Eq. (5.23),

$$\sin \bar{\mu}_{\text{max}} = 1/2 \tag{5.25}$$

$$\bar{\mu}_{\text{max}} = 30° \tag{5.26}$$

and there is no advantage to having $\bar{\mu}$ greater than 30°. In practice, when it is not necessary to observe all the possible reflections, a setting of 20° is often used, with a considerable decrease in the exposure time required. Even smaller angles may be used for special purposes.

The remaining variables are interrelated and are concerned with the setting of the level screen. They are the radius of the clear ring (r in Fig. 5.46) and the screen-to-crystal distance. It is obvious that a ring of small radius placed close to the crystal will pass the same rays as one of larger radius closer to the film. The range over which the screen position can be varied is limited, however, by the necessity that the various pieces of the apparatus clear each other as they move, so a series of screens with rings of different radii is usually provided. When these are combined with the available fore-and-aft movement of the screen holder, they allow complete coverage of possible diffraction cone sizes.

It is clear from Fig. 5.46 that the proper choice of r and the screen-to-crystal distance (CS) depends on the precession angle $\bar{\mu}$. For the zero-level case pictured,

$$\frac{r}{CS} = \tan \bar{\mu} \tag{5.27}$$

$$CS = \frac{r}{\tan \bar{\mu}} \tag{5.28}$$

so for any desired $\bar{\mu}$ and screen the proper distance can be found. For $\bar{\mu} = 20°$, a good combination is a screen of 14-mm radius and a spacing of 41.2 mm.

The scaling of the reciprocal lattice to the camera settings as described by Eq. (5.22) applies as well in the third dimension. To take photographs of the nth upper level of the reciprocal lattice, the film is set toward the crystal by

$$CF\, d_n^* \tag{5.29}$$

Here CF is again the zero-level crystal-to-film distance and d_n^* is the perpendicular distance in r.l. units between the zero and nth levels. The screen must also be positioned to limit the observed reflections to the desired set. For a screen with a ring radius r and a given setting of $\bar{\mu}$, the crystal-to-screen distance CS is given by

$$CS = r \cot[\cos^{-1}(\cos \bar{\mu} - d_n^*)] \tag{5.30}$$

There are usually several possible combinations of r and CS, and tables and graphs of solutions of Eq. (5.30) are given by Buerger.[12]

Upper-level precession photographs characteristically have a blank zone in the center where r.l. points never pass through the sphere of reflection. This reduces their utility for measuring intensities but is not a serious problem for checking on the distribution of systematic absences (see Section 5.12).

Alignment

The principal difficulty encountered with the precession technique arises from the necessity of setting the desired direct axis so that it is parallel to the beam when $\bar{\mu} = 0°$. We shall consider this process first for a crystal that has already been aligned on a Weissenberg camera, returning afterwards to the more general case.

All precession cameras are built to accept standard goniometer heads, so it is possible to transfer the mounted and aligned crystal and head from one camera to another. They are also provided with the possibility of rotating the crystal about its mounting axis (the Weissenberg rotation axis) to any desired azimuth setting.[14] Unfortunately, although the screw threads for attaching the head are standardized, the location of the indexing pin is not; so a given crystal position often may give different azimuth readings on different instruments.[15] The correction factor between cameras can be obtained by mounting a small mirror on a head with sealing wax and using this in connection with an autocollimator[16] to determine the azimuth corresponding to the position normal to the collimator.

The problems of finding the desired axis vary with the crystal class. For those in which all the angles are 90°, the direct and reciprocal axes coincide and the problem is simple. The azimuth at which any axial line crosses the center of a Weissenberg photograph is the azimuth at which the corresponding reciprocal axis is tangent to the sphere of reflection, that is, perpendicular to the beam. If the direct and reciprocal axes are coincident, a rotation of ±90° will bring both of them parallel to the beam, as is desired.

For monoclinic crystals two cases can be distinguished. In the first the crystal is mounted along the *b* axis, so the Weissenberg photograph shows the a^*c^* net. In this case both the *a* and *c* axes are perpendicular to the mounting axis, and all that is required is the proper azimuth setting. This is most easily obtained by noting that at the azimuth at which the a^* axis is tangent to the sphere the *c* axis is parallel to the beam (Fig. 5.48), and similarly for the c^* and *a* axes.[17]

In the second case the crystal is mounted about one of the inclined axes, say *a*. Under these conditions (Fig. 5.49), the coincident *b* and b^* axes are in the plane perpendicular to the mounting axis, but the *c* axis is not. If the

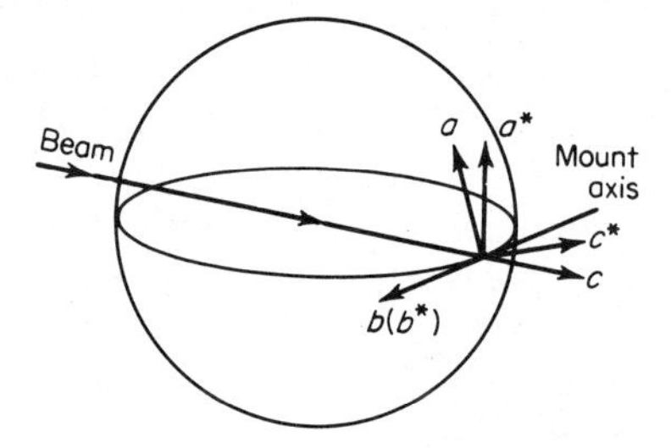

Figure 5.48. Precession orientation for a monoclinic crystal mounted about the *b* (b^*) axis, *c* axis in the X-ray beam, a^*b^* net photographed.

[14]The azimuth reading varies irregularly as the crystal precesses, so all readings and adjustments must be made with $\bar{\mu}$ set to 0°.

[15]For details on one method of correcting for such discrepancies, see D. R. Stirling and J. F. deWet, *J. Appl. Cryst.*, **9**, 339 (1976).

[16]An autocollimator is a device furnished with most precession cameras for projecting a beam of light down an oversize collimator tube and observing when it is reflected back from a crystal face perpendicular to the beam.

[17]This is obviously true as well for the orthorhombic crystal and is an alternative to the method given in the previous paragraph.

precession photographs are needed merely to provide a view of the $h00$ reflections and to tie the Weissenberg levels together, much the simplest answer is to use the b axis and set it as described for orthorhombic cells. If, on the other hand, it is desired to use c as the precession axis, the best procedure is to realign the crystal so that a^* and not a lies on the spindle axis. This change results in having both c and b directly accessible as precession axes by merely changing the azimuth.[18]

If the azimuth setting that places b^* normal to the X-ray beam is known, its use will put c in the plane defined by the spindle axis and the beam but offset from the latter by $(\beta - 90)°$ (Fig. 5.49). In order to bring it into line, the arcs of the head will have to be adjusted. If one arc is parallel to the X-ray beam, either by good luck or by intent, the correction is made on it. If this is not the case, it is easiest to use the general method of Stirling and deWet[15] to calculate the appropriate corrections in terms of the rotation needed to bring the head to a standard orientation.

A triclinic crystal can be set in the same way as a monoclinic one by choosing an azimuth that will place an r.l. axis perpendicular to the beam. The direct axis, which should be along the beam, will again be misaligned in the beam/spindle plane but will usually be close enough that the corresponding r.l. plane can be recognized in orientation photographs. Alignment is then most easily achieved by using the methods for the general case.

While the preceding discussion was based on the use of a prealigned crystal of known setting, this is not necessary since the precession camera can itself afford all the information needed for alignment. Many workers, in fact, use it routinely in preference to a Weissenberg camera, although this is largely a matter of personal taste and experience.

Photographs suitable for orientation are obtained by using *unfiltered* radiation, no screen or one with a central hole instead of an annulus, and a small precession angle ($\bar{\mu} = 5$–$10°$). Under these conditions an exposure of

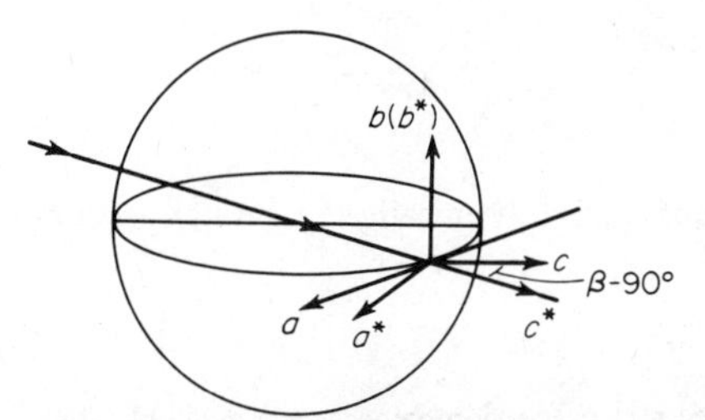

Figure 5.49. Orientation for a monoclinic crystal mounted about the a axis when transferred from a Weissenberg to a precession camera.

[18]This orientation, with a reciprocal axis as the mounting axis, is very useful if intensity data are to be collected on a diffractometer, and the methods to be described serve as a convenient means of attaining it photographically.

only a few minutes is usually enough.[19] Such a photograph, taken on an approximately aligned crystal, will show a small portion of the zero-level r.l. net near the center of the film. Upper-level reflections may also appear, but only as doubled spots outside the zero-level region (Fig. 5.50). The trace of the zero-level net is roughly circular and is outlined by dark spots[20] that mark the ends of the white radiation streaks in the net. If the direct precession axis is aligned with the X-ray beam, this circle will be centered on the image of the beam (beam stop). If the alignment is not perfect, the circle will be shifted and distorted by amounts that depend on the size of the error (Figs. 5.50 and 5.52). The corrections required can be obtained from measurements of this eccentricity.

The process of aligning a crystal first involves finding a trial orientation that will bring a direct axis within 10–15° of the X-ray beam. Errors no greater than these will usually allow the desired r.l. plane to be recognized by the experienced eye, and alignment can then be completed systematically. The obvious choices for trial are positions that bring face normals or various crystal edges into line with the beam; and only when these attempts have failed is it necessary to make a systematic search by taking photographs at azimuth intervals of 15° for one or more settings of the head arcs. It may be a consoling thought that the space groups of higher symmetry in which it is most desirable to find the direct axes parallel to symmetry elements tend to have these arcs along the obvious face normals, while in

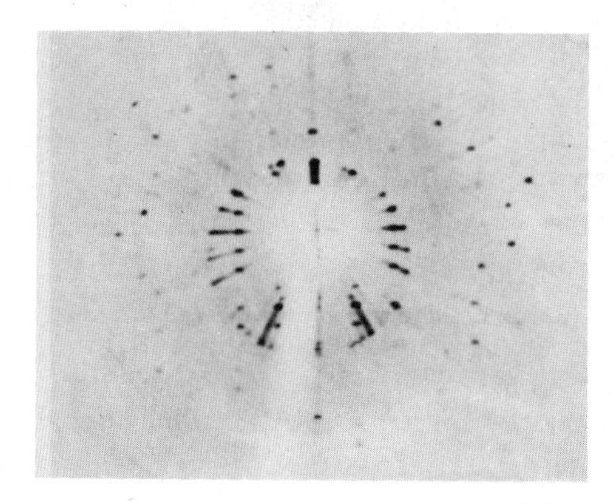

Figure 5.50. Precession setting photograph of a slightly misaligned crystal. (Courtesy Dr. L. Sieker.)

[19]Adapters allowing the use of very fast Polaroid film in precession cameras are available. The speed of the film, coupled with the very rapid and easy development, speeds the alignment process greatly.
[20]Note that these are *not* r.l. points. In Fig. 5.50 the reciprocal lattice consists of densely packed points and generates a conspicuous circle on the film. Smaller direct cells produce many fewer reflections, and it can be more difficult to recognize the pattern.

the triclinic case the normal to any r.l. plane found can, in principle, be chosen as an axis.

Once a suitable axis has been found, it is highly desirable to have the head oriented so that its arcs are reasonably close to being parallel and perpendicular to the X-ray beam. This is particularly true if the lower head arc deviates appreciably from the zero setting. Figure 5.51 shows a general, approximately oriented r.l. plane. The misalignment can be represented in terms of three errors: ϵ_H and ϵ_V relate to the horizontal and vertical angular separations of the direct axis from the beam; ψ is the amount the net is rotated about the direct axis from its desired position. If the arcs are oriented as suggested above, a change in the azimuth effects ϵ_V alone, the parallel (horizontal) arc affects ϵ_H, and the perpendicular (vertical) one affects ψ. If the orientation is not as described, the arc motions are not independent, and changing one will affect all the errors at once.[15]

An r.l. plane oriented as in Fig. 5.51 will yield an orientation photograph that may be idealized as Fig. 5.52. If the distances from the beam to the edge of the zero-level trace are measured in the two horizontal (Δx) and the two vertical (Δy) directions, the distance $\Delta\xi$ in r.l. units between the points appearing at the edges is

$$\Delta\xi_V = \frac{\| y_1| - |y_2\|}{2 \times \text{film-to-crystal distance}} \tag{5.31}$$

$$\Delta\xi_H = \frac{\| x_1| - |y_2\|}{2 \times \text{film-to-crystal distance}} \tag{5.32}$$

For small ($\leq 10°$) values of $\bar{\mu}$, the corresponding angular errors ϵ_H and ϵ_V are given by

$$\epsilon = \tfrac{1}{2} \tan^{-1} \Delta\xi \tag{5.33}$$

which for errors of a few degrees and the common camera spacing of 60 mm implies

$$\epsilon \text{ (in minutes)} \approx 14 \times \| x_1| - |x_2\| \text{ (in mm)} \tag{5.34}$$

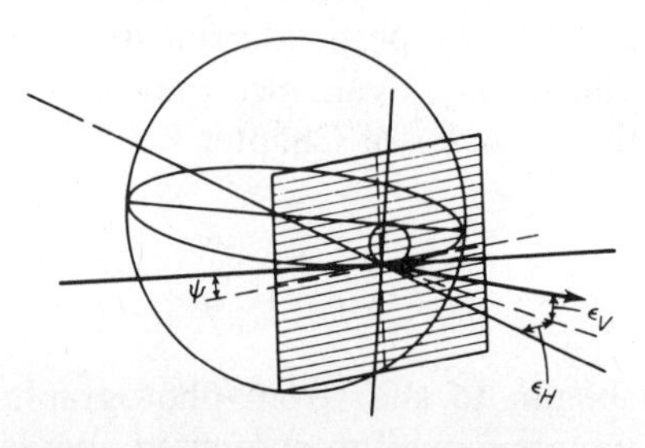

Figure 5.51. Reciprocal lattice plane misset by ϵ_H in horizontal plane, ϵ_V in vertical plane, and ψ around the X-ray beam.

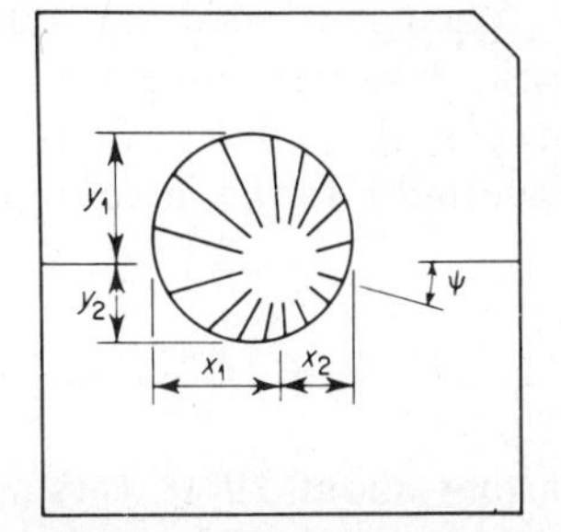

Figure 5.52. Idealized precession photograph showing misalignment corresponding to Fig. 5.51.

If the arcs are properly oriented, corrections for these errors can be made by changing the azimuth and the horizontal arc by $-\epsilon_V$ and $-\epsilon_H$, respectively. The sense of the correction is obtained by noting that the direct axis lying within the sphere of reflection is always tipped toward the *short* distances on the film (x, y) and the arcs must be changed to bring it into alignment with the beam.

The final correction—the rotation ψ needed to bring an r.l. axis onto the spindle axis—is just the angular deviation of this axis from the beam/spindle plane. It can be measured from the angle on the film between the desired axis and a horizontal line (the trace of the beam/spindle plane) passing through the direct beam point. This line is defined on most cameras by having two tiny holes in the film holder that admit light and register as dark spots near the edge of the film. These holes are located so that the line between them is parallel to the spindle axis.

Again, for properly oriented arcs, the correction ψ can be made using only the vertical arc. If the horizontal arc setting is not 0°, this correction will have some effect on the orientation of the r.l. plane as well, and a final refinement of the horizontal arc and azimuth settings (especially the latter) may be required.

If the initial errors are large, the approximations made in the derivations of these corrections and the interactions of the arc motions will usually prevent the first changes from leading to perfect alignment. They should produce a great improvement, however, and repetition of the process should give excellent results.

It should be clear that the importance of the precession method is that it permits a view of the reciprocal lattice more or less perpendicular to that given by the Weissenberg without the necessity of remounting the crystal. In this way measurements can be made that involve the reflections corresponding to the mounting axis. In particular, the angle β^* can be measured for monoclinic crystals rotated about a or c. Similarly, it is possible, using a precession camera alone, to measure all the constants necessary to characterize a triclinic crystal without remounting. Note that the spacings needed for conversion to r.l. distances by Eqs. (5.22) are the repeat distances along the lattice lines and not the perpendicular distances between these lines. Once the r.l. distances are available, they can be converted into the direct repeats using the formulas of Chapter 2.

5.10. DIFFRACTOMETER THEORY

Starting about 1960, crystallographers began to shift from photographic methods to[21] ones based on quantum counters moved to computed angular orientations. A number of different mechanical arrangements were tried,

[21]Actually the movement was "back to" since the earliest structural work, that of W. H. and W. L. Bragg, used an electrometer as the radiation detection device.

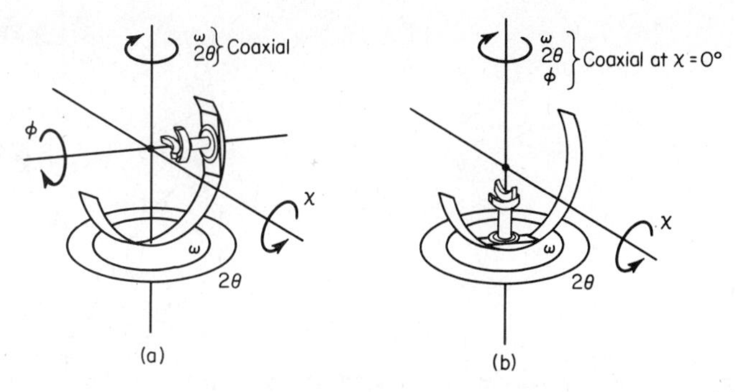

Figure 5.53. Schematic of a four-circle diffractometer. (*a*) At $\chi = 90°$; (*b*) at $\chi = 0°$.

but the field has converged on variants of the four-circle diffractometer. This possesses four arcs that can be used to adjust the orientation of the crystal and counter so as to bring any desired plane into a reflecting position and detect this reflection. These are normally divided into the crystal orienter, which contains the phi (ϕ) and chi (χ) circles, and the base, which contains omega (ω) and two theta (2θ). Figure 5.53 shows an outline of the arrangement for two values of χ.

The two base circles, ω and 2θ, are mounted about a common axis and can either be adjusted independently or be geared together so that their movements are correlated. The orienter is mounted within the ω circle and pinned so that it rotates with ω. When $\chi = 0°$ the axis of ϕ coincides with that of ω and 2θ. The ϕ circle rides on the χ arc as shown, and at $\chi = 90°$ its axis is perpendicular to that of ω and 2θ.

Although ϕ, ω, and 2θ are true circles, χ can be either a full circle (Fig. 5.54) or a segment of one with a setting range only a little more than 90° (Fig. 5.55). The original single-crystal orienter was of the latter type and had the advantage of allowing freer access to the space around the crystal. The extended range of χ setting provided by a full circle has certain

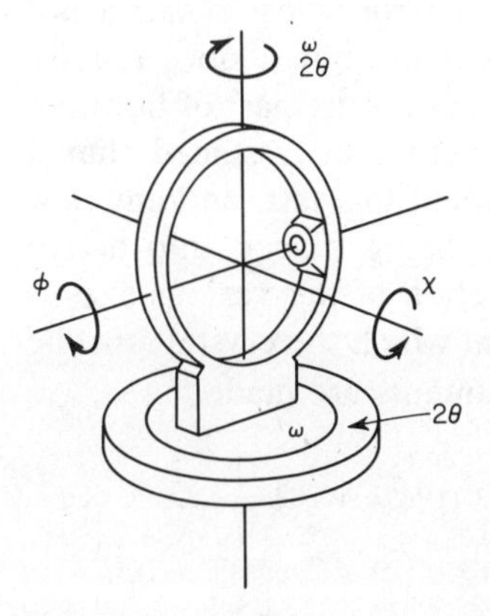

Figure 5.54. Full χ circle on a four-circle diffractometer.

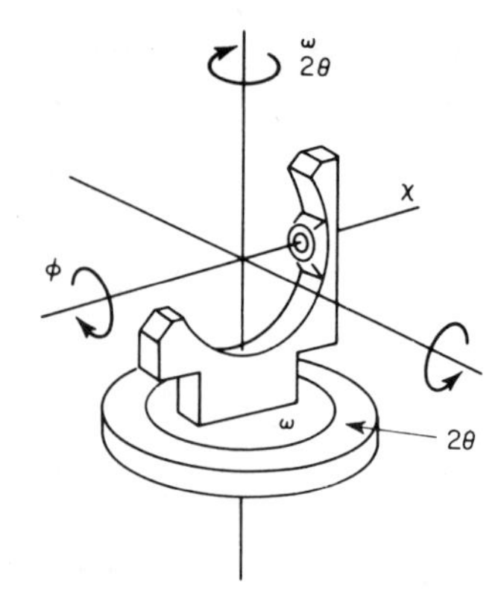

Figure 5.55. Quarter χ circle on a four-circle diffractometer.

conveniences in crystal alignment and compensates for the possible "blind spot" produced by the vertical circle. An alternative approach replaces the χ arc with an additional axis (κ, kappa) offset by 50° from the $\omega/2\theta$ axis. This provides enough additional rotational freedom to approximate a full-circle device, although at the cost of extra mathematical complexities (fortunately concealed in software) in calculating the appropriate setting angles.

The crystal is mounted in a goniometer head attached about the ϕ axis. Normally a *eucentric* head is used in which the centering adjustments are above, rather than below, the head arcs. With this arrangement, the crystal can be positioned so that it is at the point that is the common center of the two arcs and consequently will not move as they are adjusted. The allowable height of the goniometer head is determined by the geometry of the orienter and was originally standardized as 49.00 mm. In practice this proved to be somewhat restricting, and many diffractometers now accept 63-mm heads.

One of the principal mechanical difficulties of diffractometers arises from the number of axes involved in the orientation process and the necessity for their precise intersection. The requirement that is expected to be met by a properly aligned instrument is that the axes of the ω (2θ) and ϕ circles and the center of the χ arc intersect within a sphere of error whose radius is not more than 0.001 in. The requirement is not extreme, but it does require careful machining and intelligent attention to detail on the part of both the manufacturer and the user. Unfortunately, it cannot be assumed that a diffractometer as received from the maker will meet this test, and any new instrument should be carefully checked before being placed into active service. The point that is defined by the intersection of the various axes is the natural center of the system and is the point at which the crystal must be placed if it is not to move as the orienting adjustments are made.[22]

[22]S. Samson and W. W. Schuelke, *Rev. Sci. Instr.*, **38**, 1273 (1967); W. Hoppe, *Acta Cryst.*, **A24**, 67 (1969).

The remainder of the active diffraction system consists of an X-ray tube and detector. These are positioned so that the focal spot of the source, the center of the orienter system, and the center of the detector aperture all lie in a plane accurately perpendicular to the $\omega/2\theta$ axis. Because the orienter is fixed in position, adjustments for vertical and horizontal position as well as take-off angle are usually provided on the X-ray tube mount. It is significantly more convenient if the tube is mounted on the diffractometer base and not on the supporting table. It should also be held and adjusted near its head, rather than at its base, since in the latter case the position of the focal spot is very sensitive to thermal expansion and other distortions of the tube length.

The detectors used are generally scintillation counters. The Geiger–Mueller tube and proportional counters were used in the past but have been completely supplanted, largely because of their lesser counting efficiency for Mo K_α radiation.[23]

Whereas the X-ray source is set in a fixed position with respect to the crystal orienter, the detector is mounted on the 2θ circle. Thus the angle between the detector and the direct beam can be set to any desired value within the range of the instrument. The usual range is about 0–160° on one side of the beam and 0 to −10–50° on the other (Fig. 5.56). It is convenient to have as large a setting range as possible on the back side, since it simplifies crystal alignment and the measurement of accurate cell constants as well as the measurement of Friedel pairs.

In addition to the elements described, the complete diffractometer normally includes a regulated power supply for the X-ray tube, scales for counting the pulses from the detector, and usually a ratemeter for displaying on a chart recorder the rate at which pulses are being received. An additional common element is a pulse-height analyzer for discriminating

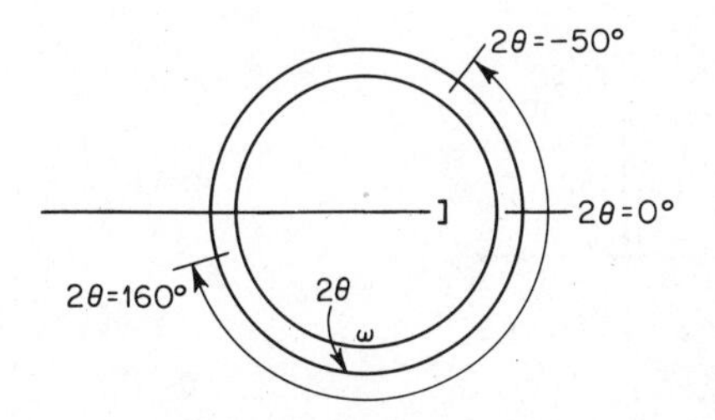

Figure 5.56. Convenient 2θ range of a diffractometer.

[23]For further discussion of detectors and associated components see *International Tables*, Vol. III, pp. 144–156; R. Rudman, *J. Chem. Ed.*, **44**, A187 (1967); and E. W. Nuffield, *X-Ray Diffraction Methods*, Wiley, New York, 1966, pp. 177–201.

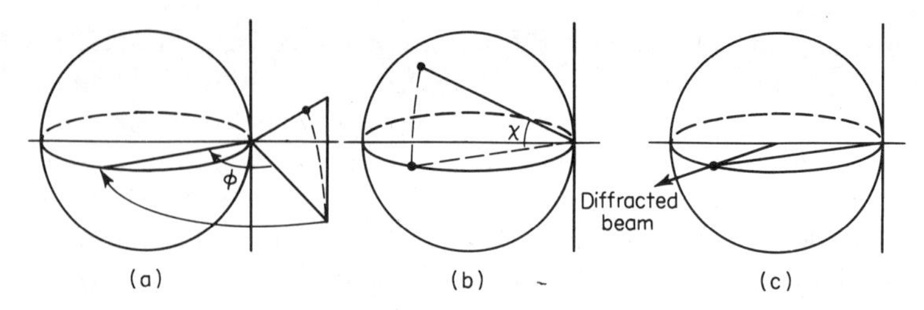

Figure 5.57. Function of ϕ and χ circles in bringing an r.l. point into the diffracting position.

against pulses that are either much weaker or much stronger than those expected for the characteristic energies being used.

It is the usual practice with an instrument of this sort to use the χ and ϕ arcs to bring any desired r.l. point into contact with the sphere of reflection in the plane defined by the source, crystal, and detector (Fig. 5.57). Thus any reflection can be recorded under conditions that are essentially those of a zero-level Weissenberg photograph. There are extra degrees of freedom in the system, and this result can be achieved in many alternative ways. In the most common arrangement (*bisecting*), $\omega = 0°$ and the χ circle bisects the angle between the incident and diffracted X-ray beams (Fig. 5.58). The combination of other values of ω with suitable settings of ϕ and χ has the effect of rotating the diffracting plane about its normal, the ψ axis. The other unique arrangement (*parallel*) has $\omega = 90°$ and the χ circle parallel to the X-ray beam at $\theta = 0°$. This tends to interfere with the beams at low θ but is more open for high values.

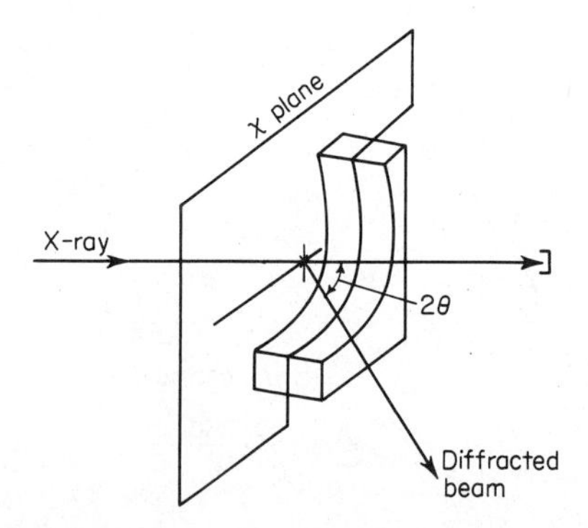

Figure 5.58. Geometry of diffraction, showing χ plane bisecting the angle between direct and diffracted beams.

5.11. DIFFRACTOMETER PRACTICE[24]

In order to measure any reflection the r.l. point must be brought onto the sphere of reflection in the equatorial plane of the diffractometer and the counter must be positioned at the proper value of 2θ. If a 2θ scan technique is to be used, ω and 2θ must be set to suitable initial values so that the scan will carry the point completely through the surface of the sphere. In order to achieve these results, the values of χ, ϕ, 2θ, and ω must be determined and set correctly for each reflection to be measured.

In the original diffractometers, various knobs or cranks were turned by hand until the correct setting was achieved. Modern instruments use electric motors and position-sensitive readout devices to set the angles automatically. In order to set these, however, the correct values must be known. Although off-line computation was used in the past, all modern instruments include built-in computers that calculate the χ, ϕ, and 2θ values necessary to obtain a reflection from each unique plane within the limiting sphere. Options are generally provided as well to calculate the initial settings for ω or 2θ scans, scan times, settings for specific reflections, or data beyond the unique portion of reciprocal space.

To calculate the settings, both the cell constants and the orientation of the crystal must be known. For use with stationary crystal, stationary counter (peak-height) counting, high precision is often required since changes of several hundredths of a degree in some settings (particularly ϕ) can produce detectable variations in intensity. These requirements are noticeably relaxed, however, when scanning techniques are used. On the other hand, the very sensitivity of the diffractometer to angle settings allows its use to achieve an accuracy of alignment and parameter measure that exceeds by an order of magnitude the accuracy usually obtained photographically.

The alignment process begins with placing the crystal, mounted on a goniometer head as usual, on the spindle. The diffractometer is more sensitive to the exact centering of the crystal than are most photographic methods, so care must be paid to locating the crystal as closely as possible at the intersection of the various axes. This is normally done by observing the crystal through a microscope mounted on the diffractometer and making translational adjustments until its apparent center does not move as ϕ and χ are changed. A small crystal can be adjusted quite precisely in this way; adjusting a large one is more difficult. Once the crystal has been centered, it is ready for the actual alignment.

Modern diffractometers come with software packages that enable them

[24]T. C. Furnas, *Single Crystal Orienter Instruction Manual*, General Electric Co., Schenectady, NY, 1957; R. A. Sparks, in *Computational Crystallography*, D. Sayre, Ed., Clarendon Press, Oxford, 1982, pp. 1–18; R. A. Sparks, in *Crystallographic Computing Techniques*, F. R. Ahmed, K. Huml, and B. Sedlacek, Eds., Munksgaard, Copenhagen, 1976, pp. 452–467.

to carry out automatic searches of a portion of reciprocal space, identify a number of reflections, center these accurately in the detector aperture and measure the setting angles, and then calculate probable cell parameters as well as the orientation matrix required to collect a data set. It is generally possible, as well as tempting, to mount a crystal in an arbitrary orientation, set it on the diffractometer, and return in a few hours to find that the machine appears to know all that is required to proceed to intensity measurements. The fact that diffractometers are frequently supplied with goniometer heads that have no adjusting arcs at all and that depend on the setting ability of the diffractometer for all angular changes shows that such behavior is expected. *This is a dangerous path to follow*. For all of its convenience, it leaves the researcher without any information about reciprocal space except in those places where the diffractometer chose to look.

We recommend strongly that crystals of new compounds be aligned photographically by the methods described above, so that at least one zero-level r.l. net (and preferably two, containing all three proposed cell axes) can be examined, and that cell constants be calculated at least roughly for comparison with those proposed by the program. If this is done, one can then choose either to leave a direct axis coincident with the diffractometer ϕ axis (the natural result of alignment on a Weissenberg camera) or bring an r.l. point onto the axis if it is not already there. This latter orientation has the virtue that if χ is set to 90°, rotation about ϕ is equivalent to rotation about the normal to the reflecting plane and provides a rough measure of the effects of absorption (see Section 7.2).[25]

As an alternative, some diffractometers provide the option of taking rotation photographs for the crystal as mounted on the orienter. These can be helpful in finding an alignment that corresponds to that of a standard camera and allows easy transfer for photography, but they do not replace Weissenberg or precession photos for giving an overall view of the lattice. In many cases, too, optical alignment of the crystal (Section 4.5) will place it in a standard orientation, suitable for both photographic and diffractometric measurement.

Once photographic alignment is complete, the diffractometer can be allowed to run through its alignment procedure, either automatically or by identifying a number of strong low-order reflections from the photographic record, finding them by manual adjustment of the setting angles, and then allowing automatic refinement of the values. This process yields a rough

[25]Such a scan shows also the variation in intensity produced by double reflections (Renninger effects; see Section 5.12). It is enlightening to observe a systematically absent low-order reflection at $\chi = 90°$ while rotating ϕ slowly and discover just how common such double reflections actually are. Their angular widths are small, however, so the intensities of only a few reflections will be affected in this way. The seriousness of these errors in affecting highly accurate structure determinations has not yet been assessed, however. See R. A. Young, *Acta Cryst.*, **A25**, 55 (1969).

orientation matrix that can be used to calculate approximate settings for a more systematic selection of reflections to be used for obtaining refined cell constants. These reflections should be chosen to be relatively intense (another reason for having a preliminary photographic view of the reciprocal lattice), of as high a 2θ value as is possible consistent with reasonable intensity, and should be distributed in more than one octant of the reciprocal lattice. Ideally, all equivalent forms of several different reflections should be measured; these should include reflections chosen from all eight octants and measured at $\pm 2\theta$. For the best results, the process can then be repeated using reflections of still higher 2θ values. Indeed, if the capability is present, it is useful to measure individual reflections at two or more settings (ψ) around the normal to the reflecting plane.

Once a number of reflections have been described in terms of measured setting angles, these are used to define two matrices, conventionally called **U** and **B**.[26] **B** uses quantities related to the direct and reciprocal cell constants to convert reflection indices to parameters (in Å^{-1}) in reciprocal space. **U** relates these reciprocal space coordinates to the coordinate system of the diffractometer to give values that can be manipulated in fairly standard ways by rotations about the various setting axes.

Various methods have been proposed for calculating these matrices independently from the observations,[26,27] which has the advantage of allowing the incorporation of known symmetry restraints into the derivation of the cell constants. Most commonly, however, the setting programs solve for the elements of the product matrix **UB** and derive the cell parameters and their standard deviations from this matrix.[28] This process has the advantage of giving the "best" orientation matrix for intensity measurements (assuming that the reflections used have been chosen appropriately) but gives parameters as though the cell is triclinic (e.g., 89.75° instead of a symmetry-constrained value of 90°), which may lead to greater errors in other parameters than are implied by the standard deviations. It also leads to questions as to what values should be reported—the observational results or the values implied (required) by the point group symmetry. Programs are available that avoid these problems, but they are not widely used.[29]

If, for some reason, one is stuck with a diffractometer that does not provide the ability to determine the orientation matrix automatically, it is necessary to carry out the alignment by hand and to determine the relationship between the crystal lattice and the zero settings of the orienting devices. The necessary procedures were discussed in our first edition,[30]

[26]W. R. Busing and H. A. Levy, *Acta Cryst.*, **22**, 457 (1967).

[27]W. Clegg, *Acta Cryst.*, **A39**, 170 (1983).

[28]K. Tichy, *Acta Cryst.*, **A26**, 295 (1970).

[29]D. P. Shoemaker and G. Bassi, *Acta Cryst.*, **A26**, 97 (1970); R. L. Ralph and L. W. Finger, *J. Appl. Cryst.*, **15**, 537 (1982).

[30]G. H. Stout and L. H. Jensen, *X-Ray Structure Determination*, Macmillan, New York, 1968, pp. 187–192.

but since they are unlikely to be widely used now we have omitted them here.

A final check on the alignment and visual evidence of the amount of latitude available can be obtained by placing a small piece of X-ray film in a black envelope just in front of the counter aperture and making a number of scans through a reflection. The film, when developed, should show a reflection spot centered in a gray area that represents the image of the receiving aperture.

5.12. SYMMETRY DETERMINATION[31,32]

The problem of defining the correct space group, that is, the correct internal symmetry, of a crystal of unknown structure is an important one, since most of the subsequent calculations will assume the relationships implied by the space group. This problem can be separated into two parts; first the assignment of the proper crystal lattice and then the selection of the proper space group from among those appropriate to the lattice.

In the past, lattice assignment was considered a routine task, one to be carried out almost automatically by inspection. Recently it has become clear that the problem is not quite so simple, and a number of examples of misassigned lattices have appeared in the literature.[33,34] In part this has resulted from the routine use of diffractometers without thoughtful inspection of diffraction photographs and with dependence on imperfect software to select the lattice. In part, however, it is the result of some subtle traps that have become more apparent with the growing number of structural examples.

The discussion of lattices in Chapter 3 suggested by implication that the process of lattice assignment consisted simply of marking in obvious axes between lattice points and checking for the presence of equal edges and special angles. In effect this is true, but in practice it can be far from simple. The problem is that in the general case one is presented with just a three-dimensional array of points in an arbitrary orientation and not with a neat assemblage of boxes to check for symmetry. There exist a vast number of possible axial choices for such an array. How does one choose the appropriate axes in all cases?

Formerly, crystallographers tended to mount crystals in ways that took advantage of the clues afforded by their exterior form. Too, alignment by oscillation photographs tended to force orientations in which a direct axis

[31] Buerger, *X-Ray Crystallography*, Chapter 4; E. W. Nuffield, *X-Ray Diffraction Methods*, Wiley, New York, 1966, pp. 87–102.

[32] *International Tables*, Vol. I, pp. 52–55, 74–352.

[33] W. H. Baur and E. Tillmanns, *Acta Cryst.*, **B42**, 95 (1986) and references cited.

[34] R. E. Marsh and F. H. Herbstein, *Acta Cryst.*, **B39**, 280 (1983); **B44**, 77 (1988); F. H. Herbstein and R. E. Marsh, *Acta Cryst.*, **B38**, 1051 (1982).

was along the mounting spindle, so the problem was limited to identifying axes in the zero-level r.l. net perpendicular to this axis. Consequently, symmetry elements were generally found if they were present and the principal concern was to ensure that a consistent unique set of axes would be found for triclinic lattices, which lack such constraints. The answer to this problem was the introduction of the *reduced cell*, a standardized cell whose edges are the three shortest noncoplanar distances between lattice points and whose angles are all acute or all obtuse.[33,35,36] The unique determination of such a cell is not always easy, especially if cell edges are nearly equal or angles are near special values (e.g., 60°, 90°). Such occurrences, when combined with experimental uncertainties, confuse the choices, but in principle it is a matter only of computation. Unfortunately, a number of conflicting conventions have been used, which have complicated the question of exactly *which* reduced cell was being discussed.

The use of reduced cells is helpful in a number of cases. It is possible, especially in triclinic crystals, for there to be a number of alternative cells with almost identical edges and different angles. In other classes it is sometimes possible to find alternative orientations that appear to have approximately the same parameters but that lead to very different indexing of reflections. If two crystals are mounted in ways that lead to these different choices, much confusion can occur unless the reduced cells are calculated and used for relating the different orientations.[37]

With the growing tendency to depend on diffractometers to identify the lattice for a crystal mounted in an arbitrary orientation, and to observe only the (supposedly) unique set of reflections, the problem has become still more complex. Assuming that the automatic search routines succeed in finding enough r.l. points to define all the lattice repeats,[38] it is a reasonably easy process to calculate a reduced cell for the lattice observed.[39] This may have cell parameters that suggest the shape of the true lattice cell, or it may be a general triclinic shape that gives no apparent clue to the possible symmetric lattices that could be assigned to the same points. It has become clear that many diffractometer setting programs do a poor job of finding the highest lattice symmetry and that it is advisable to check the results independently.[40]

[35] *International Tables*, Vol. A, pp. 737–744.

[36] A. D. Mighell, *J. Appl. Cryst.*, **9**, 491 (1976).

[37] For an example, see W. H. DeCamp, *Acta Cryst.*, **B32**, 2257 (1976).

[38] This *should* be the case, but it is not a foregone conclusion. A crystal with pseudosymmetry that causes all the reflections of odd l to be much weaker than those of even l could fail to have any of the first set found by a rapid search. The consequence would be a c axis with half its true length, loss of half the intensity data, and conversion of the psuedosymmetry into true symmetry in the resulting model.

[39] Whether this is *the* reduced cell is more of a problem.

[40] R. Harlow, *ACA Newsletter*, **18**(4), 11 (August 1987).

A number of workers have studied the problem of assigning correct lattices,[41,42] and a number of different techniques have been proposed. The method of Andrews and Bernstein[41] has the advantage of offering a quantitative measure of the degree of agreement between the observed parameters and those required for any proposed lattice.

The geometric quantities used in these methods to assign the lattice depend strictly on the values found for the distances and angles between the points in the reciprocal lattice. They ignore entirely the intensities associated with these points and the extent to which they do or do not show the symmetry associated with a particular Bravais lattice. Clearly this approach rejects a great deal of potentially useful information, information that often appears automatically in photographs. Hence, many of the problems that appear in the general case are minimized if pictures are taken of the reciprocal lattice. It is true, however, that those conditions that can lead to accidental simulation of the lattice parameters of a higher-than-true symmetry can also lead to apparent symmetry in the intensities, so photographs cannot resolve all the possible problems.

The most common difficulties that arise in lattice assignment involve lattices that can be indexed as centered cells. Probably the most frequent error has been the misidentification of rhombohedral cells (which are hard to recognize even on photographs) as C-centered monoclinic, an indexing that is always possible. This error is often indicated by the apparent occurrence of 12 molecules in a cell assigned to $C2/c$. Other common mistakes are assignment of monoclinic C-centered cells to triclinic, and centered orthorhombic cells to monoclinic.[43]

Once the lattice has been assigned, there remains the task of determining the actual space group if this can be done. The great usefulness of photographs in space group determination lies in their ability to show in a single image the systematic presence or absence of specific groups of reflections and thus imply the presence or absence of certain symmetry elements. This ability is not unique to these techniques; any method that allows the reconstruction of the reciprocal lattice point by point will give the same results. Photographs have the advantage, however, over a list of intensities that they provide a *pattern* of reflections that is more apparent to the human mind. Furthermore, a reflection observable on film generally represents a greater number of quanta than a similar observation on a diffractometer; so the random uncertainty is reduced and it is usually clearer what is or is not present.

[41]L. C. Andrews and H. J. Bernstein, *Acta Cryst.*, **A44**, 1009 (1988).

[42]W. Clegg, *Acta Cryst.*, **A37**, 913 (1981); Y. LePage, *J. Appl. Cryst.*, **15**, 255 (1982); V. L. Himes and A. D. Mighell, *Acta Cryst.*, **A43**, 375 (1987).

[43]Note, however, that the appearance of apparent orthorhombic lattice geometry does not necessarily mean that the true lattice is orthorhombic. In a number of cases, analysis of the final structure has confirmed that the monoclinic assignment was correct (see footnote 33 reference).

Unfortunately, the information provided by systematic absences is not complete; only translational symmetry elements can be detected, and there are many space groups that are not fully defined by these. Fortunately, however, two of the most common for organic compounds, $P2_1/c$ and $P2_12_12_1$, *are* uniquely determined by their systematic absences. In other cases, for example, $P2_1$ and $P2_1/m$, one is so much more common than the other that the probable—though not certain—choice is clear. Table 5.1 gives the results of a survey[44] on the distribution of space groups reported for organic crystals up to 1981. These show clearly that organic compounds are not distributed uniformly over all the space groups but rather are concentrated into a relatively small number (75% in 5 groups, 90% in 16). It is therefore possible to consider particularly those space groups and symmetry elements that are most likely to be found, while remaining aware of the other possibilities.

The *systematic extinctions* caused by each kind of translational symmetry element are given in Table 5.2. The presence of each of the symmetry elements listed in column 1 causes those reflections of the classes in column 2 to vanish when the indices are as specified in column 3. The indices h, k, and l may assume any value unless specifically restricted to zero, and n is an integer (i.e., $2n+1$ is a general representation for the set of odd numbers).

Of the elements listed, the 2-fold screws, the simple glide planes, and the side-centered lattices are very much the most common in organic crystals. The principal difficulty that confronts the beginner in practice lies in identifying their systematic absences on Weissenberg photographs of a crystal of uncertain orientation. Fortunately, however, the mirror planes and the even-fold axes are flagged in photographs by being lines of mirror symmetry and are thus relatively easy to locate.

Examination of Table 5.2 shows that screw axes cause the disappearance of certain reflections along an axial line in the reciprocal lattice, glide planes affect reflections in the corresponding reciprocal lattice planes, and centering causes systematic absences throughout reciprocal space. The 2-fold screw is much the simplest in its effects and merely causes the extinction of reflections with odd index on the corresponding reciprocal

TABLE 5.1 Percentage of Common Space Groups Found for 29,059 Organic Compounds

$P2_1/c$	36.0%
$P\bar{1}$	13.7
$P2_12_12_1$	11.6
$P2_1$	6.7
$C2/c$	6.6

[44] A. D. Mighell, V. L. Himes, and J. R. Rodgers, *Acta Cryst.*, **A39**, 737 (1983); see also A. J. C. Wilson, *Acta Cryst.*, **A44**, 715 (1988).

TABLE 5.2 Translational Symmetry Elements and Their Extinctions

Symmetry Element	Affected Reflection	Condition for Systematic Absence of Reflection
2-fold screw (2_1) along a	$h00$	$h = 2n + 1 = $ odd
4-fold screw (4_2) along b	$0k0$	$k = 2n + 1$
6-fold screw (6_3) along c	$00l$	$l = 2n + 1$
3-fold screw ($3_1, 3_2$), 6-fold screw ($6_2, 6_4$) along c	$00l$	$l = 3n + 1, 3n + 2$, i.e., not evenly divisible by 3
4-fold screw ($4_1, 4_3$) along a	$h00$	$h = 4n + 1$, 2, or 3
b	$0k0$	$k = 4n + 1$, 2, or 3
c	$00l$	$l = 4n + 1$, 2, or 3
6-fold screw ($6_1, 6_5$) along c[a]	$00l$	$l = 6n + 1$, 2, 3, 4, or 5
Glide plane perpendicular to a		
Translation $b/2$ (b glide)	$0kl$	$k = 2n + 1$
$c/2$ (c glide)		$l = 2n + 1$
$b/2 + c/2$ (n glide)		$k = l = 2n + 1$
$b/4 + c/4$ (d glide)		$k + l = 4n + 1$, 2, or 3
Glide plane perpendicular to b		
Translation $a/2$ (a glide)	$h0l$	$h = 2n + 1$
$c/2$ (c glide)		$l = 2n + 1$
$a/2 + c/2$ (n glide)		$h + l = 2n + 1$
$a/4 + c/4$ (d glide)		$h + l = 4n + 1$, 2, or 3
Glide plane perpendicular to c		
Translation $a/2$ (a glide)	$hk0$	$h = 2n + 1$
$b/2$ (b glide)		$k = 2n + 1$
$a/2 + b/2$ (n glide)		$h + k = 2n + 1$
$a/4 + b/4$ (d glide)		$h + k = 4n + 1$, 2, or 3
A-centered lattice (A)	hkl	$k + l = 2n + 1$
B-centered line (B)		$h + l = 2n + 1$
C-centered lattice (C)		$h + k = 2n + 1$
Face-centered lattice (F)		$h + k = 2n + 1$, $h + l = 2n + 1$, $k + l = 2n + 1$ (i.e., h, k, l not all even or all odd)
Body-centered lattice (I)		$h + k + l = 2n + 1$

[a]Note that in the crystal classes in which 3- and 6-fold screws occur as cell axes, these are conventionally assigned to be c, so only the $00l$ reflections need be considered.

axis. In the absence of other extinctions this effect is easily recognized. The presence of such an axis parallel to the rotation axis cannot be detected from Weissenberg photographs taken on a single mounting, since the required reflections are never visible. They will be revealed, however, by precession photographs or by Weissenberg photographs of a crystal rotating about a different axis.

The pattern of extinctions resulting from a glide plane is somewhat more complicated, and its appearance varies with the rotation axis chosen. In general, for a single glide plane (e.g., *Pc*) the r.l. absences will form the patterns, shown in Fig. 5.59, where the rotation axis is normal to the glide

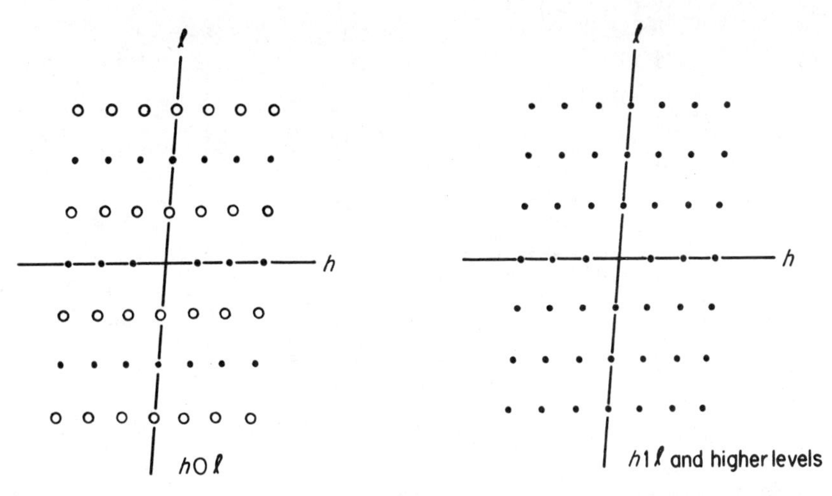

Figure 5.59. $h0l$ and $h1l$ reciprocal lattice nets for space group *Pc*. Systematic absences designated by small circles.

plane ($h0l$, $h1l$);[45] those in Fig. 5.60, where it is in the glide plane but angled to the direction of translation ($0kl$, $1kl$); and those in Fig. 5.61, where it is in the plane and parallel to the translation ($hk0$, $hk1$). If the inclined axes are chosen so that the space group is *Pn*, the appearance of the $h0l$ net is as shown in Fig. 5.62. Because a glide plane produces extinctions that are not limited to one axis, it can be detected regardless of the orientation of the crystal, although a perfectly general orientation would make it difficult to recognize the extinctions even though they could be "observed."

The extinctions resulting from centering are more extensive and affect general reflections rather than only those with at least one index zero. Centering is most easily recognized either from the "diamond pattern" produced when every festoon is composed of reflections alternating with extinctions or from the observation that when photographs of successive levels are superimposed no two reflections coincide. For side-centered lattices (A, B, C) the first case will be seen when rotation is about the axis (a, b, c) corresponding to the centering. Otherwise the second will be seen.

[45]The ability of a glide plane to extinguish entire festoons of reflections when viewed in this orientation can pose a hazard in the assignment of axes in monoclinic crystals. In the $h0l$ net of *Pc*, for example, half of the reflections (those with indices $2n$, 0, 2) of the visible $h02$ festoon lie on central lattice lines that pass through r.l. points in the $h01$ festoon and half do not. The reflection chosen to define the c^* axis (i.e., 002) must belong to the former set, but this cannot be assured since 001 is systematically absent. For this reason it is necessary to choose axes with reference to a first-level ($h1l$) film, in which the 011 reflection may be seen in the position corresponding to the absent 001.

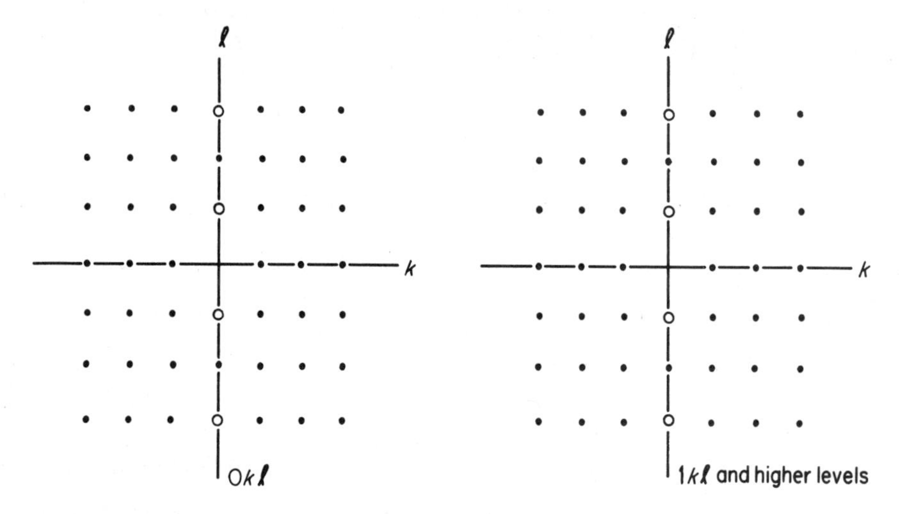

Figure 5.60. $0kl$ and $1kl$ reciprocal lattice nets for space group Pc. Systematic absences designated by small circles.

F-centering shows the latter effect for all axes, and I-centering shows both simultaneously for all axes.

As a practical matter, however, one cannot count on translational symmetry elements appearing singly, and so it is often necessary to decide whether the extinctions that are present belong to more than one set. A good general rule is to work from the most complex to the simplest, that is, to look for centering, then glides, and finally screws. Any reflection that would be extinguished by a given symmetry element occurring alone is also

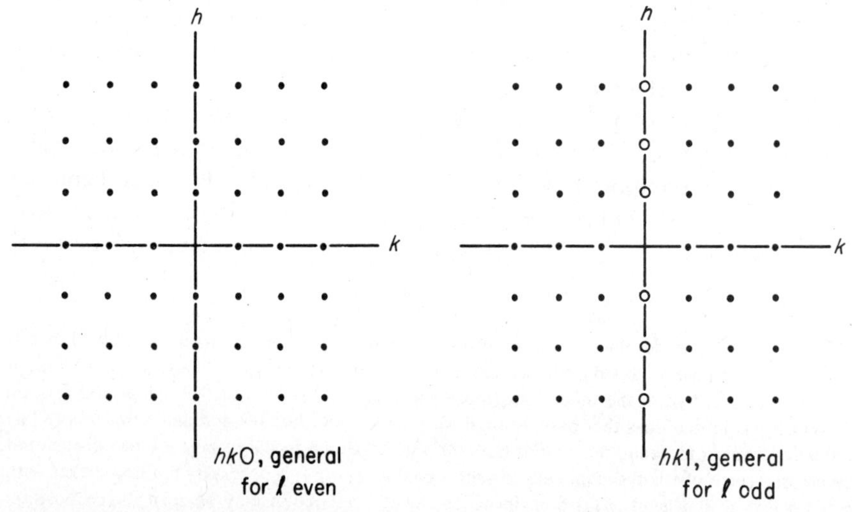

Figure 5.61. $hk0$ and $hk1$ reciprocal lattice nets for space group Pc. Systematic absences designated by small circles.

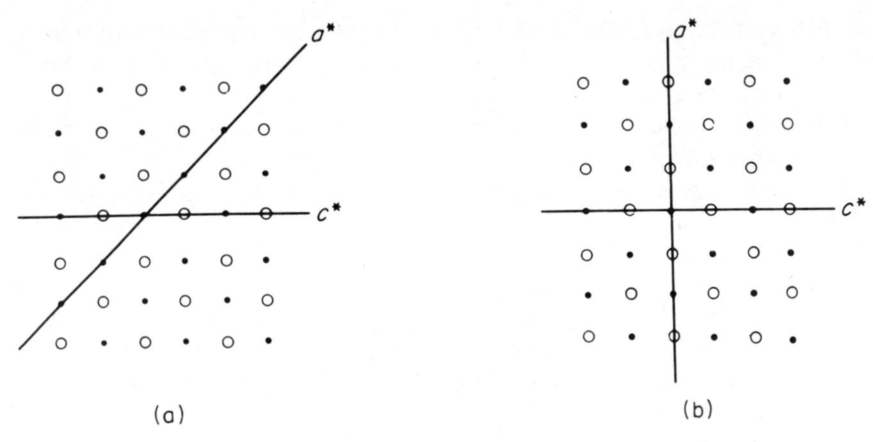

Figure 5.62. (*a*) Monoclinic a^*c^* net with absences for *c* glide. (*b*) Same net with axes chosen to show *n*-glide absences.

extinguished when this element occurs in combination. Thus the extinction patterns that appear in photographs of complex space groups are the sums or superpositions of the patterns of the individual elements. Consequently when a symmetry element has been recognized, the extinctions that it causes can be discounted and the photographs examined to see if others remain. If all are accounted for, no further translational elements are to be found.[46] If the apparent extinctions lead to a space group not in Table 5.2, a nonstandard indexing has been chosen. Although this may be advantageous in special cases, it is convenient to have a space group orientation that is one of those illustrated in *International Tables*.

The extinctions that we have been discussing are invariant and are absolutely mandatory for the symmetry elements they represent. Thus, in principle, the observation of a single reflection that should be extinguished should be sufficient grounds for rejecting a symmetry element and replacing it by "near symmetry" or "pseudosymmetry." Unfortunately, false observations can occur for reasons that do not violate space group symmetry, and it is necessary to be aware of these possibilities in order to avoid an overhasty rejection.

One possible source of false reflections is the misidentification of the K_β spot from a strong reflection as a K_α spot. This is discussed in Section 6.1 and, except for the particular reflections cited there, is rarely a problem.

[46]This does not always mean that they are not present, however. $C2/c$, for example, contains 2-fold screws produced by the combined operation of the 2-fold axis and the C centering. The reflections that would be extinguished, $0k0$ with k odd, form part of the set $hk0$ with h plus k odd, which is extinguished by the centering. Thus, despite the extra translational symmetry, $C2/c$ cannot be distinguished from Cc by systematic absences and the pair represents one of the most common sources of space group uncertainty.

Similarly, if a copper X-ray tube is even slightly contaminated with iron, as does sometimes occur, the fourth-order (e.g., 040) reflection of Fe K_α ($\lambda = 1.9373$) will fall at the expected position for the fifth-order (050) of Cu K_α. This contamination can be detected, however, by observing the Fe spot corresponding to an intense general reflection.

A second and more serious source is the possibility of double reflections (Renninger effect) in which radiation reflected in turn from two strongly reflecting planes appears to arise by single reflection from a third.[47,48] Normally errors from this source appear to be small and are ignored, but if the apparent reflection occurs at the site of a systematic absence it can be disturbing. Usually glide planes cause so many extinctions that apparent failure in one case can be ascribed to double reflection without qualms. Screw axes lead to fewer absences, however, and the failure of one of these poses more problems. Three methods are available for identifying a double reflection. The first depends on its appearance. By the nature of the double reflection process, Renninger spots tend to be more sharp-edged and well-defined than normal reflections, and with experience they are often identifiable. A more positive decision can be reached by looking for the reflection on a photograph taken of the crystal mounted in a different orientation. The occurrence of a specific double reflection requires a particular orientation of the reciprocal lattice around the vector between the r.l. origin and the apparent source of the reflection. If the crystal is remounted so that the reciprocal lattice is effectively rotated about this vector,[49] the conditions for the observed double reflection are destroyed and the Renninger spot should vanish.[50] Finally, for a given orientation, a particular Renninger effect will be observed only for a single wavelength. Thus if photographs are taken using two different characteristic radiations, any reflection that appears on both can safely be taken as real.

This discussion of Renninger effects should not be allowed to shake one's confidence in the use of extinctions in determining symmetry. In practice, Renninger reflections are found only rarely, and it is only the need to recognize their possibility that leads to their inclusion here.[51]

Once all the translational symmetry elements have been found, it is necessary to consider whether they determine the space group uniquely. Since pure rotation axes and mirror planes do not cause extinctions, no

[47]M. Renninger, *Z. Physik.*, **106**, 141 (1937).

[48]H. Lipson and W. Cochran, *The Determination of Crystal Structures*, G. Bell and Sons, London, 1957, pp. 30–32.

[49]If the vector can be made coincident with the mounting axis, this effect is achieved very simply in a series of precession photographs taken at different azimuth settings or by diffractometer ϕ scans at $\chi = 90°$.

[50]See L. Cohen, B. S. Fraenkel, and Z. H. Kolman, *Acta Cryst.*, **16**, 1192 (1963); H. Cole, F. W. Chambers, and H. M. Dunn, *Acta Cryst.*, **15**, 138 (1962).

[51]For examples in which this problem has occurred in practice, see, among others, A. Hargreaves and S. H. Rizvi, *Acta Cryst.*, **15**, 365 (1962) and J. C. Speakman, *Acta Cryst.*, **18**, 570 (1965).

deductions can be made about their presence or absence, and the extinction patterns will be the same for space groups with the same translational symmetry, regardless of what other nontranslational elements are added. Table 5.3 lists the triclinic, monoclinic, and orthorhombic space groups that will have identical extinctions. As can be seen, only a few of these possess unique diffraction patterns, although fortunately they include several of the most important. Any differentiation within the nonunique sets, however, will require further information.

The required information may in some cases be derived from chemical knowledge about the compound, even though this is very scanty. In particular, if the compound is known to be optically active, it is impossible for it to crystallize in a space group containing either a mirror (or glide) plane or a center of symmetry. These symmetry elements require the presence of both right- and left-handed asymmetric species in the crystal, a condition that cannot be fulfilled by a single enantiomorph. Obviously, however, this argument cannot be applied to a racemic mixture, nor is the absence of asymmetric centers in the molecule a guarantee against crystals of low symmetry.[52]

TABLE 5.3 Orthorhombic and Lower Space Group Sets with Identical Extinctions

$P1$,[a] $P\bar{1}$[b]	$Pmn2$, $Pmmm$[b]
	$Pba2$, $Pbam$[b]
$P2$,[a] Pm, $P2/m$[b]	$Pna2_1$, $Pnma$[b]
$P2_1$,[a] $P2_1/m$[b]	$Pnn2$, $Pnnm$[b]
Pc, $P2/c$[a]	$Cmc2_1$, $Ama2$, $Cmcm$[b]
$C2$,[a] Cm, $C2/m$[b]	$Cccc2$, $Cccm$[b]
$P2_1/c$[b]	$Abm2$, $Cmma$[b]
Cc, $C2/c$[b]	$Aba2$, $Cmca$[b]
	$Fdd2$
$P222$,[a] $Pmm2$, $Pmmm$[b]	$Iba2$, $Ibam$[b]
$P222_1$[a]	$Ima2$, $Imma$[b]
$P2_12_12$[a]	$Pnnn$[b]
$P2_12_12_1$[a]	$Pban$[b]
$C222_1$[a]	$Pnna$[b]
$C222$,[a] $Cmm2$, $Amm2$, $Cmmm$[b]	$Pcca$[b]
$F222$,[a] $Fmm2$, $Fmmm$[b]	$Pccn$[b]
$I222$,[a] $I2_12_12_1$,[a] $Imm2$, $Immm$[b]	$Pbcn$[b]
$Pmc21$, $Pma2$, $Pmma$[b]	$Pbca$[b]
$Pcc2$, $Pccm$[b]	$Ccca$[b]
$Pca21$, $Pbcm$[b]	$Fddd$[b]
$Pnc2$, $Pmna$[b]	$Ibca$[b]

[a]Possible for optically active compounds.
[b]Centrosymmetric.
Space groups occurring singly are completely determined by their systematic absences.

[52]See F. H. Herbstein and F. R. L. Schoening, *Acta Cryst.*, **10**, 657 (1957).

Since for optically active substances the space group is known to be free of reflective elements, the problem of determination is much simplified. Among the systems we are considering, that is, orthorhombic and lower, only the point groups 1, 2, and 222 are allowed. On this basis all noncentrosymmetric space groups in Table 5.3 can be assigned uniquely except for the extremely uncommon pair $I222$ and $I2_122$.

A second, less powerful guide is sometimes given by the number of molecules in the unit cell. If the molecular weight is known even approximately, comparison of this value with the total molecular weight of the cell contents (see Section 4.6) will give the number of molecules per cell, usually 1, 2, 4, or 8. Each space group has a characteristic number of asymmetric units that are to be found within its unit cell, the number depending on the point group and the presence or absence of centering. Table 5.4 lists these numbers. It is usual, although there are frequent exceptions, for an asymmetric unit to consist of one molecule, so that in most cases the number of molecules in the cell will equal the number of asymmetric units. Since the space groups with corresponding extinctions generally belong to different point groups, the number of molecules can serve as a *suggestion* as to the correct choice. Unfortunately, it is quite possible for there to be two or more molecules in an asymmetric unit, so the presence of an excessive number of molecules is no bar to groups with fewer asymmetric units. On the other hand, a unit cell can contain fewer molecules than asymmetric units only if the molecules possess one or more symmetry elements that also appear in the space group and can position themselves in the cell so that the corresponding elements coincide.[53] In this way, one part of the molecule is "produced" by a symmetry operation on another part, and the unique portion is less than a whole molecule. Because of the lack of symmetry in the most stable conformation of most large organic molecules and because of packing considerations, this effect is by no means common, but it must be considered before arbitrarily discarding a possible space group as being of too high symmetry.[54]

Once intensity data have been collected, it is often possible to prove the absence of a center from consideration of the distribution of intensity values found (see Section 7.3; but note warning). These techniques cannot prove unequivocally the presence of a center, however, although they can suggest it. In some cases, accurate collection of intensities for the entire sphere of

[53]Note that it is not possible for a nonpolymeric molecule to possess translational symmetry, so coincidences with screws and glides need not be feared and such common space groups as $P2_1$ and $P2_12_12_1$ must have at least one molecule per asymmetric unit. [See R. B. Woodward and J. Z. Gougoutas, *J. Am. Chem. Soc.*, **86**, 5030 (1964).]

[54]A further complication that can arise in nearly symmetric compounds is the possibility that through random packing they may achieve the appearance of a symmetry higher than that which they really possess. An excellent discussion of this problem is given in connection with the crystal structure of azulene [J. M. Robertson, H. M. M. Shearer, G. A. Sim, and D. G. Watson, *Acta Cryst.*, **15**, 1 (1962)].

TABLE 5.4 Asymmetric Units per Unit Cell

	Point Group							
	1	$\bar{1}$	2	*m*	2/*m*	222	*mm*2	*mmm*
P	1	2	2	2	4	4	4	8
A, *B*, *C*			4	4	8	8	8	16
F						16	16	32
I						8	8	16

data will provide useful information about the actual symmetry of the intensity-weighted reciprocal lattice, especially if a heavy atom is present.

If the space group is still in question, and if no derivative or modification free of the ambiguity is available, the best approach is to attempt the solution in each of the possible groups and see which one succeeds. The most common uncertainty is whether or not a center of symmetry exists in the crystal, and it is often possible to obtain solutions in both centrosymmetric and noncentrosymmetric groups.[55] The best choice to start with, in the absence of other information, is the most common (see Table 5.2 or footnote 38) of those space groups of the set that have the number of asymmetric units indicated by the number of molecules.

KEY FORMULAS

ALIGNMENT

Oscillation

$$\text{tilt error} = \left|\frac{\Delta r - \Delta l}{2}\right|; \qquad \text{bow error} = \left|\frac{\Delta r + \Delta l}{2}\right| \tag{5.2}$$

Weissenberg

$$\delta = \frac{|\angle A - \angle B|}{2} = |\angle C - \tau| \qquad \text{(5.11) and (5.13)}$$

$$\Delta = 1.12\delta \tag{5.12}$$

[55]Since one of the most common errors is the failure to find a center of symmetry that is actually present, it is important to consider carefully at the end of the analysis which of the two solutions is correct. See R. E. Marsh, *Acta Cryst.*, **B42**, 193 (1986). This is normally done in terms of the refinement behavior of the two models and the topological analysis of the final structures (see references listed in footnotes 33 and 34). See Chapter 19 for a further discussion.

Precession

$$\Delta\xi = \frac{||x_1| - |x_2||}{2 \times \text{film-to-crystal distance}} \tag{5.31}$$

$$\epsilon = \tfrac{1}{2} \tan^{-1} \Delta\xi \tag{5.33}$$

$$\epsilon\ (\text{min}) \approx 14 \times ||x_1| - |x_2|| \quad (\text{mm}) \tag{5.34}$$

AXIAL MEASUREMENTS

Rotation

Repeat distance on direct axis of rotation:

$$r = \frac{\lambda}{d_1^*} = \frac{n\lambda}{\sin \tan^{-1}(y_n/R)} \tag{5.5}$$

Interplanar distance in the reciprocal lattice:

$$d_n^* = \frac{y_n}{(R^2 + y_n^2)^{1/2}} = \sin \tan^{-1}\left(\frac{y_n}{R}\right) \quad \text{in r.l.u.}$$

$$= \frac{y_n}{\lambda(R^2 + y_n^2)^{1/2}} = \frac{\sin \tan^{-1}(y_n/R)}{\lambda} \quad \text{in Å}^{-1} \tag{5.3}$$

Weissenberg

Distance of a point (axial or otherwise) on zero-level Weissenberg from the r.l. origin:

$$\begin{aligned} \xi &= 2 \sin(57.30 \times 2y/4R) && \text{in r.l.u.} \\ &= (2/\lambda) \sin(57.30 \times 2y/4R) && \text{in Å}^{-1} \end{aligned} \tag{5.17}$$

Precession

Distance of a point on a zero-level precession photograph from the r.l. origin:

$$\begin{aligned} PO &= \xi = RF/CF && \text{in r.l.u.} \\ &= \xi = (1/\lambda)(RF/CF) && \text{in Å}^{-1} \end{aligned} \tag{5.22}$$

SETTING

Weissenberg

$$\mu = \sin^{-1}(d^*/2), \text{ with } d^* \text{ in r.l.u.} \tag{5.7}$$

$$\text{screen shift} = r_s \tan \mu \tag{5.8}$$

Precession

Screen-to-crystal distance for a screen of radius r and a precession angle $\bar{\mu}$:

$$CS = \frac{r}{\tan \bar{\mu}} \tag{5.28}$$

BIBLIOGRAPHY

Blundell, T. L., and L. N. Johnson, *Protein Crystallography*, Academic, London, 1976, pp. 254–309.

Buerger, M. J., *X-Ray Crystallography*, Wiley, New York, 1942.

Glusker, J. P., and K. N. Trueblood, *Crystal Structure Analysis*, 2nd ed., Oxford University Press, New York, 1985, Chapter 4.

Henry, N. F. M., H. Lipson, and W. A. Wooster, *The Interpretation of X-Ray Diffraction Photographs*, Macmillan, London, 1960.

Jeffery, J. W., *Methods in X-Ray Crystallography*, Academic, New York, 1971, pp. 134–237, 422–445.

Luger, P., *Modern X-Ray Analysis on Single Crystals*, de Gruyter, New York, 1980, pp. 52–79, 94–110, 160–195.

CHAPTER 6

INTENSITY DATA COLLECTION

In Chapter 5 we saw how the geometric properties of unit cells can be deduced from the locations of X-ray reflections. In this chapter we shall be concerned with the measurement of the relative intensities of these reflections, since it is from these intensities that we are able to deduce the electron density distribution in the crystal cell. The connections between the intensities and the electron density are described in the following chapters; here we shall be concerned only with *how*, not why, the intensities are measured.

Two general methods are available for measuring the intensities of diffracted beams. Either the beams can be detected by a quantum counting device that measures the number of photons directly (diffractometer or counter methods) or else the degree of blackening of spots on diffraction photographs can be measured and taken as proportional to the beam intensity (photographic methods). During the growth of X-ray crystallography as a structural tool, most of the work involved photographically measured intensities. Within the last twenty years, however, diffractometers have become almost universally used. They have many real advantages: If properly used they can provide more precise intensity measurements than photographic methods, they are much faster for the same accuracy, and they allow data to be collected routinely and almost automatically. On the other hand, this very ease can be seductive and deceptive; the machines appear to do everything, but they do it with the utmost stupidity, and unless the operator asks the right questions it is easy to pass over important facts.

We shall consider first the general aspects of intensity measurement, mainly in terms of film, since it is easy to see what is happening away from the reciprocal lattice point, and then some of the problems specific to diffractometers. In general, however, the basic problems are the same for all methods, and whenever a diffraction spot is mentioned in the following discussion the argument also holds for a diffracted beam as seen by a counter.

6.1. FUNDAMENTALS

The quantity with which we shall be dealing is called the *integrated intensity*. It is not a particularly easy concept to define rigorously,[1] but a definition adequate for our purposes is "a measure of the total number of photons of the characteristic wavelength being used that are diffracted in the proper direction while a reciprocal lattice point passes from the outside to the inside of the sphere of reflection or vice versa." The definition is an ideal one, and the actual methods of intensity measurement will fulfill it only more or less closely.

It is worth noting that a reflection is a three-dimensional element of intensity, a spatial image of the r.l. point as it passes through the sphere of reflection. We tend to observe it as a two-dimensional object (a spot on a film) or one-dimensional object (a scan tracing), but these actually represent one- or twofold integrations of the true intensity distribution.

There are a few general precautions that are applicable to any intensity-measuring method. The crystal should be entirely bathed in the X-ray beam and should be centered as well as possible so that it does not move from one part of the beam to another as it rotates. Correspondingly, the beam should be as uniform as possible, so that all parts of the crystal receive the same X-ray flux. The significance of these requirements is obvious; if one is to compare the relative intensities of two reflections, the incident radiation must be constant.[2]

The major difficulties in intensity measurements arise from three sources. The first of these is the scattering of radiation by processes other than Bragg reflection. The second is the spectral impurity of the incident beam, which gives rise to appreciable amounts of scattered and diffracted radiation of wavelengths other than that of the desired characteristic. The last is the need to define the boundaries of the reflection correctly regardless of variations in size, shape, and intensity.

[1]For more detailed and rigorous discussions, see R. W. James, *The Optical Principles of the Diffraction of X-Rays*, Cornell University Press, Ithaca, New York, 1965, pp. 34–52; Buerger, *Crystal–Structure Analysis*, pp. 29–48.

[2]Although the situation described is still the optimum one for most work, there are alternatives. Some workers use extremely intense, highly focused beams that are smaller than the crystal and entirely contained within it. Thus the flux distribution can be quite uneven as long as all of it sees crystal. Furthermore, a process known as local scaling (see Section 7.2) enables approximate corrections to be made for slow variations in space or time of the beam and irradiated volume.

Examination of a diffraction photograph (e.g., Fig. 5.20), shows that the diffraction spots are accompanied by other scattered radiation (*background*), which must be eliminated if true integrated intensities are to be obtained. This background can be divided into two kinds, which pose rather different problems. The first kind is caused by diffuse scattering of X-rays of all wavelengths by all objects in the beam, including air and the crystal. Although the darkening produced by this scattered radiation is not uniform over the film, it tends to vary relatively slowly and as a function of 2θ only. Thus it is relatively simple to measure the background intensity in the neighborhood of a spot to obtain a value to be subtracted from the total intensity at the spot.

A more serious background problem—if a less general one—arises from spectral impurities that lead to dark streaks (*white radiation streaks*) passing through the desired diffraction spot. The spectral distribution of copper radiation, filtered with nickel foil in the usual way, is shown in Fig. 1.4. It is obvious that considerable amounts of extraneous radiation, of both longer and shorter wavelengths, are still present. If the reciprocal lattice is defined such that the distance of any r.l. point from the origin is λ/d, the distance varies directly as the wavelength of the radiation used. (See Section 2.4). Thus for nonmonochromatic radiation an r.l. point is not a point at all but rather a sausagelike object extended along the line joining it to the origin (Fig. 6.1). The end nearest the origin corresponds to the r.l. point as defined for the shortest-wavelength component of the incident radiation; the other end corresponds to the point for the longest-wavelength component. Somewhere between is the usual r.l. point, the one calculated for the

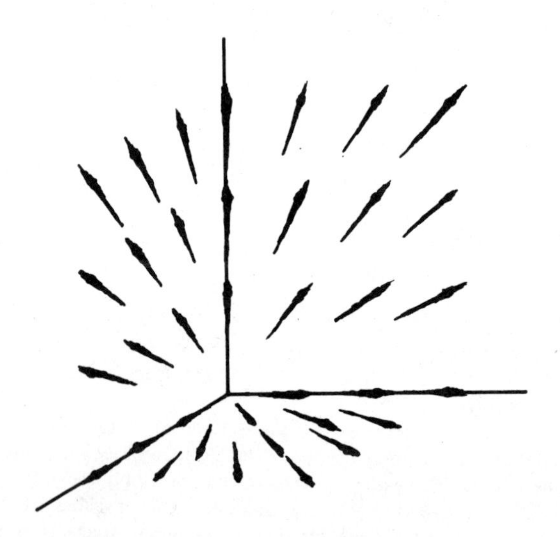

Figure 6.1. White radiation streaks in reciprocal space.

characteristic radiation being used.[3] As each portion of the elongated r.l. point passes through the sphere of reflection, the conditions of reflection are met for the corresponding wavelength. Thus a reflection appears as an elongated streak (Fig. 6.3) whose intensity varies as the intensity of the various components of the incident beam, modified by functions describing the differing efficiencies of reflection and detection for different wavelengths. It is this change in efficiency coupled with the effects of filtering that causes the dark streaks to be more pronounced on the long-wavelength side (i.e., farther from the film image of the direct beam) than on the short-wavelength side.

Since the elongated r.l. point extends radially from the origin, it is possible for it to overlap a second r.l. point that represents the next-higher order of reflection from the same family of planes; for example, the streak from 121 can overlap 242. This phenomenon is particularly apparent for reflections along the axial central lattice lines where the points tend to be the most closely spaced. The observed consequence of such overlap is that the dark streak originating from an intense axial reflection extends far enough on the film to overlap the next reflection in line and thus to provide a very significant contribution to the local background at that point. In

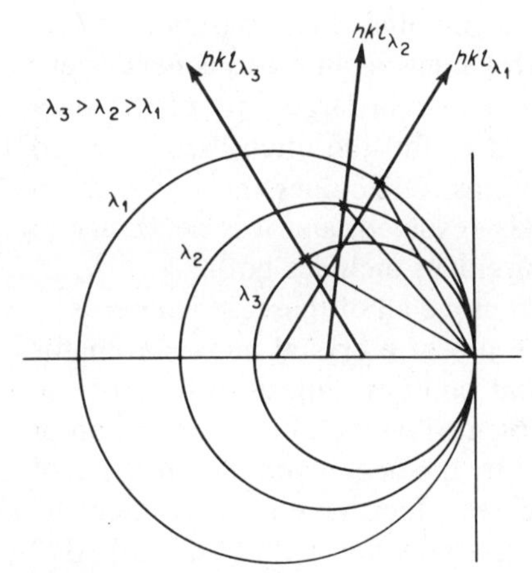

Figure 6.2. Sections of spheres of reflection for $\lambda_1 < \lambda_2 < \lambda_3$.

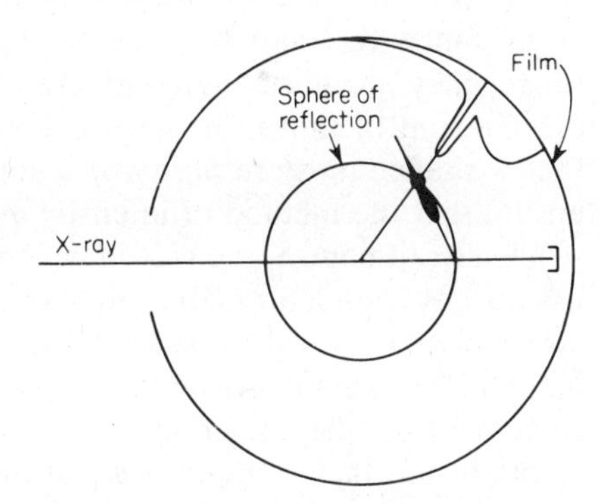

Figure 6.3. White radiation streak in reciprocal space and intensity distribution superimposed on film.

[3]The corresponding picture for the reciprocal lattice defined in terms of $1/d$ leaves the r.l. points unchanged but converts the sphere of reflection into a set of nested spheres of differing radii, all tangent at the r.l. origin. Each sphere has the radius $1/\lambda$ for some one component of the radiation, and reflection at this wavelength occurs as usual when an r.l. point passes through the sphere. Figure 6.2 shows a section through this construction. The other view is simpler, however, and is one of the principal justifications for introducing λ into the r.l. definition.

order to account for this contribution, the background should be measured at points on either side of the reflection *in the trace of the central lattice line passing through it*. Fortunately this is possible for diffractometer 2θ scans as well as for some photographs.

Two special cases of nonmonochromaticity are important. The first concerns the K_β component. If this is removed by selective filtering, the removal is rarely complete. Consequently the K_β reflection can usually be seen as a ghost near intense reflections. Normally this causes no difficulty, but as indices become higher the gap between the two reflections becomes larger and the K_β spot may be mistaken for one of interest. For example,

$$d^*_{0,10,0}(\text{Cu K}_\beta) = \frac{1.3922}{b/10} = \frac{13.922}{b} \text{ r.l.u.} \quad (6.1)$$

$$d^*_{090}(\text{Cu K}_\alpha) = \frac{1.5418}{b/9} = \frac{13.876}{b} \text{ r.l.u.} \quad (6.2)$$

so that the K_β from 0, 10, 0 will appear at very nearly the same location as the K_α reflection of 090. The latter is an odd axial reflection whose absence is often required by common space groups, so its seeming appearance could lead to a wrong space group assignment.

The second case is similar to the first but involves the separation of $K_{\alpha 1}$ and $K_{\alpha 2}$. This becomes visible only at high values of $\sin\theta$ and is never very large. Since $K_{\alpha 1}$ and $K_{\alpha 2}$ are not resolved over a large part of the data range, they are normally considered together, and the intensities recorded are the sum of those for the two wavelengths. Difficulties then arise when the two reflections are partly or completely resolved, and it is necessary to ensure that the method of intensity measurement includes both.

Ideally, of course, the way in which to remove all of these difficulties is to use strictly monochromatic radiation. The use of a crystal monochromator (see Chapter 1) will provide radiation that consists almost entirely of the $K_{\alpha 1}$ and $K_{\alpha 2}$ characteristic lines. The problems involved in the operation of such a device, the loss of intensity suffered by the beam, and the presence of errors from other sources of equal or greater effect have all prevented the general use of crystal monochromators in connection with film methods.[4] They are very commonly employed in conjunction with diffractometer measurement, however.

The problem of spot shape is simply that two reflections of the same integrated intensity may have spots that differ markedly in shape and more particularly in area. Consequently the radiation density in the two will differ, and any intensity measurement at one point only would not show them to be equal. The problem is accentuated if the density distribution over the reflections is irregular. It is for this reason that it is necessary to

[4]But see L. V. Azaroff, *Acta Cryst.*, **10**, 413 (1957).

discuss the total or integrated intensity of reflections. Admittedly, *peak height* methods, which depend on using the maximum intensity of a reflection as a measure of its integrated intensity, have been used successfully, but their success reflects rather the relative insensitivity of structural analysis to poor intensity data than any virtues of the method.

The reasons behind the variation in spot shape are numerous. Among the simplest are changes in the apparent area of the reflecting surface as a crystal rotates. The integrated intensities will not be affected by the crystal shape, however, because the same number of electrons contribute to the scattering of X-rays whether they are stacked in narrow columns or spread over broad thin sheets.

Geometric factors in the diffraction process also affect spot shapes. There is a general tendency for reflections to be larger and more diffuse at high values of sin θ than at low values. Coupled with this is the problem of the gradual separation of $K_{\alpha 1}$ and $K_{\alpha 2}$ with increasing sin θ. The angular spread over which reflection occurs also depends on the variation in orientation of the mosaic blocks (see Section 4.3) making up the crystal. In extreme cases, the crystal may be split or even multiple, leading to partially resolved peaks inside the reflection envelope. A better crystal is to be preferred, but often none is available and one must make do. If a crystal monochromator is used, it too has a mosaic spread and the two effects interact.

Unique Data

The amount of data available for any crystal is limited by the wavelength of the radiation used. Since a reciprocal lattice point must pass through the sphere of reflection in order to produce an observable reflection, only those points where the distance from the r.l. origin is less than the diameter of this sphere can be observed. Thus a sphere of radius equal to this diameter, described about the origin of the reciprocal lattice (Fig. 2.23), enclose all of the possible reflecting points for a given wavelength. Such a sphere is known as the *limiting sphere* or, since its contents vary with the radiation used, as the *copper sphere* or *molybdenum sphere*.

Not all the points within the limiting sphere represent independent observations. Because of the symmetry of the reciprocal lattice, those reflections having indices related in certain ways will have the same intensities, and only one member of the set need be measured. The portion of the data that is unique is often discussed in terms of that geometric portion of the limiting sphere that it represents (see Table 3.4).

The sphere can be considered to be cut by three planes, each containing two of the r.l. axes (Fig. 6.4). By Friedel's law, $I_{hkl} = I_{hkl}$ and the distribution of intensities is centrosymmetric (see Chapters 2 and 8). Thus it is never necessary to investigate more than a hemisphere of points cut from

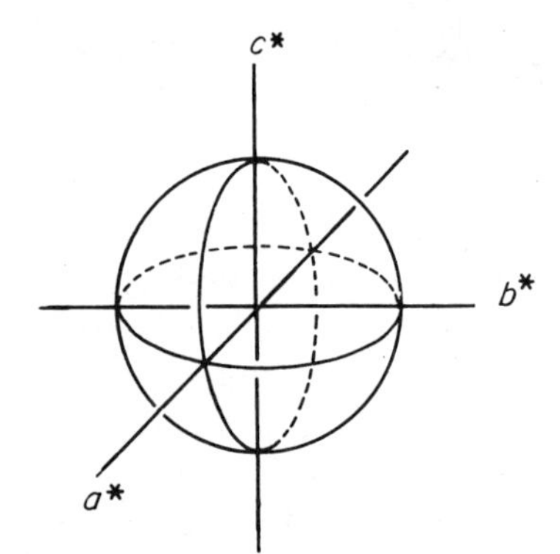

Figure 6.4. Intersection of the principal reciprocal planes with the limiting sphere.

the limiting sphere (Fig. 6.5).[5] In principle this hemisphere can be taken arbitrarily, but it is usual to select one bounded by one of the planes of Fig. 6.4.

If the symmetry of the space group is higher than triclinic, additional symmetry elements enter the picture. If Friedel's law holds, the intensity weighted reciprocal lattice of a monoclinic crystal always has the symmetry $2/m$, and the reflections with the same absolute values of h, k, and l can be divided into two sets:

$$I_{hkl} \equiv I_{\overline{hkl}} \equiv I_{\bar{h}k\bar{l}} \equiv I_{h\bar{k}l} \tag{6.3}$$

$$I_{\bar{h}kl} \equiv I_{h\overline{kl}} \equiv I_{hk\bar{l}} \equiv I_{\overline{hk}l} \tag{6.4}$$

$$I_{hkl} \neq I_{\bar{h}kl} \tag{6.5}$$

In these sets, the second intensity is related to the first by Friedel's law, the

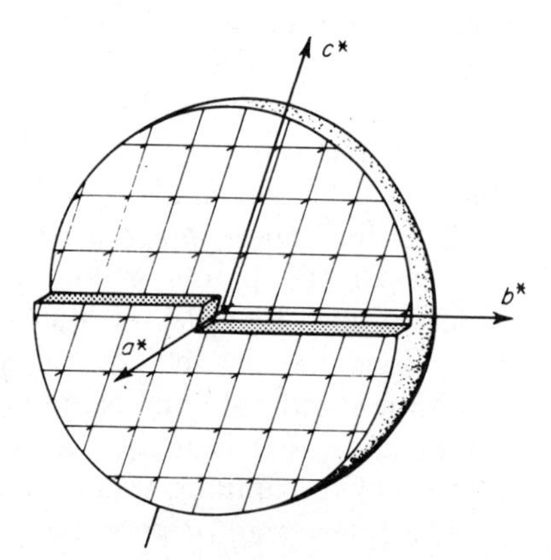

Figure 6.5. Unique volume in reciprocal space for a triclinic crystal.

[5]Except in the presence of dispersion (Section 8.13), for which Friedel's law fails.

third by rotation about the 2-fold axis parallel to b^*, and the fourth by reflection in the a^*c^* mirror plane. The corresponding fragment of the sphere of reflection is a wedge-shaped quadrant (Fig. 6.6) bounded on one side by the a^*c^* plane. Application of the various symmetry operations of $2/m$ to this quadrant can be seen to regenerate the entire sphere.

In the orthorhombic case, the symmetry is *mmm* and

$$I_{hkl} \equiv I_{\bar{h}kl} \equiv I_{h\bar{k}l} \equiv I_{hk\bar{l}} \equiv I_{hkl} \equiv I_{\bar{h}k\bar{l}} \equiv I_{hkl} \equiv I_{hkl} \tag{6.6}$$

Here only one octant is required (Fig. 6.7). For crystals of higher symmetry, further equivalences appear and the unique data may consist of even less than an octant.

As a consequence of the above arguments it is possible to see that the unique reflections whose intensities should be measured are the following:

Triclinic	One index running $0 \to \infty$
	Two indices $-\infty \to \infty$
Monoclinic	k and one of h or l running $0 \to \infty$
	The second of h or l running $-\infty \to \infty$
Orthorhombic	All indices running $0 \to \infty$

These rules hold for general (hkl) reflections, but those that have one or more indices zero may represent special cases. Thus,

Triclinic	If one index 0, one index runs $0 \to \infty$ and one $-\infty \to \infty$
	If two indices 0, the remaining index runs $0 \to \infty$
Monoclinic	For $h0l$, like triclinic
	For $hk0$ or $0kl$, both indices run $0 \to \infty$
	If two indices 0, the remaining index runs $0 \to \infty$

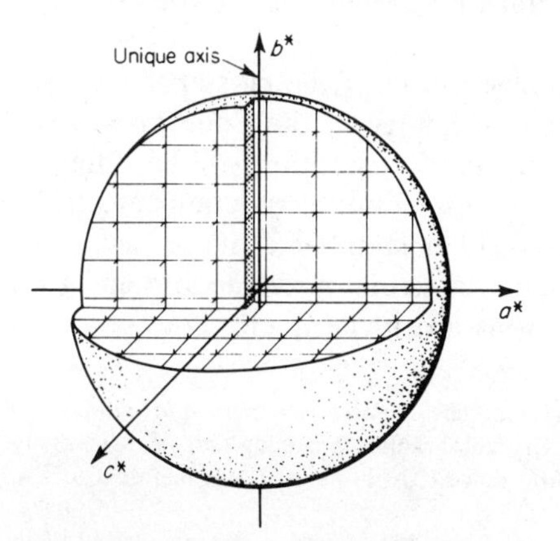

Figure 6.6. Unique volume in reciprocal space for a monoclinic crystal.

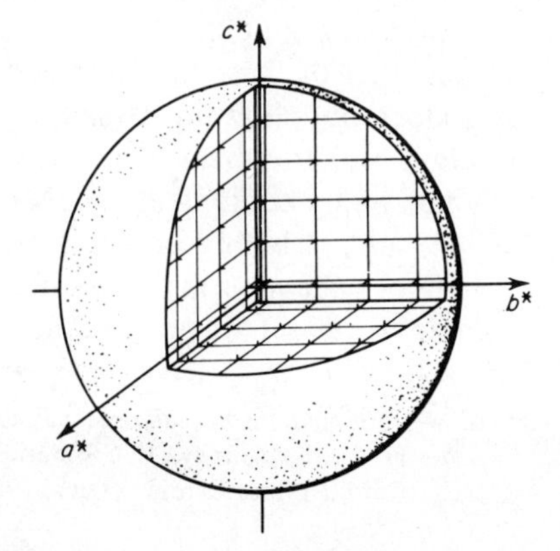

Figure 6.7. Unique volume in reciprocal space for an orthorhombic crystal.

These restrictions can be seen to follow from Friedel's law and the equalities of (6.3) and (6.4). They are illustrated in Figs. 6.5–6.7, which show the details of the exact boundaries of the regions containing the unique reflections.

For space groups of higher symmetry, the equivalent reflections are best obtained from the rotation matrices (Section 3.5) describing the appropriate point group symmetry (usually the Laue symmetry, but the true point group for cases in which Friedel's law fails). All equivalent reflections are related by

$$\mathbf{h}' = \mathbf{hR} \tag{6.7}$$

where $\mathbf{R}$ is one of the point group rotation matrices.[6] Thus for the monoclinic case (Laue group $2/m$) the equivalent reflections are

$$[h \quad k \quad l]\begin{bmatrix} 1 & 0 & 0 \\ 0 & 1 & 0 \\ 0 & 0 & 1 \end{bmatrix} = \begin{bmatrix} h \\ k \\ l \end{bmatrix} \qquad [h \quad k \quad l]\begin{bmatrix} -1 & 0 & 0 \\ 0 & -1 & 0 \\ 0 & 0 & -1 \end{bmatrix} = \begin{bmatrix} -h \\ -k \\ -l \end{bmatrix}$$

(a) identity (b) center

$$[h \quad k \quad l]\begin{bmatrix} -1 & 0 & 0 \\ 0 & 1 & 0 \\ 0 & 0 & -1 \end{bmatrix} = \begin{bmatrix} -h \\ k \\ -l \end{bmatrix} \qquad [h \quad k \quad l]\begin{bmatrix} 1 & 0 & 0 \\ 0 & -1 & 0 \\ 0 & 0 & 1 \end{bmatrix} = \begin{bmatrix} h \\ -k \\ l \end{bmatrix} \tag{6.8}$$

(c) 2-fold (c) mirror

Use of Equivalent Reflections

It would seem that obtaining a complete set of intensity data, having the minus signs restricted to the minimum number of indices, would be simplified by using the identities between various reflections. Although this is commonly done, there are potential pitfalls that might lead to difficulties in the later stages of structure refinement.

The first of these is concerned with absorption. If the crystal is irregular and does not show in its shape the full symmetry of the Laue group (see Section 3.6) to which it belongs, the effects of absorption will be different for nominally equivalent reflections. If no absorption corrections are to be made, the relabeling of a reflection as another member of its set will have no effect except to make possible systematic errors more random and thus harder to detect. If absorption corrections are to be made, however, it is

[6]For a proof and further discussion, see S. Hovmöller, *Rotation Matrices and Translation Vectors in Crystallography*, IUCR Series on Crystallographic Teaching, no. 9, University College Cardiff Press, Cardiff, Wales, 1981. Note that $\mathbf{hR}$ is, in general, not identical to $\mathbf{Rh}$. The first case requires $\mathbf{h}$ to be a *row vector* $[h \quad k \quad l]$, and the second, a *column vector* $\begin{bmatrix} h \\ k \\ l \end{bmatrix}$ in order for the proper multiplication to occur.

vital that the reflections be correctly indexed at least through the absorption calculation. Otherwise, the calculated "corrections" can actually increase, rather than lessen, the error.

Another source of difficulty, and one that is both more subtle and more important, arises when Friedel's law fails to hold and $I_{hkl} \neq I_{\bar{h}\bar{k}\bar{l}}$. In this case the symmetry of the intensity-weighted reciprocal lattice is just that of the point group to which the crystal actually belongs.

These differences between I_{hkl} and $I_{\bar{h}\bar{k}\bar{l}}$ are relatively small, usually only 5–10% at most. They are often not apparent to the eye on photographs and appear only when careful intensity measurements are made. Thus they have relatively minor effects on the results of the structural analysis (but see Section 17.6) and have often been overlooked or neglected. For the most accurate results, however, it is worth measuring both members of a Friedel pair and either using their average as a better measure of F_0 free of anomalous scattering or dealing with them independently. In this case equivalences can be assumed only between those reflections that are related in the reciprocal lattice by the true symmetry of the crystal point group. Thus for a crystal in space group $P2$ or $P2_1$ (point group 2), and hkl and $\bar{h}k\bar{l}$ reflections can be safely taken as Friedel equivalent under all conditions [see Eqs. (6.8a) and (6.8c)]. If the space group is Pc (point group m), the hkl and $h\bar{k}l$ reflections are equivalent [Eqs. (6.8a) and (6.8d)], while for the centrosymmetric $P2_1/c$ (point group $2/m$) the general equivalences of Eqs. (6.3), (6.4), and (6.8) hold.

Even if corrections for these errors are not to be made, it is probably worthwhile to measure the intensities with correct indexing,[7] changing to equivalent indices only when the data are reduced (see Chapter 7). If the data are sufficiently accurate and if various sets of *supposedly* equal reflections are measured, it may be possible to detect which ones are *actually* equal. From this information the point group symmetry of the reciprocal lattice and thus of the crystal can be inferred, and the problem of space group determination becomes greatly simplified.

Finally, it is possible at the end of the refinement of such a noncentrosymmetric crystal to determine its absolute configuration (Section 18.8). To do so, however, requires that correctly indexed experimental data be available.

Selection of Data

The range of observable intensities provided by a normal crystal is very large, sometimes as much as 100,000 to 1. This is beyond the range of linearity of any convenient measuring method, and consequently means are required to attenuate the strong reflections by known factors so that the resulting intensities fall within an easily measurable range. For diffrac-

[7]As described in Chapter 5 and by A. F. Peerdeman and J. M. Bijvoet, *Acta Cryst.*, **9**, 1012 (1956). Normally this is handled automatically by the diffractometer setting program.

tometers this is normally effected by the use of calibrated filters placed in either the incident or diffracted beam. In the case of film methods, photographs can be taken for different times. In order to reduce the number of exposures needed, however, it is common practice to use not one but several films in the camera for each exposure. In this way the film closest to the crystal removes a certain fraction of the radiation, acting as a filter before the second, which in turn reduces the intensity reaching the third, and so on.

A question that arises early in the data collection process is how many reflections are needed to ensure finding the correct structure. A normal organic crystal will usually provide about 100–150 unique reflections theoretically available to copper radiation per nonhydrogen atom. Molybdenum radiation multiplies this by a factor of 10. It is usually felt that one is well protected from forcing a grossly incorrect structure on the data by being overdetermined by a factor of 4 or 5, that is, by having four or five observations (reflections) for every adjustable parameter in the final structure. This is very much a rule of thumb, since not all parameters are equivalent, but it provides a generous margin of safety and is best adhered to if possible. The final structure will normally possess either four or nine parameters (three positional and either one or six describing the thermal motion) per atom depending on the extent to which the refinement is carried. There will also be a relatively small number of scale factors as parameters. Highly refined structures will involve parameters for hydrogen atoms as well.

Regardless of the large number of reflections theoretically available, only a fraction of these can normally be observed. The average reflection intensity decreases with increasing $(\sin\theta)/\lambda$ at a fairly rapid rate (see Section 7.3). Because of the decrease in intensity, the percentage of unobserved reflections usually rises steadily as one considers shells of increasing values of $(\sin\theta)/\lambda$. There are often very few observable reflections beyond the limit of copper data, so going to Mo K_α radiation produces little or nothing in the way of increased information. Since part of this intensity loss is due to the thermal vibration of the atoms, cooling the crystal to liquid nitrogen temperatures will often markedly increase the fraction of observed reflections, particularly for compounds with melting points not far above room temperature.[8] Also, crystals that deteriorate in the X-ray beam (e.g., proteins) generally show strikingly greater stability at low temperatures. Various devices for accomplishing this cooling have been devised,[9,10] and it is now being increasingly used, especially for unstable compounds.[11]

[8]For an example, see J. Fridrickson and A. McL. Mathieson, *Rev. Sci. Instr.*, **29**, 784 (1958).

[9]C. Altona, *Acta Cryst.*, **17**, 1282 (1964) and references cited there.

[10]R. Rudman, *Low Temperature X-Ray Diffraction*, Plenum, New York, 1976.

[11]For examples of the improvements produced by such cooling, see F. L. Hirshfeld and G. M. J. Schmidt, *Acta Cryst.*, **9**, 233 (1956).

It is customary to collect data for small molecules to approximately the limit of the copper sphere [$(\sin \theta)/\lambda \approx 0.63$] regardless of whether the data are obtained with copper or molybdenum radiation. This gives an effective resolution (see Section 2.4) of about 0.8 Å, more than enough to resolve all bonded atoms clearly. Although this is more than is needed for many purposes, modern diffractometers have reduced the effort of data collection so much that it is always preferable to collect more data rather than less. A reduced set can be selected later if desired.

Accuracy

The accuracy with which intensities can be measured depends on the methods used and the care taken. This is discussed below in connection with the various techniques available. How accurately the intensities should be measured depends, as usual, on the requirements to be fulfilled. It has been shown that even data that have been classified merely as "strong," "medium," or "weak" will give quite recognizable electron density distributions.[12] The average error in the measured intensities is reflected in the uncertainties of the parameters of the final structure, but although this is serious where accurate bond lengths are desired it is most likely to affect the gross structure. *Both theory and experience suggest, however, that the process of going from the raw data to a correct structure is made considerably more difficult by poor intensity measurements.*

6.2. FILM METHODS FOR INTENSITY MEASUREMENT

In the past, most film intensity data were obtained from measurements on Weissenberg photographs. Precession photographs have been used for providing cross scaling (Section 7.2) between Weissenberg levels.[13] More recently, protein crystallographers have returned to the older oscillation photographs, with automatic indexing being provided by a computer program.[14,15] This approach reduces the time required for collecting complete data, since all the reflections during a given oscillation are recorded, whereas in the Weissenberg method the majority are absorbed by the layer line screen. Regardless of how the photographs are obtained, however, the methods used for measuring the intensities are generally similar.

[12]J. Lukesh, *J. Appl. Phys.*, **18**, 493, (1947). For still more extreme examples, see R. Srinivasan, *Proc. Ind. Acad. Sci.*, **A53**, 252 (1961).

[13]The precession camera has also been used in protein studies for data collection since its limited range of sin θ is no deterrent in this case.

[14]Arndt and Wonacott, *The Rotation Method in Crystallography*.

[15]For an excellent exposition of the practical details, see M. F. Schmid, L. H. Weaver, M. A. Holmes, M. G. Grütter, D. H. Ohlendorf, R. A. Reynolds, S. J. Remington, and B. W. Matthews, *Acta Cryst.*, **A37**, 701 (1981); see also T. J. Greenhough and F. L. Suddath, *J. Appl. Cryst.*, **19**, 400 (1986).

Two problems arise in estimating integrated intensities from the blackening produced on diffraction photographs. The first is that of ensuring that the proportionality between the blackening and the amount of incident radiation is known, and the second lies in devising a method for measuring the total blackening produced. The first is a minor difficulty, since the optical density $[=\log(I_0/I)]$ of film treated with X-rays is linear over a useful range of exposures. This behavior, which is very different from the nonlinear response of the same film to light (Fig. 6.8), arises because the absorption of a single X-ray quantum is enough to sensitize a silver halide grain to development. Thus if the processing conditions are held constant and the exposures are kept at a level at which no grain is likely to receive more than one quantum,[16] the total amount of silver produced and hence the blackening will be directly proportional to the number of quanta diffracted.[17] The linearities of response of a large number of films have been examined,[18] and the most commonly used films are adequately linear (i.e., ±2–3%) over a range of at least 1–64.

The second problem, that of measuring the total blackening, is more severe, and a number of solutions have been proposed. Some of these, such as the direct measurement of the amount of silver in a spot by chemical or radioactivation analysis, are impractical for large numbers of reflections and are never used today.[19] The current methods all depend on light absorption by the deposited silver grains and in modern practice are always carried out with a photometer, although visual estimation of spot darkening was the

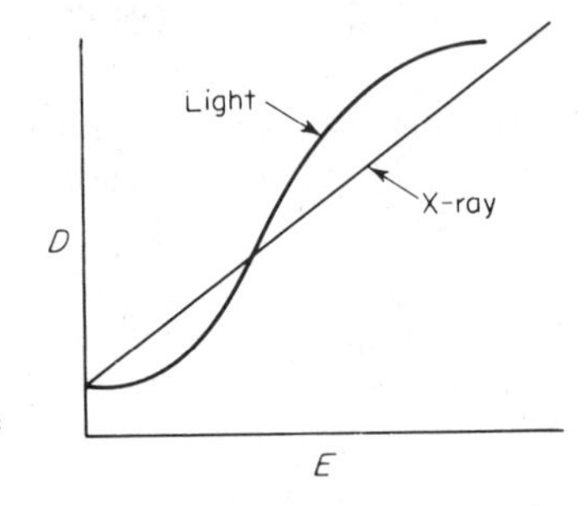

Figure 6.8. Photographic response curve, visible light, and X-rays.

[16]This restriction limits the intensity range that can be measured on a single film.

[17]For a discussion of the inherent accuracy of photographic recording and the limitations of film methods, see K. M. Rose and J. W. Jeffery, *Acta Cryst.*, **17**, 21 (1964); see also U. W. Arndt, D. J. Gilmore, and A. J. Wonacott, in Arndt and Wonacott, *The Rotation Method in Crystallography*, pp. 207–218.

[18]Commission on Crystallographic Apparatus of the International Union of Crystallography, *Acta Cryst.*, **9**, 520 (1956); H. Morimoto and R. Uyeda, *Acta Cryst.*, **16**, 1107 (1963); S. Abrahamsson, O. Lindqvist, L. Sjölin, and A. Wlodawer, *J. Appl. Cryst.*,**14**, 256 (1981); W. C. Phillips and G. N. Phillips, Jr., *J. Appl. Cryst.*, **18**, 3 (1985).

[19]For a much more extensive discussion of various methods for measuring film densities, see Buerger, *Crystal Structure Analysis*, pp. 78–111.

original technique for estimating intensities from films. Although it might at first appear to be a "peak height" method, because of the natural integrating properties of the eye it gives results that are surprisingly close to true integrated intensities.[20]

Photometric methods all replace the human eye by an optical device for measuring the extent of film blackening. Early techniques involved modifying the camera to generate reflections that would give an integrated intensity more or less directly,[21,22] but the modern approach uses an integrating densitometer that scans either the entire film or selected areas of it with a very tiny light beam (50–200 μm in diameter).[23] The output from the detector during this scan is passed through a logarithmic device to convert it to optical density and is then totaled. Individual reflections are generally identified by a computer program, which must make decisions about the boundaries of the spot and background corrections similar to those of a diffractometer area detector (see below).[24]

The accuracy of photometered intensities appears to run about 5% for the best data and 8–10% for average results. In one case the average deviation in intensities among determinations made from equivalent reflections on a zero-level Weissenberg photograph was 3%, but such a case is unusually favorable, and the precision is not likely to be maintained throughout the collection of full three-dimensional data.[25]

6.3. COUNTER METHODS FOR INTENSITY MEASUREMENT[26–31]

Single Detector

Despite the successes originally achieved by crystallographers using film, intensities are now generally measured by direct counting of the diffracted

[20]For practical details, see G. H. Stout and L. H. Jensen, *X-Ray Structure Determination*, Macmillan, New York, 1968, pp. 166–168.

[21]W. A. Wooster, *Acta Cryst.*, **17**, 878 (1964).

[22]G. H. Stout and L. H. Jensen, *X-Ray Structure Determination*, Macmillan, New York, 1968, pp. 168–176.

[23]A. J. Wonacott and R. M. Burnett in Arndt and Wonacott, *The Rotation Method in Crystallography*, pp. 119–138.

[24]M. J. Ross and R. M. Stroud, *Acta Cryst.*, **A33**, 500 (1977).

[25]See J. W. Jeffery and A. D. Whitaker, *Acta Cryst.*, **19**, 963 (1965) and J. W. Jeffery, *Acta Cryst.*, **A25**, 153 (1969) for discussions of some of the experimental requirements for accurate photographic intensity measurements.

[26]U. W. Arndt, *Acta Cryst.*, **17**, 1183 (1964).

[27]S. C. Abrahams, *Acta Cryst.*, **17**, 1190 (1964).

[28]L. E. Alexander and G. S. Smith, *Acta Cryst.*, **17**, 1195 (1964).

[29]U. W. Arndt and D. C. Phillips, *Acta Cryst.*, **14**, 807 (1961).

[30]L. E. Alexander and G. S. Smith, *Acta Cryst.*, **15**, 983 (1962).

[31]U. W. Arndt and B. T. M. Williams, *Single Crystal Diffractometry*, Cambridge University Press, Cambridge, 1966.

photons. There are two reasons for this: first, the desire to improve the accuracy of intensity measurements so as to allow the drawing of more refined structural conclusions, and second, the increasing need for automated methods for measuring very large numbers of reflections with moderate accuracy.

The desired intensity information can be obtained by a number of procedures. For example, the diffracting point can be placed on the sphere of reflection, allowed to remain there for a specified period of time, and the recorded counts taken as the intensity. For a number of reasons, among them the difficulty of obtaining the required precision of setting, this is not recommended, although it has been used. The crystal, and thus the r.l. point, might be rotated at a constant speed and the maximum intensity found used as the intensity measure. Because spot shape variation occurs in diffractometry as well as in photographic techniques, errors will arise from such peak height measurements. Finally, the crystal can be rotated at a slow enough speed[32] that a reasonable number of quanta are diffracted and detected during one pass through the reflecting position. If the counter is kept in position to receive the diffracted ray during the entire period, the conditions are approximately those appearing in our definition of an integrated intensity, and so the total number of counts should be what is desired. It is this last method that is best suited to the general operation of a diffractometer.

The scanning process can, however, be treated in two different ways, commonly called the ω and 2θ methods. The principles involved are shown in Figs. 6.9 and 6.10. In the ω scan, the counter is held stationary while the crystal, and thus the reciprocal lattice, is rotated by the ω circle to carry the lattice point into the sphere of reflection. Consequently the portion of reciprocal space that is observed as it passes through the sphere lies on an arc passing through the desired lattice point and having the lattice origin as a center (Fig. 6.9). In the 2θ technique, the counter moves as well, but at an angular rate that is just twice that of the crystal, that is,

$$\Delta 2\theta = 2\,\Delta\omega \tag{6.9}$$

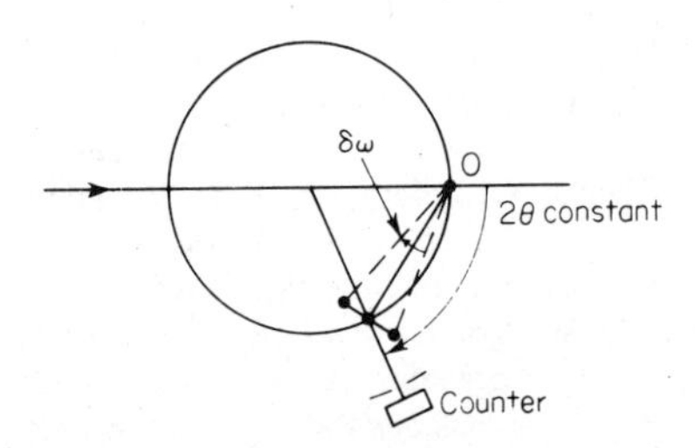

Figure 6.9. An ω scan.

[32]A variant of this method steps the crystal and counter through the desired range, pausing to count for a preset time after each angular increment.

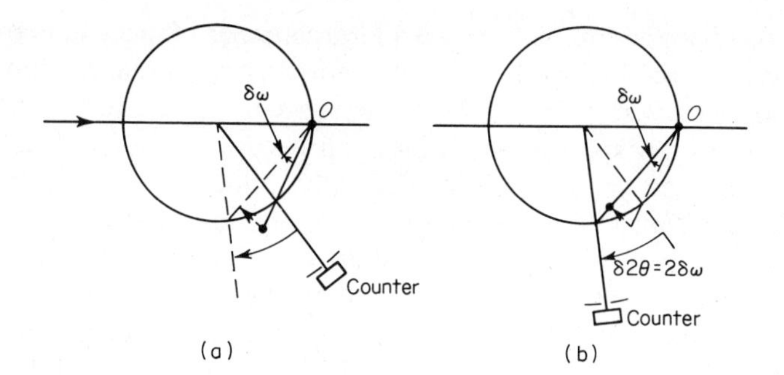

Figure 6.10. An ω–2θ scan. (*a*) Starting position; (*b*) final position.

As a result (see Section 5.4), the region scanned lies along the line connecting the lattice point with the origin (Fig. 6.10). The ω scan is similar to a photometric trace made parallel to the central line of the film on a Weissenberg photograph, while the 2θ corresponds to a similar trace made along the diagonal line passing through reflections on a common central lattice row.

Under ideal conditions and using strictly monochromatic radiation there is little to choose between the two methods. With conventional tubes and filters, however, there is always the problem of extraneous radiation to be considered and the possibility that one reflection may lie on the white radiation streak of another. An ω scan cuts across the white streaks (which, as pointed out earlier, always lie along central lattice rows) and thus does not reveal their presence. A 2θ scan, on the other hand, examines just these lines, and so the high and irregular nature of the background can be observed. For this reason the use of the 2θ scan is generally to be preferred.[33]

The proper method of carrying out the scan is open to discussion.[34] The most widely used technique is to scan through the reflection (usually over a 2θ range of 1.5–2.5°, generally increasing somewhat with 2θ to allow for the gradual resolution of the α_1 and α_2 lines) in a fixed period (often 20–150 sec) or at a constant speed (1–2°/min). The corresponding background is then obtained from the sum of two counts, each for half the scan time, taken with the apparatus stationary at the beginning and end of the scan range [*background–peak–background* (BPB) method]. It is often tempting to save time by reducing the time spent counting the backgrounds, but, especially for weak reflections, the uncertainty in the background correction contributes significantly to the uncertainty in the final intensity and becomes increasingly important as the background time is decreased (see Section 6.4 and Appendix E). An alternative method that is

[33]R. D. Burbank, *Acta Cryst.*, **17**, 434 (1964).

[34]For references and discussion, see A. McL. Mathieson, *Acta Cryst.*, **A38**, 378 (1982).

statistically advantageous is to make a large number of measurements of the background between r.l. points. These are used to generate a highly precise mapping of the average background in terms of the diffractometer setting angles, from which appropriate values can be drawn for each reflection.[35] This approach has the advantage of sharply reducing the random error in the background but at the cost of giving up any information about specific local variations (e.g., white streaks). Which method is best will depend on the actual distribution of intensities and backgrounds in the problem at hand, but in general BPB is preferable for stable small molecules with intense reflections and fewer limitations on counting time, while the averaged background offers advantages for macromolecules in which most reflections are weak and the background is a sizable fraction of many observed intensities.

A method that offers superior correction for background of all kinds but at the cost of twice the counting time per reflection is the use of *balanced filters*. In this technique, the reflection is scanned and counted in the usual way using radiation filtered through the conventional β filter (e.g., nickel for Cu K_α). A second scan is then made using an α filter, that is, one whose absorption edge is at a wavelength just longer than the K_α radiation being used (cobalt for Cu K_α). If the thicknesses of the filters are adjusted properly, their transmissions at noncharacteristic wavelengths can in principle be made nearly identical, and so the difference in counts recorded between the two scans will represent the amount of characteristic radiation diffracted (Fig. 6.11). Unfortunately, a large part of the general background consists of randomly scattered characteristic radiation, and this is also affected by the α filter. Consequently the most accurate correction for background requires not only two scans, but also four stationary half-counts (two for each scan) to determine the general background levels. Because of this doubling of time as well as the practical difficulties of adjusting the filters, they are rarely used, especially since the widespread adoption of crystal monochromators.

Other methods that are used to reduce the effects of extraneous radiation involve the recognition by the counter of photons that have too much or too little energy and their rejection from the counting process (*pulse height discrimination*). While this sort of rejection is useful for reducing the effective intensity of short-wavelength white radiation [particularly with copper radiation (Fig. 6.12); the molybdenum characteristic lines are too close to the white maximum], it is not practical to provide sharp cutoffs by this method, and it cannot replace the use of filters.

Obviously, the use of crystal-monochromatized radiation is the most elegant solution to the problem in principle, and monochromators are

[35] M. Krieger, J. L. Chambers, G. G. Christoph, R. M. Stroud, and B. L. Trus, *Acta Cryst.*, **A30**, 740 (1974).

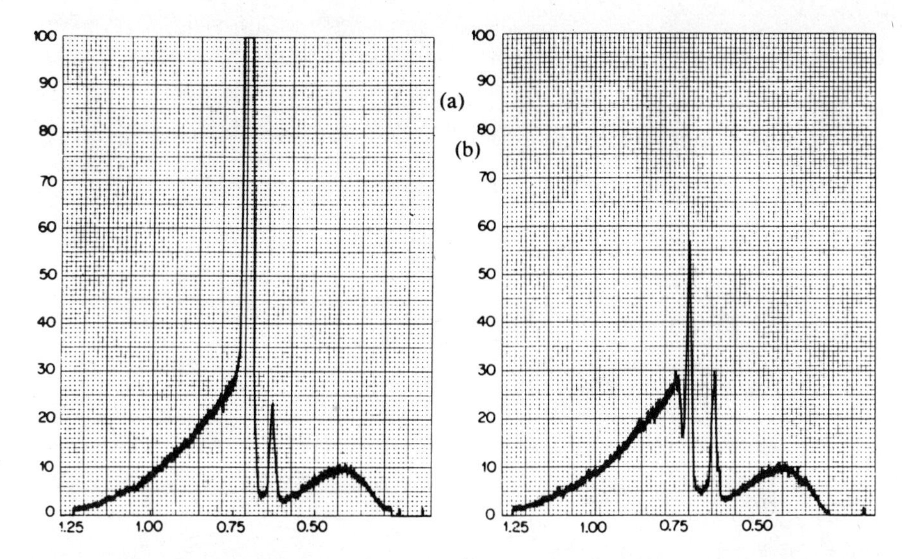

Figure 6.11. (*a*) Diffractometer tracing (2θ scan) 020 reflection of thiocytosine. Molybdenum radiation, 0.0025-in. zirconium β filter. (*b*) Same as (*a*) but with a balanced yttrium α filter.

widely used. They can produce other problems, however,[36] and alternative methods are still used as well.

Since the net count for any peak is given by the total counts measured for the peak minus the counts for the background, and since the precision of this difference decreases with increasing background intensity, it is advantageous to keep the background contribution as small as possible.[37] One way this can be done is to shrink the receiving aperture to minimize the amount of background seen around the reflection. It is vital, however, that no reflection be limited by this aperture. The size of the aperture that can be used will depend on the accuracy of alignment of the crystal and its mosaic spread, but in many cases an angular opening of 1–1.5° (as compared to a common fully opened value of ~3°) is satisfactory.

It is possible to overload a counter on intense reflections and observe only a fraction of the total diffracted photons. The counting rate at which this becomes a problem will depend on the counter used, the associated circuitry, and the error that can be tolerated, but for scintillation counters with pulse height discrimination a reasonable upper limit is ~10,000–20,000 counts/sec. Reflections whose peaks run over the value must be reduced in intensity by the use of absorbers, often one or more thicknesses of aluminum foil mounted in a holder that can be placed in front of the

[36]See S. Harkema, J. Dain, G. J. von Hummel, and A. J. Reuvers, *Acta Cryst.*, **A36**, 433 (1980).
[37]See A. J. C. Wilson, *Acta Cryst.*, **A36**, 929 (1980).

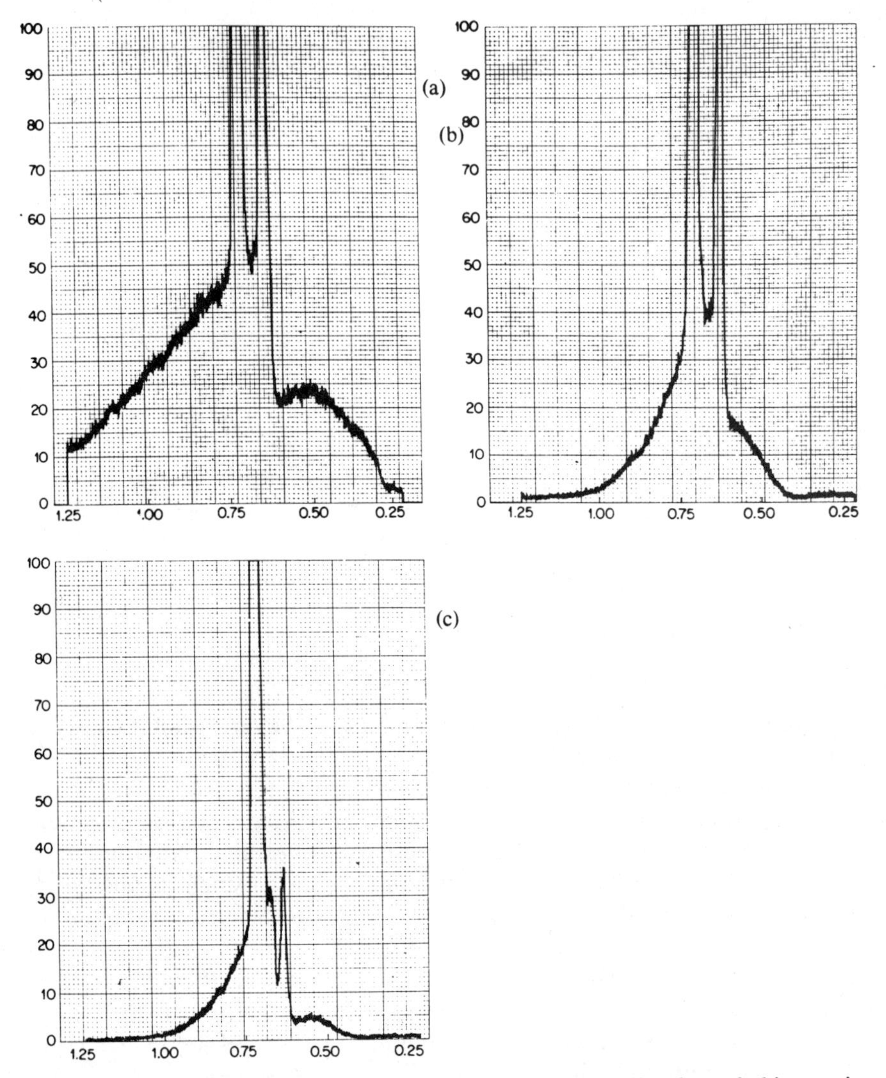

Figure 6.12. (*a*) Diffractometer tracing (2θ scan), 020 reflection of thiocytosine; molybdenum radiation. Shows characteristic peaks surrounded by background of white radiation. (*b*) Same as (*a*) but with pulse height discrimination. (*c*) Same as (*a*) but with pulse height discrimination and 0.001-in. niobium β filter. Note the great improvement in effective monochromaticity.

detector. The filter factors can be determined from the measurement of a number of reflections with and without the filter. A plot of the observed factors versus the unfiltered counting rate should be nearly constant at low count rates and then begin to fall off as counter overloading becomes serious. Such a curve can be used to decide the point at which the use of filters is needed to keep this error to a desired value.

The practical precision (*not* accuracy) of intensity measurements with a diffractometer seems to be about 1–3% for the best work and somewhat higher for routine operations.[38]

At these levels one must begin to worry about a number of effects (absorption, Renninger effects,[39] variation in crystals, etc.) that have often been ignored in the past, if full use is to be made of the possibilities for accuracy inherent in the instrument.

One means that is highly recommended for improving the absolute accuracy of a set of intensity measurements is to remeasure at least three or four selected standard reflections several times a day. The variations in the intensities found gives an indication of the working precision, and any systematic changes due to long-term drift in the equipment or changes in the crystal can be detected. If these changes are not too large (e.g., <20% of total intensity), accuracy adequate for most work can be obtained by using the standard reflections to define a normalization curve to bring reflections measured at various times onto a common basis.

Area Detector[40,41]

The rapidly increasing interest in the X-ray crystallography of macromolecules that followed the successful structure elucidation of myoglobin and hemoglobin and has now extended to intact viruses has led steadily to the study of larger and larger unit cells. These not only produce greater total numbers of reflections to be measured, but also cause more to be reflecting at any given time, greatly increasing the information loss attendant on measuring them singly. It was for this reason that many protein crystallographers returned to film methods, giving up the greater accuracy of the diffractometer for the practical advantages of measuring many reflections simultaneously.

It was widely appreciated that there would be great advantages to a device that combined the sensitivity, accuracy, and convenience of the diffractometer with the ability to measure many reflections at once. A number of different designs of detection systems have emerged, and the resulting practical instruments are beginning to be widely distributed. The currently most widespread area detector is the *multiwire proportional chamber* (MWPC)[42] derived from the standard proportional radiation

[38]See B. J. Wuensch, *Acta Cryst.*, **16**, 1259 (1963), but also S. C. Abrahams and J. L. Bernstein, *Acta Cryst.*, **18**, 926 (1965); L. E. McCandlish, G. H. Stout, and L. C. Andrews, *Acta Cryst.*, **A31**, 245 (1975); and W. A. Denne, *Acta Cryst.*, **A28**, 192 (1972).

[39]See R. D. Burbank, *Acta Cryst.*, **19**, 957 (1965); R. A. Young, *Acta Cryst.*, **A25**, 55 (1969).

[40]U. W. Arndt and A. R. Faruqi, in Arndt and Wonacott, *The Rotation Method in Crystallography*, pp. 219–226; A. R. Faruqi, *ibid*., pp. 227–243; U. W. Arndt, *ibid*., pp. 245–261.

[41]U. W. Arndt, *J. Appl. Cryst.*, **19**, 145 (1986); R. Hamlin, C. Cork, A. Howard, C. Nielsen, W. Vernon, D. Matthews, Ng H. Xuong, and V. Perez-Mendez, *J. Appl. Cryst.*, **14**, 85 (1981); R. Hamlin, *Trans. Am. Crystallogr. Assoc.*, **18**, 95 (1982).

[42]Nicolet/Xentronics, Nicolet Corp., Madison, Wisconsin; and ADSC, Area Detector Systems Corp., San Diego, California.

counter. In this device two or three grids of fine wires supported in a xenon-filled chamber serve to detect X-ray quanta passing near their crossing points and return count data together with wire identification allowing the geometric location of the reflection. A second method[43] uses a fluorescent screen together with optical fibers and photodetectors to convert the X-rays to visible light and count the resulting light. A third, which appears to hold much promise but has not yet been so widely examined, uses *imaging plates*, films in which the X-rays generate energy centers in Eu^{2+}-doped barium halide crystals suspended in a plastic matrix. These centers can later be stimulated by laser light to emit their energy as countable photons in a process that mimics the photometry of conventional films but has the advantages of greater sensitivity, much greater linear range, reusability of the film, and the absence of the chemical processing step.[44]

Regardless of the actual device used, the techniques of data collection with an area detector are generally similar and resemble those used for oscillation photographs. The crystal is mounted at an appropriate angle, generally one that places an axis along the diffractometer ω axis, and is rotated about ω (or at $\chi = 0°$, about ϕ) very slowly or in very small steps. A total rotation of a few hundredths to a few tenths of a degree constitutes a "frame" corresponding to a single photograph. The process is continued until enough frames have been accumulated to ensure that one or more measurements have been made of each set of equivalent reflections accessible at that crystal setting.

Because most area detectors are flat rather than cylindrical, they sample a smaller area than oscillation films do, and a number of scanning runs must be made to measure most of the reflections to a desired resolution.[45] This has secondary advantages, however. Area detectors suffer in general from greater variability over their active surface than does film, and the multiple scans tend to give a number of replicate measurements of most or all of the reflections. These can be used to check the accuracy of the corrections used to scale values from different parts of the detector. In addition, they allow the recognition of individual measurements that are grossly in error (e.g., by being split between one frame and the next although they were predicted to be entirely in one) and provide estimates of the accuracy (or at least the reproducibility) of the observed intensities.

Area detectors generate data at such high rates that it is easily possible to fill a 500-Mbyte disk with frame data in a few days. For this reason there

[43]FAST system, Enraf-Nonius, Delft, Holland.

[44]J. Miyahara, K. Takahashi, Y. Anemiya, N. Kamiya, and Y. Satow, *Nucl. Instr. Methods*, **A246**, 527 (1986).

[45]For discussions see Ng H. Xuong, C. Nielsen, R. Hamlin, and D. Anderson, *J. Appl. Cryst.*, **18**, 342 (1985) and H. Blum, P. Metcalf, S. C. Harrison, and D. C. Wiley, *J. Appl. Cryst.*, **20**, 235 (1987).

has been a tendency for the collection programs to do the preliminary combination of individual frame pixels into reflections on-line and discard the uninteresting portions of the frame. Although this may be necessary for reasons of storage, it is always risky to discard raw data, and it is to be hoped that the advent of storage devices with massive capacities will soon allow the archiving of frames for examination off-line.

One serious limitation of most current area detectors is that they are very sensitive to overloading and coincidence losses. This arises because the electronics have to cope not only with possibly coincident quanta arriving at a particular pixel, but also with any other quanta striking anywhere on the active surface. Thus the counting rate is limited to less than 50,000 counts/sec *over the entire detector* as well as lesser figures for any one pixel. A surprising amount of this counting limit can be lost to background, even in the absence of any particularly strong reflection in the frame.

At present, area detectors are specialist devices, extremely expensive and of interest only to those wishing to collect immense amounts of relatively weak data, but 25 years ago the same would have been said of diffractometers.

6.4. INTEGRATED INTENSITIES

When the raw intensities are obtained by visual estimation of photographic spots or by integration of continuous photometer traces through a reflection and its adjacent background, there usually appears to be a clear distinction between peak and background. If not, the peak is often defined as unobserved. With diffractometer data it is generally desirable for decisions about peak intensities to be made by computer program rather than a human.

A fundamental limitation on the accuracy of any diffractometer intensity derived by subtracting a background count from a peak count lies in the random nature of the counting process. For observations of 100 counts or more, the probability of observing a particular value is approximately normally distributed about the mean (true?) value N with a standard deviation equal to $\sqrt{N}$. This applies to the background observations as well as to the peak, and if the backgrounds were measured for the same total time,

$$I = N_{\mathrm{p}} - N_{\mathrm{b}} \tag{6.10}$$

$$\sigma_I = (\sigma_{\mathrm{p}}^2 + \sigma_{\mathrm{b}}^2)^{1/2} \tag{6.11}$$

$$\sigma_I = (N_{\mathrm{p}} + N_{\mathrm{b}})^{1/2} \tag{6.12}$$

Note that although the net counts are the *difference* between the peak and background measurements, the squared uncertainty (*variance*) is the *sum*.

Thus the greater the background contribution or smaller the net counts, the greater the percentage uncertainty. It is for this reason that it is so desirable to minimize the total background included in the volume of reciprocal space actually scanned, and that the use of precisely known, averaged backgrounds is attractive.

Many diffractometer programs offer the option of "constant precision" data measurement, in which the measurement of a single peak is continued until a chosen number of counts are accumulated, but it is questionable if this is worthwhile in most cases. Although it may provide a fixed precision for a given measurement, the net counts can still be very uncertain unless the same time is spent measuring the background as well. Furthermore, it is rarely necessary to know very weak reflections precisely; most of the information is obtained once it is known that they are indeed very weak.

More recently, the problem of selecting the limits of a reflection, assigning the proper background, and estimating the total integrated intensity has been changed by the need for computerized fitting of data collected in steps (step scans, pixels on area detectors, even digitized output from film scanners).[46] For intense reflections that have been stepped entirely across, it is simple to add the counts in the n steps, subtract n backgrounds, and use the sum as the intensity. For weak reflections or ones that have been measured only in the region of the peak (to maximize the number of counts collected in a given time), the process is more difficult.

Because of the uncertainties imposed by the $\sqrt{N}$ limitation, it is tempting to concentrate measurements in the regions near the peak, where the greater counting rate gives a reasonable number of events during a fixed counting time. If only a part of the peak is measured, however, one has to make assumptions about the missing intensity (e.g., that it is the same fraction for all peaks), and if these are wrong one risks converting random uncertainties into systematic errors. One technique that has been used repeatedly[47] and that seems to offer advantages for data sets containing a large fraction of weak reflections (e.g., proteins) is to fit the profile of the reflections with an assumed function that depends on a few parameters. These parameters, which describe the width of the peak, its exact position in the series of steps measured, and perhaps details of the function, are assumed to change only slowly with time and orientation. Thus they can be "learned," and information can be carried from stronger, precisely measured reflections to fit neighboring weaker ones.[48]

[46]For a discussion, see D. Taupin and J. Nogues, *J. Appl. Cryst.*, **14**, 478 (1981).

[47]J. C. Hanson, K. D. Watenpaugh, L. Sieker, and L. H. Jensen, *Acta Cryst.*, **A35**, 616 (1979); W. Clegg, *Acta Cryst.*, **A37**, 22, 437 (1981); S. Oatley and S. French, *Acta Cryst.*, **A38**, 537 (1982).

[48]For similar techniques extended to area detector data, see L. Sjölin and A. Wlodawer, *Acta Cryst.*, **A37**, 594 (1981) and references cited there.

BIBLIOGRAPHY

Arndt, U. W., and A. McL. Mathieson, Eds., "Conference Report on Accurate Determination of X-Ray Intensities and Structure Factors," *Acta Cryst.*, **A25**, 1–276 (1969).

Arndt, U. W., and A. J. Wonacott, Eds., *The Rotation Method in Crystallography*, North-Holland, Amsterdam, 1977.

Buerger, M. J., *Crystal–Structure Analysis*, Wiley, New York, 1960, Chapter 6.

Hope, H., in *Anomalous Scattering*, S. Ramaseshan and S. C. Abrahams, Eds., Munksgaard, Copenhagen, 1975, pp. 293–305.

Jeffery, J. W., *Methods in X-Ray Crystallography*, Academic, London, 1971, pp. 256–279.

Luger, P., *Modern X-Ray Analysis on Single Crystals*, de Gruyter, New York, 1980, pp. 196–204.

Wyckoff, H. W., C. H. W. Hirs, and S. N. Timasheff, Eds., *Diffraction Methods for Biological Macromolecules*, Part A, *Advances in Enzymology*, Vol. 114, Academic, New York, 1985, pp. 199–510.

CHAPTER 7

DATA REDUCTION

Intensity data collected as described in Chapter 6 constitute the raw material from which crystal structures are derived. In most cases they represent all the information that will be obtained from physical measurements on the crystal, and the subsequent development of a structure will depend on the skillful extraction of the information contained within the intensities. It is the preliminary manipulation of these intensities—their conversion to a corrected, more generally usable form, which is referred to as *data reduction*—with which we shall deal in this chapter. In addition, we shall consider some useful conclusions that can be drawn from these intensities without the aid of a known structure.

7.1. THEORY OF LORENZ AND POLARIZATION CORRECTIONS

As will become evident in the following chapters, the most important quantity derived from the intensities is the structure factor modulus (structure amplitude), $|F_{hkl}|$.[1] This quantity is related to the experimentally observed intensities[2]

$$|F| \propto \sqrt{I} \tag{7.1}$$

[1]The symbol $|x|$ is used to indicate "absolute value of x."

[2]Properly speaking, this proportionality holds only for crystals that have a mosaic structure (see Sections 4.3 and 17.6), but since this is so generally the case it is usually assumed without comment. Buerger, *Crystal–Structure Analysis*, pp. 195–202, and James, *The Optical Principles of the Diffraction of X-Rays*, pp. 44–66.

and at the same time can be calculated theoretically once the positions of the atoms in the cell are known. Furthermore, the structure factors are the quantities that are used in the calculation of electron density maps from which the positions of atoms can be determined. For these reasons, it is customary to convert the intensities into "observed" structure amplitudes ($|F_o| \equiv |F_{observed}|$) by a data reduction program and to use these as the observed data in subsequent calculations.

The relationship between $|F_o|$ and I depends on a number of factors, primarily geometric, that are related to the individual reflection and to the apparatus used to measure its intensity. The proportionality of Eq. (7.1) can be rewritten in a more informative fashion as[3]

$$|F_{hkl}| = (KI_{hkl}/Lp)^{1/2} \tag{7.2}$$

Here p, the *polarization factor*, is given by

$$p = (1 + \cos^2 2\theta)/2 \tag{7.3}$$

and is a simple function of 2θ, independent of the method of data collection except when a crystal monochromator is used. The *Lorentz factor* L depends on the precise measurement technique used. For the equi-inclination Weissenberg it is given by

$$p = \frac{\sin\theta}{\sin 2\theta(\sin^2\theta - \sin^2\mu)^{1/2}} \tag{7.4}$$

where μ is the equi-inclination setting angle. For zero-level reflections, Eq. (7.4) reduces to

$$L = \frac{1}{\sin 2\theta} \tag{7.5}$$

and this expression is applicable to zero-level Weissenberg photographs, the zero-layer reflections of rotation and oscillation photographs, and diffractometer data obtained by the usual 2θ or ω scans.

The term K in Eq. (7.2) depends on crystal size, beam intensity, and a number of fundamental constants. It is normally a constant for any given set of measurements and is commonly omitted from the data reduction calculations. Thus the results obtained are relative $|F|$ values ($|F_{rel}|$'s) defined by

$$|F_{rel}| = k'|F_o| = (I_{hkl}/Lp)^{1/2} \tag{7.6}$$

[3]For an analysis of the origins of these formulas, see Buerger, *Crystal–Structure Analysis*, pp. 152–194.

The scaling between $|F_{rel}|$ and $|F_o|$ is usually obtained at a later stage (see Chapter 9) by comparison of the $|F_{rel}|$'s with the $|F|$'s ($|F_c|$'s) calculated on the basis of the structure found. As this procedure adds only one additional parameter, it does not seriously affect the accuracy of the final results.

The polarization term p arises because of the nature of the X-ray beam and the manner in which its reflection efficiency varies with the reflection angle. The usual X-ray beam is unpolarized; that is, the electric vectors associated with its photons can point in any direction normal to the direction of propagation (Fig. 7.1). Each of these vectors can be considered in terms of its components in two directions, one parallel to the surface of the reflecting plane ($I_{\parallel}$ or π beam)[4] and the other at right angles to the first ($I_{\perp}$ or σ beam) (Fig. 7.2), and because of the random orientation the sums of these two components will be equal. Waves having their electric vector parallel to the reflecting plane are reflected to an extent that is determined only by the electron density in the plane and is independent of the reflection angle 2θ. The reflection of the other vector depends on both the electron density and $\cos^2 2\theta$ and therefore decreases to zero at $2\theta = 90°$. Because the initial energy is equally divided between the two electric vectors, no more than half the intensity will be lost to this effect, however.

Because of the greater efficiency of reflection of $I_{\parallel}$, it will be represented to a greater extent in the reflected beam than is $I_{\perp}$, and so the beam will be partially polarized.[5] Normally this causes no difficulty, but the partial polarization of an X-ray beam from a crystal monochromator must be taken into account because it will affect the subsequent reflection from the crystal under study. In this case the polarization correction is

$$p = \frac{1 + K\cos^2 2\theta}{1 + K} \tag{7.7}$$

where K is the ratio of the power in the π beam to that in the σ beam in the

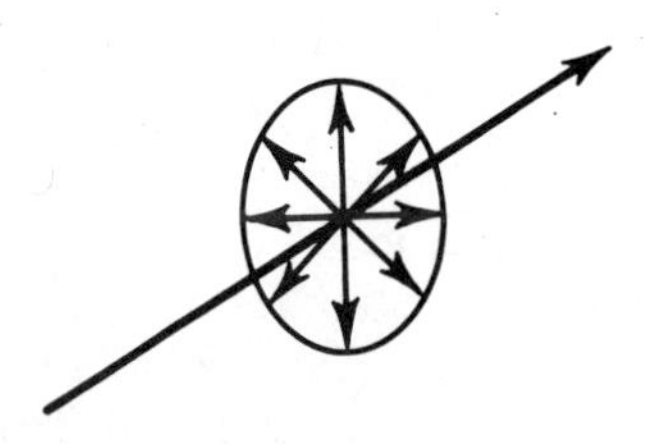

Figure 7.1. Electric vectors in an unpolarized X-ray beam.

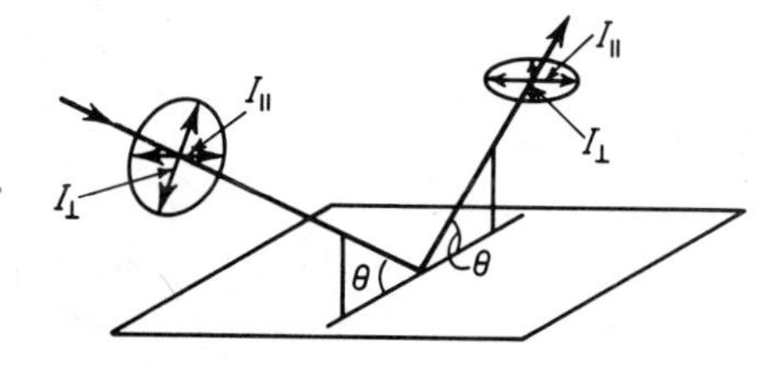

Figure 7.2. Partial polarization by reflection.

[4]L. V. Azaroff, *Acta Cryst.*, **8**, 701 (1955).
[5]The phenomenon described here is exactly the same as the well-known polarization of light by selective reflection from a smooth, shiny surface.

monochromatized source. Because the usual monochromatizing crystals show very nonideal reflection, K is not predicted accurately by theory.[6] Thus it is important to determine K for a particular apparatus,[7] although $\cos 2\theta_m$, where θ_m is the diffraction angle of the monochromator crystal, can be used as an approximation.[8]

The Lorentz factor arises because the time required for an r.l. point to pass through the sphere of reflection is not constant but rather varies with its position in reciprocal space and the direction in which it approaches the sphere. The simplest case is that of the zero-level rotation or Weissenberg photograph or a diffractometer. The crystal and the reciprocal lattice are rotated at a constant angular velocity ω, so the linear speed of an r.l. point at a distance d^* r.l.u. from the origin as it approaches the sphere is

$$v = d^*\omega \tag{7.8}$$

$$v = (2 \sin \theta)\omega \tag{7.9}$$

The time required for a point to pass through the sphere along a path of length p is

$$t = p/v \tag{7.10}$$

or

$$t = p/(2\omega \sin \theta) \tag{7.11}$$

The path length depends on the angle between the surface of the sphere and the path followed by the r.l. point. It can be shown without difficulty that

$$p \propto \frac{1}{\cos \theta} \tag{7.12}$$

and thus, dropping the constant term ω from Eq. (7.11),

$$t \propto \frac{1}{2 \cos \theta \sin \theta} \tag{7.13}$$

Since

$$2 \cos \theta \sin \theta = \sin 2\theta \tag{7.14}$$

[6]L. D. Jennings, *Acta Cryst.*, **A37**, 584 (1981); **A40**, 12 (1984).

[7]Y. Le Page, E. J. Gabe, and L. D. Calvert, *J. Appl. Cryst.*, **12**, 25 (1979); M. G. Vincent and H. D. Flack, *Acta Cryst.*, **A36**, 614 (1980) but see also A. McL. Mathieson, *Acta Cryst.*, **A38**, 739 (1982); P. Suortti, Å. Kviak, and T. M. Emge, *Acta Cryst.*, **A42**, 184 (1986).

[8]Further complications occur with the inherently polarized beam from a beam line. See R. Kahn, R. Fourme, A. Gadet, J. Janin, C. Dumas, and D. André, *J. Appl. Cryst.*, **15**, 320 (1982).

Eqs. (7.5) and (7.13) are the same except for a constant proportionality factor. The more complicated form of the general Weissenberg expression [Eq. (7.4)] arises from the fact that in general the path length is a function of both θ and μ.

7.2. DATA REDUCTION IN PRACTICE

The minimal input to the data reduction program consists of raw intensity data, each reflection intensity being identified by its indices h, k, and l, and the cell parameters. Coded information on the method of data measurement is often given to enable the proper form of the Lorentz expression to be used. Additional data such as the diffractometer setting angles and estimates of the instrumental precision may be needed as well.

Absorption

Corrections for absorption are potentially the most difficult and are the least often made.[9] To some extent they have become less significant with diffractometer data on small molecules because these are routinely collected with molybdenum radiation, for which the coefficients are usually (but not always!) much less than with Cu K_α. Nevertheless it is often enlightening to determine at least roughly the magnitude of the effect for rays passing through the maximum and minimum dimensions of the crystal in order to estimate the seriousness of the problem.[10]

The difficulty in making rigorous absorption corrections arises from the complexity of the calculations in the general case. To obtain the theoretical correction for a reflection it is necessary to calculate the absorption for the actual path length traveled within the crystal by the beam reflecting from each infinitesimal portion of the crystal and then to integrate these results over the entire volume of the crystal.[11] This problem cannot be solved explicitly for a crystal of general shape but rather requires a precise description of the crystal envelope and extensive numerical computations.

As a result, the use of empirical corrections has become common, especially in protein crystallography, where many reflections are involved and the absorption arises from both the crystal and the medium around it. Probably the most common approximation has been that of North et al.,[12] which involves setting χ to 90° and measuring the variation in the in-

[9]Buerger, *Crystal-Structure Analysis*, pp. 204–231.

[10]J. W. Jeffery and K. M. Rose, *Acta Cryst.*, **17**, 343 (1964).

[11]See W. R. Busing and H. A. Levy, *Acta Cryst.*, **10**, 180 (1957); J. de Meulenaer and H. Tompa, *Acta Cryst.*, **19**, 1014 (1965); P. Coppens, J. de Meulenaer, and H. Tompa, *Acta Cryst.*, **22**, 601 (1967) and references cited therein.

[12]A. C. T. North, D. C. Phillips, and F. S. Mathews, *Acta Cryst.*, **A24**, 351 (1968).

tegrated intensity of one or more reflections as ϕ is rotated through 360°. This process rotates the reflecting plane about its normal and rotates the crystal about the axis (***ψ axis***, in reciprocal space but ***not*** necessarily an r.l. axis) connecting the observed r.l. point with the r.l. origin. This observed variation can easily be converted to an approximate correction as a function of ϕ and applied to the entire data set. Unfortunately, a number of drastic approximations must be made, and they become increasingly serious as the Bragg angle increases. Thus the method is more suitable to low-resolution macromolecular data than to those extending to the copper limit (although the use of molybdenum radiation helps here, too). Nevertheless, it is probably better than no correction at all.

Numerous attempts have been made to generalize this approach by generating a more accurate absorption correction surface depending on both ϕ and χ (and even θ). Some[13] have depended on observing the variation in other reflections as they were rotated about their ψ axes, but this is both tedious and difficult with most diffractometers. Others[14] have used the measurement of equivalent reflections to achieve the same end, but these methods are not well suited to low-symmetry space groups.

One promising approach[15] that requires no additional measurements generates an absorption correction surface by comparing groups of observed F's with their calculated values. It is argued that although the F_c's contain model-dependent errors their effects are distributed uniformly in reciprocal space, unlike those of absorption, allowing the two problems to be separated. Because this approach requires reasonably good F_c's, it can be applied only late in the solution process. If corrections are needed earlier, the North method could be used at first, switching to this during refinement.

Deterioration

Many crystals, especially proteins and other macromolecules but also many others, show a more or less steady decrease in diffraction intensity during the process of data collection. This is difficult to recognize in film data but is shown when standard reflections are measured repeatedly at intervals with a diffractometer.[16] The origins of this deterioration are obscure but seem to be random events occurring in the crystal as a result of direct or indirect radiation damage. Limited evidence suggests that the decrease in intensity

[13]See, for example, G. Kopfmann and R. Huber, *Acta Cryst.*, **A24**, 348 (1968).
[14]C. Katayama, N. Sakabe, and K. Sakabe, *Acta Cryst.*, **A28**, 293 (1972); H. D. Flack, *Acta Cryst.*, **A30**, 569 (1974).
[15]N. Walker and D. Stuart, *Acta Cryst.*, **A39**, 158 (1983); F. Ugozzli, *Comput. Chem.*, **11**, 109 (1987); C. Katayama, *Acta Cryst.*, **A42**, 19 (1986).
[16]S. C. Abrahams, *Acta Cryst.*, **A29**, 111 (1973).

for a given reflection is approximately linearly proportional[17] to the total radiation dose received.[18]

Unfortunately, such repeated measurements also show that the change usually differs considerably among reflections. Some of this variation arises because the effect is larger for high-order reflections (which are more sensitive than low-order reflections to small structural changes) and can be approximated by a pseudo-temperature factor depending on $(\sin^2 \theta)/\lambda^2$. Unfortunately there remains a large and apparently random fluctuation such that the variability will often be about one-half the total decay. That is, if the average intensity loss for a given resolution is 20%, the individual standards will vary between 10% and 30% or even more widely. Thus the greater the deterioration at a given time of measurement, the greater the uncertainty in the individual F's, quite aside from the inherent inaccuracy in estimating the average correction factor.[19]

Because of the data errors produced in this way, it is common to use multiple crystals for the measurement of deteriorating compounds (although this introduces other problems in crystal-to-crystal scaling) and to attempt to limit the decay of an individual data set to a maximum of 20–25%. Many experimental variations have been tried to minimize the effect: Highly filtered or monochromatized radiation may help by removing the more highly absorbed white radiation; cooling the crystal, even as low as liquid nitrogen temperatures, can be very effective and also increases the average intensity by reducing the overall temperature factor.[20]

Scaling

Once corrections have been made to individual reflections, it is often necessary to bring various sets of reflections to a common basis. When data have been collected by a photographic method, they are usually divided into groups (e.g., $h0l$, $h1l$, $h2l$) obtained on different films at different camera settings. These groups (*levels*) are internally consistent but need to be brought to a common scale in order to have a uniform data set. It is possible to achieve this end approximately by making all the exposures for the same time and making no changes in the experimental conditions during

[17]For more extensive consideration, see S. C. Abrahams and P. Marsh, *Acta Cryst.*, **A43**, 265 (1987) and J. Sygush and M. Allaire, *Acta Cryst.*, **A44**, 443 (1988).

[18]Recent observations with high dose rates suggest that there are probably separate phenomena related both to the total dose and to the elapsed time since irradiation. Thus deterioration is often a less serious problem with data collected on a beam line, where the exposure time for an area detector frame may be fractions of a second, than with conventional sources requiring minutes.

[19]L. E. McCandlish, G. H. Stout, and L. C. Andrews, *Acta Cryst.*, **A31**, 245 (1975).

[20]See H. Hope, *Acta Cryst.*, **B44**, 22 (1988).

or between the exposures, but a more reliable procedure is to scale all the levels to one or more ***cross-level photographs***, that is, data collected about another crystal axis (e.g., $0kl$). Because reflections measured under different instrumental conditions have different Lorentz corrections, raw intensities cannot be compared directly. Instead they are usually converted to $|F_{rel}|$ values, which are then placed on a common scale by using the cross-level film(s) as a basis to which to relate the various levels. In any case the final scaled $|F_{rel}|$ values are usually maintained in their original level-by-level groupings so that ultimately the scale factor for each level can be refined, if desired, by comparison between the observed and calculated structure factors (see Section 9.1).

Single-counter diffractometer data are usually collected under the same conditions for each reflection and so do not require scaling. Exceptions occur when the data have been obtained from more than one crystal. In this case it is necessary to scale all the crystals together using common reflections between the sets. An extension is *local scaling*,[21] which scales one region in reciprocal space to an equivalent region in the same or a different crystal. This can be used to scale together groups of Friedel-related reflections at $\pm 2\theta$ or related crystals used for isomorphous replacement techniques (Section 13.2). The technique can minimize errors between related reflections in cases in which absorption corrections are not made or are inaccurate, or in which other experimental causes have led to reasonably smooth fluctuations in the average intensity for regions of similar $(\sin \theta)/\lambda$.

Area detectors, on the other hand, collect individual reflections under quite different conditions because the sensitivity of the detector surface is rarely uniform. Thus the raw data frequently need to be scaled pixel by pixel by reference to a uniform radiation field in order to achieve a constant response. These corrections must then be applied before the pixels are combined and reduced to reflection intensities. In addition, frames or groups of adjacent frames are usually scaled together through multiply measured reflections to form a consistent data set.

Once all the desired corrections have been made, the output from the data reduction program consists of a set of records, one to a reflection, containing the information that will be needed for subsequent calculations. This includes *hkl* and $|F_{rel}|$ as a minimum. Special flags for unobserved reflections, measures of probable error in $|F_{rel}|$, $(\sin \theta)/\lambda$, scattering factors, or weights for use in least squares refinement may also be present, as well as any other information that appeared pertinent to the programmer.

[21]B. W. Matthews and E. W. Czerwinski, *Acta Cryst.*, **A31**, 480 (1975).

7.3. INTERPRETATION OF INTENSITY DATA[22]

Uncertainties and Unobserved Reflections

One of the advantages of diffractometer data is that it is relatively easy to analyze the intensities and estimate, at least in a formal sense, their uncertainties. As was pointed out in Chapter 6, the random error in a single measurement of N counts is approximately $\sqrt{N}$, and the uncertainty in a net intensity is as given in Eq. (6.12).[23] Experiments show, however, that this expression underestimates the real random error in data, and it is usual to include an additional term that reflects a constant fractional uncertainty in the observed intensities. Thus the expression for σ_I becomes

$$\sigma_I = (T + P^2 I^2)^{1/2} \tag{7.15}$$

where T is the total counts in peak and background and P is the "instability constant." For stable crystals and a well-behaved diffractometer, P should be $\leq 1\%$. Higher values are often used but generally reflect an unjustified attempt to absorb other errors into the machine instability. One of the most serious of these is the uncertainty in the correction for deterioration in unstable crystals, and this can rapidly dominate all other errors.[19] It is much better to try to estimate the effects of other sources of error, which may depend on parameters other than intensity, and to propagate them explicitly and correctly rather than lump them as a general instrumental uncertainty.

The problem of treating extremely weak reflections is one that has produced much discussion and some strong words. On film, not all reflections will be found to be detectably different from background. Such unmeasurably weak reflections are commonly called ***unobserved reflections*** or ***unobserveds***.[24] These may at first seem to provide no information, but this is not, in fact, the case. A reflection of zero intensity is nearly as improbable an occurrence as one of high intensity, and on this basis very weak reflections are more informative than those that are nearer the average value. Of course, such a statement presupposes that the minimum observable level is carried low enough that unobserveds are at least relatively uncommon. The usual treatment of unobserveds from film was to mark them in a way that identified them to the data reduction program. They could be assigned a value just less than the weakest observable

[22]For an extensive discussion of the statistical properties of intensity data, see D. Rogers in *Computing Methods in Crystallography*, J. S. Rollett, Ed., pp. 117–148. See also H. Lipson and W. Cochran, *The Determination of Crystal Structures*, Cornell University Press, Ithaca, NY, 1966, pp. 46–65.

[23]This is true even though the distribution of the net value, being a difference of two Gaussians, is *not* normal. See A. J. C. Wilson, *Acta Cryst.*, **A36**, 929 (1980).

[24]Note that we are referring here only to reflections that are *not* systematically absent. Absent reflections are identically zero and are usually omitted entirely from the intensity data, although to do so may require hand editing of diffractometer output.

reflection (I_{min}) that appears in a region of similar background. In the absence of errors, this value should represent the maximum intensity that the reflection can have.[25] Alternatives were to assign unobserved reflections their most probable value, $\frac{1}{2}I_{min}$ often being used as an approximation, or a value of zero, but the first procedure is preferable.

This habit of defining weak reflections as unobserved has carried over to diffractometer data. There are two major reasons: First, crystallographers have become used to agreement factors obtained between observed and calculated $|F|$ values neglecting the weakest 10–40% of the data; second, if all intensity measurements are considered, some of the weakest will show negative net counts purely as a consequence of random variations. Such a result appears physically meaningless and cannot be directly converted to an $|F_o|$ by taking a square root. Consequently, it became common to call unobserved those reflections for which the intensity is less than some multiple of (often twice) its standard deviation as calculated from Eq. (7.15) or its equivalent.

Aside from the fact that omitting weak reflections gives better-looking results, it has no justification. It is theoretically inferior to using all the data and weighting the reflections by their uncertainties in the refinement process. Furthermore, although these reflections have little effect on refinement of structures for which the model fits the data well,[26] they provide extremely powerful information for improving partial structures and detecting errors in poor models.[27]

For reasonably intense reflections it is possible to estimate the standard deviation in $|F_o|$ from I_o, but the assumptions become increasingly inaccurate as the intensity decreases. The limit is reached as the intensity goes negative and even $|F_o|$ cannot be calculated. It is possible to avoid this problem during the refinement of the structures by minimizing the differences between I_o and I_c, which are not perturbed by negative values for I_o, but it is still convenient to have a measure of $|F_o|$ as well. The most elegant method for obtaining $|F_o|$ in these cases is based on an approach using Bayesian statistics and the physical expectation that a true intensity cannot be negative although it may appear so because of random deviations.[28] One consequence of this assumption is that the more negative an observation, the more likely the true intensity is to be close to zero; thus, in the absence of systematic errors a negative intensity actually serves to define the probable true value more closely than one that is slightly larger than zero. Using this method it is possible to estimate I, σ_I, $|F_o|$, and σ_{F_o} in ways that link smoothly with the results obtained for stronger reflections

[25]For a more extensive discussion with particular reference to diffractometer measurements, see F. L. Hirshfeld and D. Rabinovich, *Acta Cryst.*, **A29**, 510 (1973).

[26]But see L. Arnberg, S. Hovmöller, and S. Westman, *Acta Cryst.*, **A35**, 497 (1979).

[27]See Chapter 15. For a theoretical discussion, see G. H. Petit and A. T. H. Lenstra, *Acta Cryst.*, **A38**, 67 (1982).

[28]S. French and K. Wilson, *Acta Cryst.*, **A34**, 517 (1978).

and that go to zero in the lower limit. This is not likely to produce great changes in the final results, but it is aesthetically satisfying and may make a difference in subtle details.

Absolute Scaling and Temperature Factors

Even if nothing structural is known about the cell contents, useful information can be obtained by statistical comparison of the observed intensity data with the theoretical predictions for a crystal composed of a random assemblage of the same atoms. The most common such comparison, which allows one to put the $|F_{rel}|$ values on an approximately absolute basis and to obtain an estimate of the effects of thermal motion of the atoms, is known as a Wilson plot, after A. J. C. Wilson, who first demonstrated its utility.[29]

If one assumes spherical atoms, the scattering power of each atom is a function only of the atom type and $(\sin\theta)/\lambda$. It is independent of the position of the atom in the cell. The scattering power of a given atom for a given reflection is known as its *scattering factor* (f_o) and is expressed in terms of the scattering power of an equivalent number of electrons located at the position of the atomic nucleus.[30]

The variation of the scattering factor of carbon with $(\sin\theta)/\lambda$ (*the scattering factor curve*) is shown in Fig. 7.3. At $(\sin\theta)/\lambda = 0$ the value of the scattering factor is always equal to the total number of electrons in the atom. As $(\sin\theta)/\lambda$ increases, however, the scattering factor decreases, because X-rays scattered from an electron in one part of an atom will be to an increasing extent out of phase with those scattered in another part of the electron cloud. Thus the variation of the scattering factor is a consequence of the finite size of the atom regarded as a scattering source.[31]

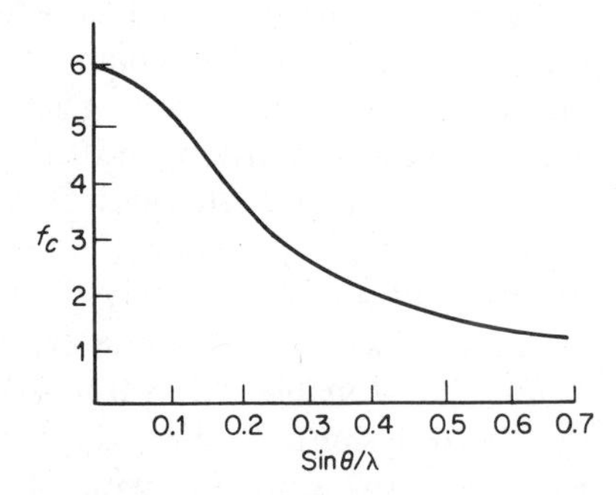

Figure 7.3. Scattering factor of carbon as a function of $(\sin\theta)/\lambda$.

[29] A. J. C. Wilson, *Nature,* **150,** 152 (1942).

[30] James, *Optical Principles . . . ,* pp. 93–134.

[31] For example, F^- contains 10 electrons in a relatively diffuse cloud and has a scattering factor curve that falls off much more rapidly than that of Mg^{2+}, which has the same number clustered more closely about the nucleus. In contrast, the scattering factors for neutrons, in which scattering actually occurs at the pointlike nucleus, are functions only of the nuclei involved and are invariant with $(\sin\theta)/\lambda$.

Thermal motion also has an effect on the X-ray intensities.[32,33] The normal scattering factor curves are calculated on the basis of the electron distribution in a stationary atom, but in fact the atoms in crystals are always vibrating about their rest points. The magnitude of the vibration depends on the temperature, the mass of the atom, and the firmness with which it is held in place by covalent bonds or other forces. In general, the higher the temperature, the greater the vibration. The effect of such *thermal motion* is to spread the electron cloud over a larger volume and thus to cause the scattering power of the real atom to fall off more rapidly than that of the ideal, stationary model. It has been shown both theoretically and practically that the change in scattering power can be given by the expression

$$e^{-B(\sin^2\theta)/\lambda^2} \tag{7.16}$$

where B is related to the mean-square amplitude $(\overline{u^2})$ of atomic vibration by

$$B = 8\pi^2\overline{u^2} \tag{7.17}$$

Thus the proper scattering factor for a real atom is not simply f_o, but rather the combined expression

$$f = f_o e^{-B(\sin^2\theta)/\lambda^2} \tag{7.18}$$

(see Figs. 7.4 and 7.5). It is convenient to have an estimate of the average value of B for the whole structure before beginning the actual analysis, and although experience can often allow one to make a reasonable guess within the usual range of 2–5 Å^2, the Wilson plot provides a more systematic working value.

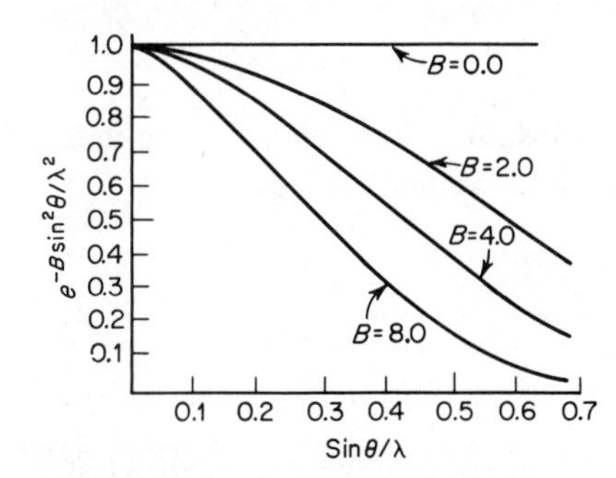

Figure 7.4. Temperature factor $e^{-B(\sin^2\theta)/\lambda^2}$ as a function of $(\sin\theta)/\lambda$.

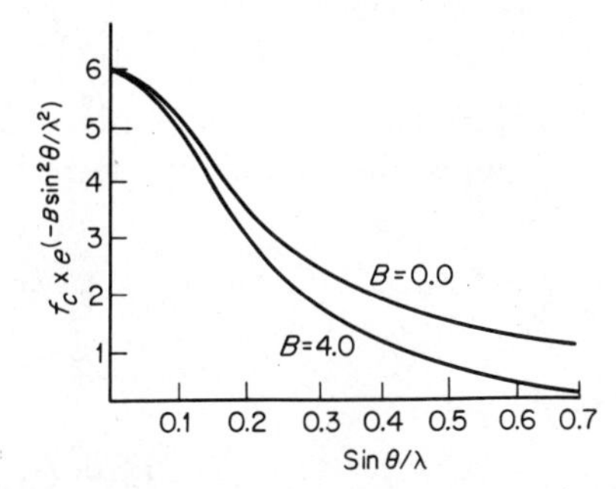

Figure 7.5. The product $f_0 e^{-B(\sin^2\theta)/\lambda^2}$ as a function of $(\sin\theta)/\lambda$.

[32]James, *Optical Principles*..., pp. 193–267.
[33]B. T. M. Willis and A. W. Pryor, *Thermal Vibrations in Crystallography*, Cambridge University Press, Cambridge, 1975.

Let us define an average observed intensity, corrected for $1/Lp$ effects, such that

$$\bar{I}_{\mathrm{rel}} = \langle |F_{\mathrm{rel}}|^2 \rangle_{\mathrm{ave}} \tag{7.19}$$

For a unit cell that contains N atoms, it can be shown fairly easily[22,29] that the theoretical average intensity is given by

$$\bar{I}_{\mathrm{abs}} = \sum_{i=1}^{N} f_i^2 \tag{7.20}$$

that is, that the *average* intensity depends merely on what is in the cell and not on where it is. Since the argument is a statistical one, difficulties arise if the contents of the cell differ too greatly from a random distribution, but most organic compounds are reasonably well behaved.

Ideally, the ratio of the I_{abs} to I_{rel} should be the scaling factor required to place the individual I_{rel} values on an absolute scale. Matters are not quite so simple, however. First, the f's are not constants but rather are functions of $(\sin\theta)/\lambda$, so that I_{abs} also varies with $(\sin\theta)/\lambda$. This is normally avoided by dividing reciprocal space into concentric shells, each thin enough that the variation of f's with $(\sin\theta)/\lambda$ within the shell can be ignored, and averaging the I_{rel} values of the reflections within each shell. This I_{rel} can then be compared with I_{abs} calculated from the f's appropriate to the shell.

The second, and more serious, problem is that the f's required in Eq. (7.20) are those that describe the atoms as they are in the crystal, that is with thermal motion included, and not the ideal f_{o}'s. Only the f_{o}'s are known a priori, so Eqs. (7.20) and (7.18) must be combined as

$$\bar{I}_{\mathrm{abs}} = \sum_{i=1}^{N} f_{\mathrm{o}i}^2 e^{-2B(\sin^2\theta)/\lambda^2} \tag{7.21}$$

where B remains to be determined. If it is assumed to have the same value for all atoms, the exponential term is the same for all $f_{\mathrm{o}i}$'s and

$$\bar{I}_{\mathrm{abs}} = e^{-2B(\sin^2\theta)/\lambda^2} \sum_{i=1}^{N} f_{\mathrm{o}_i}^2 \tag{7.22}$$

Now if

$$\bar{I}_{\mathrm{rel}} = C\bar{I}_{\mathrm{abs}} \tag{7.23}$$

$$\bar{I}_{\mathrm{rel}} = Ce^{-2B(\sin^2\theta)/\lambda^2} \sum_{i=1}^{N} f_{\mathrm{o}_i}^2 \tag{7.24}$$

$$\frac{\bar{I}_{\mathrm{rel}}}{\sum_{i=1}^{N} f_{\mathrm{o}_i}^2} = Ce^{-2B(\sin^2\theta)/\lambda^2} \tag{7.25}$$

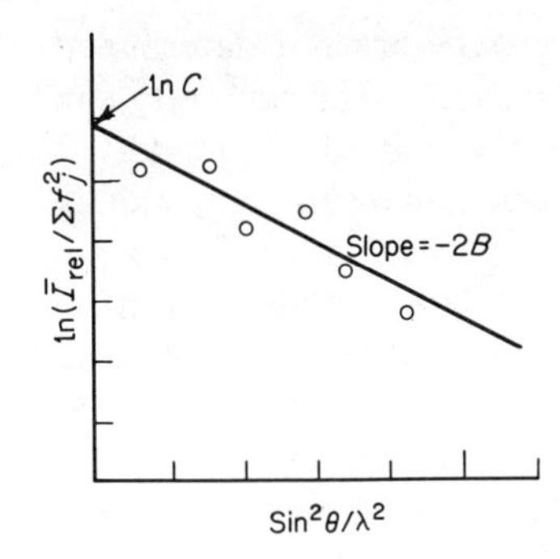

Figure 7.6. Wilson plot for determining scale and thermal parameters.

and taking the natural logarithm of both sides,

$$\ln\left(\frac{\bar{I}_{\text{rel}}}{\sum_{i=1}^{N} f_{o_i}^2}\right) = \ln C - \frac{2B(\sin^2\theta)}{\lambda^2} \tag{7.26}$$

Thus if the left side of Eq. (7.26) is evaluated for each of the shells of constant f, and the values are plotted against $(\sin^2\theta)/\lambda^2$, the result should be a straight line in which the extrapolated intercept at $(\sin^2\theta)/\lambda^2 = 0$ is $\ln C$ and the slope is $-2B$. B can thus be obtained directly from the slope, and C is related to the scale constant k needed to convert $|F_{\text{rel}}|$ to $|F_{\text{abs}}|$ by

$$k = 1/\sqrt{C} \tag{7.27}$$

where

$$|F_{\text{abs}}| = k|F_{\text{rel}}| \tag{7.28}$$

In practice it is necessary to balance the need for a shell large enough to give a good average over I_{rel} against the desire to keep the variation in f as small as possible. Usually a shell containing 50–100 reflections is a reasonable compromise. For various reasons it is advisable to omit from consideration the reflections lying closest to the r.l. origin, that is, those with all indices 0 or 1. The points obtained often show considerable scatter, and it is necessary to fit them by some best line. This can be done systematically by least squares, but the reliability of the results is probably not much improved over those from an estimated best fit. The accuracy of the values found is only fair (±10–15% in k and B), but the results are quite adequate for preliminary working values. Figure 7.6 shows an example of a Wilson plot.

Symmetry

A second useful result that can be obtained from a study of the intensities is some additional information about the symmetry of the crystal under study.

As was discussed in Section 5.12, most space groups are not determined completely by their systematic extinctions. If, however, extra symmetry can be found in the intensity set, additional useful distinctions can be made.

If the crystals being studied contain atoms heavier than sulfur, it is likely that there will be deviations from Friedel's law if the true space group is not centrosymmetric. Careful comparison of accurate intensities obtained from Laue equivalent reflections may succeed in demonstrating that true equivalence holds between some but not all. In this case, the observed equivalences will identify those symmetry elements that are actually present and restrict the space group possibilities. Unfortunately, it is just such crystals that may suffer most heavily from absorption errors, and it is necessary to consider other sources of nonequivalence rather than accepting the results blindly.

There also occur cases in which the cell constants appear to have, seemingly within experimental error, values that suggest a higher lattice symmetry than is actually present. In some cases, this near symmetry extends to the general pattern of intensities, and it is only by careful comparison that it can be seen that supposedly equivalent reflections are not.

These both represent rather special cases but may be common enough to justify collecting routine data sets rapidly over the entire sphere of data rather than more slowly over the unique set. If there are no problems, equivalent reflections can be averaged to improve the statistical precision; if there are questions, the data are at hand to examine.

Additional symmetry information can arise because although the *average* reflection intensity depends in principle only on what is in the unit cell and not on where it is, the same is not true of the distribution of the individual intensities about this average. Wilson[33] and others have shown that the intensities from a noncentrosymmetric crystal tend to be bunched more tightly about their mean than are those from a centrosymmetric one. A consequence of this fact is that centrosymmetric crystals tend to have relatively more weak or unobserved reflections than noncentrosymmetric crystals.

A large number of quantitative tests have been devised for comparing an observed distribution of intensities with those predicted theoretically.[34] Probably the test most commonly used today is based on the values of various quantities related to the normalized structure factors E (see Section 11.2). An alternative test that has the advantage of giving a comparison between two distribution curves rather than merely between two numbers is that of Howells, Phillips, and Rodgers.[35] This consists of determining the

[33] A. J. C. Wilson, *Acta Cryst.*, **2**, 318 (1949).

[34] R. Srinivasan, *Acta Cryst.*, **13**, 388 (1960); R. Srinivasan and E. Subramanian, *Acta Cryst.*, **17**, 67 (1964); F. Foster and A. Hargreaves, *Acta Cryst.*, **16**, 1133 (1963).

[35] E. R. Howells, D. C. Phillips, and D. Rogers, *Acta Cryst.*, **3**, 210 (1950).

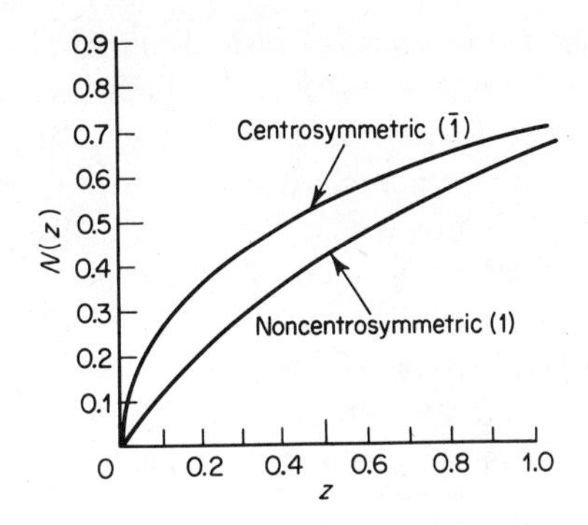

Figure 7.7. Distribution plot of X-ray intensities. $N(Z)$ is the fraction of the reflections with $I/\langle I \rangle$ less than Z.

fraction $N(Z)$ of the reflections (excluding those systematically absent) less than a specified fraction of the average intensity and plotting these quantities against each other. Figure 7.7 shows the theoretical curves for centric ($\bar{1}$) and acentric (1) distributions, while Table 7.1 gives $N(Z)$ as a function of $Z = I/\langle I \rangle$. It can be seen that there is a considerable difference between the two curves, and in practice it is usually sufficient to distinguish between the two distributions. Since the differences occur predominantly in the weakest reflections, however, it is very important that these actually be measured and not merely lumped together as unobserveds less than some cutoff based on an arbitrary (large) number of standard deviations.[36]

Unfortunately, the presence of certain symmetry or near symmetry in the cell, whether or not it is among the elements of the space group, can cause the distribution to assume the ($\bar{1}$) form even in a noncentrosymmetric crystal.[37] Thus the observation of a curve resembling that of a centric distribution, although indicative, is not absolute proof of a centrosymmetric crystal. If an acentric distribution is found, the implication is stronger that the space group does not have a center of symmetry, although even here the test can fail.

This test for a center of symmetry is usually applied to ***hkl*** data divided into shells of $(\sin\theta)/\lambda$ as in the Wilson plot. The distributions found for each shell are averaged to give the overall result. An alternative method is to correct the observed intensities for the decrease in average intensity that occurs with $(\sin\theta)/\lambda$ and then treat all of the reflections together.

TABLE 7.1 Theoretical Intensity Distributions

	Z										
	0	0.1	0.2	0.3	0.4	0.5	0.6	0.7	0.8	0.9	1.0
($\bar{1}$)	0	0.248	0.345	0.419	0.479	0.520	0.561	0.597	0.629	0.657	0.683
(1)	0	0.095	0.181	0.259	0.330	0.394	0.451	0.503	0.551	0.593	0.632

[36]For examples of problems caused by omitting weak reflections, see R. E. Marsh, *Acta Cryst.*, **B37**, 1985 (1981).

[37]D. Rogers and A. J. C. Wilson, *Acta Cryst.*, **6**, 439 (1953).

Similar tests have been devised along the same lines for detecting other symmetry elements. These usually depend on the fact that when the elements are seen in projection they resemble centers and thus cause the intensities of specific layers of r.l. points to have a centric distribution. For example, when the space group $P2_1$ is viewed down the *b* axis with the contents projected onto the *ac* plane, the 2-fold screw seen end on appears to be a center. As a consequence, the intensity distribution of the $h0l$ reflections[38] is centric while that of the general (hkl) reflections is not. On the other hand, *Pm* contains no 2-fold axis, and the projection of the unit cell contents on *ac* is noncentrosymmetric. Thus the $h0l$ and hkl reflections have the same acentric distribution.

An additional and often complementary effect can be detected in particular projections of some space groups. Thus, for example, although the projection of the unit cell of *Pm* onto the *ac* plane is noncentrosymmetric, each atom appears on top of its mirror-related image (Fig. 7.8), so the cell seems to contain half as many atoms, each with twice the number of electrons. This pattern is very definitely not a random distribution of N atoms as required by the derivation of Eq. (7.20), but rather a distribution of $N/2$ atoms, each with scattering factor $2f_i$. Thus for the $h0l$ reflections,

$$\bar{I}_{\text{abs}} = \sum_{i=1}^{N/2} (2f_i)^2 \tag{7.29}$$

$$\bar{I}_{\text{abs}} = 4 \sum_{i=1}^{N/2} f_i^2 \tag{7.30}$$

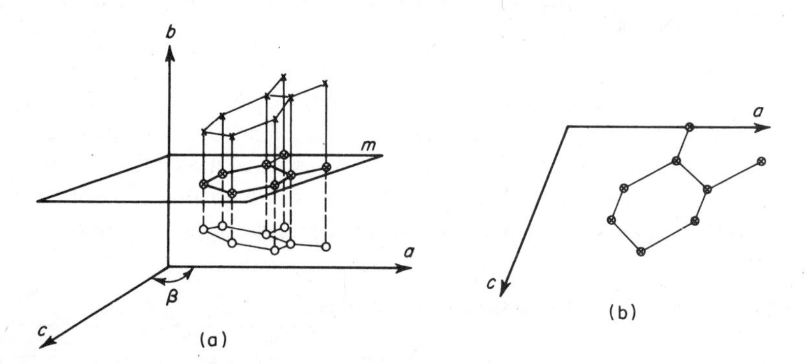

Figure 7.8. (*a*) Two molecules in space group *Pm*. (*b*) Their superimposed projections onto the *ac* plane.

[38]Note that since the planes producing these reflections are parallel to the *b* axis, their intensities are unaffected by the *y* coordinates of the scattering atoms and thus depend only on their *x* and *z* coordinates, that is, on their projections on the *ac* plane.

while for the general hkl it can be seen by dividing the atoms into two mirror-related groups with $f_i = f_{i+N/2}$.

$$\bar{I}_{abs} = \sum_{i=1}^{N/2} f_i^2 + \sum_{i=N/2+1}^{N} f_i^2 \tag{7.31}$$

$$\bar{I}_{abs} = 2 \sum_{i=1}^{N/2} f_i^2 \tag{7.32}$$

Thus,

$$\frac{(I_{abs})_{h0l}}{(I_{abs})_{hkl}} = 2 \tag{7.33}$$

so the presence of the mirror plane can be detected from the greater average intensity of the $h0l$ reflections and the results used as a check in distinguishing *Pm* from *P*2. Certain other symmetry elements produce similar effects in various classes of reflections, and these are tabulated in *International Tables*.[39]

It is worth repeating that the application of such tests, depending as they do on the use of a random model for their theory, is subject to unexpected disturbances, and the results should be treated with caution. With this proviso, however, they are extremely useful and often provide convenient guideposts as to how to attack a structural problem.

BIBLIOGRAPHY

Blundell, T. L., and L. N. Johnson, *Protein Crystallography*, Academic, London, 1976, pp. 310–336.

Buerger, M. J., *Crystal-Structure Analysis*, Wiley, New York, 1960, pp. 152–241.

Gabe, E. J., and Y. Le Page, in *Computational Crystallography*, D. Sayre, Ed., Clarendon Press, Oxford, 1982, pp. 41–55.

James, R. W., *The Optical Principles of the Diffraction of X-Rays*, Cornell University Press, Ithaca, NY, 1965, Chapters 3 and 5.

Jeffery, J. W., *Methods in X-Ray Crystallography*, Academic, London, 1971, pp. 115–126, 279–283.

Luger, P., *Modern X-Ray Analysis on Single Crystals*, de Gruyter, New York, 1980, pp. 205–216.

Prout, C. K., in *Computing Methods in Crystallography*, J. S. Rollett, Ed., Pergamon, Oxford, 1965, pp. 89–95.

Rogers, D., in *Computing Methods in Crystallography*, J. S. Rollett, Ed., Pergamon, Oxford, 1965, pp. 117–148.

[39] *International Tables*, Vol. II, pp. 355–357.

CHAPTER 8

THEORY OF STRUCTURE FACTORS AND FOURIER SYNTHESES

Our purpose in this chapter is to develop two expressions of fundamental importance in crystal structure determinations. One of these is the structure factor F_{hkl}, which is a measure of the amplitude of the reflection from the set of planes h, k, l. It is a function of the reflection indices and the positions of the atoms in the unit cell. The second expression gives the electron density $\rho(xyz)$ as a Fourier series involving the structure factors. To develop these expressions, we first derive equations describing simple harmonic motion.

8.1. SIMPLE HARMONIC MOTION[1]

Consider the point A in Fig. 8.1a moving on a circle at constant angular velocity. Its projection on the x axis, the point B, is said to execute simple harmonic motion. A plot of the linear displacement of B as a function of the angle ϕ of the radius vector[2] $\mathbf{f}$ is simply the cosine function, $OB = f \cos \phi$ (Fig. 8.1b). The maximum displacement, equal to the magnitude of $\mathbf{f}$, is defined as the *amplitude* of the wave.

[1] J. Strong, *Concepts of Classical Optics*, Freeman, San Francisco, 1958, pp. 22–34.

[2] We shall use here and later the symbolism that a vector quantity (magnitude and direction) is indicated by boldface type. The magnitude alone of such a vector is usually described by the same symbol in normal or italic type. Vector quantities can also be represented, however, as complex numbers (see Section 8.2) and as such are not printed in boldface type. In this case the magnitude (modulus) is indicated by placing the symbol between two vertical lines. Many crystallographic quantities can be treated in either way, but complex notation is more common, and we shall use it except in special cases.

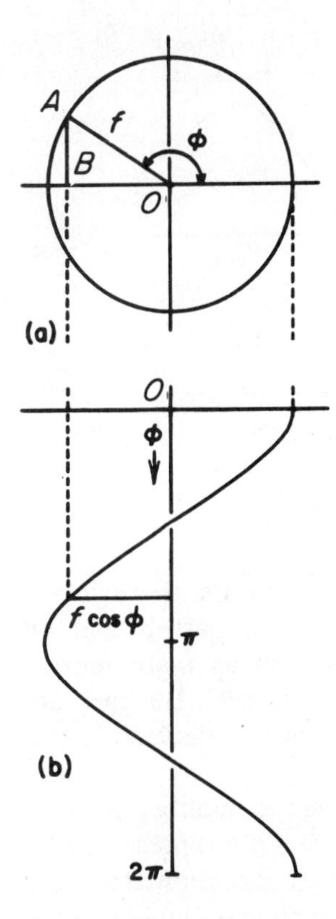

Figure 8.1. Cosine function as representation of simple harmonic motion.

If the angular velocity ω of A is constant, ϕ is proportional to ωt. Hence a plot of the linear displacement of B as a function of time will appear similar to Fig. 8.1*b*. The frequency of the wave is the number of revolutions made by A in unit time, that is, in cycles per second [properly, but less informatively, hertz (Hz)].

8.2. WAVES, VECTORS, AND COMPLEX NUMBERS

Figure 8.2 is a plot of linear displacement as a function of ϕ for a simple harmonic wave, similar to the one in Fig. 8.1*b* except that it is shifted along the ϕ axis by the angle δ. A wave of constant wavelength λ such as this one is characterized by two quantities, the amplitude f and the phase angle δ. The same two quantities characterize a vector of length f in the complex plane making an angle δ with the positive real axis x (Fig. 8.3). Therefore the vector in this figure can be taken to represent the wave in Fig. 8.2, enabling one to describe that wave by the expression

$$f = a + ib \tag{8.1}$$

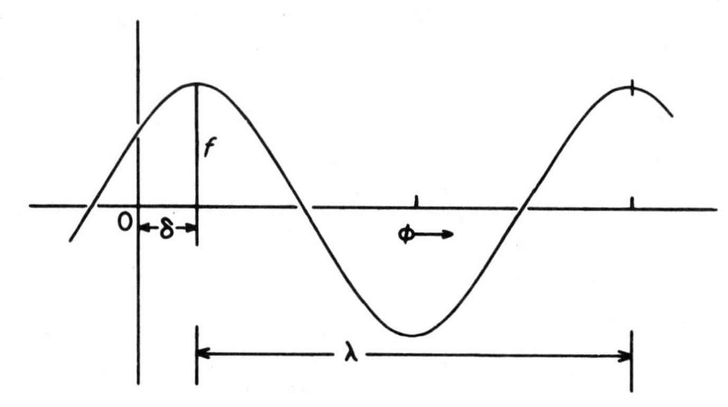

Figure 8.2. Wave with amplitude f and phase angle δ.

where a and b are real numbers and $i = \sqrt{-1}$. Numbers such as $a + ib$ are termed complex because they are composed of a real part a and an imaginary part ib. It is unfortunate that numbers such as i are termed imaginary since it perpetuates the notion that they are unreal. But they are just as "real," for example, as negative numbers, and are particularly useful in representing waves.

Some understanding of the basic meaning of complex numbers can be gained from the following considerations. Real numbers are represented by points on a straight line infinitely extended in either direction from the origin. Since imaginary numbers cannot be expressed as real numbers, they belong nowhere on this line. A clue as to how they can be represented is gained by noting that multiplication twice by i is equivalent to multiplication by -1. But multiplying by -1 corresponds to a change of direction on a straight line, that is, rotation by 180°. Hence multiplication once by i can be interpreted as a rotation of 90°. Thus in a system of orthogonal axes a real number is represented by a point on one axis, and an imaginary number by a point on the other. The complex number $a + ib$ is therefore represented by a point in the *complex plane*, a units along x, the real axis, and b units along i, the imaginary axis (Fig. 8.3).

A complex number can equally well be represented by polar coordinates

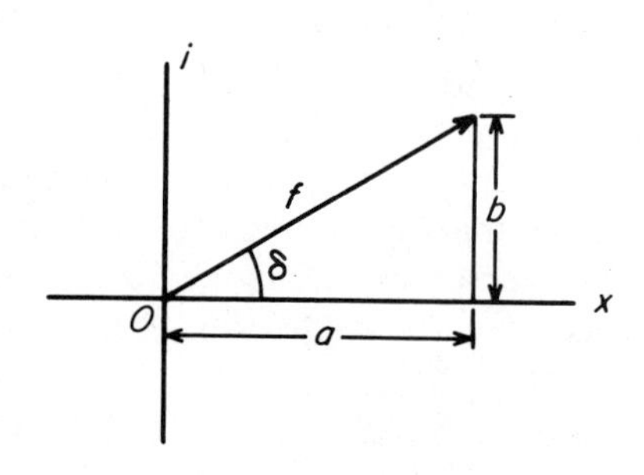

Figure 8.3. Vector **f** in complex plane with modulus $|f|$ and phase angle δ.

r, δ. In Fig. 8.3 it is clear that

$$a + ib = f(\cos \delta + i \sin \delta) \tag{8.2}$$

The form $a + ib$ is termed the *rectangular form*, and $f(\cos \delta + i \sin \delta)$ the *polar* or *trigonometric form*. We shall often discuss complex values in terms of the vector with length (amplitude, absolute value, or *modulus*) $|f| = \sqrt{a^2 + b^2}$ and phase angle (argument) $\delta = \tan^{-1} b/a$ (Fig. 8.3).

The number $a - ib$ is termed the *complex conjugate* of $a + ib$. The square of the absolute value of a complex number is defined as the product of the number and its complex conjugate since $(a + ib)(a - ib) = a^2 + b^2$.

8.3. SUPERPOSITION OF WAVES[1]

When a crystal is subjected to the electromagnetic field of an X-ray beam, each electron in the structure scatters a small fraction of the incident energy. The principle of superposition tells us how such scattered waves are to be combined to give the resultant wave reflected from each set of planes *hkl*. It states that the amplitude resulting from the simultaneous action of several waves at a point is the sum of the displacements of the individual waves. The principle is general, holding for any number of waves regardless of phase, amplitude, or frequency, but since we wish to apply it to the scattering of essentially monochromatic X-rays we shall assume a constant frequency in the following discussion.

The principle of superposition amounts to adding the vectors representing the individual waves. This is illustrated in Fig. 8.4 for the case of three waves with amplitudes f_1, f_2, f_3 and phase angles $\delta_1, \delta_2, \delta_3$. The vectors representing the individual components are joined successively as shown in Fig. 8.4*b*, and the resultant is drawn from the initial point of the first vector to the terminal point of the last. Note that the same amplitude and phase are obtained for the resultant regardless of the order in which the vectors are added. Thus in Fig. 8.4*c* the component vectors are laid out in the inverse order, beginning at the origin with $\mathbf{f}_3$. The resultant is identical.

A very useful alternative method of adding vectors is shown in Fig. 8.5, where the projections of the resultant vector on the x and y axes are simply the sums of the component projections on the same axes. Hence we may write for the x component of the resultant vector

$$x = f_1 \cos \delta_1 + f_2 \cos \delta_2 + f_3 \cos \delta_3 = \sum_j f_j \cos \delta_j \tag{8.3}$$

and correspondingly for the y component

$$y = f_1 \sin \delta_1 + f_2 \sin \delta_2 + f_3 \sin \delta_3 = \sum_j f_j \sin \delta_j \tag{8.4}$$

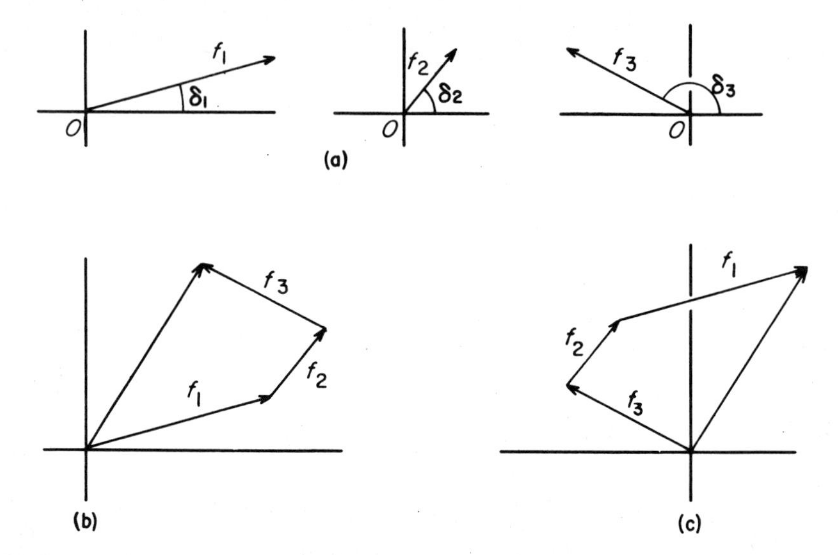

Figure 8.4. (*a*) Vectors representing three waves of different amplitude and phase. (*b*) Resultant. (*c*) Resultant after addition in reverse order.

In Fig. 8.5 it is clear that the magnitude or absolute value of the resultant $|F|$ is

$$|F| = (x^2 + y^2)^{1/2} = \left[\left(\sum_j f_j \cos \delta_j\right)^2 + \left(\sum_j f_j \sin \delta_j\right)^2\right]^{1/2} \tag{8.5}$$

and the phase angle is

$$\alpha = \tan^{-1} \frac{\sum_j f_j \sin \delta_j}{\sum_j f_j \cos \delta_j} \tag{8.6}$$

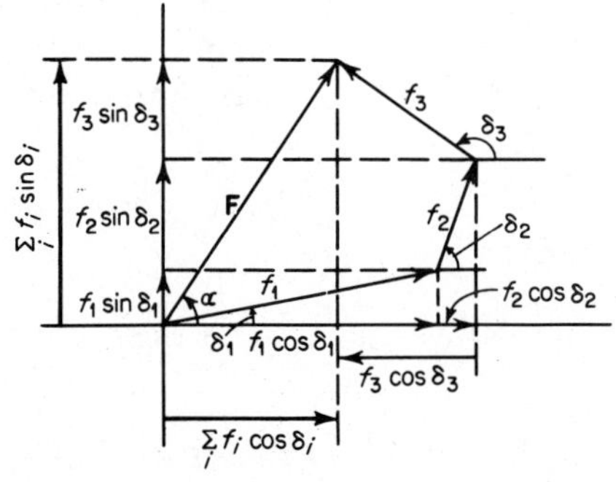

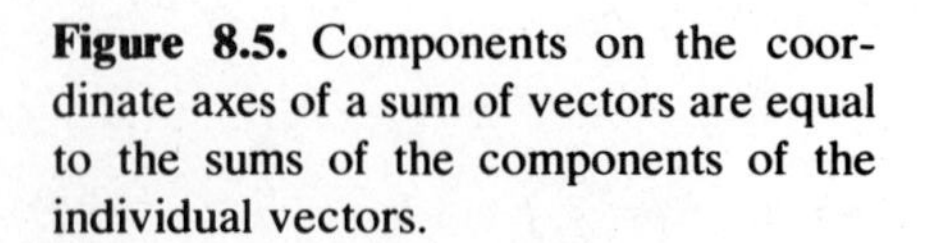

Figure 8.5. Components on the coordinate axes of a sum of vectors are equal to the sums of the components of the individual vectors.

8.4. THE STRUCTURE FACTOR

The structure factor is the resultant of N waves scattered in the direction of the reflection hkl by the N atoms in the unit cell. Each of these waves has an amplitude proportional to f_j, the scattering factor of the atom (see Chapter 7), and a phase δ_j with respect to the origin of the unit cell. Before the structure factor can be calculated, an expression for the phases in terms of the positions of the atoms and the indices of the reflection is needed. Such an expression can be derived by reference to Fig. 8.6. From the definition of the indices (see Chapter 2), the set of planes hkl cuts $\mathbf{a}$ into h, $\mathbf{b}$ into k, and $\mathbf{c}$ into l divisions. Since there is a phase difference of one cycle (2π radians, or 360°) between reflections from successive planes of any given set hkl (see the derivation of Bragg's law), it is clear that the phase differences for unit translations along the axes or along any lines parallel to these axes are $2\pi h$, $2\pi k$, and $2\pi l$ radians, respectively. For a fraction of a unit translation the phase difference will be that fraction of the phase difference for a unit translation. In Fig. 8.6 it can be seen that the phase difference in radians between the two points 0, 0, 0 and x, y, z for the set of planes hkl is the sum of the phase differences between the ends of vectors parallel to the axes and joining the two points. Thus if coordinates are expressed in fractions of a unit cell edge, the phase difference between 0 and $x, 0, 0$ is $(2\pi h)x$, that between $x, 0, 0$ and $x, y, 0$ is $(2\pi k)y$, and that

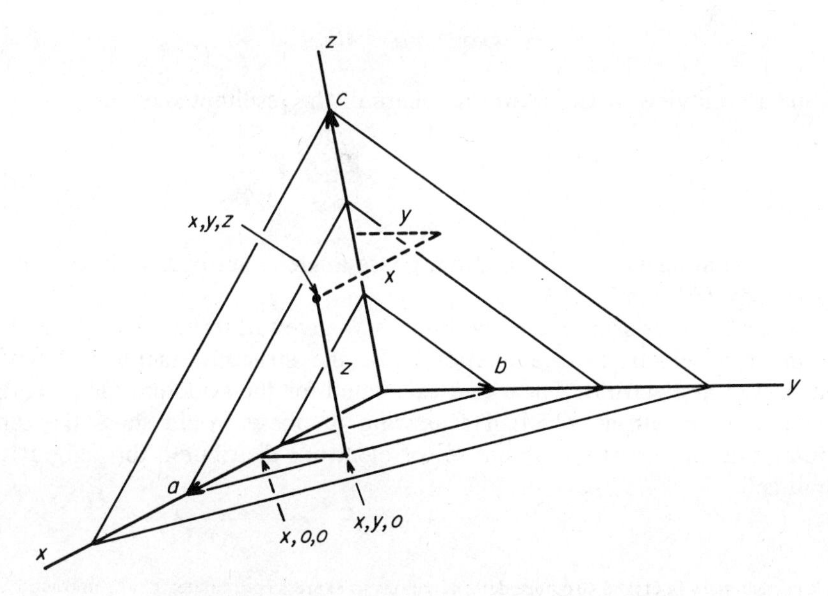

Figure 8.6. Axial components of the point x, y, z.

between $x, y, 0$ and x, y, z is $(2\pi l)z$.[3] Hence the total phase difference in radians between the origin and the point x, y, z is

$$\delta = 2\pi(hx + ky + lz) \tag{8.7}$$

Substitution in Eq. (8.5) gives for the magnitude of the structure factor

$$|F_{hkl}| = \left\{\sum [f_j \cos 2\pi(hx_j + ky_j + lz_j)]^2 + \left[\sum f_j \sin 2\pi(hx_j + ky_j + lz_j)\right]^2\right\}^{1/2} \tag{8.8}$$

$$|F_{hkl}| = (A_{hkl}^2 + B_{hkl}^2)^{1/2} \tag{8.9}$$

where

$$A_{hkl} = \sum f_j \cos 2\pi(hx_j + ky_j + lz_j) \tag{8.10}$$

and

$$B_{hkl} = \sum f_j \sin 2\pi(hx_j + ky_j + lz_j) \tag{8.11}$$

Comparison with Eq. (8.5) and reference to Fig. 8.5 will show that A and B are the projections of $\mathbf{F}$ on the x and y axes of a Cartesian coordinate system, so we may write

$$F_{hkl} = A_{hkl} + iB_{hkl} \tag{8.12}$$

Similarly, in view of Eq. (8.6), the phase of the resultant wave is

$$\alpha_{hkl} = \tan^{-1} \frac{B_{hkl}}{A_{hkl}} \tag{8.13}$$

In computational work, $|F|$ and the phase angle α are ordinarily calculated from Eqs. (8.8) and (8.13).

Since the scattering factor of an atom is given in terms of an equivalent number of electrons (see Section 7.3), the structure factor is likewise measured in electrons. The value calculated for the structure factor is that number of electrons which if scattering in phase would show the same diffracting power as the actual set of electrons distributed throughout the unit cell.

[3]It is customary in crystal structure determinations to express coordinates x, y, z in fractions of the unit-cell edges.

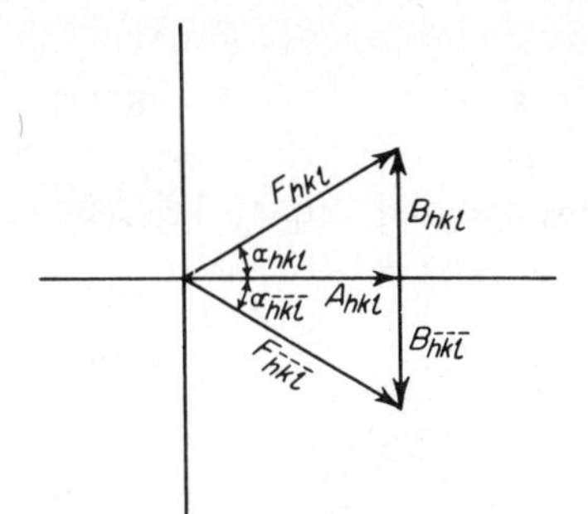

Figure 8.7. Vector representation of F_{hkl} and $F_{\overline{hkl}}$.

8.5. FRIEDEL'S LAW

In Chapter 3, Friedel's law, $I_{hkl} = I_{\overline{hkl}}$, was justified in terms of reflection from opposite sides of the same set of planes. It can now be derived in a straightforward way from the expression for the structure factor, Eq. (8.12),

$$F = A + iB \tag{8.14}$$

From Eqs. (8.10) and (8.11) it is evident that

$$A_{hkl} = A_{\overline{hkl}} \tag{8.15}$$

and

$$B_{hkl} = -B_{\overline{hkl}} \tag{8.16}$$

Since $I = A^2 + B^2$, it follows directly that

$$I_{hkl} = I_{\overline{hkl}} \tag{8.17}$$

It should be noted, however, that Friedel's law does not require that $F_{hkl} = F_{\overline{hkl}}$, but only that $|F_{hkl}| = |F_{\overline{hkl}}|$. Thus Fig. 8.7 gives the phase diagram for these reflections and shows that when Friedel's law holds,

$$\alpha_{hkl} = -\alpha_{hkl} \tag{8.18}$$

8.6. THE STRUCTURE FACTOR IN EXPONENTIAL FORM[4]

It is well known that e^x, $\cos x$, and $\sin x$ can be expressed by

$$e^x = 1 + x + x^2/2! + x^3/3! + \cdots \tag{8.19}$$

$$\cos x = 1 - x^2/2! + x^4/4! - \cdots \tag{8.20}$$

[4]Mellor, *Higher Mathematics* . . . , pp. 280–286.

and

$$\sin x = x - x^3/3! + x^5/5! - \cdots \tag{8.21}$$

Substituting $x = i\delta$ in the expression for the exponential, Eq. (8.19), and multiplying both sides of the equation by f, we have

$$fe^{i\delta} = f(1 + i\delta - \delta^2/2! - i\delta^3/3! + \delta^4/4! + \cdots) \tag{8.22}$$

$$fe^{i\delta} = f[1 - \delta^2/2! + \delta^4/4! + \cdots + i(\delta - \delta^3/3! + \delta^5/5! - \cdots)] \tag{8.23}$$

or

$$fe^{i\delta} = f(\cos\delta + i\sin\delta) \tag{8.24}$$

The right-hand side of Eq. (8.24) is simply a complex number in polar form and represents a wave with amplitude f and phase angle δ. The left-hand side of Eq. (8.24) is thus a compact representation of the same wave and enables us to write an expression for the structure factor:

$$F = \sum_j f_j e^{i\delta_j} \tag{8.25}$$

where f_j is, as previously, the scattering factor of the jth atom and δ_j is the phase with respect to the origin of the unit cell. Substituting in Eq. (8.25) the phase difference as given in Eq. (8.7),

$$F_{hkl} = \sum f_j e^{2\pi i(hx_j + ky_j + lz_j)} \tag{8.26}$$

This is the exponential form of the structure factor.

8.7. THE STRUCTURE FACTOR IN VECTOR FORM

Equation (8.26) represents the structure factor in a way that gives direct access to the three coordinates in direct space and the three indices in reciprocal space. A more compact, if less explicit, notation takes the three indices as characterizing a reciprocal space vector $\mathbf{h}$ and the three coordinates as defining a direct space vector $\mathbf{r}$. The *dot product* of these two vectors is defined as

$$\mathbf{h} \cdot \mathbf{r} = hx + ky + lz \tag{8.27}$$

so Eq. (8.26) can be rewritten as

$$F_{\mathbf{h}} = \sum f_j e^{2\pi i(\mathbf{h} \cdot \mathbf{r})} \tag{8.28}$$

This notation can be carried on to trigonometric expressions as well. Thus,

$$F_{\mathbf{h}} = \sum f_j\{\cos 2\pi(\mathbf{h}\cdot\mathbf{r}) + i \sin 2\pi(\mathbf{h}\cdot\mathbf{r})\} \tag{8.29}$$

We shall use this form where compactness is more important than explicit reference to the individual vector components.

8.8. THE GENERALIZED STRUCTURE FACTOR

Thus far the structure factor has been considered as the resultant of adding the waves scattered in the direction of the *hkl* reflection from the *N* atoms in the unit of structure. This approach was based on the assumption that the scattering power of the electron cloud surrounding each atom could be equated to that of the proper number of electrons concentrated at the atomic center (Chapter 7). But the structure factor can equally well be considered as the sum of the wavelets scattered from all the infinitesimal elements of electron density in a unit cell, with no assumptions being made about the distribution of this density. Since electron density ρ is defined as the number of electrons per unit volume, it follows that the number of electrons in any volume element dv is

$$\rho(x, y, z)\, dv = \rho(\mathbf{r})\, dv \tag{8.30}$$

In the exponential form the wavelet scattered by this element is

$$\rho(x, y, z)e^{2\pi i(hx+ky+lz)}\, dv = \rho(\mathbf{r})e^{2\pi i(\mathbf{h}\cdot\mathbf{r})}\, dv \tag{8.31}$$

The resultant is the sum of all the elements in the unit cell, that is, the integral over its volume:

$$F_{hkl} = \int_v \rho(x, y, z)e^{2\pi i(hx+ky+lz)}\, dv = \int_v \rho(\mathbf{r})e^{2\pi i(\mathbf{h}\cdot\mathbf{r})} dv \tag{8.32}$$

While structure factors are usually calculated in terms of scattering from discrete atoms, modern computers are sufficiently fast that calculations by Eq. (8.32) are feasible for a reasonable size of the elementary volume.[5] However, introducing this equation here does not stem from practical numerical considerations but from its use in the following section to derive an expression for the electron density in crystalline solids.

[5]This equation is very similar to the one that is used to obtain scattering factors from theoretical electron distributions for an atom.

8.9. FOURIER SYNTHESES

In the preceding sections of this chapter, we have seen how it is possible to calculate structure factors for a given electron distribution, either atomic or continuous. It is also necessary to perform the inverse operation to obtain an electron distribution given a set of structure factors. Since crystals are periodic structures, they are most naturally described by periodic functions. Of these, a series of cosine and sine terms with appropriate coefficients and with arguments that are successive multiples of x has proved most useful. Such series are termed *Fourier series* in honor of the French mathematician who made extensive use of them. They have important theoretical implications and find wide application in practical problems arising in science and engineering.[6]

One form of a general one-dimensional Fourier series can be written as follows:

$$f(x) = a_0 + a_1 \cos 2\pi x + a_2 \cos 2\pi(2x) + \cdots + a_n \cos 2\pi(nx) + b_1 \sin 2\pi x + b_2 \sin 2\pi(2x) + \cdots + b_n \sin 2\pi(nx) \quad (8.33)$$

$$= a_0 + \sum_{h=1}^{n} (a_h \cos 2\pi hx + b_h \sin 2\pi hx) \quad (8.34)$$

where the h's are integers, a and b are constants, and x is a fraction of a period. For the proof that a function, periodic or otherwise, can be fitted by such a series, the reader is referred to more specialized mathematical works.[7,8]

A simple example of a Fourier series is the four-term cosine approximation to the periodic step function of Fig. 8.8a:

$$y = \pi/4 + \cos 2\pi x - \tfrac{1}{3}\cos 2\pi(3x) + \tfrac{1}{5}\cos 2\pi(5x) \quad (8.35)$$

Successive terms of this series are shown in Figs. 8.8b–e, and their sum, the approximation to the function, is shown in Fig. 8.8f. Only the constant term y has a mean value different from zero; hence the mean value of the series is the value of this term, $\pi/4$. In this case just four terms give a fair approximation. Additional terms would give rapid improvement, and by taking a sufficient number of terms, the function, except at discontinuities, could be fitted to any desired degree of accuracy.

In this example, the series involves only cosine terms because the

[6]For examples, see Mellor, *Higher Mathematics* . . . , pp. 477–497.

[7]R. V. Churchill, *Fourier Series and Boundary Value Problems*, 2nd ed., McGraw-Hill, New York, 1963, Chapters 4–6.

[8]P. Franklin, *An Introduction to Fourier Methods and the Laplace Transform*, Dover, New York, 1958, Chapter 2.

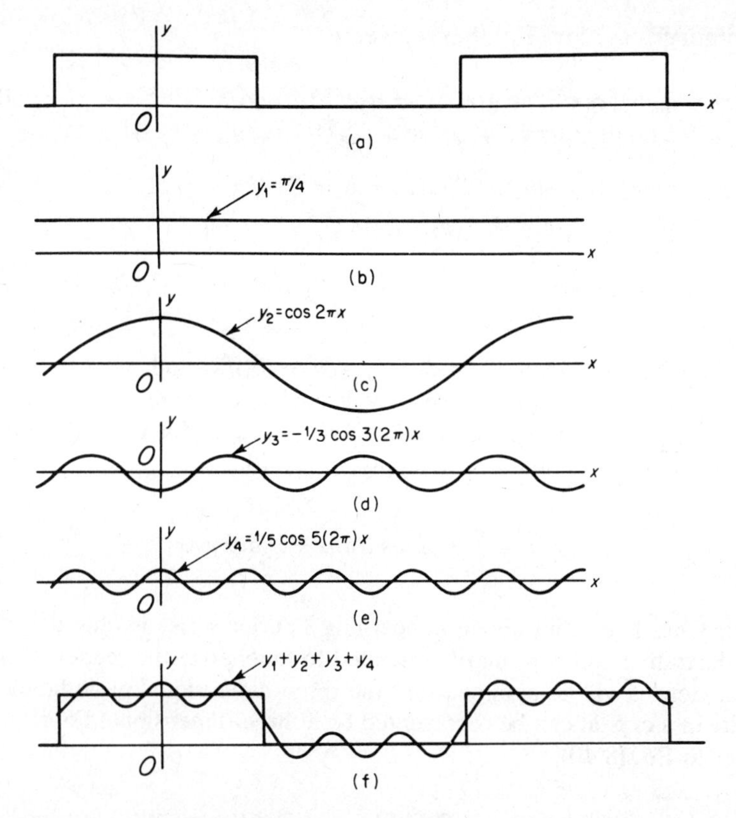

Figure 8.8. (*a*) A periodic step function. (*b*)–(*e*) Graphs of the first four terms of the Fourier series representing (*a*). (*f*) Sum of terms represented by (*b*)–(*e*) as an approximation to the function.

function to be fitted is centrosymmetric, that is, $f(x) = f(-x)$, and the cosine function has this property. If the function in Fig. 8.8*a* were modified so as to destroy the center of symmetry, the series representing it would include sine terms as well as cosine terms.[9]

It is often convenient to represent the Fourier series in exponential form. By an appropriate substitution in Eq. (8.19) and in view of Eqs. (8.20) and (8.21), we find

$$\cos x = \frac{e^{ix} + e^{-ix}}{2} \tag{8.36}$$

$$\sin x = \frac{-i(e^{ix} - e^{-ix})}{2} \tag{8.37}$$

[9]Mellor, *Higher Mathematics . . .*, pp. 473–476.

Substituting these into Eq. (8.34) gives

$$f(x) = a_0 + \tfrac{1}{2}[a_1 e^{2\pi ix} + a_1 e^{-2\pi ix} + a_2 e^{2\pi i(2x)} + a_2 e^{-2\pi i(2x)} + \cdots] \qquad (8.38)$$
$$- (i/2)[b_1 e^{2\pi ix} - b_1 e^{-2\pi ix} + b_2 e^{2\pi i(2x)} - b_2 e^{-2\pi i(2x)} + \cdots]$$

$$f(x) = a_0 + \tfrac{1}{2}[(a_1 - ib_1)e^{2\pi ix} + (a_2 - ib_2)e^{2\pi i(2x)} + \cdots$$
$$+ (a_1 + ib_1)e^{-2\pi ix} + (a_2 + ib_2)e^{-2\pi i(2x)} + \cdots] \qquad (8.39)$$

$$f(x) = \sum_{-n}^{n} C_h e^{2\pi ihx} \qquad (8.40)$$

where we have defined $C_0 = a_0$, $C_h = (a_h - ib_h)/2$, and $C_{\bar{h}} = (a_h + ib_h)/2$. This is a general form of a one-dimensional Fourier series in exponential form. An alternative expression, which is often used for both manipulation and computation, is obtained by applying Eq. (8.24):

$$f(x) = \sum_{-n}^{n} C_h(\cos 2\pi hx + i \sin 2\pi hx) \qquad (8.41)$$

What has been said above concerning Fourier series is only descriptive and illustrative and is primarily intended to familiarize the reader with such series. Here it will be assumed that the three-dimensional periodic electron density in a crystal can be represented by a three-dimensional Fourier series similar to Eq. (8.40),

$$\rho(x, y, z) = \sum_{h'} \sum_{k'} \sum_{l'} C_{h'k'l'} e^{2\pi i(h'x+k'y+l'z)} \qquad (8.42)$$

where h', k', l' are integers between $-\infty$ and ∞.

In order to determine the coefficients $C_{h'k'l'}$ in the three-dimensional Fourier series representing electron density, we substitute from Eq. (8.42) into (8.32) to give

$$F_{hkl} = \int_v \sum_{h'} \sum_{k'} \sum_{l'} C_{h'k'l'} e^{2\pi i(h'x+k'y+l'z)} e^{2\pi i(hx+ky+lz)}\, dv \qquad (8.43)$$

$$F_{hkl} = \int_v \sum_{h'} \sum_{k'} \sum_{l'} C_{h'k'l'} e^{2\pi i[(h+h')x+(k+k')y+(l+l')z]}\, dv \qquad (8.44)$$

The exponential is periodic [see the trigonometric form of a complex number in Eq. (8.24)], and the integral over one period is zero for all terms except that one for which $h' = -h$, $k' = -k$, $l' = -l$. In this case the periodicity disappears and

$$F_{hkl} = \int_v C_{\overline{hkl}}\, dv = VC_{\overline{hkl}} \qquad (8.45)$$

$$C_{\overline{hkl}} = (1/V)F_{hkl} \tag{8.46}$$

In Eq. (8.42) substituting $\bar{h}, \bar{k}, \bar{l}$ for h', k', l' and $(1/V)F_{hkl}$ for $C_{\overline{hkl}}$, and remembering that summing over $\bar{h}, \bar{k}, \bar{l}$ carries the same meaning as summing over h, k, l lead to the series

$$\rho(x, y, z) = \frac{1}{V}\sum_h \sum_k \sum_l F_{hkl} e^{-2\pi i(hx+ky+lz)} \tag{8.47}$$

Comparison of the expression for the electron density, Eq. (8.47), with that for the structure factor, Eq. (8.26), shows their marked similarity. That they should be closely related is suggested by the fact that Eq. (8.47) is an expression for the electron density in direct space in terms of the structure factors in reciprocal space, while Eq. (8.26) represents the structure factors in terms of electron density. Put another way, the electron density is the *Fourier transform* of the structure factors, while the structure factors are in turn the Fourier transform of the electron density. It should be noted, however, that this does not imply that the equations relating the two quantities are identical. Indeed, the transformation in one direction is always associated with a positive exponential term [e.g., Eq. (8.26)] and the reverse transformation with a negative term [Eq. (8.47)]. It is a matter of convention which sign is used for which equation, but the distinction must be maintained.

In the same way that the structure factor expression can be written in terms of a dot product of two vectors [Eq. (8.28)], Eq. (8.47) can be written as

$$\rho(\mathbf{r}) = \frac{1}{V}\sum F_{\mathbf{h}} e^{-2\pi i(\mathbf{h}\cdot\mathbf{r})} \tag{8.48}$$

This notation is more compact and convenient for general discussions of Fourier transforms.

An alternative expression for a three-dimensional Fourier series can be obtained by noting that the structure factor can be written in the form

$$F_{hkl} = |F_{hkl}|e^{i\alpha_{hkl}} = |F_{hkl}|e^{2\pi i\alpha'_{hkl}} \tag{8.49}$$

where α_{hkl} is the phase angle in radians and α'_{hkl} is the phase angle in cycles. Substitution into Eq. (8.47) gives

$$\rho(x, y, z) = \frac{1}{V}\sum_h \sum_k \sum_l |F_{hkl}|e^{2\pi i\alpha'_{hkl}} e^{-2\pi i(hx+ky+lz)} \tag{8.50}$$

$$\rho(x, y, z) = \frac{1}{V}\sum_h \sum_k \sum_l |F_{hkl}|e^{-2\pi i(hx+ky+lz-\alpha'_{hkl})} \tag{8.51}$$

Expanding Eq. (8.51) in terms of sine and cosine, and then assuming that Friedel's law holds so that the sine terms cancel for pairs of F_{hkl} and $F_{\overline{hkl}}$ (see Sections 8.5 and 8.11) leads to

$$\rho(x, y, z) = \frac{1}{V} \sum_h \sum_k \sum_l |F_{hkl}| \cos 2\pi(hx + ky + lz - \alpha'_{hkl}) \qquad (8.52)$$

This form of the three-dimensional Fourier series is sometimes advantageous because the phase angle $2\pi\alpha'_{hkl}$ for each coefficient appears explicitly.

8.10. STRUCTURE FACTOR AND FOURIER EXAMPLE

The necessary basic formulas for determining crystal structures are now in hand, and their use will be illustrated by a simple example.

Consider a hypothetical one-dimensional structure of period 10 Å. Using Cu K_α radiation ($\lambda = 1.5418$ Å) and applying Bragg's law (Section 2.2), we find that there will be 25 orders of h ranging from -12 through 0 to $+12$ (Table 8.1). The values of h and $\sin\theta$ are listed for each diffracted ray, together with the structure factors F_h for a structure with carbon atoms at $x = 1.833$ Å and 8.167 Å. Because of the assumed periodic nature of the structure, this is equivalent to having atoms at $+1.833$ Å and -1.833 Å, that is, the structure is centrosymmetric. From Eq. (8.26) we have, then, for

TABLE 8.1 Data for a One-Dimensional Hypothetical Structure with $a = 10$ Å

h	$\sin\theta$	F_h
0	0.000	12
±1	0.077	5
±2	0.154	−7
±3	0.231	−8
±4	0.308	−1
±5	0.385	5
±6	0.462	4
±7	0.539	−1
±8	0.616	−4
±9	0.693	−2
±10	0.770	2
±11	0.847	3
±12	0.924	1

the structure factor

$$F_h = \sum_{j=1}^{2} f_j e^{2\pi i h x_j} \tag{8.53}$$

$$F_h = \sum_{j=1}^{2} f_j(\cos 2\pi h x_j + i \sin 2\pi h x_j) \tag{8.54}$$

Since $x_{C2} = -x_{C1}$,

$$F_h = f_C[\cos 2\pi h x_{C1} + \cos 2\pi h(-x_{C1}) + i \sin 2\pi h x_{C1} + i \sin 2\pi h(-x_{C1})] \tag{8.55}$$

Since $\cos(-x) = \cos x$, and $\sin(-x) = -\sin x$,

$$F_h = 2 f_C \cos 2\pi h x_{C1} \tag{8.56}$$

The simplification resulting from the cancellation of the imaginary terms and the combination of the real terms is due to the centrosymmetry of the structure, that is, $x_{C2} = -x_{C1}$. In general for centrosymmetric structures, the imaginary terms cancel in pairs and the structure factor is simply a sum of cosine terms, one term for one atom of each centrosymmetric pair multiplied by 2 to account for the contribution of the other atom of the pair. Since

$$\sum f_C i \sin 2\pi h x = 0 \tag{8.57}$$

$\tan \alpha = 0$ and $\alpha = 0$ or π. These two phase angles correspond to the assignment of a + or a − sign to the magnitude $|F|$ in Table 8.1.

The column headed F_h lists "observed" amplitudes along with the phases (± 1) as calculated from Eq. (8.56). It is the intensities that are the experimental observables, and from them the amplitudes are derived. But, in general, it is not possible to observe the phases. This lack of information gives rise to the *phase problem*, which will be dealt with in Chapters 11–14. For the present we will take the phases as known in order to illustrate how electron density can be computed and how it can differ from that actually present.

By analogy with Eq. (8.47) we write for the one-dimensional Fourier series representing the electron density

$$\rho(x) = \frac{1}{L} \sum_h F_h e^{-2\pi i h x} \tag{8.58}$$

For a centrosymmetric structure, Eq. (8.58) can be simplified in a manner similar to that used for the structure factor expression. Expanding in terms of cosine and sine,

$$\rho(x) = \frac{1}{L} \sum_{-12}^{12} F_h[\cos 2\pi hx - i \sin 2\pi hx] \tag{8.59}$$

It can be seen from Eq. (8.56) that $F_h = F_{\bar{h}}$. Hence

$$\rho(x) = \frac{1}{L} \left[\sum_{-12}^{-1} F_h(\cos 2\pi hx - i \sin 2\pi hx) + F_0 + \sum_{1}^{12} F_h(\cos 2\pi hx - i \sin 2\pi hx) \right] \tag{8.60}$$

or

$$\rho(x) = \frac{1}{L} \left[F_0 + \sum_{1}^{12} F_h(\cos 2\pi hx + i \sin 2\pi hx) + \sum_{1}^{12} F_h(\cos 2\pi hx - i \sin 2\pi hx) \right] \tag{8.61}$$

Combining corresponding terms in Eq. (8.61),

$$\rho(x) = \frac{1}{L} \left[F_0 + 2 \sum_{1}^{12} F_h \cos 2\pi hx \right] \tag{8.62}$$

Substitution of the structure factors from the column headed F_h of Table 8.1 gives for the series representing the electron density at any point x,

$$\rho(x) = \tfrac{1}{10}[12 + 2(5) \cos 2\pi x - 2(7) \cos 2\pi(2x) - 2(8) \cos 2\pi(3x) - \cdots + 2(1) \cos 2\pi(12x)] \tag{8.63}$$

The 13 terms of this equation appear in the columns of Table 8.2 at values of $x = 0.00, 0.02, \ldots, 0.24$. Evaluation to $x = 0.24$ is sufficient to illustrate the behavior of the electron density in the region of an atom. The assumed electron density in the atom and that reconstructed from the Fourier series are shown in Fig. 8.9.

The maximum electron density from the Fourier series is almost, but not quite, coincident with the assumed position of the atom. Both this inexactness and the ripples in the curve from the Fourier series away from the atom arise largely from series termination errors due to truncation at $h = 12$. To feed back exactly the electron density that was assumed would require a series with an infinite number of terms and error-free coefficients, ideals impossible to attain in practice. How errors due to these effects may be minimized will be dealt with in a subsequent chapter.

TABLE 8.2 $2F_h \cos 2\pi h(x)$

h	F_h	x												
		0.00	0.02	0.04	0.06	0.08	0.10	0.12	0.14	0.16	0.18	0.20	0.22	0.24
0	12	12.0	12.0	12.0	12.0	12.0	12.0	12.0	12.0	12.0	12.0	12.0	12.0	12.0
1	5	10.0	9.9	9.7	9.3	8.8	8.1	7.3	6.4	5.4	4.3	3.1	1.9	0.6
2	−7	−14.0	−13.6	−12.3	−10.2	−7.6	−4.3	−0.8	2.7	6.0	9.0	11.3	13.0	13.8
3	−8	−16.0	−14.8	−11.7	−6.9	−1.0	4.3	10.2	14.1	15.8	15.5	12.9	7.5	3.0
4	−1	−2.0	−1.8	−1.1	−0.1	0.8	1.6	2.0	1.9	1.3	0.4	−0.6	−1.5	−1.9
5	5	10.0	8.1	3.1	−3.1	−8.1	−10.0	−8.1	−3.1	3.1	8.1	10.0	8.1	3.1
6	4	8.0	5.8	0.5	−5.1	−7.9	−6.5	−1.5	4.3	7.8	7.0	2.5	−3.4	−7.4
7	−1	−2.0	−1.3	0.4	1.8	1.9	0.6	−1.1	−2.0	−1.5	0.1	1.6	1.9	0.9
8	−4	−8.0	−4.3	3.4	7.9	5.1	−2.5	−7.8	−5.8	1.5	7.4	6.5	−0.5	−7.0
9	−2	−4.0	−1.7	2.6	3.9	0.8	−3.2	−3.5	0.2	3.7	2.9	−1.2	−4.0	−2.2
10	2	4.0	1.2	−3.2	−3.2	1.2	4.0	1.2	−3.2	−3.2	1.2	4.0	1.2	−3.2
11	3	6.0	1.1	−5.6	−3.2	4.4	4.9	−2.6	−5.8	0.4	5.9	1.9	−5.3	−3.8
12	1	2.0	0.1	−2.0	−0.4	1.9	0.6	−1.9	−0.9	1.8	1.1	−1.6	−1.3	1.5
1/10	Σ	0.6	0.1	−0.4	0.3	0.2	1.0	0.5	2.1	5.4	7.5	6.2	3.0	0.9

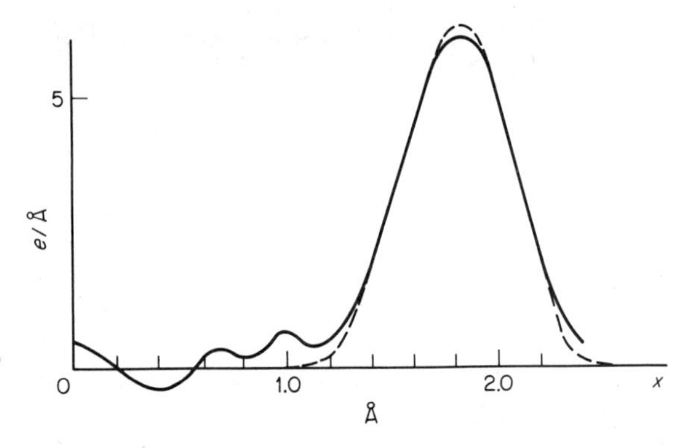

Figure 8.9. Dashed line, electron density in a carbon atom projected onto a line through its center; solid line, its representation by a 13-term Fourier series.

8.11. SPECIAL STRUCTURE FACTOR EXPRESSIONS

Although structure factors can always be calculated by using the general expression [Eq. (8.26)], there are many cases in practice in which it is more convenient to reduce this expression by mathematical manipulations to special forms. These can be used to save computing time, but more important they can point up phenomena such as systematic extinctions that are not otherwise immediately obvious. The special forms are obtained by modifying the general structure factor expression to include explicitly the effects of the symmetry elements. Thus for N atoms in the cell,

$$F_{hkl} = \sum_{j}^{N} f_j e^{2\pi i(hx_j+ky_j+lz_j)} \tag{8.64}$$

$$F_{hkl} = \sum_{n} f_n \left(\sum_{m} e^{2\pi i(hx_{m,n}+ky_{m,n}+lz_{m,n})} \right) \tag{8.65}$$

where the N atoms are now divided into n unique atoms (one asymmetric unit), each of which is converted by the symmetry operations of the cell into a set of m symmetry-related atoms (m = the number of asymmetric units).[10] Consequently,

$$F_{hkl} = \sum_{n} f_n (E_{hkl})_n \tag{8.66}$$

[10]Clearly, caution is required if some atoms in an asymmetric unit lie in special positions (see Section 3.4), since such positions are characteristically unchanged by some symmetry operation(s) of the space group. An atom placed at such a location will be transformed into itself and will appear with a multiple weight. The simplest, although not the only, solution is to multiply the scattering factors used for an atom in a special position by the ratio of the number of its equivalent locations to m and allow the normal symmetry generation process to bring the atoms back to full weight.

and

$$(E_{hkl})_n = \sum_m e^{2\pi i(hx_{m,n}+ky_{m,n}+lz_{m,n})} \tag{8.67}$$

Since Eq. (8.66) holds for an atom placed anywhere in the cell, we can drop the subscript n in Eq. (8.67) and consider E_{hkl} as a function of x, y, z for the space group described by the symmetry elements relating the m equivalent positions. The most important such element is a center of symmetry, for which the equivalent positions are x, y, z and $\bar{x}, \bar{y}, \bar{z}$. Using the notation of Eq. (8.67) and expressing the equivalent atoms explicitly,

$$E_{hkl} = e^{2\pi i(hx+ky+lz)} + e^{-2\pi i(hx+ky+lz)} \tag{8.68}$$

Expressing the exponentials in trigonometric form [cf. Eq. (8.24)],

$$\begin{aligned} E_{hkl} &= \cos 2\pi(hx+ky+lz) + i \sin 2\pi(hx+ky+lz) \\ &\quad + \cos 2\pi(hx+ky+lz) - i \sin 2\pi(hx+ky+lz) \end{aligned} \tag{8.69}$$

$$E_{hkl} = 2 \cos 2\pi(hx+ky+lz) \tag{8.70}$$

$$F_{hkl} = 2 \sum_n f_n \cos 2\pi(hx_n+ky_n+lz_n) \tag{8.71}$$

Thus for any centrosymmetric model with the atoms in general positions, the summation can be carried out over one atom of each related pair, and doubled. No sine terms need be calculated.

A graphical representation of this result can be obtained by noting that Eq. (8.69) implies

$$E_{hkl} \propto A + iB + A - iB \tag{8.72}$$

Figure 8.10*a* shows this expression in terms of the components A and B, and Fig. 8.10*b* for the corresponding **f**'s. It is clear that for centrosymmetric crystals the resultant of $\mathbf{f}_1$ and $\mathbf{f}_{\bar{1}}$ lies along x, that is, that α is 0 or π.

Similar reductions can be made for any other symmetry element or elements. In particular, those involving translation lead to the extinction rules listed in Table 5.2. As examples to illustrate the methods, we consider three cases: (1) a 2-fold screw axis along [001], (2) a net glide n reflecting in (010), and (3) an A-centered lattice.

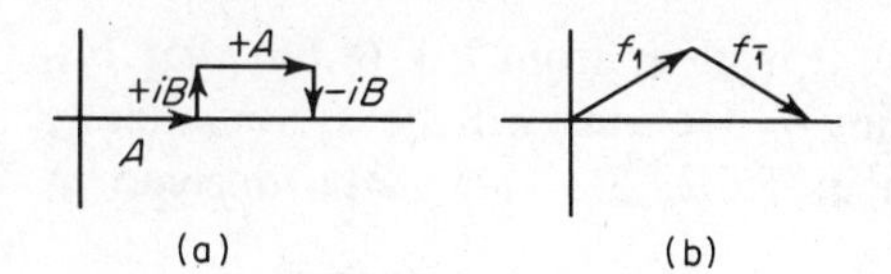

Figure 8.10. Graph showing that the vector sum in a centrosymmetric structure must lie along the real axis.

1. 2_1: Without loss of generality, we take the 2-fold screw axis coincident with the c axis of the unit cell, equivalent positions: x, y, z; $-x, -y, z+\frac{1}{2}$. Again using Eq. (8.67),

$$E_{hkl} = e^{2\pi i(hx+ky+lz)} + e^{-2\pi i(hx+ky-lz-l/2)} \tag{8.73}$$

$$E_{hkl} = e^{2\pi i(hx+ky)}e^{2\pi ilz} + e^{-2\pi i(hx+ky)}e^{2\pi ilz}e^{2\pi il/2} \tag{8.74}$$

$$E_{hkl} = e^{2\pi ilz}[e^{2\pi i(hx+ky)} + e^{-2\pi i(hx+ky)}e^{\pi il}] \tag{8.75}$$

In general, $E_{hkl} \neq 0$ except in the case of the $00l$ reflections, when

$$E_{00l} = e^{2\pi ilz}(1+e^{\pi il}) \tag{8.76}$$

$$E_{00l} = 2e^{2\pi ilz} \qquad \text{for } l=2n \tag{8.77}$$

$$E_{00l} \equiv 0 \qquad \text{for } l=2n+1 \tag{8.78}$$

2. An n-glide reflecting in (010): Without loss of generality we take the glide plane as $y=0$: equivalent positions x, y, z; $x+\frac{1}{2}, -y, z+\frac{1}{2}$. As in the preceding case, using Eq. (8.67),

$$E_{hkl} = e^{2\pi i(hx+ky+lz)} + e^{2\pi i(hx+h/2-ky+lz+l/2)} \tag{8.79}$$

$$E_{hkl} = e^{2\pi i(hx+lz)}e^{2\pi iky} + e^{2\pi i(hx+lz)}e^{-2\pi iky}e^{2\pi i(h/2+l/2)} \tag{8.80}$$

$$E_{hkl} = e^{2\pi i(hx+lz)}[e^{2\pi iky} + e^{-2\pi iky}e^{\pi i(h+l)}] \tag{8.81}$$

In general $E_{hkl} \neq 0$ except in the case of the $h0l$ reflections, when

$$E_{h0l} = e^{2\pi i(hx+lz)}[1+e^{\pi i(h+l)}] \tag{8.82}$$

$$E_{h0l} = 2e^{2\pi i(hx+lz)} \qquad \text{for } h+l=2n \tag{8.83}$$

$$E_{h0l} \equiv 0 \qquad \text{for } h+l=2n+1 \tag{8.84}$$

3. An A-centered lattice: equivalent positions x, y, z; $x, y+\frac{1}{2}, z+\frac{1}{2}$. As in the preceding cases, Eq. (8.67) gives

$$E_{hkl} = e^{2\pi i(hx+ky+lz)} + e^{2\pi i(hx+ky+k/2+lz+l/2)} \tag{8.85}$$

$$E_{hkl} = e^{2\pi i(hx+ky+lz)}[1+e^{\pi i(k+l)}] \tag{8.86}$$

$$E_{hkl} = 2e^{2\pi i(hx+ky+lz)} \qquad \text{for } k+l=2n \tag{8.87}$$

$$E_{hkl} \equiv 0 \qquad \text{for } k+l=2n+1 \tag{8.88}$$

Under the conditions when $E_{hkl} \equiv 0$, it is clear from Eq. (8.66) that F_{hkl} will also be zero and the corresponding reflections will be systematically absent. The various extinctions given in Table 5.2 can all be derived in similar fashion.

Reduced expressions in trigonometric form, together with the implied extinctions and other relevant information are listed in the *International Tables*[11] for all space groups. Although for the more involved symmetry elements and combinations of symmetry elements the process of reduction can become difficult, it is a rewarding experience to derive the simplified expressions for oneself before checking against the tables.

8.12. FAST FOURIER TRANSFORM[12]

For many years the actual calculation of Fourier transforms was carried out with factored forms of Eqs. (8.47) and (8.52). Starting in the 1960s it became widely appreciated[13] that it was possible to speed the process greatly by suitably separating the terms in the summation, doing a number of shorter and faster transformations on the separated elements, and then combining these transforms at the end. The fundamental ideas are most easily seen in a one-dimensional example.

Given a series of the form

$$\rho(x) = \sum_{0}^{H} F_h e^{2\pi i h x} \tag{8.89}$$

this can be factored into two summations over alternating terms in even and odd h.

$$\rho = \sum_{0}^{H/2} F_{2h} e^{2\pi i(2h)x} + \sum_{0}^{H/2} F_{2h+1} e^{2\pi i(2h+1)x} \tag{8.90}$$

Removing the common term in the odd summation gives

$$\rho = \sum_{0}^{H/2} F_{2h} e^{2\pi i(2h)x} + e^{2\pi i x} \sum_{0}^{H/2} F_{2h+1} e^{2\pi i(2h)x} \tag{8.91}$$

In Eq. (8.91) the exponential terms are the same in both summations and so need to be calculated only once and used twice. This factorization corresponds to separating the data into two similar lots with a constant phase shift $e^{2\pi i x}$ between them, carrying out the same summation with different coefficients (F's) on each, and then combining them with the aid of the phase shift.

[11] *International Tables*, Vol. I, pp. 368–525.

[12] E. Oran Brigham, *The Fast Fourier Transform*, Prentice-Hall, Englewood Cliffs, NJ, 1974.

[13] This fast Fourier calculation is often referred to as the Cooley–Tukey algorithm, though the underlying ideas go back at least as far as Danielson and Lanczos in 1942. In fact, similar concepts were used by Beevers and Lipson in 1936 in the construction of calculating strips for the hand computation of Fourier transforms.

The most important quality of this derivation is that it can be applied repeatedly to the individual summations in Eq. (8.91) as long as the number of terms in each series is evenly divisible into even and odd. It is for this reason that many general-purpose FFT programs (as well as the corresponding chips that perform the process in hardware) expect the number of data points to be transformed to be an integral power of 2 (e.g., 512 or 1024). The bookkeeping necessary to apply the proper phase shifts at the right time and to combine the appropriate summations is a little complicated but is easy for a computer to keep track of, and the net result is a transform in a time proportional to $H \log_2 H$ rather than H^2. For large H the saving in time is immense.

The process of factorization is not limited to division by 2 but can be applied to any factor, although with greater complexity. Similarly, the process can be generalized to two or three dimensions. Most crystallographic FFT programs now handle all prime factors up to some specified number but can be expected to run faster if the number of data points[14] in each dimension is cleanly factorable into a collection of small primes.

8.13. ANOMALOUS SCATTERING AND ITS EFFECTS

In the treatment of the structure factor, we have assumed that the scattering factors f_j are represented by real numbers. This is true for the values in common use, since they are calculated on the assumption that the frequency of the incident radiation differs widely from that of any natural absorption frequency of the atoms. Although this is generally the case for light atoms and the radiations commonly used in X-ray diffraction, it is often not true for heavier atoms. It is certainly not justified for atoms with Z near that of the element used to generate the incident X-rays. If the frequency of the incident beam does fall near a natural frequency of some atom type, an anomalous phase change occurs on scattering by atoms of that type, resulting in what is termed *anomalous scattering* or *dispersion*.[15] The scattering itself is anomalous in the sense that correction terms must be applied to the normal scattering factors. Thus they are represented by the expression

$$f_o^{\text{anom}} = f_o + \Delta f' + i\,\Delta f'' = f' + i\,\Delta f'' \qquad (8.92)$$

[14]Most FFTs store the F values in the array that will ultimately hold the final Fourier map, padding the extra spaces (the number of F's must be less than the number of map points) with zeros, so that the formal number of data points is the number of divisions assigned to each cell edge. Thus one can often choose "good" (i.e., factorable) values that give approximately the desired map resolution. See L. F. TenEyck, *Acta Cryst.*, **A33**, 486 (1977); **A29**, 183 (1973).

[15]Although this phenomenon is commonly referred to as "anomalous" dispersion, it is closely related to the more familiar change of refractive index for light as a function of wavelength (as demonstrated by dispersion by a prism) and is not anomalous at all.

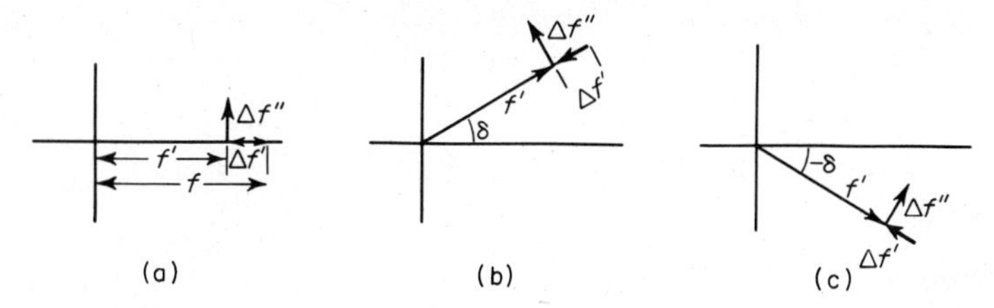

Figure 8.11. Vector representation of f, $\Delta f'$, and $\Delta f''$ for an atom showing anomalous scattering. (*a*) At the origin; (*b*), (*c*) at other positions in the unit cell.

where f_o is the normal scattering factor, $\Delta f'$ is a real correction term (usually negative), and $\Delta f''$ is the imaginary component.[16] Figure 8.11*a* shows the effect of this complex scattering factor on the wave scattered anomalously by an atom at the origin, while Figs. 8.11*b* and *c* show that the relationship between f' and $\Delta f''$ remains the same as the phase angle changes, that is, as the anomalous scatterer is moved around the cell.

The magnitude of dispersion effects is evident from inspecting Table 8.3, which gives $\Delta f'$ and $\Delta f''$ for selected atoms and radiations at $(\sin\theta)/\lambda = 0$. Figure 8.12 shows the variation of $\Delta f'$ and $\Delta f''$ with atomic number for copper radiation and nearby elements. It is evident that the effect can be appreciable even for elements well removed in the periodic table from the element used to generate the X-ray beam. It can be seen too that the effects are more pronounced for long-wavelength radiation.

The correction terms $\Delta f'$ and $\Delta f''$ that arise from anomalous scattering are almost independent of $\sin\theta$. This occurs because the effect involves primarily the inner electrons of the atom. Hence for a given atom the effects of dispersion are relatively greater at high $\sin\theta$ than at low. For example, at $(\sin\theta)/\lambda = 0$, $(\Delta f''_{\text{Co}}/f'_{\text{Co}})_{\text{Cu K}_\alpha} = 3.6/24.5$, while at $(\sin\theta)/\lambda = 0.6$ (approximately the limit of Cu K_α data) the ratio is 3.6/7.8, an increase by a factor of 3. Figure 8.13 shows plots of f_{Co}, $-\Delta f'_{\text{Co}}$, $\Delta f''_{\text{Co}}$ and $f_{\text{Co}} + \Delta f'_{\text{Co}} = f'_{\text{Co}}$, as a function of $(\sin\theta)/\lambda$.

TABLE 8.3 Dispersion Effects on Scattering Factors

	Cr K_α		Cu K_α		Mo K_α	
	$\Delta f'$	$\Delta f''$	$\Delta f'$	$\Delta f''$	$\Delta f'$	$\Delta f''$
K	0.1	2.1	0.4	1.1	0.2	0.3
Br	−0.2	2.6	−0.8	1.3	−0.4	2.5
Ag	−0.9	8.2	−0.1	4.3	−1.1	1.1
I	−5.9	12.8	−0.6	6.8	−0.7	1.8
Hg	−4.8	14.1	−5.0	7.7	−3.1	9.2

[16]These quantities are tabulated for various elements and radiation in *International Tables*, Vol. IV, pp. 148–151; but see also M. J. Cooper, *Acta Cryst.*, **A33**, 229 (1977).

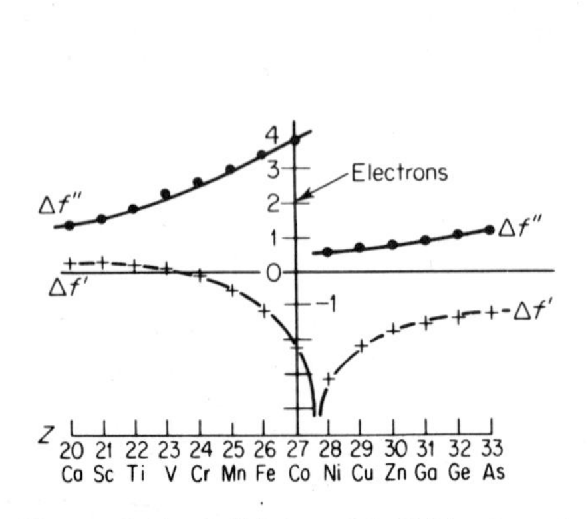

Figure 8.12. $\Delta f'$ (+) and $\Delta f''$ ($\cdot$) plotted as a function of atomic number for Cu K_α radiation.

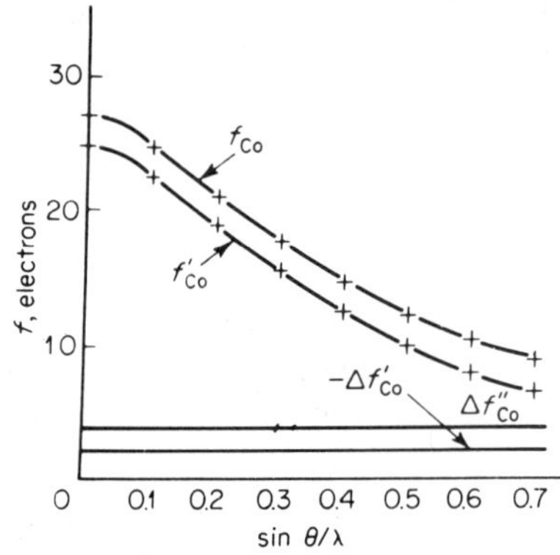

Figure 8.13. f_{Co}, f'_{Co}, $-\Delta f'_{\text{Co}}$, and $\Delta f''_{\text{Co}}$ for Cu K_α radiation as functions of $(\sin\theta)/\lambda$.

The effects of dispersion on F_c can be illustrated by a series of Argand diagrams. Figure 8.14 shows the case of a centrosymmetric structure that consists of a number of atoms that *do not* exhibit dispersion and a pair of atoms that *do*. F_w is the resultant of scattering from the atoms without dispersion, and f' and $\Delta f''$ are the real and imaginary parts of the scattering factors for each atom with dispersion. It can be seen that the phases of reflections, though equal, now differ from 0 or π and that the F's are complex; nevertheless, $|F_{hkl}| = |F_{\bar{h}\bar{k}\bar{l}}|$, and Friedel's law holds. It is evident, however, that unless the proper scattering factor is included for the atoms with dispersion, the magnitude of F_c will be in error.

The case for a noncentrosymmetric structure is shown in Fig. 8.15. As before, F_w is the resultant of scattering from atoms without dispersion, but f' and $\Delta f''$ are now the resultants of scattering with dispersion from one or more identical atoms. In Fig. 8.15*a* is shown the case for F_{hkl}, and in Fig. 8.15*b* that for $F_{\bar{h}\bar{k}\bar{l}}$. Figure 8.15*c* repeats 8.15*a* and reflects 8.15*b* across the real axis so that the difference between $|F_{hkl}|$ and $|F_{\bar{h}\bar{k}\bar{l}}|$ is readily seen. It is clear that for this case Friedel's law does not hold and that $\alpha_{hkl} \neq -\alpha_{\bar{h}\bar{k}\bar{l}}$.

If the scattering factors are not handled properly, dispersion introduces

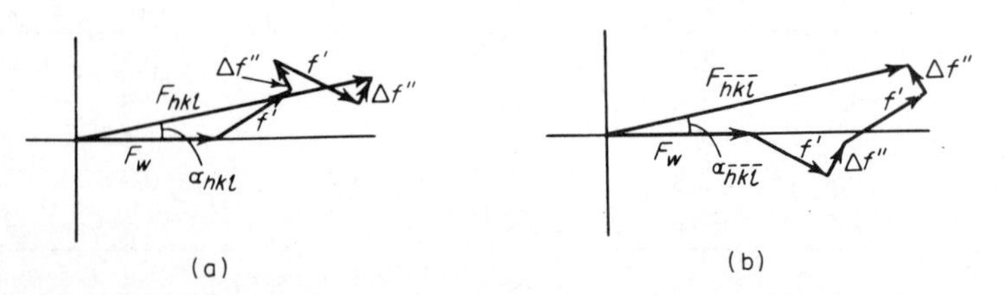

Figure 8.14. Vector representation showing anomalous scattering for a Friedel pair in a centrosymmetric structure; $F_{hkl} = F_{\bar{h}\bar{k}\bar{l}}$.

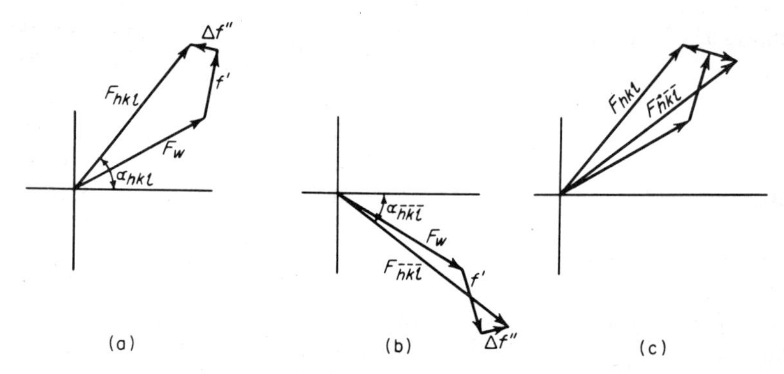

Figure 8.15. Vector representation showing anomalous scattering effects for a Friedel pair in a noncentrosymmetric structure; $F_{hkl} \neq F_{\bar{h}\bar{k}\bar{l}}$.

error. This can be minimized, however, by an appropriate choice of radiation. On the other hand, as will be shown later (Chapters 13 and 18), the effects can be useful, particularly in phase estimation and for determining absolute molecular configurations. In such cases, the radiation should be chosen to enhance the effects so that they are measurable with greater accuracy. One of the attractions of synchrotrons as storage rings is that the emission can be tuned to any desired wavelength and so selected to maximize dispersion effects if desired.

KEY FORMULAS

Structure Factor

$$A_{hkl} = \sum_j f_j \cos 2\pi(hx_j + ky_j + lz_j) \qquad (8.10)$$

$$B_{hkl} = \sum_j f_j \sin 2\pi(hx_j + ky_j + lz_j) \qquad (8.11)$$

$$|F_{hkl}| = \sqrt{A_{hkl}^2 + B_{hkl}^2} \qquad (8.9)$$

$$\alpha_{hkl} = \tan^{-1}\left(\frac{B_{hkl}}{A_{hkl}}\right) \qquad (8.13)$$

Fourier Series

$$\rho(x, y, z) = \frac{1}{V}\sum_h \sum_k \sum_l F_{hkl} e^{-2\pi i(hx+ky+lz)} \qquad (8.47)$$

$$\rho(x, y, z) = \frac{1}{V}\sum_h \sum_k \sum_l |F_{hkl}| \cos 2\pi(hx + ky + lz - \alpha'_{hkl}) \qquad (8.52)$$

Anomalous Scattering

$$f_o^{anom} = f_o + \Delta f' + i\,\Delta f'' \tag{8.92}$$

BIBLIOGRAPHY

Buerger, M. J., *Crystal–Structure Analysis*, Wiley, New York, 1960, pp. 7–24, 259–272, 352–406.

Glusker, J. P., and K. N. Trueblood, *Crystal Structure Analysis*, 2nd ed., Oxford University Press, New York, 1985, Chapters 5 and 6.

James, R. W., *The Optical Principles of the Diffraction of X-Rays*, Cornell University Press, Ithaca, NY, 1965, Chapter VII.

Jeffery, J. W., *Methods in X-Ray Crystallography*, Academic, London, 1971, pp. 93–115.

Ladd, M. F. C., and R. H. Palmer, *Structure Determination by X-Ray Crystallography*, Plenum, New York, 1985, pp. 143–179.

Lipson, H., and W. Cochran, *The Determination of Crystal Structures*, Cornell University Press, Ithaca, NY, 1966, pp. 9–16, 66–96.

Luger, P., *Modern X-Ray Analysis on Single Crystals*, de Gruyter, New York, 1980, pp. 148–160.

Mellor, J. W., *Higher Mathematics for Students of Chemistry and Physics*, Dover, New York, 1955.

Nyburg, S. C., *X-Ray Analysis of Organic Structures*, Academic, New York, 1961, pp. 96–111.

Press, W. H., B. P. Flannery, S. A. Tenkolsly, and W. T. Vetterling, *Numerical Recipes*, Cambridge University Press, New York, 1986, Chapter 12.

Ramachandran, G. N., and R. Srinivasan, *Fourier Methods in Crystallography*, Wiley, New York, 1970.

CHAPTER 9

CALCULATION OF STRUCTURE FACTORS AND FOURIER SYNTHESES

In our first edition it was explicitly assumed that anyone planning to undertake X-ray crystallographic studies of organic molecules would use a high-speed computer. Now it goes without saying, and the choices come between alternative modes of computation. Advances have occurred in computer practice as in all other areas of crystallography, and the actual methods used in programs are often more sophisticated and less easily understood than those outlined in Chapter 8. This is particularly true for Fourier calculations, which have been revolutionized by the fast Fourier transform (FFT) technique. On the other hand, the power of modern computing allows the calculation of structure factors by pure brute force methods that are inelegant but easy to program and follow.

The exact data that are required in practice and the manner in which they are handled will depend on the particular programs being used. There are, however, a number of general considerations that are more or less independent of program details. It is these that we shall treat in the present chapter.

9.1. STRUCTURE FACTORS[1–3]

The normal methods for calculating structure factors are all based on the approximation of atoms that are spherically symmetric. Given this con-

[1]Lipson and Cochran, *The Determination of Crystal Structures*, pp. 66–82.
[2]Buerger, *Crystal-Structure Analysis*, pp. 259–282.
[3]Rollett, *Computing Methods in Crystallography*, pp. 38–46.

dition, it can be shown, as pointed out in Chapter 7, that the effects of finite atomic size are the same as those produced by assuming point atoms of suitably varying scattering power located at the positions of the real atomic centers. Thus all that is needed for the calculation of structure factors based on Eq. (8.8) is the location of the atoms to be used.

The scattering factor curves for other atoms resemble that of carbon (Fig. 7.4), differing only in the value at $(\sin\theta)/\lambda = 0$ and in the fine details of the shape of the curve. These curves have been calculated for all important atomic species using theoretical electron distributions of varying degrees of elegance. The results are available in tabular form in *International Tables*.[4]

As a first approximation, the relative scattering powers of different kinds of atoms are the same regardless of the radiation used. Secondary effects, generally lumped together under the title of *anomalous scattering*, arise, however, when the wavelength of the incident beam lies near an absorption edge in the scattering element. The proper treatment of dispersion involves the introduction of two scattering factor corrections, $\Delta f'$ and $\Delta f''$, for each kind of atom that will show the effect. Of these, $\Delta f'$ is simply a term to be subtracted (usually) from the normal f_o, and its use requires very little additional computation. The application of $\Delta f''$ is more complex (see Section 8.13), but its effects are often smaller, at least for centrosymmetric crystals.

Because structure factors are usually calculated numerous times during the course of a structural analysis, it is convenient to have the scattering factors available for each reflection without having to obtain them each time from tables. For this reason, the input to the data reduction program may include tabulated scattering factor data for the atom types expected to be found. The output then associates with each reflection interpolated scattering factors that correspond to its value of $(\sin\theta)/\lambda$ and that can be read and used directly by the structure factor program.[5]

We shall assume here that the structure analysis has proceeded to the point that a molecular model, described in terms of the coordinates of its component atoms, is available. How this model was obtained and what its shortcomings may be do not concern us at present; all that is required is that atoms be placed within the unit cell. The natural coordinate system to use in describing the atomic locations in the cell is that of the cell axes, and the natural units are fractions of a unit translation along these axes (see Section 8.4). Thus an atom may be described as being at $x = 0.0550$, $y = 0.1667$, $z = 0.2576$.

[4] *International Tables*, Vol. IV, pp. 71–147.

[5] The method of interpolation in tabulated data is not the only way in which scattering factors can be obtained. Analytic expressions involving a very few empirical parameters have been devised that fit the scattering factor curves of most atoms surprisingly well, and these are increasingly widely used. See *International Tables*, Vol. IV, pp. 99–101.

Since all cells are the same, the addition of ± 1.0000 to any of these coordinates merely corresponds to shifting attention from one atom to an identical one in another cell and can be done at will. The input to a structure factor program normally consists of a list of atomic positions and an indication for each atom of its kind so that the program can select the proper scattering factor.

The usual list of atoms consists of only those in one asymmetric unit, with the program accounting for the others in the cell on the basis of symmetry information given it. The problem of incorporating symmetry is normally handled in one of two ways. The general structure factor expression can be reduced as described in Chapter 8 to a set of simplified equations that take into account the atomic positions generated from the fundamental set. If structure factors are now calculated with these expressions, using the atoms of the asymmetric unit only, the symmetry of the space group is allowed for automatically. An alternative is to generate the remaining atoms by applying the symmetry operations to those of the asymmetric unit and then carry out the structure factor calculation in the most general manner by using all the atoms in the cell in Eq. (8.26).[6]

In general, the first method minimizes the number of operations and is more efficient. For this reason, special forms of the structure factor expression have been calculated for each space group and are tabulated in *International Tables*.[7] For calculations to be done by computer, however, this method has the disadvantage that the program must be changed for each different space group. The problem is not very difficult, but it does reduce the attractiveness of the method for the nonprogramming user. This approach was used more in early programs, when computers were small and slow, than it is at present.

The second, generative, method is now the common technique. It provides more or less complete generality without program modification, and if cleverly programmed it will handle a large number of atoms at high speed. A few simplifications based on the special structure factor equations are generally available even with this sort of program, however, and their use saves computing time. The simplest and most general of these is the classification of a given space group as centrosymmetric or noncentrosymmetric. As shown in Chapter 8, the combined effects of two centrosymmetrically related atoms is to multiply the cosine terms of the structure factor expression Eq. (8.8) by 2 and eliminate the sine terms [Eq. (8.71)]. In general for such a space group it is necessary to calculate with only half the atoms, those that are not centrosymmetrically related, and account for the others by doubling the cosine terms. The sine terms need

[6]For reasons of efficiency, the programs sometimes do not generate the symmetry-related atoms but rather perform equivalent operations on the indices of the reflections. The actual method used is of no concern in practice, however.

[7]*International Tables*, Vol. I, pp. 353–525.

not be computed. Both of these omissions, but particularly the former, save significant amounts of time, and nearly all structure factor programs operate centrosymmetrically or not, as required.

A second useful but less common saving is available when dealing with lattice-centered space groups. Every centered cell corresponds to a primitive cell with one or more identical cells superimposed onto it with an offset that is the translation of the centering operation. The effect of adding the extra atoms to the primitive set is to extinguish one-half (A, B, C, or I centering—one additional set added) or three-fourths (F centering—three additional sets added) of the reflections and multiply the remaining ones by 2 or 4 as the case may be [see Eqs. (8.87)–(8.88)]. Thus, if the data for which structure factors are being calculated do not include the systematically absent reflections, as is usually the case, it is not necessary to calculate over all the atoms in the cell but only over those in the primitive unit (or half of these if the space group is centrosymmetric). The centering can then be accounted for by multiplication by the proper factor.

Other parameters usually provided with the input data are those describing the thermal motion of the atoms. As was discussed in Section 7.3, the usual scattering factor curves are calculated on the assumption of atoms at rest, but in fact atoms in crystals oscillate with appreciable amplitudes about their rest positions. The effect of this motion is to modify the scattering factors of the atoms by multiplying them by the expression

$$e^{-B(\sin\theta)^2/\lambda^2} \tag{9.1}$$

where B is an adjustable parameter related to the magnitude of vibration. Values of B for organic molecules commonly lie between 2 and 5 but can go as low as 1 or above 10.

The correction for thermal motion can be applied to structure factor calculations at several levels of approximation. The simplest of these, that of *overall isotropic* vibrations, assumes that all the atoms are vibrating with the same amplitude and that their motions, like their shapes, are spherically symmetric. This approximation makes for the fastest computation, since expression (9.1) need be evaluated only once per reflection and the result used to multiply the calculated structure factor. Since the temperature factor B provided by the Wilson plot of the intensity data (Section 7.3) is an average factor for the whole molecule, this approach is generally used in the early stages of a structural analysis.

It is obvious that the approximation of equal amplitudes is not a very good one. An atom at the end of a long aliphatic chain can reasonably be expected to be less firmly braced by its neighbors than, say, a quaternary carbon in a rigid ring structure. The *individual atom isotropic* assumption permits the assignment of temperature factors to each atom but retains the idea of spherical symmetry. The change from overall isotropic calculation generally improves the fit between the observed and calculated data markedly, and the individual temperature factors can usually be seen to have physical significance. This added refinement increases calculation time, although not greatly, because expression (9.1) must now be evaluated for each atom that contributes to each reflection and used to modify the scattering factor for that atom.

The final level of elegance is reached with the *individual atom anisotropic* approximation. Here the assumption of spherical symmetry is abandoned and the single atomic thermal parameter is replaced by six parameters that describe the size and orientation of the vibration ellipsoid (see Appendix B). Because of the loss of symmetry this approximation cannot be used with programs using factored expressions for the structure factor calculations. Instead, each atom must be treated individually. The only factorizations that may be used are those accounting for centers of symmetry and lattice centering. The greatly increased number of terms that must be handled can increase the computing significantly, and as anisotropic temperature factors are introduced only in the late stages of analysis we shall postpone their consideration to Chapter 16.

Once the parameters characterizing the model have been provided, the program can turn its attention to the reflection data. The input format for these data to the structure factor program should properly be the same as the output format from the data reduction. That is, the two programs should be part of a set such that the output of one can serve directly as the input for the next.[8] The programs used will determine how much information is read in and how much is generated internally, but a common input list would include the indices, the observed structure factor, and generally $(\sin\theta)/\lambda$ and perhaps a set of scattering factors for each reflection. Commonly programs read a reflection record, perform the calculations, and put out the results before turning to the next reflection. In this way an unlimited number of structure factors can be calculated without requiring excessive storage.

The output from the structure factor program normally includes as a minimum for each reflection the indices, the scaled observed structure amplitude $|F_o|$, the calculated structure amplitude $|F_c|$, and the phase of $|F_c|$. The difference in magnitude (ΔF) between F_o and F_c can also be included, where

$$\Delta F = |F_o| - |F_c| \tag{9.2}$$

as well as any further information that the programmer felt would be useful.

The phase of F_c can be represented in a number of ways. In the most

[8]It is this problem of matching input/output formats that frequently makes borrowing miscellaneous programs so unsatisfactory. It is highly advisible to obtain the major tools (data reduction, structure factor, Fourier, and a least squares or other refinement program) from the same source in order to have them properly linked.

general case, $|F_c|$ is obtained [Eq. (8.9)] from

$$|F_c| = \sqrt{A^2 + B^2} \tag{9.3}$$

where A and B are, respectively, the cosine and sine parts of the structure factor expression, Eqs. (8.10) and (8.11). By definition $|F_c|$ is taken to be the positive root and is merely the modulus of the structure factor vector. Where no ambiguity is likely, the modulus signs are often omitted and $|F_c|$ is written as F_c. The phase angle α can be obtained from

$$\alpha = \tan^{-1}(B/A) \tag{9.4}$$

by considering the signs of A and B individually, and can be reported in degrees or millicycles. A form that is slightly less compact but more useful for subsequent calculations involves reporting the cosine and sine of α:

$$\cos \alpha = \frac{A}{|F_c|} \tag{9.5}$$

$$\sin \alpha = \frac{B}{|F_c|} \tag{9.6}$$

As has already been discussed, for centrosymmetric space groups the sine terms (B) become identically zero. In these cases,

$$|F_c| = |A| \tag{9.7}$$

$$\cos \alpha = \pm 1 \tag{9.8}$$

$$\sin \alpha = 0 \tag{9.9}$$

that is, α is restricted to either 0° or 180°.[9] Since the quantities used for most subsequent calculations are actually A and B obtained from $|F_c| \cos \alpha$ and $|F_c| \sin \alpha$, it is common practice to speak of F_c in the centrosymmetric case as the signed quantity defined as

$$F_c = A \tag{9.10}$$

After all the reflections have been processed, the structure factor program usually puts out some summary information about the agreement between the observed and calculated data. The general agreement in magnitude is given by some sort of rescale factor. This can be calculated in

[9]Properly speaking this is true only when the origin of the system is placed at the center of symmetry and in the absence of dispersion. Both of these conditions are generally assumed to be fulfilled, in the absence of contrary comment.

several ways, but the simplest is the *linear rescale factor* (LRS), which is

$$\text{LRS} = \sum |F_c| \Big/ \sum |F_o| \tag{9.11}$$

that is, it is derived from the condition that regardless of individual variations the *average* of the observed and calculated reflections should be the same. If the model used for the calculations contains all of the atoms in the cell, this rescale factor is usually more nearly correct than the one obtained from the Wilson plot, and its product with the input scale factor will usually be used as the scale factor for the next cycle of calculations. Since the $|F_c|$'s are based on an absolute quantity, the amount of scattering material in the cell, the $|F_o|$'s, which are derived from the relative observed intensities, are normally scaled to match them. Some programs do scale the $|F_c|$'s, but this procedure is less desirable.

Many programs have provisions for calculating rescale factors for various portions of the data, particularly for individual levels. This can be useful for film data in which level-to-level scale factors were obtained from cross-level films (see Section 7.2). As a general rule, though, if reasonably good measured scale factors are available, they should not be changed until late in the refinement, if at all, and initially only the overall scale should be adjusted to bring $\sum |F_o|$ and $\sum |F_c|$ into agreement.

It is also possible at this stage to obtain a refined value for the overall temperature factor B. This can be done by dividing the data into small ranges on $(\sin^2 \theta)/\lambda^2$ and plotting $\ln(|F_o|/|F_c|)$ versus $(\sin^2 \theta)/\lambda^2$ in the manner of a Wilson plot (Section 7.3). The slope of the resulting line is $-\Delta B$, that is, the correction required in B. Analogous computations are carried out by some F_c programs, which offer directly a value for ΔB.

When the scale factor has been adjusted to approximately 1, the question arises as to the reflection-by-reflection agreement between the observations and the calculations. It is customary to describe this in terms of the *residual index* (or, less desirably, *reliability index*) R, defined as

$$R = \sum |\Delta F| \Big/ \sum |F_o| = \sum \Big| |F_o| - |F_c| \Big| \Big/ \sum |F_o| \tag{9.12}$$

Once R has been calculated, it is necessary to interpret its value in terms of the probable correctness of the model used. The R function is by no means the perfect guide to correctness of fit, but it has been so widely used that to neglect it would be to ignore a large amount of accumulated experience, which does not exist for the other proposed measures.

It has been shown[10] that the theoretical values for R that would be obtained by using a model consisting of the proper kind and number of

[10] A. J. C. Wilson, *Acta Cryst.*, **3**, 397 (1950).

atoms placed randomly in the cell are

$$R_{\text{random, centric}} = 0.83 \tag{9.13}$$

$$R_{\text{random, acentric}} = 0.59 \tag{9.14}$$

Note that the results are different for centric and acentric space groups. This behavior continues as R falls, and in general acentric structures will show a lower value of R at a given stage of refinement than will centric ones.

The values given in (9.13) and (9.14) serve to set an upper bound on our interest.[11] Any model that gives results approaching these is obviously little better than a random collection of atoms, and in the absence of very strong indications otherwise it has little probability of refining into the true structure without major modifications. Since R for the most thoroughly refined organic structures now can fall below 0.02, it is clear that such a model can be expected to contain a large number of major and minor errors. The problem of interpreting intermediate values is more difficult and is one to which we shall return frequently. *Very* roughly, we can say that an R around 0.45 is not hopeless but suggests that considerable changes are needed; 0.35 is quite hopeful, and the solution can often be obtained by more or less routine refinement; and below 0.25 the structure is probably correct except for small (<0.1 Å) atomic shifts and changes in thermal parameters. With modern techniques, any well-behaved structure should be capable of refinement to an R below 0.1, and most below 0.06.

On the other hand, in the early stages of structure analysis when the models used lack one or more of the atoms of the real molecule, R is a particularly feeble guide and may run at disturbingly high values until all of the atoms have been inserted. An alternative, which is often more informative with structures giving relatively high R's, is the correlation coefficient (CC). This can be expressed in general form as

$$\text{CC} = \frac{\sum ab - (\sum a \sum b)/N}{\{[\sum a^2 - (\sum a)^2/N]^{1/2}\}\{[\sum b^2 - (\sum b)^2/N]^{1/2}\}} \tag{9.15}$$

and appears in various programs with either $a = |F_o|$, $b = |F_c|$ or $a = |F_o|^2$, $b = |F_c|^2$. The correlation coefficient runs from -1 (perfect inverse correlation), through 0 (no correlation), to $+1$ (perfect correlation). It has the advantages of being almost independent of scaling between $|F_o|$ and $|F_c|$ and of being much more sensitive than R in the region where R approaches its random limits. Conversely, the coefficient approaches 1.0 closely as R goes below 0.2, and so becomes of limited value.

A more subjective indication, which resembles the correlation coefficient, is simply the side-by-side comparison of the $|F_o|$ and $|F_c|$ lists. If

[11]But see also A. S. Douglas and M. M. Woolfson, *Acta Cryst.*, **7**, 517 (1954).

these tend to go up and down together in roughly comparable fashion, the signs are hopeful regardless of the poor agreement of the actual values. The only sure evidence, however, is the success of attempts to refine the model.

9.2. FOURIER SYNTHESIS[12]

The operation of calculating Fourier series is used for several purposes in crystallographic studies and can lead to a number of different results depending on the coefficients used. The method is very similar in all cases, however; so we shall first consider the most common use, that of calculating electron density maps, in some detail and afterwards we shall look at some of the other possibilities.

The general expression [Eq. (9.16)] for the calculation of the electron density in the cell has been derived [Eq. (8.47)].

$$\rho(x, y, z) = \frac{1}{V}\sum_h \sum_k \sum_l F_{hkl} e^{-2\pi i(hx+ky+lz)} \qquad (9.16)$$

The difficulty in its application arises from the fact that the coefficients in this series, the F_{hkl}'s, are not merely the moduli of the structure factors, which are available as the $|F_o|$'s, but rather the moduli plus the unobservable phases. This can be expressed in several ways, of which the most compact is

$$F_{hkl} = |F_{hkl}| e^{i\alpha} \qquad (9.17)$$

and the one actually used for the computation is

$$F_{hkl} = |F_{hkl}| \cos \alpha + i|F_{hkl}| \sin \alpha \qquad (9.18)$$

or

$$F_{hkl} = A_{hkl} + iB_{hkl} \qquad (9.19)$$

Thus before an electron density map can be calculated it is necessary to be able to assign to each modulus a phase that it is hoped, is more or less correct. Some of the means by which these phases can be obtained in the opening stages of a structure analysis are considered in Part II.

Once the first plunge has been made, however, the most common way of obtaining the phases is to use the ones given by structure factors calculated

[12]Lipson and Cochran, *The Determination of Crystal Structures*, pp. 83–108; Buerger, *Crystal–Structure Analysis*, pp. 370–508; Rollett, *Computing Methods in Crystallography*, pp. 82–88; J. M. Robertson, *Organic Crystals and Molecules*, Cornell University Press, Ithaca, NY, 1953, pp. 101–108.

on the basis of some approximate model. If these phases and *their* moduli, the $|F_c|$'s, were used as coefficients, the results would be simply a density map of the model. When the calculated phases are combined with the observed moduli, as a better approximation to the true structure factors, it is expected that the calculated electron distribution will resemble the true molecule more closely than did the original model. On the basis of this distribution, new atomic parameters can be assigned that are more accurate than before; new structure factors, with improved phases, can be calculated; and the process can be repeated as needed until a satisfactory fit is reached. It is not obvious, perhaps, that this sequence of operations should work and particularly that it should converge at a reasonable rate, but in fact it is remarkably powerful and is the mainstay of X-ray structural analysis.

Because crystallographic Fourier series are calculated on digital computers, the electron density cannot be generated as a continuum but rather must be evaluated at discrete sampling points within the cell. Fortunately, density distributions are relatively smooth and without discontinuities; so if the spacing between the points is chosen properly, little information is lost by the digital method. Although it is not mandatory, convenience in both computing and interpreting the results is served if the sampling points are arranged in a regular lattice with uniform spacings along each crystallographic axis.

It has been found empirically that for data extending to the edge of the copper sphere a grid that provides a separation of about 0.3 Å between points is satisfactory. This is consistent with the results from information theory, since $d_{min}(\text{Cu}) = 0.77$ Å, which is approximately the limit of resolution (see Section 2.4), and a sampling at least twice as fine is needed to avoid the loss of information. In the early stages of analysis, when one is more concerned with locating new atoms than with obtaining particularly accurate positions for those already found, this spacing can be increased to 0.5 or 0.6 Å with a saving in time. A coarse grid is useful, too, with large molecules, since it produces smaller maps, which are often easier to draw and interpret as representations of a connected molecule. Correspondingly, if the data do not extend to the copper limit, as is generally the case with macromolecules, there is no point in calculating maps on a finer scale than about one-half the actual resolution. The grid used can also be selected for convenience in interpreting the output, for example, it is particularly easy to place atomic coordinates on a grid marked out in multiples of 1/100, or to compensate for the characteristics of a printer when one wishes to obtain maps that are true to scale. If the program uses the FFT method, there may be requirements that the number of grid points on each cell edge be divisible by specific factors.

Three-dimensional Fourier syntheses are calculated in the form of sections through the cell and usually parallel to a cell face. All of the points on one section are printed on a page of output, and successive pages are viewed as though stacked on top of one another. Because it is easier to visualize structures that are presented on the minimum number of sheets, sections chosen to cut the shortest axis will usually give the clearest view (but see below about contouring and display). Some programs are restricted to calculating sections about particular axes, but it is usually possible then to construct views sliced in other directions (or even in planes that are not parallel to cell faces).

It is obvious that the minimum volume over which the Fourier must be calculated in order to be able to study the entire cell contents is one asymmetric unit. One cannot choose just any block of the proper volume, however. The essential characteristic of an asymmetric unit is that successive application of the space group symmetry operations to it will generate *all* of the unit cell. Most arbitrary volumes, when subjected to the symmetry operations, will generate only part of the cell, overlapping with themselves at some points.

In general, an asymmetric unit will not neatly contain a single molecule but rather will cut through parts of several.[13] These can be related to a single molecule only by considering their symmetry-related images. It is not difficult to generate the related portions, but it does require some time and offers opportunities for making mistakes. In the past the extra costs of whole-cell calculations compelled crystallographers to compute over an asymmetric unit and twist and turn the output to generate the equivalent parts. Now Fourier calculations have become so fast that it is usually easier to compute an entire cell and select only those parts needed.

The input data for Fourier calculations consist of the observed structure factors (F_o) and some source of phases for these. In the usual case, the phases are those calculated for some kind of trial structure, and so, as in the case of data reduction and structure factor calculations, it is most desirable that the structure factor and Fourier programs be linked so that the output from the first may be used as the input to the second.

The amount of data used can be varied over a wide range. The usual upper limit is all of the unique reflections (see Section 6.1) for the space group being studied. Given these, the program either generates the equivalent reflections to provide data for an entire hemisphere of reciprocal space,[14] making use of symmetry information given it, or uses space-group-specific program changes to carry out the calculation by means of a reduced

[13]This is often true as well for a unit cell since the molecules can stray over the cell boundaries, but in this case only unit translations are needed to connect the parts, rather than rotations and reflections.

[14]As long as Friedel's law holds, the general Fourier expression, Eq. (9.16), can always be reduced in a manner similar to that used for structure factor calculations in centrosymmetric space groups (see Section 8.12). Consequently, the computation need be carried out using only half the points in reciprocal space and the results multiplied by 2. If Friedel's law fails, the full data set is required and the resulting density is complex.

form of the general Fourier expression. As in the case of structure factors, operational simplicity is favored by the generative approach.

No lower limit can be set on the number of reflections required to compute a Fourier map. Most programs do not require that the reflection list be complete, so any arbitrary selection of reflections can be omitted without disturbing the operation of the program. Such omissions may affect the results of the Fourier synthesis, however.

In general, the appearance of an F_o map will be very little affected by omission of the unobserved and weak reflections. In fact, experience has shown that a synthesis calculated using only the strongest 10% of the reflections, correctly phased, will give an entirely recognizable molecular structure. It is this fact that is responsible for much of the success of calculations based on direct determination of the phases of the strongest reflections (see Chapter 11).

If the number of reflections is reduced by omitting all those with $(\sin \theta)/\lambda$ greater than some specified value, the effect is to spread the atoms somewhat and ultimately to cause them to overlap. The loss of data corresponds to the introduction of an abnormally high temperature factor. Since the phases calculated for reflections of low $(\sin \theta)/\lambda$ on the basis of an inaccurate model are generally more nearly correct than those of higher $(\sin \theta)/\lambda$, the use of data limited in this way is often advantageous early in an analysis. Unfortunately, the sharp cutoff also produces diffraction ripples, particularly about heavy atoms, that can obscure structural details.

Whereas structure factors can usually be calculated in any random order with no effect on the program efficiency, the speed of a given Fourier summation can be affected to a marked degree by the sequence in which the reflections are presented. This sensitivity to order will vary considerably from program to program and will usually be discussed in the program write-up. It can range from a flat requirement that, for example, hkl must be followed by $hk\bar{l}$ for the program to operate at all to a slight increase in operating speed for properly sorted data as compared to a random selection.

The output of the Fourier program usually appears in intermediate form on a disk file or magnetic tape. This can then be converted to a useful product in various ways. The most direct is to print rows and columns of numbers, each proportional to an electron density, together with some indication of the coordinates of each value in terms of the grid system used. Most listings are right-angled, although some programs have provision for offsetting successive lines so as to approximate the angles of a nonorthogonal cell. For most work done on Fourier listings the small errors in appearance that are produced by treating nonorthogonal cells as orthogonal are not serious and can be neglected.

The output of some Fourier programs is not directly in electrons per cubic angstrom, the units of electron density, but in some quantity related to the density. This relationship depends on two quantities, the scaling used, directly or indirectly, and the magnitude of the F_{000} reflection.

The F_{000} reflection is not an observable datum, corresponding as it does to a reflection in the line of the direct beam, but it is necessary as the constant term of the three-dimensional Fourier series. As was pointed out in the example of Chapter 8, the average value of a trigonometric function over one full cycle is identically zero. Consequently, the average over one cycle of a Fourier series consisting only of trigonometric terms is also zero. In order to have a nonzero average, a constant term, F_{000}, equal to this average must be added to the series. It can be shown easily from the structure factor expression that the proper value of F_{000} is merely the total number of electrons in the cell,

$$F_{000} = \sum_{j=1}^{N} Z_j \tag{9.20}$$

Equation (9.20) is true only when the F_o's to be used with F_{000} are on an absolute scale (see Section 7.3). If they are on a relative scale, F_{000} must be increased or decreased by the ratio of the relative scale to an absolute scale. Small errors in the relative scales of F_{000} and the F_o's are comparatively unimportant, however.

Depending on the program, F_{000} can be supplied either as a constant in the operating instructions or as a pseudoreflection with the indices 0, 0, 0 and a phase of 0° ($\cos \alpha = 1$).

The situation with regard to scaling varies from program to program. Normally no notice is taken of the $1/V$ term in Eq. (9.16). For programs that generate a hemisphere of data, there is an additional factor of 2 that is also often omitted.

Thus the printed output is actually

$$(V/2) \times \text{scaling factor} \times \rho \tag{9.21}$$

and consequently a value of

$$(V/2) \times \text{scaling factor} = 1\ \text{e/Å}^3 \tag{9.22}$$

For programs that use data from only one-fourth (monoclinic) or one-eighth (orthorhombic) of reciprocal space, using program modifications rather than the generative method, the 2 of Eqs. (9.21) and (9.22) can be replaced by 4 or 8. Normally this point will be discussed in the program description.

As a rough check on scaling, one can make use of the fact that the usual density of carbon atoms at their centers is about 4–8 e/Å^3. Nitrogen and oxygen are usually about 7–10 and 8–12 e/Å^3, respectively. These values are quite sensitive, however, to the temperature factors of the atoms involved and to the amount of data used in the calculation.

To increase the readability of density maps, it is customary to *contour* them by drawing closed curves through points of constant electron density. The results are similar to a topographic map, with atoms appearing as maxima surrounded by several concentric contour lines. For preliminary work, the values at which the contours are drawn can be chosen arbitrarily, but it is usually just as easy and a bit more informative to use Eq. (9.22) to calculate the values corresponding to integral values of electrons per cubic angstrom. A convenient choice, which gives neat drawings of organic molecules, is to contour at intervals of $1\ e/Å^3$ starting at $2\ e/Å^3$. Occasionally it may be of interest to contour along the line $\rho = 1\ e/Å^3$, but early in a structure determination this merely adds spurious peaks without being particularly useful. Heavy atoms, of course, must be contoured at more widely spaced intervals if the lines are to be kept to a reasonable number.

The modern approach to the problem of contouring involves the use of automatic curve-drawing equipment to read the file produced by the computer and generate a contour map from it. If the contours are being drawn automatically, it is probably worthwhile to use a finer interval than $1\ e/Å^3$, especially if the actual values are not listed simultaneously. Systems that are prepared to contour maps automatically usually also contain programs that scan the entire three-dimensional array and attempt to identify those maxima that might be atoms. These are presented as a list in order of decreasing peak height and greatly simplify the process of reading off coordinates from the map. It is often necessary to select from a larger list those peaks that best constitute a molecule, but this is generally easy with the map in hand.

In order to study molecules in three dimensions, some sort of "three-dimensional drawing board" is needed. Various model-building and other representational schemes have been tried, but the most common has been to use transparent sheets of glass or plastic onto which the contours of the Fourier are traced. These sheets are then held in order in some way that provides the proper vertical spacing, and a three-dimensional contour map results (Fig. 9.1). Plastic (Lucite or Plexiglas) is more expensive than glass and is more easily scratched, but its lighter weight and nonbreakable nature recommend it. Increasingly, however, this method is being supplemented or supplanted by graphic display units in which the Fourier map, generally contoured in mesh (cage) form (Fig. 9.2), is shown on an interactive graphics terminal so that the density can be rotated, moved, studied, and fitted with more or less ideal models. As the price of such displays falls, it is probable that they will become the method of choice, although contours on sheets are often advantageous for whole-molecule views of complex structures.

If the Fourier map has been calculated over only one asymmetric unit, it is often useful to carry out the tracing process so as to see the whole of a

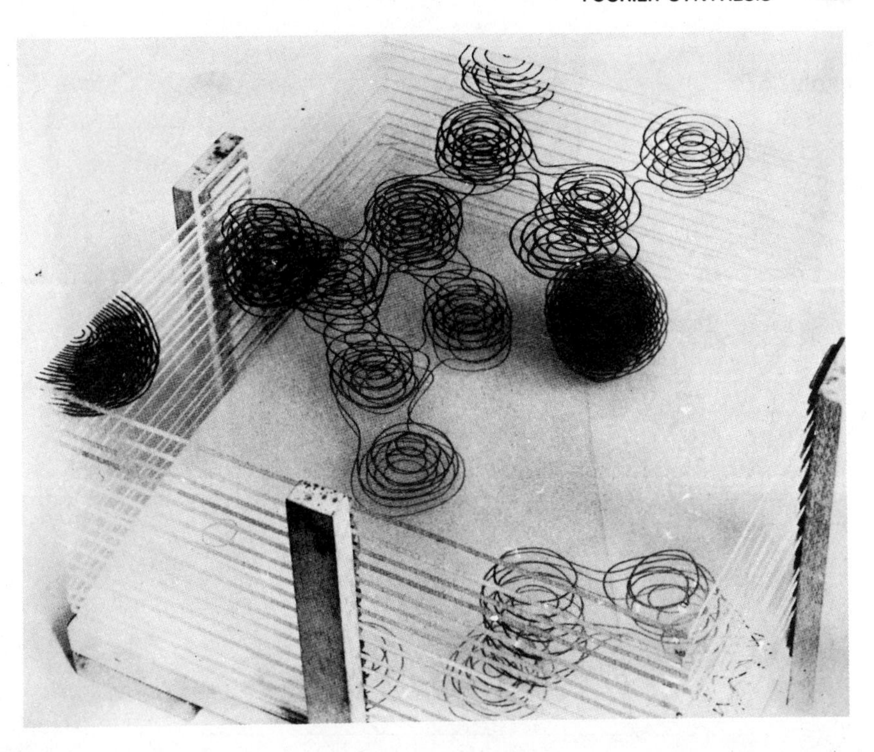

Figure 9.1. Photograph of a three-dimensional electron density map of potassium benzyl penicillin, showing the use of stacked plastic sheets to represent sections. (From D. Crowfoot, C. W. Bunn, B. W. Rogers-Low, and A. Turner-Jones, in *The Chemistry of Penicillin*, H. T. Clarke, J. R. Johnson, and R. Robinson, Eds., Princeton University Press, Princeton, NJ, 1949, p. 327. Reproduced by permission).

single molecule. This can be done by performing the symmetry operations of the space group on the output in hand. For example, if a $P2_1$ map were calculated over a block $1 \times \frac{1}{2} \times 1$ with the sections perpendicular to the b axis, the sections with $y = \frac{1}{2} \rightarrow 1$ would be just those with $y = 0 \rightarrow \frac{1}{2}$ rotated 180° in their plane, that is, about the 2-fold screw that passes through $x = \frac{1}{2}$, $z = \frac{1}{2}$ (Fig. 9.3). On the other hand, if the sections cut the a axis, the operations are a little more complicated. The section x, $y = 0 \rightarrow \frac{1}{2}$, $z = 0 \rightarrow 1$ is transformed by the screw at $x = \frac{1}{2}$, $z = \frac{1}{2}$ to one running $-x$, $y = \frac{1}{2} \rightarrow 1$, $z = 1 \rightarrow 0$ (Fig. 9.4). The x and y coordinates are no problem, corresponding as they do to a simple translation and an interchange of sections on either side of $x = \frac{1}{2}$. The change in direction along the z axis resulting from the rotation of the section parallel to the 2-fold screw is equivalent to turning the page over and recontouring it from behind. This is most easily

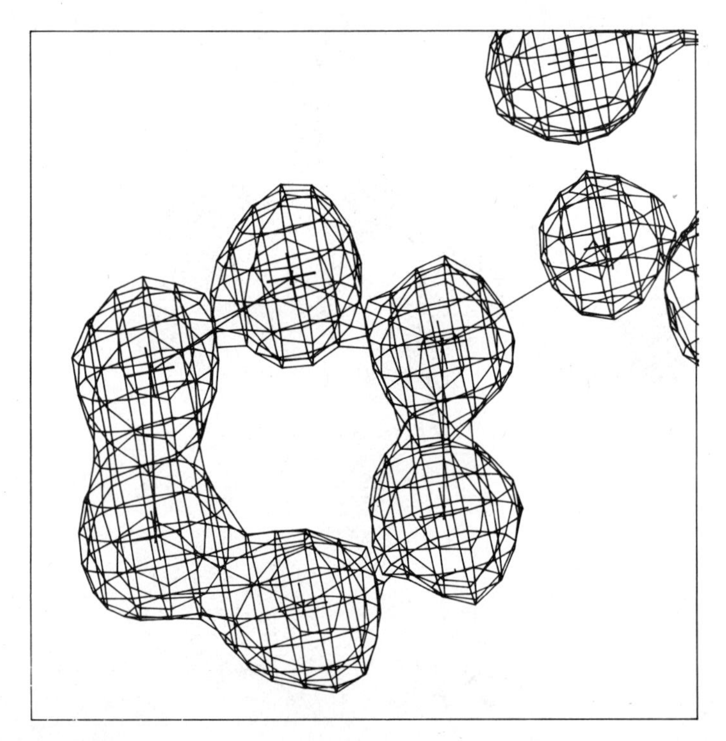

Figure 9.2. Graphic display of cage contours. (Courtesy of Prof. R. E. Stenkamp and Dr. M. A. Holmes.)

done by tracing the original contours onto the back of the transparent sheet. If the sheets are thin compared to the spacings between them, they can be used in this form. If not, it is best to copy the contours onto the front face and then clean the backs.

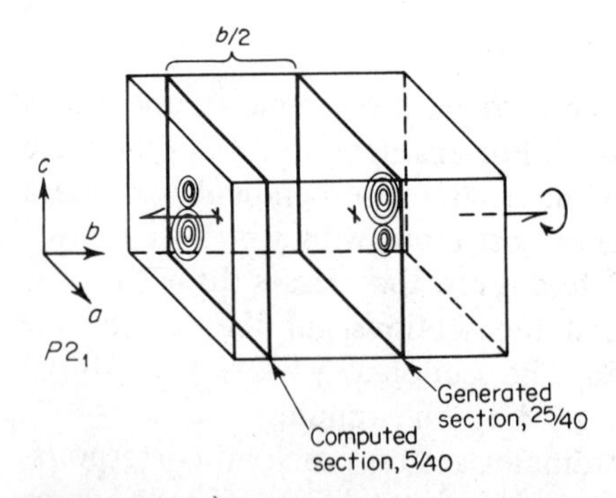

Figure 9.3. Generation of a section outside the unique volume calculated in space group $P2_1$. Sections calculated normal to b.

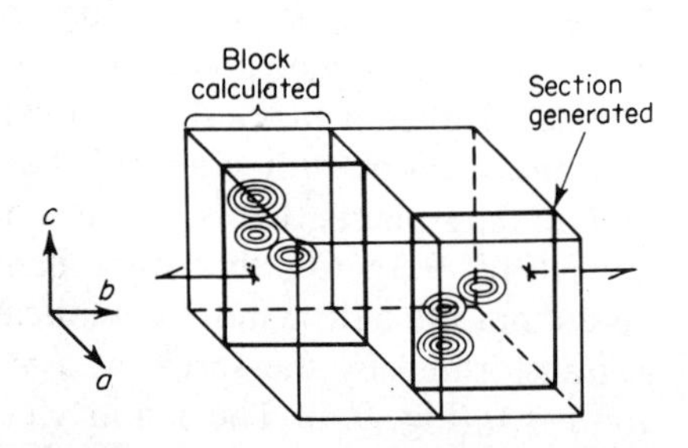

Figure 9.4. Generation of a section outside the unique volume calculated in space group $P2_1$. Sections calculated cutting a, parallel to bc.

9.3. OTHER COEFFICIENTS

The general principles described above apply to crystallographic Fourier series calculated with any sort of coefficients. A large number of such series have been used for one purpose or another, but two have become particularly important and deserve a brief mention at this time.

The first of these is the Patterson or $|F|^2$ function. Its uses are discussed in later chapters, but here we are concerned only with the outlines of its calculation. In this case, the coefficients of the Fourier series are the squares of the $|F_o|$'s, that is, the observed intensities after corrections for the geometric factors involved in data collection. Since the coefficients are squares, they are phaseless; but more commonly they are treated by the program as though they all had phases of 0°, so that

$$\cos\alpha = +1; \qquad \sin\alpha = 0; \qquad A = |F|^2; \qquad B = 0 \tag{9.23}$$

The ability to calculate Patterson functions is usually provided as an option on the standard Fourier program. Setting a control in the input data causes the program to square the input $|F_o|$'s and treat them as phaseless quantities. It is often possible, too, to use either $|F_{rel}|$ data (i.e., the usual input to the structure factor program) or the structure factor output as the input to a Patterson calculation, as either set contains the observed F's.

The symmetry of the Patterson function is related to, but may be different from, the symmetry of the space group. This point is discussed in Chapter 12, but it is noted here to warn that the size and location of asymmetric units may not be the same in both cases.

Because of the squaring of the coefficients, the average value of the Patterson function is very much higher than that of the corresponding F_o Fourier. A scale factor of 10^{-3}–10^{-4} is often used to keep the peaks from becoming too large. One particular problem arises from the fact that the simple Patterson function generates a very large peak at the origin, much larger than any other peak in the function. This peak, which provides little information, is often removed by modifying the coefficients (see Chapter 12), but if it is present the scaling should be adjusted so that the rest of the Patterson map shows a reasonable range between the highest and lowest points. The origin peak need be considered only when the program will not allow any numbers to overflow—that is, to become larger than the printing space available—without stopping.

Patterson functions are contoured in the usual way, usually at arbitrary contour intervals. If a correct I_{000} ($= F^2_{000}$) datum is included, the function should be everywhere positive, and negative regions can be regarded as equal to zero.

The second Fourier function of importance is the so-called *difference synthesis* or *difference Fourier*. In this case the coefficients are the ΔF's, that is, the quantities $|F_o| - |F_c|$, calculated on the basis of some model. The ΔF's and the Fourier series calculated from them are obviously related to the errors in the model as compared to the true structure. In fact, difference syntheses have proved to be an exceedingly valuable tool both for locating new atoms and for correcting the positions of those already present. These uses are discussed in Chapters 15 and 16.

The calculation of difference syntheses, like that of F_o syntheses, requires using the phases of the F_c's from the model. Thus the usual input is the output from the structure factor program. Most or all Fourier programs can be switched to calculate difference syntheses, and often several options are available as to how the coefficients are to be further modified or chosen (see Chapter 15). When the F_c's are calculated from a model including all the electrons in the unit cell, ΔF_{000} is zero and is not needed. If the model is incomplete, however, ΔF_{000} is simply the number of missing electrons.

The difference synthesis can be shown to be equal to an F_o synthesis calculated as usual with the F_c phases, minus an F_c synthesis calculated with F_c phases *and* F_c moduli. The natural units, therefore, are differences in electrons per cubic angstrom (e/Å^3), and these can be obtained by scaling the output data exactly as described for the F_o synthesis.

The heights of the peaks of a difference map will vary from a few e/Å^3 in the early stages of an analysis to a few tenths near the end of the refinement in the absence of heavy atoms. For this reason and also because one is often concerned with gradients as well as peaks, it is generally advisable to use a larger scaling factor for difference Fourier maps. The contour interval should also be adjusted to the size of the maximum excursions.

If no constant term is included, the mean value of a properly scaled difference synthesis is zero, and the resulting map is as likely to have negative holes as positive peaks. These holes can be just as important as the peaks and should be contoured in the same way. To distinguish the two it is useful to contour the negative regions in one color and the positive ones in another.

KEY FORMULAS

Structure Factor

$$\Delta F = |F_o| - |F_c| \tag{9.2}$$

$$\cos \alpha = A/|F_c| \tag{9.5}$$

$$\sin \alpha = B/|F_c| \tag{9.6}$$

$$R = \frac{\sum ||F_o| - |F_c||}{\sum |F_o|} = \frac{\sum |\Delta F|}{\sum |F_o|} \tag{9.12}$$

CORRELATION COEFFICIENT

$$CC = \frac{\sum ab - (\sum a \sum b)/N}{\{[\sum a^2 - (\sum a)^2/N]^{1/2}\}\{[\sum b^2 - (\sum b)^2/N]^{1/2}\}} \quad (9.15)$$

FOURIER SYNTHESIS

$$F_{000} = \sum_{j=1}^{N} Z_j \quad (9.20)$$

PATTERSON SYNTHESIS

$$P(x, y, z) = \frac{1}{V} \sum_h \sum_k \sum_l |F_{hkl}|^2 \cos 2\pi(hx + ky + lz)$$

DIFFERENCE SYNTHESIS

$$\Delta\rho(x, y, z) = \frac{1}{V} \sum_h \sum_k \sum_l \Delta F e^{-2\pi i(hx+ky+lz)}$$

BIBLIOGRAPHY

Buerger, M. J., *Crystal–Structure Analysis*, Wiley, New York, 1960, pp. 259–282, 370–477.

Lipson, H., and W. Cochran, *The Determination of Crystal Structures*, Cornell University Press, Ithaca, NY, 1966, pp. 66–108.

Nyburg, S. C., *X-Ray Analysis of Organic Structures*, Academic, New York, 1961, pp. 111–119.

Pepinsky, R., J. M. Robertson, and J. C. Speakman, Eds., *Computing Methods and the Phase Problem in X-Ray Crystal Analysis*, Pergamon, Oxford, 1961.

Rollett, J. S., *Computing Methods in Crystallography*, Pergamon, Oxford, 1965, pp. 38–46, 82–88.

PART II

THE PHASE PROBLEM

CHAPTER 10

THE PHASE PROBLEM

The elements of the phase problem—*the* problem of X-ray crystallography—have been touched on in the preceding chapters. If the structure factors and phases are known, the electron density distribution of the unit cell can be calculated; if the electron density is known, structure factors can be obtained. Structure solutions would thus be trivial exercises in computation except for the fundamental difficulty that the crystallographically available data consist only of the structure factor magnitudes and not their phases. It is the necessity for supplying the missing information that is the source of the phase problem.[1]

One important question is whether there is necessarily a one-to-one relationship between structure factor magnitudes and electron density, that is, whether a given set of magnitudes *must* correspond to one and only one electron distribution. If the question is phrased mathematically in its most general form, the answer is no. Although the operation of going from any arbitrary electron density to structure amplitudes is unique, the return path is not. That this should be so is suggested by the fact that the number of observed structure factors is finite while the number of points in the unit cell whose electron density can be varied arbitrarily is infinite.

On the other hand, the practical success of crystallographic methods indicates that there is apparently a significant difference between the general theory and experience. The answer to this difference lies in the fact

[1]Direct experimental determination of the signs of structure factor triples of the sort discussed in Chapter 11 have appeared in the last few years. These offer an indirect means of obtaining phases for individual reflections. See B. Post, *Acta Cryst.*, **A39**, 711 (1983) and F.-S. Han and S.-L. Chang, *Acta Cryst.*, **A39**, 98 (1983).

that the density distributions with which we are dealing are very far from being arbitrary. Since we are concerned with real objects—crystals—we expect the electron density to be everywhere real, positive, and continuous. On the basis of other evidence, we also expect the density to be concentrated into more or less spherical regions that we call atoms. These restrictions are quite severe, and they form the theoretical basis for the direct methods of phase determination discussed in Chapter 11. If we also require that the atoms be located so as to form chemically reasonable molecules in which the interatomic distances and angles vary only between rather restricted limits, and that the molecules not approach each other too closely, the restrictions become so severe that the solutions appear to be unique for practical purposes.

It has been shown theoretically[2] that the first restrictions are not sufficient to guarantee uniqueness of solution. The effects of the second are harder to judge, but as a practical matter no case has yet been found of a false structure that fit the observed data as well or that gave bond lengths and angles as consistent with expectations as did the true one. This is not to say that false structures do not appear during structure analysis. Any incorrect model that gives a reasonable fit to the $|F_o|$'s is in a sense a false structure, and some of these persist for disturbing lengths of time. Indeed, structures varying from mildly to grossly wrong have been reported in the literature. In all known cases, however, they contained telltale signs of error, even if those signs were missed by the original authors, and the corrected structures fit both observations and expectations more closely.[3] Unfortunately, the increased power of modern computing has made it easier to force a structure into "acceptable" agreement by excessive manipulation of parameters, and the growing automation of the process has reduced the extent of critical review of the details.

Although the large number of $|F_o|$ values obtained for the average structure serve to overdetermine the atomic positions greatly,[4] attempts to obtain the coordinates of these positions by solving the structure factor expressions analytically have not been successful, and there are good theoretical reasons for believing that this will remain the case.[5] Consequently the practical process of advancing from the observed structure magnitudes to the final structure will be considered in three stages. The first

[2]A. L. Patterson, *Phys. Rev.*, **65**, 195 (1944).

[3]There do exist cases where uncertainties remain, especially with regard to structures that are either approximately or exactly symmetric, but these differences are very small and to some degree philosophical. In any case, nature is a continuum and any attempt to draw a dividing line will always come upon ambiguous questions.

[4]This is true for small molecule structures but generally not for macromolecular ones (e.g., proteins), for which implicit or explicit reference to assumed molecular geometries is necessary to allow atomic solutions.

[5]See H. Lipson and W. Cochran, *The Determination of Crystal Structures*, G. Bell and Sons, London, 1957, pp. 246–250.

and most critical of these involves the development of a phase set and associated structural model sufficiently close to correct for Fourier methods to be used. We shall call this model a *phasing model*. It is often incomplete in the sense of not representing the entire chemical structure and may not even be recognized when it first appears, but its existence is shown by the success of the second stage. Here, the phasing model is elaborated, usually by successive cycles of structure factor and Fourier calculation, until all of the atoms in the molecule are found and placed in reasonably correct positions. By this time the chemical structure should be clear except possibly for distinctions about atom types and bond orders. At the end of this stage the phases should be fairly well fixed and should not vary a great deal during the last step, the final refinement. In refinement the atomic positions and temperature factors are adjusted, generally by nearly automatic procedures, to obtain the best fit between $|F_c|$ and $|F_o|$ and the best approximation to the true structure. Of course, the three steps are not entirely independent, particularly the second and third, but in general they use different methods and can be considered separately.

In the material that follows, Chapters 11–14 are concerned with various methods of obtaining phasing models. There exist two main lines of approach: the first by a direct calculation of initial phases, which then allow the deduction of an atomic model from a Fourier map; the second by the direct location of enough atoms in the cell to give approximate but adequate phases on an F_c calculation. Which of these should be used in a given case depends on the nature of the problem and on the data available.

Once a phasing model has been obtained, Chapter 15 discusses the general techniques involved in expanding it into a full molecular skeleton containing all of the atoms of the unit cell. Chapter 16 describes the techniques of systematic refinement, and Chapter 17 considers the problem of describing the probable accuracy of the results. Chapter 18 examines some useful results that can be obtained from an analysis of the final structural parameters, while Chapter 19 reviews the reliability of the final results.

BIBLIOGRAPHY

Bunn, C. W., *Chemical Crystallography*, 2nd ed., Oxford University Press, London, 1961, Chapter X.

CHAPTER 11

DIRECT METHODS

At the time of our first edition, the major method of solving the phase problem in practice involved introducing a heavy atom into the molecule of interest, locating this atom by methods involving the Patterson function, and using it as an initial phasing model by which to bootstrap to the entire structure. Methods of direct calculation of phases were known and discussed but were regarded as experimental, constituting a significant research undertaking in themselves. Furthermore, although a few noncentrosymmetric structures had been solved by direct phasing,[1,2] the methods had in general been restricted to centrosymmetric space groups.

Twenty years later this picture has changed entirely. Direct phasing is now the method of choice, with the result that 80–90% of reported small-molecule structures are solved in this way. The methods have been automated to the extent that they are commonly used as "black box" techniques in which the raw data go in at one end and the essentially solved structure appears at the other. As much as any other advance, direct methods have been responsible for the widespread use of X-ray crystallography as a routine technique for structure determination, attested by the 1985 award of the Nobel prize in chemistry to Jerome Karle and Herbert Hauptman, two of the pioneers and steady trail breakers in the field.

Although most direct solutions are performed with one of a few widely distributed programs, we believe that it is important for the user to have some understanding of the principles behind these programs. Although

[1]I. L. Karle and J. Karle, *Acta Cryst.*, **17**, 835 (1964).

[2]I. L. Karle, *Abstracts of American Crystallographic Association Meeting*, Austin, Texas, February–March, 1966, p. 18.

many techniques have been proposed, the field has converged to a few basic methods. These do have their differences, however, and in some cases one will work where another fails. Thus it is useful to be aware of the distinctions and their implications.

11.1. INEQUALITIES

One of the earliest attempts to relate phases to intensities led to the development of the Harker–Kasper inequalities.[3] These expressions, resulting from the combination of structure factor expressions with certain classical inequalities, provided the first means of determining the phase of one reflection in terms of its magnitude and those of others. Thus using the general structure factor expression the trivial result

$$|F_{hkl}|^2 \leq F_{000}^2 \tag{11.1}$$

could be derived. If the requirement of centrosymmetry was added, however, a significant result was obtained.

$$F_{hkl}^2 \leq F_{000}[\tfrac{1}{2}F_{000} + \tfrac{1}{2}F_{2h,2k,2l}] \tag{11.2}$$

The meaning of Eq. (11.2) can be clarified by dividing both sides by F_{000}^2 and defining

$$u_{hkl} = F_{hkl}/F_{000} \tag{11.3}$$

Equation (11.2) then becomes

$$u_{hkl}^2 \leq \tfrac{1}{2} + \tfrac{1}{2}u_{2h,2k,2l} \tag{11.4}$$

The importance of this equation lies in the fact that both the magnitude and sign (+) of u_{hkl}^2 are known while the phase of $u_{2h,2k,2l}$ remains the only unknown. Thus

$$u_{hkl}^2 \leq \tfrac{1}{2}(\pm\tfrac{1}{2}|u_{2h,2k,2l}|) \tag{11.5}$$

If the magnitudes of u_{hkl} and $u_{2h,2k,2l}$ are sufficiently large, they may require the selection of the positive sign for the latter in order to ensure that the inequality holds.

Table 11.1 shows some examples of the use of inequalities as well as their

[3]D. Harker and J. S. Kasper, *Acta Cryst.*, **1**, 70 (1948). These inequalities reflect the restriction that the electron density must be everywhere real and nonnegative. See J. Karle and H. Hauptman, *Acta Cryst.*, **3**, 181 (1950).

TABLE 11.1 Examples of Phase Determination by an Inequality

u_{hkl}^2	$\|u_{2h,2k,2l}\|$	Phase +	Phase −	Comment
0.60	0.20	0.60	0.40	$u(2h, 2k, 2l)$ must be +
0.50	0.10	0.55	0.45	Must be +
0.40	0.10	0.55	0.45	Could be either
0.40	0.30	0.65	0.35	Must be +
0.25	0.50	0.75	0.25	Almost certainly +
0.25	0.30	0.65	0.35	Could be either

weakness; in order to obtain definite results the reflections must have amplitudes that are large fractions of F_{000}; that is, they must represent the in-phase scattering of most of the electrons in the cell. Unfortunately, such reflections are rare in normal crystals, even if allowance is made for the decline in average scattering with increasing sin θ (see below). For this reason, inequalities rarely are useful for determining phases for complex structures. Furthermore, as we shall see, the inequalities represent limiting cases of the more general probability relationships and merely define the point at which the probably determined phase becomes certain.

11.2. *F*'S, *U*'S, AND *E*'S

One of the obvious difficulties with applying inequalities to ordinary $|F|$'s or $|U|$'s is that the decline in $|F|$ with sin θ will rapidly lower the $|F|/F_{000}$ ratios below the level at which there is any chance of obtaining phase information by this method. The derivation of the inequalities makes no assumption about the shape of the atoms in the cell, however, and they can be applied equally well to the F's (F_{point}) that would be found if the real atoms were replaced by point atoms, that is, atoms in which all of the scattering power was concentrated at the nucleus.

One of the characteristics of such a point atom is that its scattering power is not a function of $(\sin\theta)/\lambda$ but is a constant equal to the atomic number Z, that is, to the usual scattering factor at $(\sin\theta)/\lambda = 0$ (Fig. 11.1).

As was discussed in Chapters 8 and 9, the structure factor is given by an expression of the form

$$F = \sum_{j}^{N} f_j e^{2\pi i(hx_j+ky_j+lz_j)} = \sum_{j}^{N} f_j e^{2\pi i(\mathbf{h}\cdot\mathbf{r})} \tag{11.6}$$

For a cell containing only one kind of atom with scattering factor f, this can be rewritten as

$$F = f\sum_{j}^{N} e^{2\pi i(\mathbf{h}\cdot\mathbf{r})} \tag{11.7}$$

$$F = fE \tag{11.8}$$

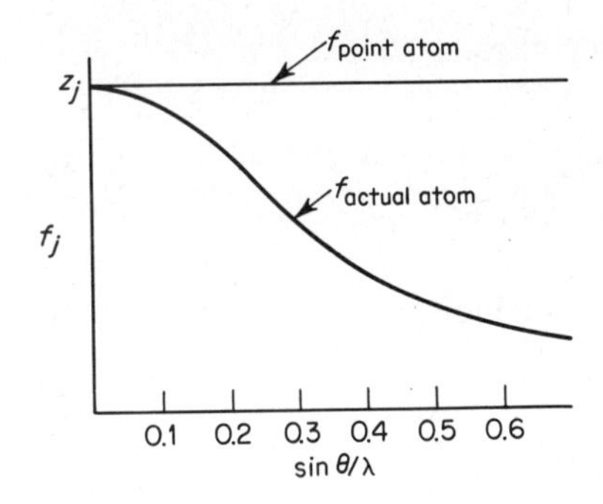

Figure 11.1. Scattering factor for an actual atom and for a point atom.

where E is the sum of the exponential terms. For real atoms, which possess thermal motion,

$$f = f_o e^{-B(\sin^2 \theta)/\lambda^2} \tag{11.9}$$

whereas for stationary point atoms,

$$f = Z \tag{11.10}$$

Thus,

$$\frac{F_{\text{point}}}{F_{\text{real}}} = \frac{ZE}{Ef_o e^{-B(\sin^2 \theta)/\lambda^2}} \tag{11.11}$$

and

$$F_{\text{point}} = \frac{Z}{f_o e^{-B(\sin^2 \theta)/\lambda^2}} F_{\text{real}} \tag{11.12}$$

Most crystals, however, contain more than one kind of atom, so the problem is not quite so simple. Nor, in fact, do all atoms even of one kind have the same temperature factor. Thus it is not possible to obtain exact F_{point} values, but some average value of Z/f must be used. A common approximation is

$$F_{\text{point}} = \frac{\sum_j^N Z_j}{(e^{-B(\sin^2 \theta)/\lambda^2}) \sum_j^N f_{oj}} F_{\text{real}} \tag{11.13}$$

and this will serve for most purposes.

In practice it is more common to define the *unitary structure factor* U such that

$$U_{hkl} = F_{hkl,\text{ point}}/F_{000} \tag{11.14}$$

or, using Eq. (11.13) and noting that $\sum^N Z_j = F_{000}$,

$$U_{hkl} = \frac{F_{hkl}}{(e^{-B(\sin^2\theta)/\lambda^2}) \sum_j^N f_{oj}} \tag{11.15}$$

If the more general scattering factors, which include the temperature effect [see Eq. (7.18)], are used, then

$$U_{hkl} = \frac{F_{hkl}}{\sum_j^N f_j} \tag{11.16}$$

Thus U is a structure factor that has the same phase as F but whose absolute value ranges from 0 to 1, the maximum value corresponding to the case in which all of the atoms scatter in phase. The importance of this result is that complete in-phase scattering can occur only if all the atoms are located at maxima (or minima) of a Bragg–Lipson chart (see Appendix D) or its three-dimensional equivalent. This fact alone will usually not locate the atoms explicitly, but it does serve to reduce the number of possible sites markedly. $|U|$'s of 1 are rarely if ever found, but the larger the value observed the greater the constraints placed on the atomic positions.

The information content of a reflection is determined by its intensity relative to the average of its neighbors. In line with this view, it will very often be found that the highest $|U|$'s are obtained for reflections of high $\sin\theta$ that are relatively very strong but much smaller in $|F|$ than many at low $\sin\theta$. Although the question of which reflections are the most important can be settled by ranking them in order of decreasing $|U|$, it is desirable to have some knowledge of the expected values for any structure so as to be able to judge significance on an absolute scale. This can be obtained by noting that the U's can also be calculated by summation over the atoms in exactly the same way as the ordinary structure factors. Thus for the centrosymmetric case [see Eq. (8.71)],

$$U_{hkl} = 2 \sum_j^{N/2} n_j \cos 2\pi(hx_j + ky_j + lz_j) \tag{11.17}$$

where n_j, which corresponds to the usual scattering factor, is

$$n_j = f_j \Big/ \sum_j f_j \tag{11.18}$$

that is, the fraction of the scattering power represented by the jth atom. Since the scattering factor curves do not all show the same shape, n_j is a function of θ if there is more than one kind of atom in the cell. If all the atoms are alike, however, it becomes

$$n_j = 1/N \tag{11.19}$$

In exactly the same way that the expected (average) value of F^2 is given by

$$\langle F^2 \rangle = \sum_j^N f_j^2 \tag{11.20}$$

(see Section 7.3 and references cited there), so the expected value of U^2 is given by

$$\langle U^2 \rangle = \sum_j^N n_j^2 \tag{11.21}$$

Thus the root-mean-square value of U is

$$U_{\text{rms}} = \left(\sum_j^N n_j^2\right)^{1/2} \tag{11.22}$$

and making the approximation, usually fairly good for organic molecules without heavy atoms, that all of the atoms are alike,

$$U_{\text{rms}} \approx 1/\sqrt{N} \tag{11.23}$$

It is obvious from Eq. (11.23) that there do not have to be many atoms in the unit cell before the average $|U|$ drops well below the values that make the use of inequalities practical. While some reflections may show large $|U|$ even for large N, the number of these decreases rapidly to the point where the structure cannot be determined by inequalities alone. As we shall see, the methods that are effective for such large systems are based on probabilities rather than certainties. For these relationships, the relative magnitudes of $|U_{hkl}|$ and U_{rms}, as well as their absolute sizes, are important in determining the reliability of the indicated phases. For this reason, among others, Karle and Hauptman[4] introduced a *normalized structure factor* E_{hkl} given by

$$E_{hkl}^2 = U_{hkl}^2 / \langle U^2 \rangle \tag{11.24}$$

The E's have certain mathematical conveniences for use in probability calculations, but their particular advantage is that they allow the normalization of all classes of reflections to a common basis. In this way it is

[4]J. Karle and H. Hauptman, *Acta Cryst.*, **9**, 635 (1956).

possible to avoid a rather subtle source of error in the comparison of special sets of reflections.

The significant factor in the calculation of E's is that a given U must be related to the true $\langle U^2 \rangle$ for the class of reflections to which it belongs. For general (hkl) reflections, this value will be that given by Eq. (11.24). As was pointed out in Chapter 7, however, in some space groups certain classes of reflections have average intensities that are multiples of that of the general set. Likewise, the $\langle U^2 \rangle$ given by Eq. (11.21) represents an average over all reflections including those systematically absent. If, as is usually the case, only the nonextinguished reflections of a set are considered, they are intensified by a factor that depends on the fraction of reflections omitted (see Sections 7.3 and 8.11).[5] Thus combining Eqs. (11.24) and (11.21) into a general form

$$E_{hkl}^2 = \frac{U_{hkl}^2}{\varepsilon \sum_j^N n_j^2} \tag{11.25}$$

or equivalently, and more commonly used for calculation,

$$E_{hkl}^2 = \frac{|F_{hkl}|^2}{\varepsilon \sum_j^N f_j^2} \tag{11.26}$$

In these expressions, ε is an integer that is generally 1 but may assume other values for special sets of reflections in certain space groups. Thus in $P2_1/c$, $\varepsilon = 2$ for the $h0l$ and $0k0$ reflections and 1 for all others. For any space group, the proper value of ε and the affected reflections can be obtained from *International Tables*.[6] Here ε ($= S/\phi$) is given for all symmetry elements (although in only one orientation; the others can be derived by analogy) and for the classes of observed reflections that they affect.

The effect of including ε in the calculation of $|E|$ is to reduce the significance of reflections that have large values of $|U|$ but that also belong to a class having, for reasons connected with the space group symmetry, an abnormally large $\langle U^2 \rangle$. Thus the improbability of these reflections, and the information carried by them, is not as great as would at first appear.

In practice, $|E|$'s are often calculated as described, based on the assumption of randomly distributed atoms. If, however, something is known of the geometry of the cell contents, this information can be applied to obtain a more specific model of the expected intensity distribution. The result is generally a better fit to the observed distribution and yields better values for the scale and temperature factors. Even better results, in terms of estimat-

[5]For a more general discussion, see A. J. C. Wilson, *Acta Cryst.*, **17**, 1591 (1964).

[6]*International Tables*, Vol. II, pp. 355–356; see also H. Iwosaki and T. Ito, *Acta Cryst.*, **A33**, 227 (1977).

TABLE 11.2 Theoretical Values Related to $|E|$'s

	Centrosymmetric	Noncentrosymmetric
Average $\|E\|^2$	1.0000	1.0000
Average $\|E^2-1\|$	0.968	0.736
Average $\|E\|$	0.798	0.886
$\|E\|>1$	32.0%	36.8%
$\|E\|>2$	5.0%	1.8%
$\|E\|>3$	0.3%	0.01%

ing $|E|$'s that lead to solutions, are often obtained by using local values of $\langle F^2\rangle$ corresponding to the averages of the observations over small ranges of $(\sin^2\theta)/\lambda^2$. This method does not give single values for scale and temperature factors but does adapt to the variations of the actual structure.

The distribution of the $|E|$ values is, in principle and often in practice, independent of the size and content of the unit cell. It does depend, however, on the presence or absence of a center of symmetry in the space group. Table 11.2 gives some useful values describing these distributions.[7] As can be seen, these values provide yet another statistical test for centric and acentric distributions of intensities (see Section 7.3). Like all tests, however, they are subject to being disturbed by particular atomic distributions in the cell and should be treated with caution.

11.3. STRUCTURE INVARIANTS AND SEMINVARIANTS

As we shall see here and in Chapter 12, one of the recurring difficulties in solving the phase problem involves defining the origin of the unit cell properly with respect to the symmetry elements and all the contents of the cell. Shifting the origin arbitrarily does not affect the structure amplitudes but may change the phases drastically. Consequently, the selection of phases is intimately associated with the choice of origin. There exist, however, certain combinations of phases that do not change regardless of the arbitrary assignment of cell origins (*structure invariants*), or that do not change with shifts among alternative origins that have the same symmetry characteristics (*structure seminvariants*).[8,9]

Among the structure invariants, the most important ones have the form

$$\phi_n = \phi_{\mathbf{h}1} + \phi_{\mathbf{h}2} + \cdots + \phi_{\mathbf{h}n} \tag{11.27}$$

[7] I. L. Karle, K. S. Dragonette, and S. A. Brenner, *Acta Cryst.*, **19**, 713 (1965).
[8] H. Hauptman and J. Karle, *Solution of the Phase Problem. I. The Centrosymmetric Crystal*, ACA Monograph 3, Polycrystal Book Service, Pittsburgh, PA, 1953.
[9] H. Schenk, in *Computational Crystallography*, D. Sayre, Ed., Clarendon Press, Oxford, pp. 65–74; H. Hauptman, *ibid.*, pp. 75–91.

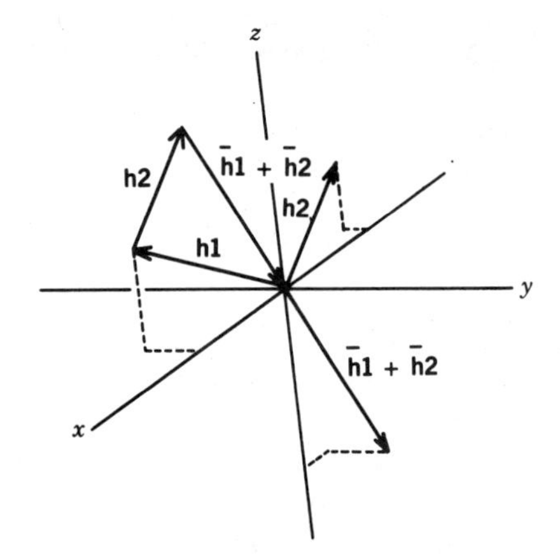

Figure 11.2. Triplet of vectors forming a closed set in the reciprocal lattice.

where

$$\mathbf{h1} + \mathbf{h2} + \cdots + \mathbf{hn} = \mathbf{0} = (0, 0, 0) \tag{11.28}$$

In other words, the sum of the phases of n reflections whose indices sum to zero is a constant for a given structure regardless of the choice of origin in the structure. Trivial examples of this are F_{000}, which always has the phase of 0, and F_{hkl} and $F_{\overline{hkl}}$, whose indices add to (000) and whose phases, in the absence of dispersion, add to $2\pi = 0$ (see Fig. 8.10). The first significant example is the triple of phases

$$\Phi_3 = \phi_{\mathbf{h1}} + \phi_{\mathbf{h2}} + \phi_{\overline{\mathbf{h1}}+\overline{\mathbf{h2}}} \tag{11.29}$$

Regardless of the magnitudes of the F's, and independent of the choice of origin, these phases will sum to a constant value. In general it is not possible to specify this value a priori, but the cases in which the Harker–Kasper inequalities hold are equivalent to demonstrating that Φ_3 must be 0.

For centrosymmetric space groups, Eq. (11.29) can be rewritten in terms of the product of the signs of the structure factors. Thus,

$$S_3 = s_{\mathbf{h1}} \cdot s_{\mathbf{h2}} \cdot s_{\overline{\mathbf{h1}}+\overline{\mathbf{h2}}} = \pm 1 \tag{11.30}$$

A convenient visualization of these structure invariants is as a closed polygon of vectors in reciprocal space (Fig. 11.2). This polygon starts and ends at the reciprocal lattice (r.l.) origin and defines a set of reflections whose relative phasing remains constant regardless of the choice of origin.

11.4. PROBABILITY METHODS

Once structures reach such a size that inequalities are ineffective, it is necessary to find another approach to the problem of direct phase determination. As we shall see, in the range of intensities that are too small for inequalities but still relatively large, it is possible to set up "equations" that are *probably* true, and from these to extract phase information. These methods were first introduced practically for problems in centrosymmetric space groups and only later extended to noncentrosymmetric groups. For clarity we shall follow the same route.

Centrosymmetric Methods

The easiest approach for the methods to be described is a 1952 paper by Sayre,[10] although mathematically equivalent results had appeared earlier.[11] It can be shown that, subject to certain restrictions,

$$F_{hkl} = \Omega_{hkl} \sum_{h'} \sum_{k'} \sum_{l'} F_{h'k'l'} F_{h-h',k-k',l-l'} \tag{11.31}$$

where Ω_{hkl} is simply a calculable scaling term. The implication of Eq. (11.31) is that any structure factor F_{hkl} is determined by the products of all of the pairs of structure factors whose indices add to give (hkl). Thus F_{213} depends on the products of F_{322} and $F_{\bar{1}\bar{1}1}$, F_{604} and $F_{\bar{4}1\bar{1}}$, and so on.

On the surface, Eq. (11.31) appears rather useless, since to determine one F it is necessary to know the magnitudes and phases of all others. Sayre pointed out, however, that for the case where F_{hkl} is large, the series must tend strongly in one direction (+ or −) and that this direction is generally determined by the agreement in sign among products between large F's. Thus for the case of all three reflections large,

$$s(F_{hkl}) \approx s(F_{h'k'l'}) \cdot s(F_{h-h',k-k',l-l'}) \tag{11.32}$$

or, as it is sometimes given,

$$s(F_{hkl}) \cdot s(F_{h'k'l'}) \cdot s(F_{h-h',k-k',l-l'}) \approx +1 \tag{11.33}$$

where s means "sign of" and $\approx$ means "is probably equal to." The quantities s( ) may be considered as ± 1 and will often be used in the form $s(hkl)$.[12]

Equations (11.32) and (11.33) are the probability equations derived from Eq. (11.27) and serve to favor one of the possible choices of the sign of the structure invariant [Eq. (11.30)]. Since the phases of U's and E's are the

[10] D. Sayre, *Acta Cryst.*, **5**, 60 (1952). This analysis adds explicitly the requirement that the electron density be composed of resolved atoms.
[11] J. Karle and H. Hauptman, *Acta Cryst.*, **3**, 181 (1950).
[12] In centrosymmetric space groups, $s(hkl) = s(\overline{hkl})$ $[s(\mathbf{h}) = s(-\mathbf{h})]$ so Eq. (11.33) can be expressed equally well as $s(\mathbf{h}) \cdot s(\mathbf{h}') \cdot s(\mathbf{h} - \mathbf{h}')$ or $s(\mathbf{h}) \cdot s(\mathbf{h}') \cdot s(\mathbf{h} + \mathbf{h}')$. In the noncentrosymmetric case more care is needed to keep track of index signs.

same as those of the corresponding F's, they also apply to these modified structure factors. In addition, Eq. (11.32) holds in those cases in which inequalities also apply, that is, the inequalities simply represent the cases in which the probability becomes certainty. For example,

$$s(2h, 2k, 2l) \approx s(hkl) \cdot s(hkl) \tag{11.34}$$

so that regardless of the sign of F_{hkl}, $F_{2h,2k,2l}$ will probably be positive $[=(+1)^2$ or $(-1)^2]$ if the reflections are sufficiently large. This is the same result that is given with certainty when the inequality of Eq. (11.4) is applicable. For this reason, inequalities are generally ignored in practice, since the same results will appear with very high probabilities from Eq. (11.34).

The question of the exact probabilities of Eq. (11.32) and (11.34) is obviously an important one, and many analyses of the problem have appeared. The result that is most commonly used, although it is not absolutely rigorous, is that of Cochran and Woolfson.[13] According to this,

$$P = \tfrac{1}{2} + \tfrac{1}{2} \tanh\{(\sigma_3/\sigma_2^3)|U_{hkl} U_{h'k'l'} U_{h-h',k-k',l-l'}|\} \tag{11.35}$$

where P is the probability that Eq. (11.32) will hold, and

$$\sigma_3 = \sum_j^N n_j^3 \tag{11.36}$$

$$\sigma_2 = \sum_j^N n_j^2 \tag{11.37}$$

the n_j being those defined by Eq. (11.18). If all the atoms of the cell are equal, it is easily shown that

$$\frac{\sigma_3}{\sigma_2^3} = \frac{N}{N^3}\left(\frac{N}{N^2}\right)^{-3} = N \tag{11.38}$$

so

$$P = \tfrac{1}{2} + \tfrac{1}{2} \tanh\{N|U_{hkl} U_{h'k'l'} U_{h-h',k-k',l-l'}|\} \tag{11.39}$$

By a simple conversion, the corresponding formulas for E can be found to be

$$P = \tfrac{1}{2} + \tfrac{1}{2} \tanh\{(\sigma_3/\sigma_2^{3/2})|E_{hkl} E_{h'k'l'} E_{h-h',k-k',l-l'}|\} \tag{11.40}$$

[13] W. Cochran and M. M. Woolfson, *Acta Cryst.*, **8**, 1 (1955).

and since

$$\frac{\sigma_3}{\sigma_2^{3/2}} = \frac{N}{N^3}\left(\frac{N}{N^2}\right)^{-3/2} = \frac{1}{\sqrt{N}} = N^{-1/2} \tag{11.41}$$

we have

$$P = \tfrac{1}{2} + \tfrac{1}{2}\tanh\{N^{-1/2}|E_{hkl}E_{h'k'l'}E_{h-h',k-k',l-l'}|\} \tag{11.42}$$

Equations (11.40) and (11.42) are safer to use than (11.35) and (11.39) because the peculiarities of the special classes of reflections have been acounted for in the calculation of the E's, while in Eq. (11.35) they should properly be considered by using different values for σ_3 and σ_2 in certain cases. Since the calculated probabilities are more indications than rigorous evaluations in practical cases, we shall assume the equal atom case from now on to keep the equations as simple as possible. In general, however, equations involving a triple product of U's can be generalized by replacing N with σ_3/σ_2^3, and those with E's by replacing $N^{-1/2}$ with $\sigma_3/\sigma_2^{3/2}$.

Table 11.3 shows P for sign relationships involving various values of $|E_{av}| = |(E_{hkl}E_{h'k'l'}E_{h-h',k-k',l-l'})|^{1/3}$ for crystals containing N equal atoms.

Probabilities calculated with the equations given hold only if there is no systematic relationship between the indices hkl and $h'k'l'$. In particular, for the case given by Eq. (11.34) the proper formulas are

$$P_+(U_{2h,2k,2l}) = \frac{1}{2} + \frac{1}{2}\tanh\left\{\frac{\sigma_3}{2\sigma_2^3}|U_{2h,2k,2l}|(U_{hkl}^2 - \sigma_2)\right\} \tag{11.43}$$

$$P_+(E_{2h,2k,2l}) = \frac{1}{2} + \frac{1}{2}\tanh\left\{\frac{\sigma_3}{2\sigma_2^{3/2}}|E_{2h,2k,2l}|(E_{hkl}^2 - 1)\right\} \tag{11.44}$$

where P_+ is now the probability that $U_{2h,2k,2l}$ has a positive sign. Probabilities calculated with these equations are always smaller than those that

TABLE 11.3 Phase Probabilities as a Function of $|E_{av}|$ and N

		N					
$\|E_{av}\|$	$\|E_{av}\|^3$	20	40	60	80	100	120
3.0	27.0	1.0	1.0	1.0	1.0	0.99	0.99
2.8	22.0	1.0	1.0	1.0	0.99	0.99	0.98
2.6	17.4	1.0	1.0	0.99	0.98	0.97	0.96
2.3	12.2	1.0	0.98	0.96	0.94	0.92	0.90
2.0	8.0	0.97	0.93	0.89	0.86	0.83	0.81
1.8	5.8	0.93	0.86	0.81	0.78	0.76	0.74
1.5	3.4	0.82	0.74	0.71	0.68	0.66	0.65

would be obtained using Eqs. (11.35) and (11.40). The implication is that Eq. (11.34) is less powerful than Eq. (11.32) as a phase-determining tool and should be used with caution, particularly if probabilities are not being calculated explicitly.

It will often occur in the later stages of the phasing process that there are a number of relationships of the form of Eq. (11.32), all relating to a given U_{hkl}, but none having sufficiently high probability to be compelling. It is obvious, however, that the expression[14]

$$S(hkl) = \sum_{h'k'l'} S(h'k'l')S(h-h', k-k', l-l') \tag{11.45}$$

more nearly resembles the full Sayre equation [Eq. (11.31)] than does Eq. (11.32) and should have a higher probability of being correct than any of the individual terms of the summation. Under these conditions[15]

$$P_+(U_{hkl}) \approx \tfrac{1}{2} + \tfrac{1}{2}\tanh\left\{N|U_{hkl}| \sum_{h'k'l'} U_{h'k'l'} U_{h-h',k-k',l-l'}\right\} \tag{11.46}$$

or its equivalent for E's,

$$P_+(E_{hkl}) \approx \tfrac{1}{2} + \tfrac{1}{2}\tanh\left\{N^{-1/2}|E_{hkl}| \sum_{h'k'l'} E_{h'k'l'} E_{h-h',k-k',l-l'}\right\} \tag{11.47}$$

where $P_+(U_{hkl})$ is the probability that U_{hkl} is positive.

Since the argument of the hyperbolic tangent in Eqs. (11.46) and (11.47) can be either positive or negative, depending on the sign of the summation of the E or U products, the values of P_+ from these formulas can range between 0 ($\sum$ = large negative, tanh = −1) and 1 ($\sum$ = large positive, tanh = +1). Values of P_+ less than $\frac{1}{2}$ are thus indications that the sign of F_{hkl} is negative with a probability P_- given by

$$P_- = 1 - P_+ \tag{11.48}$$

Noncentrosymmetric Methods

Although it was not appreciated that practical use could be made of the information, it was known by 1955[16,17] that expressions equivalent to Eq.

[14]This expression is frequently called the Σ_2 (sigma two) relationship after a notation used in H. Hauptman and J. Karle, *Solution of the Phase Problem*. I. *The Centrosymmetric Crystal*, A.C.A. Monograph No. 3, Polycrystal Book Service, Pittsburgh, 1953. Σ_1 (sigma one) was used in this reference for Eq. (11.34).

[15]M. M. Woolfson, *Direct Methods in Crystallography*, Clarendon Press, Oxford, 1961, p. 54; see also A. Klug, *Acta Cryst.*, **11**, 515 (1958) for a more precise analysis.

[16]W. Cochran, *Acta Cryst.*, **8**, 473 (1955).

[17]J. Karle and H. Hauptman, *Acta Cryst.*, **3**, 181 (1950).

(11.32) could be derived for noncentrosymmetric space groups as well. These appear as

$$\phi(\mathbf{h}) \approx \phi(\mathbf{h}') + \phi(\mathbf{h} - \mathbf{h}') \tag{11.49}$$

or

$$\phi(\mathbf{h}) + \phi(\mathbf{h}') + \phi(\bar{\mathbf{h}} + \bar{\mathbf{h}}') = \Phi_3 \approx 0 \tag{11.50}$$

where ϕ is the phase angle expressed in fractions of a cycle. As before, the probability that Eq. (11.49) holds or that $\Phi_3 = 0$ increases with the magnitude of the reflections involved. Because the phases can now assume general values, the probabilities assume the form of a distribution that gives the likelihood of various degrees of error. For Eq. (11.49) this distribution is given by[16]

$$P\{\Delta\phi(\mathbf{h})\} = \frac{e^{\kappa(\mathbf{h},\mathbf{h}')\cos\{\phi(\mathbf{h}) - [\phi(\mathbf{h}') + \phi(\mathbf{h}-\mathbf{h}')]\}}}{2\pi I_0[\kappa(\mathbf{h}, \mathbf{h}')]} \tag{11.51}$$

where

$$\kappa(\mathbf{h}, \mathbf{h}') = 2(N^{-1/2})|E(\mathbf{h})E(\mathbf{h}')E(\bar{\mathbf{h}} + \bar{\mathbf{h}}')| \tag{11.52}$$

and I_0 is a modified Bessel function.[18] Figure 11.3 shows how this probability varies for triples with different values of κ, while Table 11.4 shows

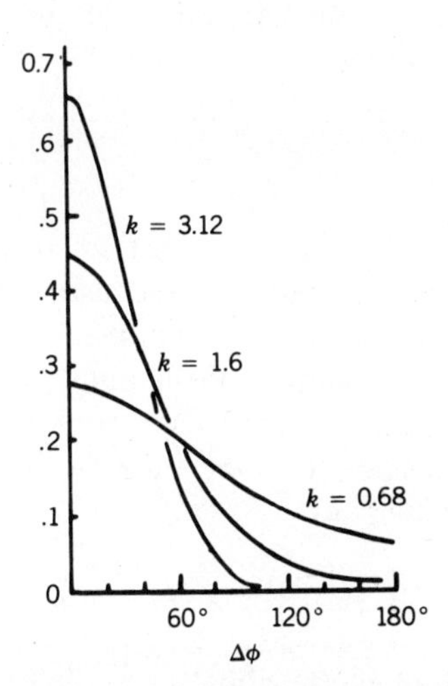

Figure 11.3. Probability distributions of ϕ_3 for three values of κ. See text.

[18] Although frequently described in the literature as a modified Bessel function of the second kind, this is, in fact, a function of the first kind by standard definitions. It is unfortunately true that many errors and discrepancies of notation have been propagated in the secondary literature and that supposedly identical equations from different sources sometimes give widely discordant results when used to calculate numerical values.

TABLE 11.4 **κ's and E's for Various Numbers of Atoms**

	$\langle E\rangle = (E_1E_2E_3)^{1/3}$	
κ	$N=100$	$N=30$
0.68	1.5	1.2
1.60	2.0	1.6
3.12	2.5	2.0

that these correspond to different average values of E for different numbers of atoms in the cell. A similar expression gives the distribution of probable values for $\phi(\mathbf{h})$ when $\phi(\mathbf{h}')$ and $\phi(\mathbf{h}-\mathbf{h}')$ and the E's are known.

If there is more than one triple pointing to the phase of a particular reflection, the expression equivalent to Eq. (11.45) is the *tangent formula* of Karle and Hauptman.[3]

$$\tan[\phi(\mathbf{h})] = \frac{\sum_{h'} \kappa(\mathbf{h}, \mathbf{h}') \sin[\phi(\mathbf{h}') + \phi(\mathbf{h}-\mathbf{h}')]}{\sum_{h'} \kappa(\mathbf{h}, \mathbf{h}') \cos[\phi(\mathbf{h}') + \phi(\mathbf{h}-\mathbf{h}')]} \tag{11.53}$$

Although this formula has flaws, it has been the most important single element in the practical extension of direct methods to noncentrosymmetric crystals. It is often used in a weighted form in which the contributions to the sums are weighted in terms of the probability of the relationships.[19] The probability distribution of the $\phi(\mathbf{h})$ predicted by Eq. (11.53) can be calculated from equations similar to Eq. (11.51) but is more often described in terms of its variance, the square of its standard deviation.[20]

11.5. MODERN EXTENSIONS[21]

The formulas detailed in Section 11.4 form the basis for most of the practical methods of phase determination currently in widespread use. A great many extensions have been considered theoretically and used in experimental cases. Some of these add to our understanding of the underlying theory of direct methods while others appear in particular program systems or hold promise for the future. Here we shall consider a few of the developments that seem to us to be the most significant.

Determinantal Forms

One of the earliest[8,11] and most important theoretical contributions to direct methods was the Karle–Hauptman determinant. Time after time it has been

[19] S. E. Hull and M. J. Irwin, *Acta Cryst.*, **A34**, 863 (1978).
[20] J. Karle and I. L. Karle, *Acta Cryst.*, **21**, 849 (1966).
[21] For discussion and references see M. M. Woolfson, *Acta Cryst.*, **A43**, 593 (1987).

shown to contain information that was first uncovered by other derivations and only later traced back to the determinant. For many reasons it did not generate the mainstream of practical structural methods (except in the able hands of the Karles themselves), but it continues to provide insight. In addition, as computer power grows it becomes increasingly possible to take advantage of some of the more powerful practical possibilities.

The Karle–Hauptman determinant (hereafter KHdet) is a determinant of order $n+1$ of the form

$$\begin{vmatrix} F_0 & F_{-\mathbf{h}1} & F_{-\mathbf{h}2} & \cdots & F_{-\mathbf{h}n} \\ F_{\mathbf{h}1} & F_0 & F_{\mathbf{h}1-\mathbf{h}2} & \cdots & F_{\mathbf{h}1-\mathbf{h}n} \\ F_{\mathbf{h}2} & F_{\mathbf{h}2-\mathbf{h}1} & F_0 & \cdots & F_{\mathbf{h}2-\mathbf{h}n} \\ & & & \vdots & \\ F_{\mathbf{h}n} & F_{\mathbf{h}n-\mathbf{h}1} & F_{\mathbf{h}n-\mathbf{h}2} & \cdots & F_0 \end{vmatrix} \geq 0 \qquad (11.54)$$

in which the fact that the electron density is everywhere nonnegative leads to the requirement that the determinant be itself nonnegative. The leading column of the determinant consists of a set of arbitrary (different) phased structure factors (real or complex). The top row consists of their Friedel mates, that is, those reflections with the signs inverted on all indices. The body of the array contains the phased structure factors for all those reflections with indices that are the sum of those at the left edge of the row and top of the column. Hence the diagonal terms are simply F_{000}.

Since the derivation of the KHdet makes no assumption about the shapes of the electron density, the formulation holds as well for determinants written in U's or E's. Using centrosymmetric U's for which $U_{\mathbf{h}} = U_{-\mathbf{h}}$ and restricting the determinant to small orders leads to familiar results.

$$|U_0| = 1 \geq 0 \qquad (11.55)$$

$$\begin{vmatrix} U_0 & U_{-\mathbf{h}} \\ U_{\mathbf{h}} & U_0 \end{vmatrix} = U_0^2 - U_{\mathbf{h}}U_{-\mathbf{h}} = 1 - U_{\mathbf{h}}^2 \geq 0 \qquad (11.56)$$

$$\begin{vmatrix} 1 & U_{-\mathbf{h}} & U_{2\mathbf{h}} \\ U_{\mathbf{h}} & 1 & U_{-\mathbf{h}} \\ U_{2\mathbf{h}} & U_{\mathbf{h}} & 1 \end{vmatrix} = 1 + 2U_{\mathbf{h}}^2 U_{2\mathbf{h}} - U_{2\mathbf{h}}^2 - 2U_{\mathbf{h}}^2 \geq 0 \qquad (11.57)$$

$$= (1 + U_{2\mathbf{h}})(1 - U_{2\mathbf{h}}) - 2U_{\mathbf{h}}^2(1 - U_{2\mathbf{h}}) \geq 0 \qquad (11.58)$$

since $(1 - U_{2\mathbf{h}})$ must be >0, this yields

$$\tfrac{1}{2}(1 + U_{2\mathbf{h}}) \geq U_{\mathbf{h}}^2 \qquad (11.59)$$

Here Eq. (11.56) is equivalent to the trivial Harker–Kasper result of Eq. (11.1), while (11.59) is the classic Harker–Kasper inequality, Eq. (11.4).

Manipulation of the more general third-order KHdet leads to expressions that contain products of $U_{\mathbf{h}}$, $U_{\mathbf{h}'}$, $U_{\mathbf{h}-\mathbf{h}'}$, and $U_{-\mathbf{h}+\mathbf{h}'}$. These can be connected with the results obtained by Sayre and others.

Extension of the KHdet to higher orders leads to the theory of quartets, quintets, and higher-order forms (see below), but more significant and specific are those techniques that make use of the capabilities of very large determinants. Obviously, as the order increases, the number of possible phase permutations becomes immense, and general techniques do not exist for finding reliably those combinations of phases that meet the requirements imposed by the determinantal inequality. There are many in any real case, but they constitute only a small fraction of those possible. Tsoucaris[22] demonstrated that the most probable set of phases is the one that maximizes the KHdet (*maximum determinant method*) and has proposed solution approaches for attacking that problem. Other attempts have used the KHdet and the maximum determinant to extend and refine phases in macromolecular problems.

Quartets and Higher Products[23,24]

Although most of the practical work with direct phasing has involved triples of reflections whose indices sum to zero, there has been increasing interest in larger sets of reflections related in the same way. Thus quartets of the form

$$\Phi_4 = \phi_{\mathbf{h1}} + \phi_{\mathbf{h2}} + \phi_{\mathbf{h3}} + \phi_{\overline{\mathbf{h1}}+\overline{\mathbf{h2}}+\overline{\mathbf{h3}}} \tag{11.60}$$

are also structure invariants and have been found to be of real practical utility. In particular, they can be related to the KHdet, and it can be shown[23,24] that for the important case in which the product $E_{\mathbf{h1}}E_{\mathbf{h2}}E_{\mathbf{h3}}E_{\overline{\mathbf{h1}}+\overline{\mathbf{h2}}+\overline{\mathbf{h3}}}$ is large, the value of Φ_4 may be distributed around 0 (*positive quartets*), around $\pi/2$, or around π (*negative quartets*) depending on whether the *cross terms* $E_{\mathbf{h1}+\mathbf{h2}}$, $E_{\mathbf{h1}+\mathbf{h3}}$, and $E_{\mathbf{h2}+\mathbf{h3}}$ are large, medium, or small. Formulas have been derived for estimating the probable value of Φ_4 and its uncertainty, but they are still a field of active study. Nevertheless, positive quartets have proved useful for confirming triple relationships in the early stages of tree building, while negative ones serve as a selection test among alternative solutions.[25] They are also effective for breaking the tendency of certain space groups (e.g., $P1$ and $P\bar{1}$) to give falsely consistent

[22] G. Tsoucaris, *Acta Cryst.*, **A26**, 492 (1970).

[23] H. Schenk in *Crystallographic Computing*, 3: *Data Collection, Structure Determination, Proteins, and Data Bases*, G. M. Sheldrick, C. Krüger, and R. Goddard, Eds., Clarendon Press, Oxford, 1985, pp. 190–199.

[24] H. Schenk, *Acta Cryst.*, **A29**, 77 (1973).

[25] G. T. DeTitta, J. W. Edmonds, D. A. Langs, and H. Hauptman, *Acta Cryst.*, **A31**, 367 (1975).

solutions in which all phases are 0. Quartets whose values tend to $\pi/2$ can be used to select enantiomorphs.

Still higher order polygons of r.l. vectors such as quintets and sextets have been examined and occasionally used, but their application is not yet widespread. It would appear likely, however, that a general approach based on the KHdet would be more useful than dealing with each individual increase in complexity.

11.6. PHASE DETERMINATION IN PRACTICE

Although the basic theory of direct phase determination is now nearly 40 years old, it has been only in the last decade that its application has undergone explosive growth to its present dominance. This is partially due to the growth in computing power and the realization that brute force is much more effective here than is logical elegance. In part, too, it reflects the time taken for practical understanding to grow from the somewhat idealized theory.

The practical objective of direct methods is to phase a sufficient number of reflections to give an identifiable Fourier representation of the molecule being studied. The number required will depend on various factors, among them the nature of the molecular system and how much is already known of its structure. Roughly, however, 10 reflections per atom in the asymmetric unit seems quite satisfactory, and in some cases as few as three to five per atom have served.

Symbolic Addition[20]

The method most used for the application of direct methods in their early years traces back to Zachariasen[26] in a 1952 paper accompanying the pioneering one of Sayre. Subsequently it was applied intermittently by various workers and was finally codified under the name of the *symbolic addition method*.[27]

The method, like so many other crystallographic techniques, is a bootstrapping operation. Here one starts with a very limited number of phases and uses them in connection with Eq. (11.32) or Eq. (11.45) to pyramid to a number large enough to give a recognizable Fourier representation of the structure. The obvious danger in this approach is that one is building one's pyramid the wrong way to; it is balanced on its point, and if one of the stones in the bottom course is faulty it will come down. Fortunately, it turns out that the labor invested is not very great and it is possible to improve

[26] W. H. Zachariasen, *Acta Cryst.*, **5**, 68 (1952).
[27] I. L. Karle, K. Britts, and P. Gum, *Acta Cryst.*, **17**, 496 (1964).

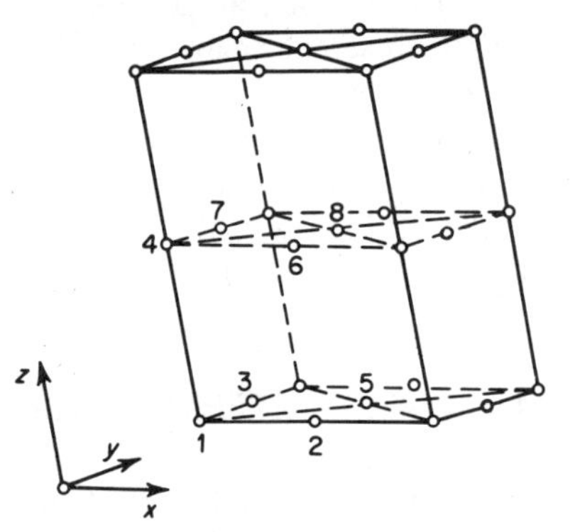

Figure 11.4. Unit cell of a centrosymmetric structure showing centers of symmetry and alternative origins.

one's chances by constructing a number of pyramids and seeing which one stands.

The first problem to be met in practice is that of obtaining some initial phases to work with. Fortunately, a limited number, usually three but sometimes more or fewer, can be assigned arbitrary values, subject to certain restrictions. These arbitrarily assigned phases constitute the initial set.

The question of what reflections can be assigned phases in various space groups has been studied by many workers.[28–30] We shall outline only the simplest case, that of the primitive centrosymmetric space groups in the triclinic, monoclinic, and orthorhombic classes. Any such space group can be generalized in the form shown in Fig. 11.4, that is, as a unit cell with centers of symmetry as shown, all other symmetry being omitted. In dealing with a centrosymmetric space group by this method, it is necessary to place the origin at a center of symmetry to keep all the phases real, but in these cases there are no further restrictions on which of the eight centers marked with numbers is chosen.[31] As discussed in Section 11.3, a change of origin from one center to another will affect only the phases and not the magnitudes of any calculated structure factors, and so no one is to be preferred. The justification of this statement, and the clue as to the effect of such an origin change on the phases, is gained by considering the general structure factor expression for a centrosymmetric crystal [Eq. (8.71)].

$$F_{hkl} = 2 \sum^{N/2} f_j \cos 2\pi(hx + ky + lz) \tag{11.61}$$

Suppose that the origin is shifted from point 1 to point 2, that x becomes

[28]K. Lonsdale and H. J. Grenville-Wells, *Acta Cryst.*, **7**, 490 (1954).

[29]H. Hauptman and J. Karle, *Acta Cryst.*, **12**, 93 (1959).

[30]J. Karle, in *International Tables*, Vol. IV, pp. 339–349.

[31]The others need not be considered since they are all related to some one of the fundamental eight by unit translations.

$x - \frac{1}{2}$. Then

$$F'_{hkl} = 2 \sum^{N/2} f_j \cos 2\pi(hx - h/2 + ky + lz) \tag{11.62}$$

$$F'_{hkl} = 2 \sum^{N/2} f_j \cos[2\pi(hx + ky + lz) - \pi h] \tag{11.63}$$

$$F'_{hkl} = 2 \sum^{N/2} f_j[\cos 2\pi(hx + ky + lz) \cos(-\pi h) + \sin 2\pi(hx + ky + lz) \sin(-\pi h)] \tag{11.64}$$

And, since $\cos n\pi = (-1)^n$, $\sin n\pi = 0$ for any integer n,

$$F'_{hkl} = 2 \sum^{N/2} f_j \cos 2\pi(hx + ky + lz)(-1)^h \tag{11.65}$$

$$F'_{hkl} = (-1)^h F_{hkl} \tag{11.66}$$

Thus the shift of an origin by $a/2$ results in a change of sign for all reflections of odd h but no changes in $|F|$. Similar derivations could be carried out for each of the centers in Fig. 11.4 with corresponding results; a change by $\frac{1}{2}$ along any axis produces a change of sign in the reflections with the corresponding indices odd. The reflections with even indices are unaffected. If the shift involves translation along two axes, for example, $a/2$ and $b/2$ to get to origin 5, the result would be

$$F_{hkl} = (-1)^{h+k} F_{hkl} \tag{11.67}$$

and a sign change would occur only if h or k but not both were odd. Table 11.5 gives the complete set of sign changes of an initially positive set of reflections for all possible combinations of even/odd indices and origins.

Table 11.5 provides a basis for deciding how arbitrary phases can be

TABLE 11.5 Relative Sign Relationships for Possible Origins

		Reflection Kind							
Origin	Shift	eee	oee	eoe	eeo	ooe	oeo	eoo	ooo
1	0	+	+	+	+	+	+	+	+
2	$a/2$	+	−	+	+	−	−	+	−
3	$b/2$	+	+	−	+	−	+	−	−
4	$c/2$	+	+	+	−	+	−	−	−
5	$(a+b)/2$	+	−	−	+	+	−	−	+
6	$(a+c)/2$	+	−	+	−	−	+	−	+
7	$(b+c)/2$	+	+	−	−	−	−	+	+
8	$(a+b+c)/2$	+	−	−	−	+	+	+	−

assigned. Since the phases of eee reflections never change for shifts among these standard origins (***structure seminvariants***), it is clear that they cannot be given values at will. All of the other classes are positive for four origins and negative for four. As long as both + and − signs appear in the table for a class of reflections, some member of it can be assigned a phase arbitrarily, since this merely corresponds to selecting one origin from the two possible sets. Thus if, for example, an oee reflection such as 742 is defined as being +, this reduces the possible number of origins to four (1, 3, 4, or 7). The phases of *all* the oee reflections are now fixed, and no further choices can be made from that set. All of the remaining classes, however, show two +'s and two −'s for the four origins that make the oee group +. Thus any reflection of these classes, for example, 516, an ooe, may be set +. This narrows the choice to origins 1 and 4, and one more choice can be made to select one of these. This choice is more restricted than before, however. Not only are the oee and ooe groups barred since they have already been fixed, but also eoe is forbidden as a choice. Inspection of Table 11.5 shows that eoe is + for both origins 1 and 4, that is, the phases for this class of reflections have been set by the choices already made and none can be chosen arbitrarily. The other classes all have a + and a − sign for these origins, and any of these can be assigned the third phase to finally fix the origin. It is clear that the choice of the first two phases is easy but that some care must be given to the third in order to avoid choosing one from a class that has become invariant.

Within the limits imposed above, any three reflections can be assigned phases. Selection of these also determines the phase of those other reflections that are related to the choices by the symmetry of the reciprocal lattice. As was discussed in Section 6.1, the concept of a unique data set arises because of the identity of the intensities from various symmetrically related reflections. Thus for monoclinic crystals, reflections with the same absolute values of h, k, and l can be divided into two groups [see Eqs. (6.3)–(6.5)],

$$|F_{hkl}| = |F_{\overline{hkl}}| = |F_{h\bar{k}l}| = |F_{\bar{h}k\bar{l}}| \tag{11.68}$$

$$|F_{\bar{h}kl}| = |F_{h\overline{kl}}| = |F_{hk\bar{l}}| = |F_{\overline{hk}l}| \tag{11.69}$$

$$|F_{hkl}| \neq |F_{\bar{h}kl}| \tag{11.70}$$

The phase relationships among reflections within one set are determined by the space group and can be calculated from the general structure factor expression by expanding it in terms of the equivalent positions. Fortunately, the results have been tabulated for all space groups in *International Tables*.[32] Thus for the common case of $P2_1/c$ we find that for all reflections

[32] *International Tables*, Vol. I, pp. 374–525.

$$F_{hkl} = F_{\bar{h}\bar{k}\bar{l}} \qquad (11.71)$$

which implies as well

$$F_{\bar{h}kl} = F_{h\bar{k}\bar{l}} \qquad (11.72)$$

The effect of changing the sign of k depends, as is generally the case, on the even or odd character of an index or sum of indices. For $P2_1/c$ this test is on $k+l$. Thus,

$k+l$ even:

$$F_{hkl} = F_{h\bar{k}l} = F_{\bar{h}k\bar{l}} \qquad (11.73)$$

$$F_{\bar{h}kl} = F_{\bar{h}\bar{k}l} = F_{hk\bar{l}} \qquad (11.74)$$

where the second equalities in Eqs. (11.73) and (11.74) follow from the application of Eq. (11.71) to the first. On the other hand,

$k+l$ odd:

$$F_{hkl} = -F_{h\bar{k}l} = -F_{\bar{h}k\bar{l}} \qquad (11.75)$$

$$F_{\bar{h}kl} = -F_{\bar{h}\bar{k}l} = -F_{hk\bar{l}} \qquad (11.76)$$

Using these relationships or the corresponding ones for other space groups, it is possible to obtain the proper phase for any member of an index set once that of one member of the set is known.

Origin assignment for noncentrosymmetric crystals is similar to that described but is complicated by the complex phases that occur for general reflections.[29,30,33] In many cases there exist special classes of reflections (e.g., $h0l$ in $P2_1$) for which the allowed phases are restricted to one of two values, but it is usually helpful to have at least one general reflection in the initial set. One additional starting phase can be assigned for many noncentrosymmetric cases in order to select the enantiomer.[34] These phases are assigned to structure invariants or seminvariants (which may be single reflections that become invariant after the origin has been fixed). In some space groups such as $P2_12_12_1$ there exist reflections for which $A \equiv 0$ and whose phases are therefore $\pi/2$ or $3\pi/2$. An appropriate one of these can be assigned $\pi/2$ (say) arbitrarily to define the enantiomorph. If no such limited reflections exist, the problem is more difficult, since it is necessary to

[33] H. Hauptman and J. Karle, *Acta Cryst.*, **9**, 45 (1956); J. Karle and H. Hauptman, *Acta Cryst.*, **14**, 217 (1961).

[34] This is *not* true for the enantiomeric space groups such as $P3_1$ and $P3_2$ for which the enantiomer is chosen by the assignment of the space group and its associated phase relationships.

select a reflection for which the correct phase is near $\pi/2$ or $3\pi/2$ (rather than 0 or π) in order to choose a hand. Often the assignment of an enantiomorph-determining reflection is omitted in the early stages of phase determination and is made only at the end when there is evidence to favor a particular choice.

The choice of origin-defining reflections is normally made by algorithms encoded in the package of phase-determining programs, but it is simple to check the correctness of the set chosen with a procedure of Hovmöller.[35]

The preliminary stage[36] of the symbolic addition method consists of calculating E's for the entire data set. For all the E's greater than some chosen minimum, often about 1.5, a list is compiled of all the triples of reflections belonging to this set for which the indices sum to zero. This is the most tedious part of the exercise and the hardest to carry out without a computer. Practical programs often also compute probabilities [see Eq. (11.42)] that $s_{\mathbf{h1}} s_{\mathbf{h2}} s_{\overline{\mathbf{h1}}+\overline{\mathbf{h2}}} = +1$ for a centrosymmetric problem or the probable variance of the Φ_3 [Eq. (11.51)] for a noncentrosymmetric one. These aid in selecting the *probably* most reliable relationships for the next step.

The list of strongest (most probable) relationships is used to select those reflections that are most often and most reliably interconnected, and appropriate ones of these are chosen for origin determination. Often the three starting reflections will combine to relate to two new ones and imply their phases by Eq. (11.45). Perhaps one of these will feed back with the original set to give yet another. Nevertheless it is generally the case that the process will terminate early by running out of new combinations. The crucial step of the method then involves bringing in a new, strong reflection as a variable, historically represented by a letter. This variable stands for either a general phase in the noncentrosymmetric case or a sign in the centrosymmetric case. Using this variable, the process is continued until all possible triples have been found, and the cycle is then started again with another variable. This series of operations is repeated until a sufficient number of reflections have been phased, either absolutely or in terms of one or more variables.

In order to outline the phase-determining process, an actual structure determination will be used as an example.[37] The crystals were those of a natural product ($C_{17}H_{16}O_7$) in space group $P2_1/c$. The indices of reflections with $|E| > 3.0$ were used to generate the triples of reflections in which the indices of two added to give those of the third. With $|E| > 3.0$ and with 96 nonhydrogen atoms in the cell, the probability that the signs of these reflections were related as given in Eq. (11.32) or (11.33) could be calculated [Eq. (11.40)] to be greater than 0.99 (see Table 11.3).

[35]S. Hovmöller, *Acta Cryst.*, **A37**, 133 (1982).

[36]For detailed examples of the solution of phase problems by symbolic addition, see I. L. Karle in *Crystallographic Computing Techniques*, F. R. Ahmed, K. Huml, and B. Sedlack, Eds., Munksgaard, Copenhagen, 1976, pp. 27–70; M. M. Woolfson, *Acta Cryst.*, **A43**, 593 (1987); and Luger, *Modern X-Ray Analysis on Single Crystals*, pp. 253–266.

[37]G. H. Stout, T. S. Lin, and I. Singh, *Tetrahedron*, **25**, 1975 (1969).

The origin was fixed by assigning + phases to the reflections $3\ 1\ \overline{17}$ (ooo), 3 4 11 (oeo), 5 0 14 (oee). These appeared in two relationships:

$$s(3\ 1\ \overline{17}) \cdot s(3\ 4\ 11) = s(6\ 5\ \bar{6}); \qquad s(6\ 5\ \bar{6}) = + \tag{11.77}$$

$$s(3\ 1\ \overline{17}) \cdot s(5\ 0\ 14) = s(8\ 1\ \bar{3}); \qquad s(8\ 1\ \bar{3}) = + \tag{11.78}$$

The high probabilities of the relationships suggested that these signs might be accepted as correct and used as the basis for further sign determinations.

Combination of one of the new reflections with one of the original set leads to a new relationship;

$$s(3\ \bar{4}\ 11) \cdot s(5\ 3\ \overline{14}) = s(8\ \bar{1}\ \bar{3}) \tag{11.79}$$

Since the space group is $P2_1/c$,

$$s(3\ \bar{4}\ 11) = -s(3\ 4\ 11) = - \tag{11.80}$$

and

$$s(8\ \bar{1}\ \bar{3}) = s(8\ 1\ \bar{3}) = + \tag{11.81}$$

$k + l$ being odd in the first case and even in the second [see Eqs. (11.73)–(11.76)]. Consequently Eq. (11.79) may be written

$$- \cdot S(5\ 3\ \overline{14}) = +; \qquad s(5\ 3\ \overline{14}) = - \tag{11.82}$$

No further relationships exist relating two of the reflections of known sign with one of the remaining reflections. To continue the process, a variable that represents the unknown + or − phase is assigned as the sign of some useful reflection. The phases of other reflections are then worked out in terms of the known signs and this variable. Thus, if by definition

$$s(6\ 5\ \overline{12}) \equiv a \tag{11.83}$$

the relationship

$$s(2\ \bar{6}\ 9) \cdot s(6\ 5\ \overline{12}) = s(8\ \bar{1}\ \bar{3}) \tag{11.84}$$

implies

$$s(2\ \bar{6}\ 9) \cdot a = + \tag{11.85}$$

or since

$$a \cdot a = + \cdot + = - \cdot - = + \tag{11.86}$$

then

$$s(2\ \bar{6}\ 9) = a \tag{11.87}$$

regardless of the actual sign corresponding to a. Since $k + l$ is odd,

$$s(2\ 6\ 9) = -s(2\ \bar{6}\ 9) = -a \tag{11.88}$$

In the same way,

$$s(3\ \bar{1}\ \overline{17}) \cdot s(3\ 6\ 5) = s(6\ 5\ \overline{12}) \tag{11.89}$$

$$+ \cdot s(3\ 6\ 5) = a; \qquad s(3\ 6\ 5) = a \tag{11.90}$$

$$s(3\ \bar{4}\ 11) \cdot s(6\ 5\ \overline{12}) = s(9\ 1\ \bar{1}) \tag{11.91}$$

$$- \cdot a = s(9\ 1\ \bar{1}); \qquad s(9\ 1\ \bar{1}) = -a \tag{11.92}$$

At this point a checking relationship becomes available, all of whose signs are known. Thus,

$$s(2\ \bar{6}\ 9) \cdot s(3\ 6\ 5) = s(5\ 0\ 14) \tag{11.93}$$

$$a \cdot a = + \tag{11.94}$$

which is correct and lends confidence to the assignments that have been made.

Using the extra phases now determined in terms of a, additional relationships can be found. For example,

$$s(2\ \bar{6}\ 9) \cdot s(7\ 7\ \overline{10}) = s(9\ 1\ \bar{1}) \tag{11.95}$$

$$a \cdot s(7\ 7\ \overline{10}) = -a; \qquad s(7\ 7\ \overline{10}) = - \tag{11.96}$$

so the sign of $7\ 7\ \overline{10}$ is determined absolutely even though the two other reflections in the equation are known only in terms of the variable. The list of signs can be extended by repeating the process until a point is reached at which no more relationships involving two known and one unknown signs can be found. In this case 11 further signs can be found. Since a number of equations remain unused, another arbitrary sign can be defined and the process repeated, new signs being found in terms of either or both of the variables.

With this set of determined signs as a basis it is now possible to extend the process to reflections of lower E value. The list of unknowns is extended by adding the set of the next most intense reflections (in terms of E), and the relationships are found that connect two reflections of known phase with a member of this set. In the actual solution of the example being used, the list was extended in steps, adding first those reflections with $|E| > 2.5$, then $|E| > 2.0$, and finally $|E| > 1.5$.

As the relationships begin to include reflections of lower $|E|$ values, their probabilities begin to fall from the very high values that allow one to accept a single sign indication as determinative. To make up for this, however, many unknown signs will be determined by more than one relationship. Thus 2 3 $\overline{20}$ ($E = 2.53$) is given by

$$s(\bar{4}\ \bar{2}\ \overline{13}) \cdot s(6\ 5\ \bar{7}) = s(2\ 3\ \overline{20}) \tag{11.97}$$

$$s(\bar{4}\ 8\ \bar{8}) \cdot s(6\ \bar{5}\ \overline{12}) = s(2\ 3\ \overline{20}) \tag{11.98}$$

$$s(\bar{6}\ 2\ \overline{17}) \cdot s(8\ 1\ \bar{3}) = s(2\ 3\ \overline{20}) \tag{11.99}$$

All these imply that

$$S(2\ 3\ \overline{20}) = -a \tag{11.100}$$

and the agreement reinforces the strength of the implication [see Eq. (11.47)].

As the phase-determining process continues, two additional phenomena occur. First, contradictions will begin to appear. The set of relationships defining a given sign may contain several that indicate it to be ab and one indicating $-ab$. Usually, the contradictory relationship is one of relatively lower probability and can be discounted against one of opposite implication. Such disagreements are to be expected, and, in fact, their failure to appear is a warning that the solution may be headed for the "uranium catastrophe" in which all the phases are consistent and all the electron density is concentrated at a single massive peak at an origin.[38]

The second phenomenon is the appearance of sets of relationships suggesting more than one variable combination as appropriate for a given reflection. These often serve to suggest relationships between the phases corresponding to some of the variables, either absolutely or in terms of other variables. This serves to reduce the number of unknowns, but usually not to zero, and often suggests the most probably successful order of phase assignment to the variables.

Symbolic addition in the noncentrosymmetric case is carried out in a similar way, but the process is complicated by the need to add phases modulo 2π rather than multiplying signs and by the accumulation of errors in slow stages even for approximately correct relationships.

When several phase indications are combined by averaging the implications of individual triples, problems arise from the step that occurs between 359° and 1° ($2\pi - \delta$ and $0 + \delta$) or between $\pm\pi$ (+179° and −179°). If a given average includes contributors from the two sides of a step, the

[38]This effect occurs when all the phases are 0 or π (+ or − signs). It may not be apparent for the origin selected, but examination of the signs produced by the changes described in Table 11.4 will show that for some origin all the signs are the same. The phenomenon is a particular problem in $P1$ and $P\bar{1}$ where triplet sign relationships alone do not provide any means of breaking out of an overconsistent set with all the signs the same.

numbers do not average correctly and the apparent distribution of values around the mean is excessively broad. One solution[20] is to calculate each average for both of the ranges $\pm\pi$ and 0–2π and select the average that gives the tighter distribution. Once an extended set of starting phases, often 100–200, has been obtained, the tangent formula [Eq. (11.53)] can be used by cycling repeatedly through the triples to obtain the most consistent phases for the original set as well as to extend the phasing to reflections with lower E values.

Multiple Solution Methods

In the early days of direct methods, computers were used to generate the lists of triples, and the subsequent symbolic addition was carried out by hand. This was fairly easy in the centrosymmetric case, somewhat more difficult in the noncentrosymmetric one. Nevertheless, it was obviously desirable to computerize the entire process, and various programs were written to carry out symbolic addition by machine. Computers are not particularly adept at quasi-algebraic manipulation, however, and the standard approach to symbolic addition does not make the best use of their abilities. They are best at keeping track of numbers, and it was on this basis that Woolfson and his collaborators built their direct methods program MULTAN,[39–42] which is extremely widely used either directly or in the form of other programs derived from it.[43] In fact, MULTAN has become almost a generic term for multisolution programs.

MULTAN in its earliest forms was a mechanization of various elements of the symbolic addition method, with the critical addition that symbolic variables were replaced by a list of trial numeric phase possibilities adequate to hit sufficiently close to the true phase of the reflection. For general noncentrosymmetric reflections these were originally 45°, 135°, 225°, and 315°, which guarantee that one trial must be within 45° of the correct phase (at worst; the average error is much less). For centrosymmetric reflections they are 0° and 180° (or other special values for particular cases). Of course, each of these trials corresponds to a distinct alternative solution and as such must be kept separate, leading to rather complex bookkeeping but well within the power of the computer. Each additional variable causes another

[39]P. Main, in *Crystallographic Computing*, 3: *Data Collection, Structure Determination, Proteins, and Data Bases*, G. M. Sheldrick, C. Krüger, and R. Goddard, Eds., Clarendon Press, Oxford, 1985, pp. 206–215.

[40]P. Main, *MULTAN 11/82 Manual*, University of York, England, 1982.

[41]M. M. Woolfson, in *Computational Crystallography*, D. Sayre, Ed., Clarendon Press, Oxford, 1982, pp. 110–125.

[42]L. Lessinger, *Acta Cryst.*, **A32**, 538 (1976).

[43]For example, MITHRIL, C. J. Gilmore, *J. Appl. Cryst.*, **17**, 42 (1984). For a list and description of available direct methods packages, see C. J. Gilmore in *Computational Crystallography*, D. Sayre, Ed., Clarendon Press, Oxford, 1982, pp. 126–140.

set of branches to sprout from each existing stem, leading to a very large tree structure of possible phase sets.

The actual number of trial sets was reduced by the introduction of *magic integer* methods.[44,45] These are sets of integers that serve to link a moderate number of unknown phases to two or three variables in a way that provides uniform coverage of approximate alternative phases. Various methods were then used to obtain the probably best values for the remaining variables. Although widely used as part of program systems, their importance has probably been lessened by the growth of purely random methods (see below).

MULTAN also formalized in CONVERGENCE[46] the process described above of working back from the entire set of triples to the most reliable and productive set of reflections on which to build the tree. This is the weakest part of the procedure, since if an early triple fails to be true, the entire process may fail, but it is no more fallible than doing the same thing by hand and is certainly more systematically complete.

Since each branch of the phase development tree is worked out from the initial set and the list of triples in terms of numeric phases, it is possible to apply the tangent formula [Eq. (11.53)] directly to the process of phase extension as well as to subsequent refinement. Unfortunately, the tangent formula sometimes fails to refine an initial set of phases properly even if the phases are correct.[42] In fact, cases are known where an apparently erroneous starting set will refine to the true solution while a seemingly much closer set will fail. On the other hand, it is the strength of the multisolution approach that it provides a large number of opportunities to arrive at the true solution.

Random Phase Methods

In the course of work with MULTAN it became clear that the most common source of difficulty was the occurrence of failures early in the phase deduction as a result of triples whose true value was grossly different from that predicted. Some of these failures could be avoided by applying constraints to the CONVERGENCE process and forcing alternative developments that accidentally placed the (unknown) triple at a less critical point, but some could not be remedied in this way.

The simplest means of avoiding the risks of depending on single triplet indications is to do away with the tree development entirely. Various approaches were made to increasing the size of the starting set of triples so as to reduce the effect of a single error,[47] and the process was carried to its

[44]P. S. White and M. M. Woolfson, *Acta Cryst.*, **A31**, 53 (1975).

[45]P. Main, *Acta Cryst.*, **A33**, 750 (1977).

[46]G. Germain, P. Main, and M. M. Woolfson, *Acta Cryst.*, **B26**, 274 (1970).

[47]For example, R. Baggio, M. M. Woolfson, J. P. Declercq, and G. Germain, *Acta Cryst.*, **A34**, 883 (1978).

logical extreme by Yao[48] in his program RANTAN, in which random phases are assigned to all strong reflections except for those defining the origin and enantiomorph and the tangent formula is then applied carefully to refine them to the most consistent possible set. The method does not guarantee to yield a successful result for any single set, but when done repeatedly it is possible to obtain and identify good solutions in a relatively small number of trials (tens, not hundreds, let alone the astronomical number of possible combinations of phases). This appears at present to be the strongest practical method and has been incorporated into MULTAN and other programs as an option, often the default option.

Getting a Model

Once the analysis is completed, the result will be a number of possible phase sets corresponding either to multiple solutions or to assignment of phases to the variables in a single symbolic list. It is now perfectly feasible to calculate three-dimensional Fourier syntheses for each of these and to examine the maps in turn to see which appear to yield promising molecular systems. The limiting factor is much more one's disinclination to plot and examine a large number of syntheses than the time required to compute them. In order to avoid wasting time studying unneccessary syntheses, it is worthwhile to consider the combinations in an order that corresponds at least roughly to their probability of being correct.

Numerous approaches have been used to try to select the best combination without actually calculating Fourier syntheses. Thus an approximation to the full Sayre summation [Eq. (11.31)] can be calculated for some or all reflections using the phases determined for each case. The agreement between the magnitudes ($|E|$ or $|F|$) predicted by the summation and those actually observed can be measured by a quantity similar to R,[49] and in principle the set of phases that gives the best agreement should be correct. In practice the results are not so clean-cut, but they are certainly helpful as a guide. Both very strong and very weak (*psi-zero test*) reflections have been used in these comparisons, but neither has been shown to be obviously superior. Often both are calculated and are combined to give a better test than either alone.[50]

A further method that has been used involves the use of other phase-determining relationships to obtain absolute phases for certain reflections

[48] J.-X. Yao, *Acta Cryst.*, **A37**, 642 (1981); **A39**, 35 (1983).

[49] This is often cited as the Karle R (R_{Karle}) after J. Karle and I. L. Karle, *Acta Cryst.*, **21**, 849 (1966), but similar formulations had appeared earlier. See W. Cochran and A. S. Douglas, *Proc. Roy. Soc. (London)*, **243A**, 281 (1951).

[50] Many other tests have been proposed and are used in various programs. For further discussion, see G. Cascareno, C. Giacovazzo, and D. Viterbo, *Acta Cryst.*, **A43**, 22 (1987).

already known in terms of variables. In this way the correct values of the variables are found.[51–53]

Once a set or sets of phases have been found, the remaining task is to calculate Fourier syntheses. The majority of reflections phased will have large indices and thus correspond to moderate to high values of sin θ. For this reason their $|F|$ values are small even though the $|E|$ values are large. If the $|F|$'s are used as the coefficients in the Fourier synthesis, many of the weaker ones tend to be swamped by a few, more intense, low-order reflections. It has been found that this can be avoided by calculating an *E-map*,[52,53] that is, a synthesis in which the coefficients are E's rather than F's. Since the E's correspond to completely sharpened atoms, the peaks on the map tend to be very sharp, but nevertheless they can easily be found on a grid with a spacing of about 0.5 Å.

Because of the high degree of sharpening present in an E-map and because the reflections involved are often specially selected by the existence of their interrelations, a considerable number of spurious peaks may be observed. Benzene rings, for example, are often found to have peaks in their centers. On the other hand, groups that do not fit into the main pattern of the molecular structure, for example, side chains attached to rings, may not show up at all. Similarly, little or no significance can be attached to the relative heights found for the various peaks. It is for this reason that a knowledge of the probable gross molecular structure can be of help in trying to visualize the way in which atoms are to be assigned to the map.[54]

If a reasonable set of atomic positions can be found, they can be used for structure factor calculation, and the subsequent development will follow the lines described in Chapter 15. If the distribution of peaks looks possible, but a molecule cannot be made out because of spurious peaks, the development of more phases and the addition of more reflections will, if they are correct, enhance the correct peaks and diminish the false ones. If only a portion of the molecule can be recognized, structure factors can be calculated from these atoms and the phases used as a starting set to feed back into tangent formula refinement (*Karle recycling*).[55] Often this process will reveal additional atoms and can be repeated to build the entire structure.

If no solution can be found, it is necessary to force the system to follow some other path of phase determination to avoid errors that may be appearing too early in the process and that trap the solution into a false result. This can sometimes be done by trying another phasing program, but it is necessary to be careful since many programs have been adapted from

[51] I. L. Karle, H. Hauptman, J. Karle, and A. B. Wing, *Acta Cryst.*, **11**, 257 (1958).

[52] I. L. Karle, K. Britts, and S. Brenner, *Acta Cryst.*, **17**, 1506 (1964).

[53] S. Block and A. Perloff, *Acta Cryst.*, **16**, 1233 (1963).

[54] For a discussion of the effects of *random* errors in phases on the appearance of E-maps, see A. M. Silva and D. Viterbo, *Acta Cryst.*, **A36**, 1065 (1980).

[55] J. Karle, *Acta Cryst.*, **B24**, 182 (1968).

others and the internal workings may be very similar. Too, some crystals prove to be very resistant to direct methods, for reasons that are not well understood.

BIBLIOGRAPHY

Dunitz, J., *X-Ray Analysis and the Structures of Organic Molecules*, Cornell University Press, Ithaca, NY, 1979, pp. 148–182.

Giacovazzo, C., *Direct Methods in Crystallography*, Academic, London, 1980.

Karle, J., in *Advances in Structure Research*, R. Brill, Ed., Interscience, New York, 1964, pp. 55–89.

Karle, J., and I. L. Karle, in *Computing Methods in Crystallography*, J. S. Rollett, Ed., Pergamon, Oxford, 1965, pp. 151–165.

Ladd, M. F. C., and R. H. Palmer, *Structure Determination by X-Ray Crystallography*, Plenum, New York, 1985, pp. 271–287.

Lipson, H., and W. Cochran, *The Determination of Crystal Structures*, Cornell University Press, Ithaca, NY, 1966, Chapter 9.

Luger, P., *Modern X-Ray Analysis on Single Crystals*, de Gruyter, New York, 1980, pp. 229–266.

Schenk, H., A. J. C. Wilson, and S. Parthasarathy, Eds., *Direct Methods, Macromolecular Crystallography, and Crystallographic Statistics*, World Scientific, Singapore, 1987.

Woolfson, M. M., Direct Methods—From Birth to Maturity, *Acta Cryst.*, **A43**, 593–612 (1987).

CHAPTER 12

PATTERSON METHODS

Although the majority of structural problems are now attacked first by direct methods, these methods sometimes fail. In addition, there exist problems, especially with macromolecules, that are too large for current direct methods. The alternative approaches generally involve analyzing the intensity data by way of the Patterson function.

12.1. THE PATTERSON FUNCTION[1–3]

It was mentioned in Chapter 9 that a Fourier calculation can be carried out using the phaseless quantities $|F|^2$ as the coefficients. In 1935 A. L. Patterson published a classic paper that pointed out the crystallographic usefulness of the results. He showed that whereas the usual synthesis with F's as coefficients showed the distribution of atoms in the cell, the map calculated with $|F|^2$ gave peaks corresponding to all of the interatomic vectors. Thus a peak at the point uvw in a Patterson map indicates that atoms exist in the crystal at x_1, y_1, z_1 and x_2, y_2, z_2 such that

$$u = x_1 - x_2 \qquad (12.1)$$

$$v = y_1 - y_2 \qquad (12.2)$$

$$w = z_1 - z_2 \qquad (12.3)$$

[1] A. L. Patterson, *Z. Krist.*, **A90**, 517 (1935).
[2] Lipson and Cochran, *The Determination of Crystal Structures*, pp. 12–15, 144–193.
[3] Buerger, *Vector Space*, pp. 5–29, 41–64.

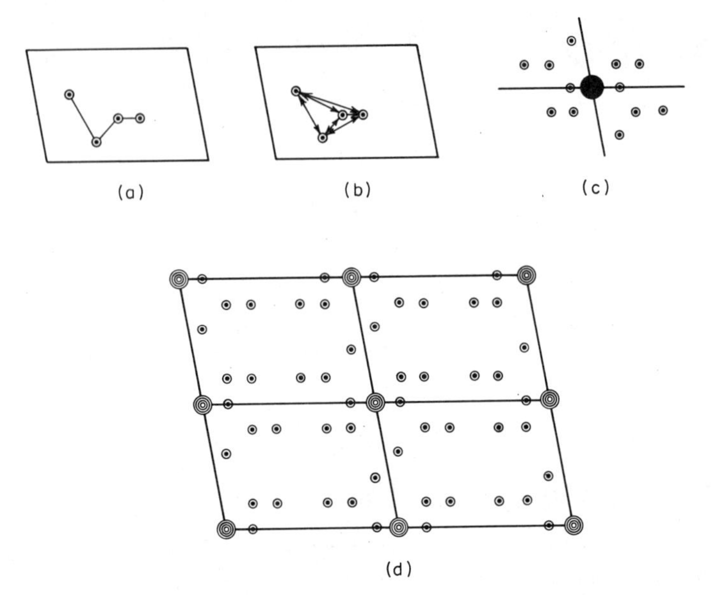

Figure 12.1. (*a*) Set of points. (*b*) Interatomic vectors. (*c*) Patterson peaks about the origin from one set; (•) origin peak. (*d*) Patterson peaks in four unit cells.

For a molecule containing N atoms in a unit cell (see Fig. 12.1 for a two-dimensional example), the Patterson synthesis will show N^2 peaks, corresponding to the N possible vectors that can be drawn from each of the N atoms. Of these, N will be vectors of zero length from each atom to itself and will be concentrated as a very large peak at the origin. The remaining $N^2 - N$ will be distributed throughout the cell. Since the cell of the Patterson synthesis is the same size as that of the crystal, the peaks are obviously much more densely packed and there is generally a great deal of overlap among them. This overlap is accentuated by the greater intrinsic breadth of Patterson peaks as compared to Fourier peaks. Figure 12.2 shows the cause of this effect. The distribution of vector density over a

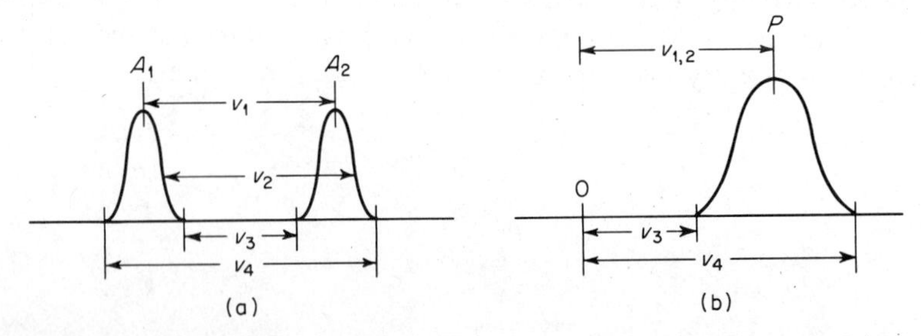

Figure 12.2. (*a*) Two atoms. (*b*) Patterson peak resulting from atoms in (*a*), showing spread of vector peak (arbitrary height). Origin peak not shown.

Patterson peak P corresponds to a summation of all the vectors between elements of electron density in the two atomic peaks A_1 and A_2. The vectors 1 and 2 have the same length and contribute to the maximum of P, but the vectors 3 and 4, which differ in length by twice the atomic width, determine its outer limits. Thus the maximum width of P is twice that of A, and the tendency to overlap is correspondingly enhanced.

For both of these reasons, the Patterson map of a molecule with a moderate number of atoms tends to be an almost featureless distribution of vector density that scarcely suggests $N^2 - N$ discrete interatomic vectors. It is possible, however, to modify the map to approach this ideal more closely by a process known as sharpening,[1,2] which greatly reduces the overlap caused by peak broadening. This process is entirely equivalent to the conversion to F_{point} (Section 11.2) that is part of the generation of U's and E's. In fact, most modern calculations of *sharpened Patterson functions* are carried out with E's rather than with specially prepared coefficients.

Once $|F_{\text{point}}|$ or $|E|$ values are obtained from the measured $|F_{\text{real}}|$ values, they can be squared and used as the coefficients for a sharpened Patterson function. The resulting map is not a true point-vector representation since the sharpening cannot be completely effective when applied to a finite number of reflections,[4] but it shows many more relatively discrete peaks than an unsharpened one (Figs. 12.3*a* and *b*).

A less serious but sometimes annoying difficulty with the Patterson function arises from the great size of the origin peak. Since it does not provide useful structural information (but see below), and since it may cause trouble in scaling the computer output, it is often eliminated during the sharpening operation.

Patterson[1] showed that it was possible to subtract the origin peak from a vector map by subtracting the *average* value of $|F|^2$ from each term in the summation. Thus for F's that have been placed on an absolute scale,

$$|F'_{hkl}|^2 = |F_{hkl}|^2 - \sum_{j}^{N} f_j^2 \tag{12.4}$$

For a sharpened Patterson function, the f_j's are all constants equal to Z_j, and the origin removal process simply consists of subtracting the quantity

$$\sum_{j}^{N} Z_j^2 \tag{12.5}$$

[4]The perfect representation of a point atom by a Fourier series requires in theory an infinite amount of data. If only a limited number of terms are included in the series, as is always the case in practice, the maxima will be surrounded by secondary ripples. The extent to which data are sharpened usually represents a personal judgment of the relative disadvantages of peak broadening (from undersharpening) and ripples (from oversharpening). One common compromise is to use Eq. (11.13) but to set B somewhat smaller than its true value, often zero. See Lipson and Cochran, *The Determination of Crystal Structures*, pp. 165–169, for further discussion.

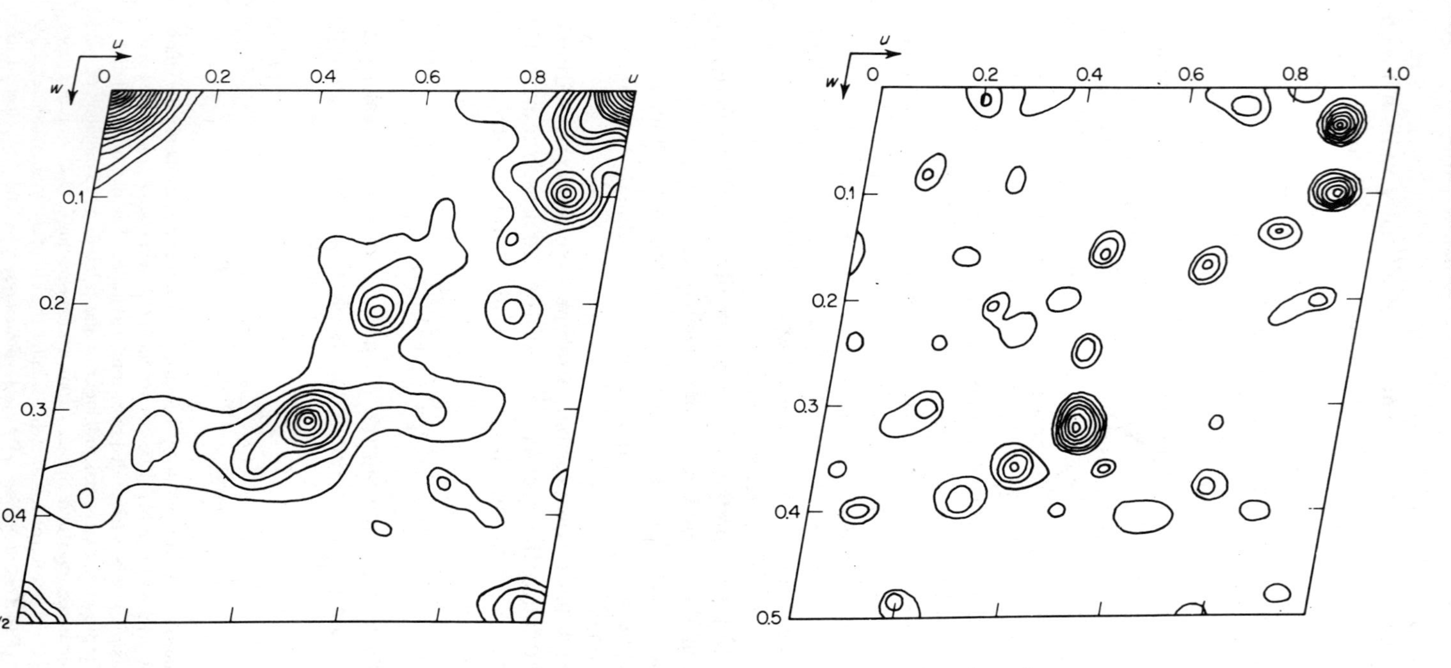

Figure 12.3. (*a*) Asymmetric part of the *u*0*w* section of the Patterson map of a natural product. (*b*) Same section after sharpening and removal of the origin peak.

from each $|F_{\text{point}}|^2$. With computers using stored files of reflection data, it is often easier and more accurate to compute the average of the actual $|F|^2$ values first and then calculate the coefficients. If $|E|$'s are used, the calculation is particularly simple since by definition the average value of $|E|^2 = 1$ and the coefficient $E^2 - 1$ is often provided directly by programs that generate E values. Figure 12.3 shows part of the zero section of a Patterson map with and without original removal. The process is not likely to reveal any totally hidden peaks near the origin, but it can convert shoulders into distinct maxima.

The weight of a Patterson peak depends on the numbers of electrons in the atoms between which the vector occurs, and can be shown to be proportional to the products of their atomic numbers. Properly speaking, this is true only of the integral over the entire assembly of vector density under the peak, but the assumption is often made that the shapes of atoms of similar atomic number are sufficiently alike that it can be applied to the peak heights as well. The proportionality constant k needed to relate a peak to the corresponding atomic number product is usually obtained from the origin peak, whose height is

$$\rho(000) = k \sum_{j}^{N} Z_j^2 \tag{12.6}$$

The major importance of this peak height relationship is that the vectors between heavy atoms appear with weights corresponding to the square of their atomic numbers and thus stand out strongly against the background of heavy–light and light–light atom vectors. It is this fact that allows the heavy atom vectors to be recognized in the Patterson map, usually without difficulty.

12.2. PATTERSON SYMMETRY

Although the atomic positions in the real crystal can be distributed in any of the 230 space groups, the vector peaks in the corresponding Patterson map are limited to only 24 space groups. This simplification reflects the loss of information suffered when proceeding from the phased F's to the phaseless $|F|^2$'s. Put in another way, it represents the symmetry of the intensity-weighted reciprocal lattice, the possible arrangements being the space groups, primitive and centered, corresponding to the 11 Laue groups (Section 3.6). The symmetry of the Patterson group is thus often higher than that of the crystal itself.

That this should be the case can be seen easily from the nature of a vector map. Any pair of atoms A and B give rise to both vectors AB and BA. These are equal in magnitude but opposite in direction and so will appear as peaks at u, v, w and $-u, -v, -w$. Consequently any Patterson map will be centrosymmetric, regardless of the space group of the atomic distribution from which it is derived. By similar but more involved arguments it can be shown that the following are true of Patterson functions[5]:

1. All Patterson functions are centrosymmetric.
2. Their lattice type (*P*, *C*, *F*, etc.) is the lattice type of the original space group.
3. Their space group is derived from the original space group by replacing all translational symmetry elements (screws, glides) by the corresponding nontranslational elements (axes, mirrors) and by adding a center of symmetry if it is not already present.

The simplification produced by these rules is particularly marked in the classes of lower symmetry with which we are concerned. The Patterson maps derived from all triclinic crystals show $P\bar{1}$ symmetry; those from monoclinic crystals, $2/m$ ($P2/m$ or $C2/m$, depending on the lattice type); and those from orthorhombic, *mmm* (*Pmmm*, *Cmmm*, *Immm*, or *Fmmm*).

An alternative expression of the above is that the symmetry of the Patterson function is represented in matrix form (see Section 3.5) by the same rotation matrices as the space group but with the translation vectors all set to 0. If there is no $\bar{1}$ rotation matrix in the space group, it is added as well.

One consequence of the change of symmetry in going from the density to the vector distributions may be a decrease in the volume of the asymmetric unit. In particular, if the crystal is noncentrosymmetric, it will contain half as many asymmetric units in the same volume as does the corresponding Patterson function (see Table 5.4). This loss of unique volume in the Patterson map reflects the inability of the intensity distribution to dis-

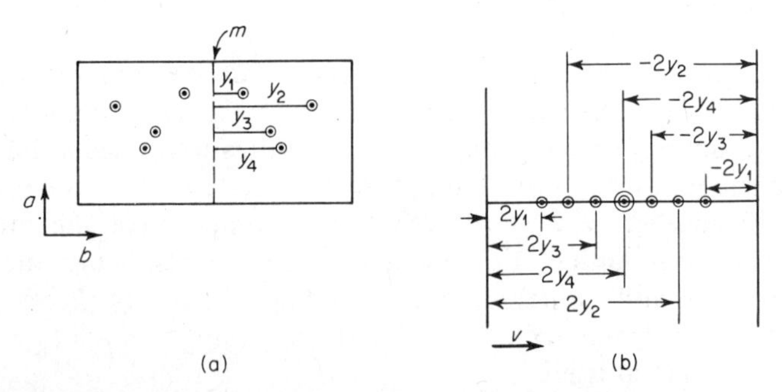

Figure 12.4. (*a*) Projection of a structure in space group *Pm* on the *ab* face. (*b*) Harker peaks produced along the *v* axis by the atoms in (*a*).

[5]Buerger, *Vector Space*, pp. 198–200.

TABLE 12.1 Some Harker Lines and Planes

2-Fold axis ∥ a, b, c	$0vw$; $u0w$; $uv0$
2-Fold screw ∥ a, b, c	$\frac{1}{2}\,u\,w$; $u\,\frac{1}{2}\,w$; $u\,v\,\frac{1}{2}$
m plane ⊥ a, b, c,	$u00$; $0v0$; $00w$
a glide ⊥ b, c	$\frac{1}{2}\,v\,0$; $\frac{1}{2}\,0\,w$
b glide ⊥ a, c	$u\,\frac{1}{2}\,0$; $0\,\frac{1}{2}\,w$
c glide ⊥ a, b	$u\,0\,\frac{1}{2}$; $0\,v\,\frac{1}{2}$

tinguish under normal conditions between the right- and left-handed members of an enantiomeric pair. This increases the crowding in the vector map and makes its interpretation more difficult.

Although the symmetry elements of the crystal space group do not necessarily appear as such in the Patterson map, they leave their traces in the form of particular concentrations of vector points. These concentrations, known as *Harker lines and planes*,[6] correspond in the Patterson map to the unusually high average intensity of certain regions of reciprocal space (see Section 7.3) and can be used in the same way to provide space group information. They arise because the vectors between corresponding atoms of molecules related by symmetry elements other than centers have one or two constant coordinates. For example, in *Pm* there is a mirror plane perpendicular to the b axis (Fig. 12.4), and thus for every atom with coordinates x, y, z there is another at $x, \bar{y}, z$. The vectors between these atoms all have the coordinates $0, 2y, 0$, that is, they are concentrated on the Harker line that is the $v(y)$ axis in the Patterson map. Likewise, $P2_1$ contains pairs of atoms of the form x, y, z, and $\bar{x}, y+\frac{1}{2}, \bar{z}$. The corresponding vectors are $2x, \frac{1}{2}, 2z$ and appear in the plane $v=\frac{1}{2}$. Table 12.1 lists the Harker sections corresponding to some of the more common symmetry elements. Those for any other symmetry element can be derived along the lines given above.

12.3. ATOM FINDING IN PRACTICE

In order to place an atom in the cell, it is merely necessary to locate it with respect to the symmetry elements of the cell and to know the relationship of these to the conventional origin of the cell. To demonstrate how this is done, we shall take some particular examples, considering for the moment only the case of a single atom per asymmetric unit. In general, what is discussed below for a particular space group is true for the other members of the same point group.

[6]D. Harker, *J. Chem. Phys.*, **4**, 381 (1936); see also Buerger, *Vector Space*, pp. 132–180, and Lipson and Cochran, *The Determination of Crystal Structures*, pp. 151–154, 175–178.

1. $P1$: Triclinic, no symmetry, equivalent positions only x, y, z. Any point can be taken as the origin, so the atom can be assigned arbitrarily the coordinates 0, 0, 0. The phase angles calculated from this model will all be 0° ($\cos \alpha = +1$) [see Eq. (8.26)]; so the Fourier map will look much like the Patterson map. Extra symmetry will be present in the Fourier map (see $P2_1$ below), so the remainder of the molecule must be picked from the two superimposed images.

2. $P\bar{1}$: Triclinic, centrosymmetric, equivalent positions x, y, z; $\bar{x}, \bar{y}, \bar{z}$. An atomic peak will appear in the Patterson map at the general position u, v, w equal to $2x, 2y, 2z$. This determines the unique atomic position with respect to the center of symmetry as an origin.

3. $P2_1$: Monoclinic, one 2-fold screw parallel to b, equivalent positions x, y, z and $x, y+\frac{1}{2}, z$. The atomic vector is to be found on the Harker plane $u, \frac{1}{2}, w$ and at the position $2x, \frac{1}{2}, 2z$. This peak determines the x and z coordinates of the atom but leaves its y value unspecified. In this space group, however, there is no unique origin point along the b axis, so the y coordinate can be assigned arbitrarily. It is convenient to give it the value $\frac{1}{4}$, since the serious problem that remains then becomes obvious. If one atom is located at $x, \frac{1}{4}, z$, the other will be at $\bar{x}, \frac{3}{4}, \bar{z}$ or equivalently at $\bar{x}, -\frac{1}{4}, \bar{z}$. It is obvious that these two atoms are related by a center of symmetry (Fig. 12.5) even though the space group is noncentrosymmetric. As a result, the structure factors calculated from the single-atom model will all have phase angles 0 or π. When these are used to phase an F_o Fourier series, the result will be an electron density map containing peaks for both the true structure and its mirror image.

The other space groups of point group 2 differ from $P2_1$ only in the location of the Harker plane in which the atomic vector is found.

4. Pc: Monoclinic, one c glide in the ac plane, equivalent positions x, y, z and $x, \bar{y}, \frac{1}{2}+z$. In this case the atomic vector must be located in the Harker line $0, v, \frac{1}{2}$ and falls at $v = 2y$. As a rule, Harker lines are less satisfactory to work with than Harker planes, since the same amount of vector information is compressed into a smaller compass, but a heavy atom peak can generally be identified regardless. Only one coordinate is determined by the Patterson synthesis, but the others are arbitrary. If the values $x = 0$, $z = \frac{1}{4}$ are assigned, it is clear that Pc, and the other members of point group m, suffer from the same false centrosymmetry problem as $P2_1$.

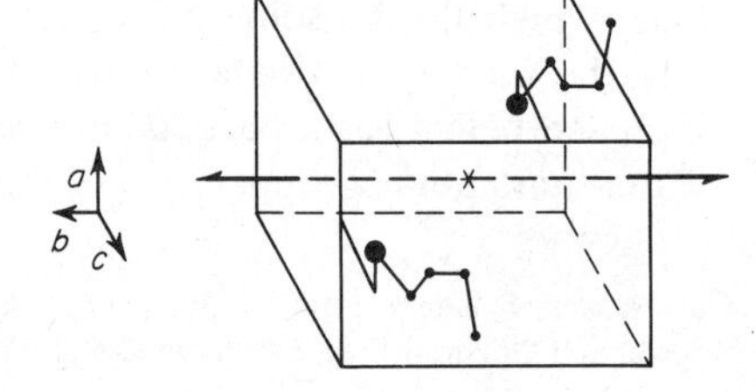

Figure 12.5. All screw-related atoms in $P2_1$ are also related by centers of symmetry lying on the screw axes. For example, the two heavy atoms (•) in the unit cell are related by a center (×) on the screw axis at $x = \frac{1}{2}$, $z = \frac{1}{2}$.

5. $P2_1/c$: Monoclinic, centrosymmetric with a 2-fold screw parallel to b and a c-glide perpendicular to it. Equivalent positions (1) x, y, z; (2) $\bar{x}, \bar{y}, \bar{z}$; (3) $\bar{x}, \frac{1}{2}+y, \frac{1}{2}-z$; (4) $x, \frac{1}{2}-y, \frac{1}{2}+z$. In this space group the screw and glide do not pass through the center of symmetry, which is taken as the origin, but are offset from it. As a result the Harker peaks fall at positions that are not simply twice the corresponding atom coordinates. Instead, while the vector between two screw-related atoms (numbers 1 and 3 above) occurs as usual in the $u, \frac{1}{2}, w$ plane, it is located at $-2x, \frac{1}{2}, \frac{1}{2}-2z$. The reverse vector $3 \rightarrow 1$ produces a second Harker peak at $2x, -\frac{1}{2}, 2z-\frac{1}{2}$, and since $v=\frac{1}{2}$ and $v=-\frac{1}{2}$ refer to the same plane, the two peaks are merely related by a center of symmetry in the $u, \frac{1}{2}, w$ plane. One might expect two more Harker peaks since there is another pair of screw-related atoms, but the vectors between these coincide with those between the first two and only serve to increase the size of the peaks.[7] Since only half the Harker plane is unique, the other half being generated by the action of the center, a Fourier synthesis calculated as usual over an asymmetric unit shows only a single peak in its output. This peak may be arbitrarily considered to represent either of the vectors described above and the x and z coordinates of the atom deduced from it. The choice between the vectors merely selects which atom of the symmetry-related set is to be characterized as xyz and which regarded as the screw image.

The y coordinate can be obtained either from the Harker line resulting from the c glide, on which the peaks appear at $0, \frac{1}{2} \pm 2y, \frac{1}{2}$ or from the peaks between centrosymmetrically related atoms. Such a peak occurs at the general location uvw (and $\bar{u}v\bar{w}$, $u\bar{v}w$, $\overline{uvw}$ derived from it by the $2/m$ symmetry of the Patterson) and represents one of the set

$$\begin{array}{ccc} -2x, & -2y, & -2z \\ 2x, & 2y, & 2z \\ 2x, & -2y, & 2z \\ -2x, & 2y, & -2x \end{array}$$

If the x and z coordinates have already been assigned from the Harker plane, a given uvw peak can be correlated with either of two of these vectors, the arbitrary choice between them merely deciding the sign of y. If no previous assignments have been made, then all the coordinates can be obtained from one uvw peak. In this case, however, it is advisable to check the Harker peaks as well, since they are twice as large as the general peaks and provide useful confirmation.

6. $P2_12_12_1$: Orthorhombic, noncentrosymmetric with three nonintersecting 2-fold screw axes. Equivalent positions (1) x, y, z; (2) $\frac{1}{2}-x, \bar{y}, \frac{1}{2}+z$;

[7]It is recommended that the reader work out the complete set of vectors for $P2_1/c$ to demonstrate the truth of this statement.

(3) $\frac{1}{2}+x, \frac{1}{2}-y, \bar{z}$; (4) $\bar{x}, \frac{1}{2}+y, \frac{1}{2}-z$. As in $P2_1/c$, the screws are offset from the origin of the cell and the atomic vectors occur on Harker planes at $\frac{1}{2}\pm 2x, \pm 2y, \frac{1}{2}$; $\frac{1}{2}, \frac{1}{2}\pm 2y, \pm 2z$; $\pm 2x, \frac{1}{2}, \frac{1}{2}\pm 2z$. There are four symmetry-related peaks in each plane, one for each combination of signs in these expressions. As before, two coordinates can be found by the arbitrary identification of one peak with any convenient sign combinations. Using one of these coordinates, the corresponding two combinations for any peak of another Harker plane can be found, and a choice between them provides the sign and value of the third coordinate. The third plane is redundant but is useful as a check on the results.

7. *Pbca*: Orthorhombic, centrosymmetric with three nonintersecting screws with glide planes perpendicular to them. Equivalent positions are

$$x, y, z; \quad \tfrac{1}{2}+x, \tfrac{1}{2}-y, \bar{z}; \quad \bar{x}, \tfrac{1}{2}+y, \tfrac{1}{2}-z; \quad \tfrac{1}{2}-x, \bar{y}, \tfrac{1}{2}+z$$
$$\bar{x}, \bar{y}, \bar{z}; \quad \tfrac{1}{2}-x, \tfrac{1}{2}+y, z; \quad x, \tfrac{1}{2}-y, \tfrac{1}{2}+z; \quad \tfrac{1}{2}+x, y, \tfrac{1}{2}-z$$

This space group is merely $P2_12_12_1$ with a center of symmetry added. The atoms can be located from the Harker planes or as in $P2_1/c$ from centrosymmetrically related atom pairs. The ambiguity of two possible atom sets that occurs in $P2_12_12_1$ does not exist in *Pbca* since the equivalent positions in this space group correspond to the simultaneous presence of *both* mirror images in the cell (see below).

As a practical example of locating an atom from Patterson data, the bromine atom in methyl micromerol bromoacetate[8] (space group $P2_12_12_1$) was located from three Harker sections, each calculated over only its unique area. Thus

(a) Section at $u = 48/96$, with $v = 0 \rightarrow 50/100$, w; $0 \rightarrow 50/100$.
Peak found at 48/96, 11/100, 21/100.
Assigned $\frac{1}{2}, \frac{1}{2}-2y, 2z$, so $y = 19.5/100$, $z = 10.5/100$.

(b) Section at $v = 50/100$, with $u = 0 \rightarrow 48/96$, $w = 0 \rightarrow 50/100$.
Peak found at 34/96, 50/100, 29/100.
Assigned $2x, \frac{1}{2}, \frac{1}{2}-2z$, so $x = 17/96$ ($\frac{1}{2}-2z = 29/100$ as required).

(c) Section at $w = 50/100$, with $u = 0 \rightarrow 48/96$ and $v = 0 \rightarrow 50/100$.
Peak found at 14/96, 39/100, 50/100.
General form $\frac{1}{2}\pm 2x, \pm 2y, \frac{1}{2}$. The peak found corresponds to $\frac{1}{2}-2x, 2y, \frac{1}{2}$.

The complete set of symmetry-related bromine atoms for this example is

Br_1	17/96, 19.5/100, 10.5/100	x, y, z
Br_2	31/96, −19.5/100, 60.5/100	$\frac{1}{2}-x, -y, \frac{1}{2}+z$
Br_3	65/96, 30.5/100, −10.5/100	$\frac{1}{2}+x, \frac{1}{2}-y, -z$
Br_4	−17/96, 69.5/100, 39.5/100	$-x, \frac{1}{2}+y, \frac{1}{2}-z$

[8] G. H. Stout and K. L. Stevens, *J. Org. Chem.*, **28**, 1259 (1963).

Any member of this set can be chosen to be x, y, z and the observed vectors will be produced, although the expressions characterizing the vector peaks will have different sign combinations.[9]

On the other hand, if the first bromine atom is placed at

$$Br_1 \quad -17/96, -19.5/100, -10.5/100$$

exactly the same set of vectors can be obtained by changing all the signs in the vector expressions. Thus there are two different sets of bromine atoms that satisfy the Harker peaks. The difference between these sets is simple and, to a crystallographer (but not a chemist), trivial—they are mirror images. Since they cannot be distinguished at this point, either can be chosen. If the structure that results proves to have the wrong conventional or absolute configuration, inversion is easily achieved by changing all the coordinate signs (or all the signs on one coordinate only, e.g., x).

More systematic and programmable means have been used for finding a general atomic position. Perhaps the most general is that known as the *symmetry minimum function*,[10] *symmetry map*,[11] or *continuous multiple implication function*.[12] To obtain such a map, each point in a unique volume of the crystal cell[13] is considered as a possible atomic location and the vectors from an atom at this point to its symmetry-related images are calculated. The Patterson map is inspected to learn if peaks are found at the predicted locations. If they are, the location is a possible one; if not, it can be rejected. The effect of this process is usually to eliminate a large fraction

[9]It is a worthwhile exercise to show that the positions corresponding to Br_2 and so on can be derived by the proper choice of signs in the vector expressions from the Patterson peaks given. Note that $\pm\frac{1}{2}$ can be added at will to any of the coordinates of the atom located since this merely corresponds to shifting the cell origin to another, equivalent position. Thus Br_1 could equally well be placed at 17/96, 19.5/100, 60.5/100 or 65/96, 19.5/100, −39.5/100, etc. The positions of the symmetry-related atoms will, of course, change as well.

[10]P. G. Simpson, R. D. Dobrott, and W. N. Lipscomb, *Acta Cryst.*, **18**, 169 (1965).

[11]B. T. Gorres and R. A. Jacobson, *Acta Cryst.*, **17**, 1599 (1964).

[12]R. D. Ellison and H. A. Levy, *Acta Cryst.*, **19**, 260 (1965).

[13]Note that this volume need not correspond to the asymmetric unit. For those triclinic, monoclinic, and orthorhombic space groups in which a unique origin position can be specified with reference to the symmetry elements of the cell (those of point groups $\bar{1}$, $2/m$, 222, and mmm among others), there are always alternative origin sites at translations of $\frac{1}{2}$ along the various axes (cf. the centrosymmetric case, Section 11.6). The symmetry map does not distinguish among these and returns the possible locations of atoms with respect to all alternative origins. Thus for the cases cited, the repeating unit of the symmetry map is always $\frac{1}{2}\times\frac{1}{2}\times\frac{1}{2}$ of the crystal cell. In addition this unit has the symmetry of the Patterson function; so the actual volume over which the computation must be made is less. Since so much less of the symmetry map than the actual crystal cell is unique, however, it suffers correspondingly from overlap and ambiguity. For space groups of higher symmetry, the number of alternative origin sites decreases, and both the volume of the repeating unit and the clarity of the resulting map increase. For a systematic analysis in terms of *Cheshire groups*, see F. L. Hirshfeld, *Acta Cryst.*, **A24**, 301 (1968).

of the cell volume from consideration but to give back many more possible atomic sites than will be actually occupied, at least if one is searching for the sites of light atoms.

Although any of these methods can be used in principle and often in practice to locate a single atom in the cell, adding a second requires some attention to detail if the two atoms are to be properly situated relative to each other. We shall consider the case of $P2_1$, but the arguments are similar for other space groups. The selection of one Harker peak as $2x, 2z$ defines the origin of the crystal cell, and any further atoms must be correctly placed with respect to this origin. Figure 12.6 shows an example in which each of the four Harker peaks resulting from two heavy atoms leads to a different origin when it is regarded as $2x\ 2z$. Fortunately, all the peaks can be related to a single origin by noting that because of the lattice nature of the Patterson function and its unit repeats, a peak of $u, \frac{1}{2}, w$ can equally well be considered as arising from a vector of length $u+1, \frac{1}{2}, w$, or $u, \frac{1}{2}, w+1$, or $u+1, \frac{1}{2}, w+1$ (Fig. 12.7). For each Harker peak, one of these combinations will correspond to the vector to one atom in the cell from the atom related by the 2-fold screw passing through a predetermined cell origin. Thus in practice a Harker peak for one atom is selected as $2x_1, 2z_1$ and one for the other is regarded as locating an atom at one of the locations

$$u/2, \ldots, w/2$$
$$u/2+\tfrac{1}{2}, \ldots, w/2$$
$$u/2, \ldots, w/2+\tfrac{1}{2}$$
$$u/2+\tfrac{1}{2}, \ldots, w/2+\tfrac{1}{2}$$

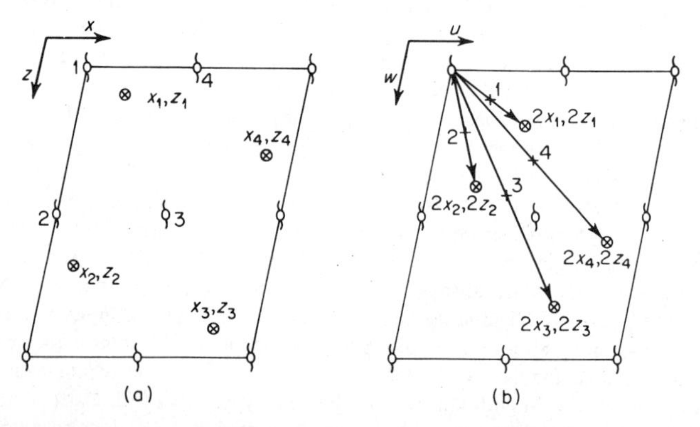

Figure 12.6. (*a*) Two heavy atoms at x_1, z_1 and x_2, z_2 and their screw-related atoms at x_3, z_3 and x_4, z_4. In projection. (*b*) Harker section $P(u, \frac{1}{2}, w)$ showing the possible atomic locations 1, 2, 3, 4 from peaks $2x_1$, $2z_1$; $2x_2$, $2z_2$; $2x_3$, $2z_3$; and $2x_4$, $2z_4$. Each location corresponds to choosing the origin on a different screw axis 1, 2, 3, or 4 in (*a*).

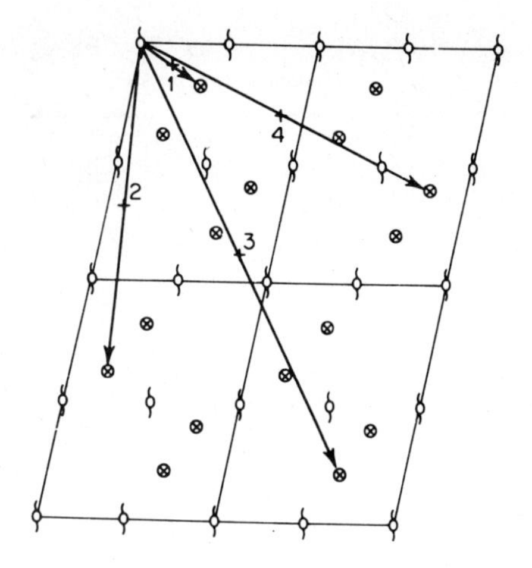

Figure 12.7. Harker sections of four adjacent cells (cf. Fig. 12.6*b*) showing the choice of peaks that give the four correct atomic positions within the unit cell.

Which of these is actually correct is determined by calculating the two *cross vectors* (atom 1 → atom 2; atom 1 → atom 2 screw image) for each of the combinations of positions and searching the Patterson map for peaks at the predicted locations. Only one set of correspondences should be found. Since y coordinates are not determined by the Harker peaks, y_1 can be set at 0 and y_2 determined from the v coordinates of the cross peaks. In general, when a set of atomic positions have been derived, the best test of their consistency is to calculate the positions for the set of unique vectors (those not related by the Patterson symmetry) and to assure oneself that suitable peaks are found at all the predicted locations.

BIBLIOGRAPHY

Buerger, M. J., *Vector Space*, Wiley, New York, 1959.

Ladd, M. F. C., and R. H. Palmer, *Structure Determination by X-Ray Crystallography*, Plenum, New York, 1985, pp. 218–252.

Lipson, H., and W. Cochran, *The Determination of Crystal Structures*, Cornell University Press, Ithaca, NY, 1966, pp. 12–15, 144–193

CHAPTER 13

HEAVY ATOM METHODS

Although the previous chapter discussed the use of the Patterson function for locating an atom in the general sense, historically one of its most important uses was for obtaining the position of a single heavy atom in a cell full of much lighter ones. As we shall see below, such an atom has a surprisingly large effect on the phases of structure factors, even though it represents a relatively small fraction of the total electron count of the cell. Thus for years the most effective means for solving the phase problem consisted in introducing a suitable heavy atom, finding its location, and then pyramiding phases based on it into a complete structure. Similar, but more complicated methods have also been responsible for the solution of protein and other macromolecular structures.

13.1. HEAVY ATOM PHASES

The use of the Patterson function in the determination of structures containing heavy atoms depends on the application of the properties described in Chapter 12, particularly the ability to resolve vectors between heavy atoms from all the others. The need for such resolution raises the question of how heavy an atom is required for a structure to be solved easily by these methods. The proper choice of the atom requires the balancing of a number of factors. On one side, the heavier an atom the easier it is to locate by means of the Patterson map and the more it tends to determine the phases and intensities of all the reflections, that is, the better a phasing model it is. On the other hand, when the atom is very heavy compared to the others in the structure, its dominance becomes too great and the comparison of $|F_o|$ and $|F_c|$ becomes relatively insensitive to the positions of the light atoms. Thus the uncertainty in the light atom coordinates increases, the reliability of bond-length measurements decreases, and in extreme cases the light atoms may not be found at all. Furthermore, absorption errors in the data are increasingly significant as heavier atoms are added. For these reasons the desire for an easy solution to the phase problem must be weighed against the accuracy required from the final results.

A convenient rule of thumb that has been used as a guide in the selection of a heavy atom is

$$\sum Z^2_{\text{heavy}} \Big/ \sum Z^2_{\text{light}} \approx 1 \tag{13.1}$$

This relation is merely a guide, however, and large deviations from it can often be tolerated.[1] Thus iodine has often been used as a heavy atom for compounds containing 20–30 carbon atoms, where Z^2 ratios fall in the range 4 to 2.5. The phasing is thereby simplified, but the precision of bond-length measurements is reduced and it has sometimes been necessary to use chemical evidence to distinguish between double and single bonds. At the other end of the scale, vitamin B_{12} ($C_{63}H_{88}N_{14}O_{14}PCo \cdot 8H_2O$) was solved,[2,3] although admittedly with great difficulty, using initially the phases from the cobalt atom (Z^2 ratio ≈ 0.17). On the whole, it is better to err on the side of too light an atom (Z^2 ratio ≤ 1).

Once the heavy atom has been located, it is the phasing model for the structure and can be elaborated by repeated cycles of structure factor and Fourier calculations as discussed in Chapter 15. If the atom is fairly heavy, ΔF syntheses are an excellent supplement to the more obvious F_o syntheses for finding additional atoms.

Although the location of the heavy atom from the Patterson map normally proceeds easily, the further expansion of the structure can prove difficult if the first Fourier maps, calculated on the basis of the heavy atom phases, have additional symmetry compared to the true space group. As has already been described, this arises from the nature of point groups 1, 2, and m when there is only one atom in the model asymmetric unit. It can also occur in other point groups when the heavy atoms are approximately or exactly located at special positions in the cell so that their symmetry is higher than that of the whole molecules.

There are a number of alternatives for one faced with this problem. One

[1]For more quantitative discussions, see G. A. Sim, *Acta Cryst.*, **10**, 177 (1957) and S. Parthasarathy, *Acta Cryst.*, **18**, 1022, 1028 (1965).

[2]J. G. White, *Proc. Roy. Soc.* (*London*), **A266**, 440 (1962).

[3]D. C. Hodgkin, J. Lindsey, M. Mackay, and K. N. Trueblood, *Proc. Roy. Soc.* (*London*), **A266**, 475 (1962).

approach uses direct methods, either alone or in conjunction with Patterson location of the heavy atom. Although the analysis of direct methods is often phrased in terms of structures composed of equal atoms, the techniques generally work as well or better if heavy atoms are present. Thus in complicated cases it may not be necessary to carry out the analysis of the Patterson map to locate the heavy atom (although it is an excellent check), and the problem of separating two overlapping images can be handled by choosing an enantiomorph-determining phase rather than by fitting a Fourier map.[4]

Alternatively, an effort can be made to sort out the true structure from its image. The method is the same whether the extra symmetry is inherent in the space group or a consequence of the special positions of the heavy atoms. What is required is the addition to the phasing model of a suitable number of atoms belonging to one molecule and so located that the atoms related by the true symmetry of the space group do not appear to be related by the false symmetry as well. The effect of this addition is to shift the calculated phases from the special values produced by the heavy atoms to more general values. When these phases are now used in a Fourier synthesis they will tend to favor the atoms in one molecule and weaken those in the image. By comparing the relative heights of corresponding peaks, it is possible to select the correct member of each pair, although it may be necessary to repeat the F_c and Fourier process several times to get them all.

The difficulty with this method lies in the first step. A single atom can always be added at the position of one of the peaks of the heavy atom Fourier map, since this merely represents the selection of the right-handed image over the left-handed one or vice versa. Unfortunately, a single carbon atom in the presence of bromine is probably not sufficient to produce an obvious effect on the heights of the remaining peaks, although it might be possible to "bootstrap" oneself along from this beginning. If, however, the atom is heavier than carbon the prognosis is more hopeful. For example, a lazy, if roundabout, approach to the question of locating two heavy atoms in an asymmetric unit is to simply find one, calculate F_c's and a Fourier synthesis and pick out the second from among the collection of image peaks. This method can also be used if the second atom is a medium-weight one (e.g., S with Br and C), whose location from the Patterson map is more difficult. It is possible for false medium-weight peaks to arise by the accidental overlap of light atoms from the two images, but consideration of peak shape and the surrounding peaks, or trial and error, should resolve the difficulty. Note that double-weight light atom peaks occur systematically on the false-symmetry elements (see below), where atoms overlap their false images.

In the most general case, however, it is necessary to break the deadlock using light atoms alone. For this purpose it is most useful to locate some

[4]R. Karlsson, *Acta Cryst.*, **A34**, 698 (1978).

molecular fragment, for example, a ring, *that allows the addition of a number of atoms that have a high probability of belonging to the same molecule*. This restriction is important because although the first atomic position can be chosen from any of the peaks, the later ones must belong to the same image if the analysis is to converge to a single molecule. Once the atoms have been added, their effect on the relative heights of the members of the uncertain pairs of peaks is used to select the other members of the correct molecular set.

In selecting the atoms to be added, some attention should be paid to their position with respect to the false-symmetry elements. The effect of the near or exact symmetry between the heavy atoms results in a Fourier synthesis that belongs more or less closely to a space group of symmetry higher than the real one. Thus $P2_1$ with two atoms at $y = \frac{1}{4}$ and $\frac{3}{4}$ has the symmetry $P2_1/m$ and mirror planes appear at $y = \frac{1}{4}$ and $\frac{3}{4}$ in the Fourier map. The atoms that fall on or near these planes overlap with their (false) mirror-related images and tend to appear as large peaks drawn out parallel to y. Although it is possible to work out reasonable positions for the atoms from the amount of elongation, the closer the atoms are to the mirrors the more their addition will reinforce rather than suppress the false symmetry. For this reason it is best to search for recognizable fragments in the region between the mirrors, in the expectation that, if found, they will be more effective in defining the rest of the associated molecule. Similar policies should be followed in other space groups.

In practice the details of the methods used in unraveling the $P2_1$ ambiguity will vary from case to case and from worker to worker. In some molecules for which the chemical structure is known, it may be possible to pick out the entire molecule from the double-image Fourier map. In others, a laborious trial-and-error process may be needed. Programs are available either for the searching of a Fourier synthesis for a known molecular fragment (see Chapter 14 for a similar process in the Patterson map) or for expanding a few atoms into possible molecules by making use of known bond lengths and angles. In the meantime, there is still some art left to crystal structures analysis, and the best guide is past experience. Previous workers[5–9] give a number of examples of structure determinations in which the problems of extra symmetry were overcome and the methods used are discussed in some detail.

[5] N. E. Taylor, D. C. Hodgkin, and J. S. Rollett, *J. Chem. Soc.*, **1960**, 3685.
[6] W. A. C. Brown and G. A. Sim, *J. Chem. Soc.*, **1963**, 1050.
[7] G. Kartha and D. J. Haas, *J. Am. Chem. Soc.*, **86**, 3630 (1964).
[8] A. M. Duffield, P. R. Jefferies, E. N. Maslen, and A. I. M. Rae, *Tetrahedron*, **19**, 593 (1963).
[9] D. C. Hodgkin, B. M. Rimmer, J. D. Dunitz, and K. N. Trueblood, *J. Chem. Soc.*, **1963**, 4945.

13.2. OTHER HEAVY ATOM METHODS

Although heavy atoms are most commonly used as phasing models for Fourier calculations, other methods are available by which the remainder of the structure can be developed once the positions of the heavy atoms are known. Two such methods that have been used with success are superposition and isomorphous replacement.

Superposition methods are discussed in Chapter 14 with particular reference to systems containing only light atoms, but the same techniques can be applied to the Patterson functions of molecules with heavy atoms. Since the method involves the isolation from the Patterson map of the vectors between known atoms and unknown ones, its application is much simplified by the assurance with which the heavy atom positions are known and by the greater weight of the heavy–light vector peaks as compared with the light–light background.

Isomorphous replacement techniques require that one have in hand two or more crystalline derivatives that have the same crystal structure but contain different heavy atoms. Classic small-molecule cases have been isomorphous salts of alkaloids,[10] and porphyrins with different metal atoms (or none at all) in the chelated positions.[11] The use of isomorphous replacement methods for general organic structures has essentially vanished as direct methods and the increasing speed of computations have made alternative techniques much easier. On the other hand, the technique is the only one that has so far been successful in the solution of totally unknown protein structures.

The principles of the method are most easily seen in the centrosymmetric case. The observed structure factors for the two isomorphs A and B can be regarded as consisting of two parts. The light part (L) is the same for both since it represents the scattering effects of the light atoms, which are the same in the two crystals. The variable part (R) reflects the scattering contributions of the replaceable atoms. Since these occupy the same positions in their respective crystals, their contributions will have identical phases but will differ in magnitude as the scattering factors differ. Thus,

$$F_A = F_L + F_{AR} \tag{13.2}$$

$$F_B = F_L + F_{BR} \tag{13.3}$$

and

$$F_A - F_B = F_{AR} - F_{BR} \tag{13.4}$$

[10]For example, strychnine; C. Bakhoven, J. C. Schoone, and J. M. Bijvoet, *Acta Cryst.*, **4**, 275 (1951).

[11]See J. M. Robertson, *Organic Crystals and Molecules*, Cornell University Press, Ithaca, NY, 1953, pp. 262–270.

Since the contributions F_{AR} and F_{BR} can be calculated from the known heavy atom positions, the right-hand side of Eq. (13.4) is known. Since only the magnitude of the observed F's are available, the left-hand side is one of

$$(\pm|F_A|) - (\pm|F_B|) \tag{13.5}$$

The values corresponding to the four possible sign combinations can be calculated in turn, and if one combination gives a clearly superior agreement to Eq. (13.4) it is accepted.[12]

The basis of the isomorphous replacement method can also be seen in terms of phase diagrams. Figure 13.1 shows three of the possible cases for two isomorphs with B containing the heavier atom. In each, the structure factor vectors $\mathbf{F}_A$ and $\mathbf{F}_B$ are shown as the summation of $\mathbf{F}_L$ and $\mathbf{F}_{AR}$ or $\mathbf{F}_{BR}$. The magnitudes and phases of the observed **F**'s depend on the magnitudes and phases of these components, which can be combined in various ways. For the centrosymmetric case, however, the vectors will always lie along the horizontal axis, that is, the phase angles are 0 or π.[13]

The problem of isomorphous replacement in noncentrosymmetric space groups can best be considered in terms of a generalization of these diagrams. This is shown in Fig. 13.2.

For this case, Eqs. (13.2) and (13.3) can be rewritten in terms of vector

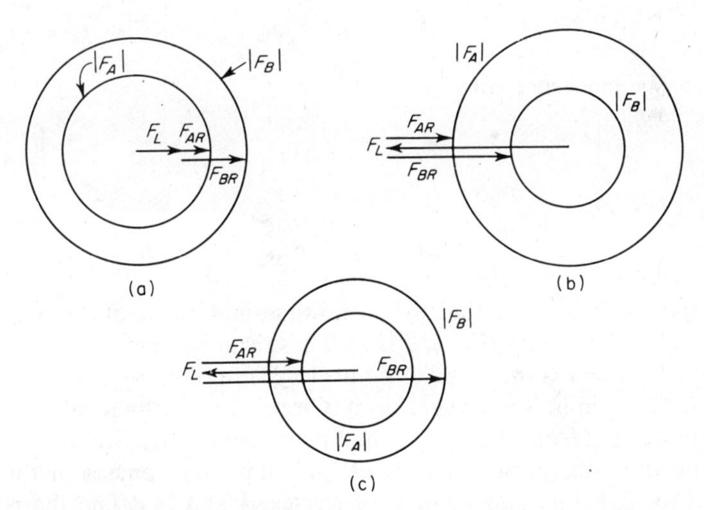

Figure 13.1. Isomorphous replacements AR and BR in a centrosymmetric structure for three reflections.

[12]For examples, see Buerger, *Crystal-Structure Analysis*, pp. 521–529.
[13]For a discussion of the practical details of isomorphous replacement in the centrosymmetric case, see A. Hargreaves, *Acta Cryst.*, **10**, 196 (1957).

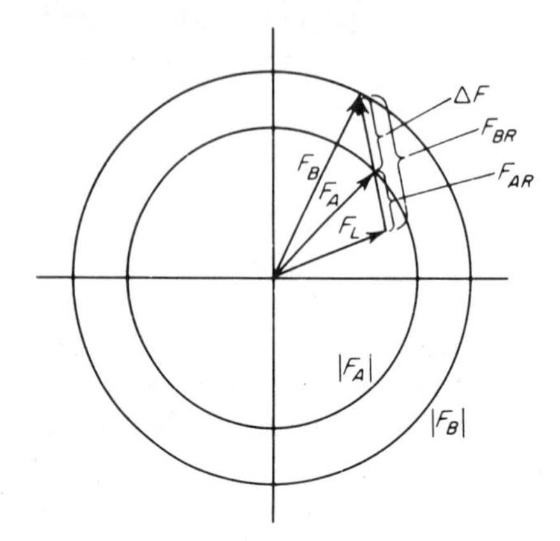

Figure 13.2. Isomorphous replacements *AR* and *BR* in a noncentrosymmetric structure.

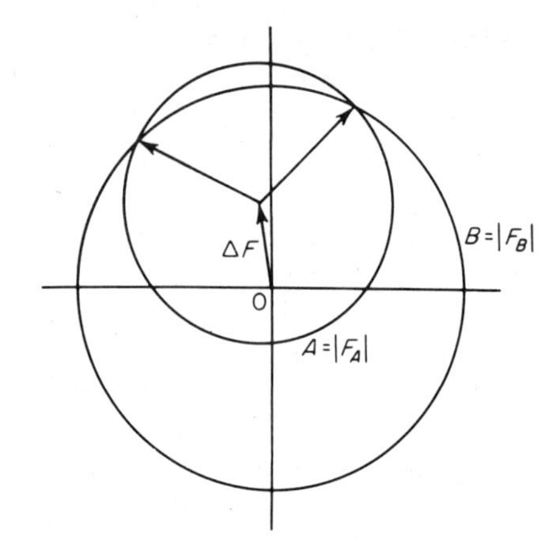

Figure 13.3. Ambiguity in phase determination for a noncentrosymmetric structure when *AR* and *BR* occupy the same location.

addition of the parts. Thus,

$$\mathbf{F}_A = \mathbf{F}_L + \mathbf{F}_{AR} \tag{13.6}$$

$$\mathbf{F}_B = \mathbf{F}_L + \mathbf{F}_{BR} \tag{13.7}$$

These can be combined into

$$\mathbf{F}_B = \mathbf{F}_A + \Delta\mathbf{F} \tag{13.8}$$

where

$$\Delta\mathbf{F} = \mathbf{F}_{BR} - \mathbf{F}_{AR} \tag{13.9}$$

The practical situation is that the magnitude and phase of $\Delta\mathbf{F}$ are known, but only the magnitudes $|F_A|$ and $|F_B|$. These can be combined as shown in Fig. 13.3. Circle B is the locus of points $|F_B|$ from the origin O. The end of the vector $\mathbf{F}_B$ will lie somewhere on it. Circle A is the locus of points lying at a distance $|F_A|$ from the end of the known vector $\Delta\mathbf{F}$. The points on this circle constitute the vector sums of $\Delta\mathbf{F}$ and all possible phases that might be assigned to $|F_A|$. The intersections of circles A and B define the points at which Eq. (13.8) holds and thus the phases of $\mathbf{F}_A$ and $\mathbf{F}_B$. Unfortunately, as can be seen from Fig. 13.3, there are two such points and consequently two sets of phases that satisfy Eq. (13.8). Only one of these is correct; so the solution of Eq. (13.8) does not completely determine the structure.

Several solutions to this problem have been proposed. If a third isomorph can be obtained that differs from one of the first pair in the change of an atom *at a different location*[14] from that varied between the first two, these two will form another isomorphic pair.[15] Using these, another comparison can be set up. This will again lead to two possible solutions, but of these only one will in general coincide with that found from the first pair, and the phases for all three isomorphs will be determined.

A second method, which has sometimes been used, is based on using only two isomorphs and a Fourier synthesis with slightly modified coefficients that includes *both* possible phases for each reflection.[16] The suggestion is that the contributions with the correct phases will tend to reinforce and give peaks at atomic locations, while those with the false phases will approximate random errors and merely add to the background "noise." Structures have been solved by this method, but it is difficult to tell whether the additional effort involved in collecting data on the second isomorph was repaid by an equivalent improvement in the case of obtaining the final structure.

A combination of the isomorphous and heavy atom methods can be used, in which that phase is chosen for $\mathbf{F}_B$ that is closer to the one suggested by the heavy atom alone, that is, by $\mathbf{F}_{BR}$. Not all of these choices will be correct, but the majority will be and the resulting phases will be more accurate than those based on the heavy atom method alone.

Finally, there is increasing interest in the combination of isomorphous replacement with direct methods and tangent refinement.[17] This appears promising as a means of obtaining a unique phase indication from a single native/derivative pair, but it has not yet been thoroughly tested in practice on unknown cases.

Of these approaches the use of several different isomorphs (MIR, *multiple isomorphous replacement*) is the most powerful. The obvious difficulty is that of obtaining three isomorphs with two different replaceable atoms. For normal organic molecules the problems are such that the method is scarcely if ever used, but in protein work, where heavy atoms can be introduced by "staining" methods into various sites in preformed crystals, it has proved most valuable. A great deal of effort is required to prepare the specimens and collect the data, but fortunately it is possible to screen prospective crystals rapidly by comparing precession photographs of the native and treated crystals. They should be similar enough geometrically to assure probable isomorphism and at the same time show sufficient differences

[14]If the three isomorphs all have the replaceable atom in the same location, the heavy atom contributions will all have the same phase and any pair will lead to the same ambiguity as any other pair. The requirement of different locations usually eliminates such apparent candidates as the HCl, HBr, and HI salts of optically active alkaloids and is so restrictive that the method is unlikely to be applicable to any but very large molecules.

[15]D. Harker, *Acta Cryst.*, **9**, 1 (1956).

[16]G. Kartha, *Acta Cryst.*, **14**, 680 (1961); see also D. M. Blow and M. G. Rossmann, *Acta Cryst.*, **14**, 1195 (1961).

[17]D. A. Langs, *Acta Cryst.*, **A42**, 362 (1986) and references cited there.

(5–10% overall) in intensities to indicate the absorption of enough heavy atoms to produce a useful effect.

Once the data are in hand, the heavy atoms are found by Patterson methods, and their positions and occupancies are adjusted to get the best fit with the observations. The possible phases can be calculated for each derivative with the native, and these sets, each of them with its 2-fold ambiguity, can be combined to give the best (most probable) overall result. Because the method involves small differences between quantities with considerable experimental error, the phase calculations are often highly uncertain, and the combination process generally does not give a single clear result. Instead, the results are described in terms of a probability distribution of values, often still centered on two possible phases, one of which is more probable. In the past the more probable value was often used, but the preferred method is to use a weighted average of the phases and to weight the Fourier coefficients by a figure of merit that reflects the angular spread between these values and the extent to which the phasing is concentrated at these points rather than being spread uniformly around the phase circle.

13.3. ANOMALOUS SCATTERING METHODS[18,19]

A further method of phasing, which has attracted increasing interest as the precision of intensity measurements has improved, is based on the use of the anomalous scattering properties of certain atoms under the proper conditions. As was discussed in Chapter 8, this effect occurs when the wavelength of the incident X-rays is reasonably near that of an absorption edge of the irradiated element. Since sulfur is about the lightest element for which such dispersion can be observed with common radiations, it is clear that, at least in organic compounds, the use of this effect is normally combined with the heavy atom method.[20] For inorganic structures, however, there may be cases in which the anomalous scatterer is not significantly heavier than the other atoms in the structure and phasing must be based solely on the anomalous effects.

In general, the practical effects of anomalous scattering are similar to those of isomorphous replacement. In fact, if conditions can be found such that the scattering factor changes associated with dispersion are reflected only in the real term $\Delta f'$ (see Section 8.13), the treatment is identical. Figure 13.4 shows the difference that will be observed if a centrosymmetric

[18] S. Ramaseshan, in *Advanced Methods of Crystallography*, G. N. Ramachandran, Ed., Academic, New York, 1964, pp. 67–95.

[19] Ramaseshan and Abrahams, *Anomalous Scattering*.

[20] Or with the isomorphous replacement technique. Cf. A. C. T. North, *Acta Cryst.*, **18**, 212 (1965) and B. W. Matthews, *Acta Cryst.*, **20**, 82 (1966).

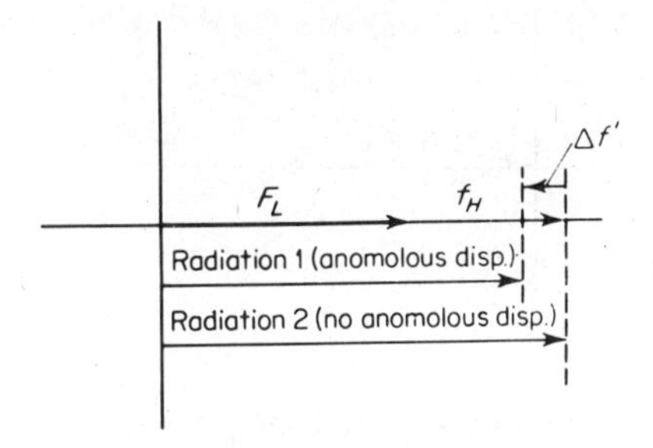

Figure 13.4. Effect of anomalous scattering in a centrosymmetric structure when $\Delta f''$ is negligible.

structure containing a heavy atom is studied with one radiation such that $\Delta f'$ is appreciable so that[21]

$$\mathbf{F}'_{\text{heavy}} = \mathbf{F}_{\text{heavy}} + \Delta\mathbf{F}' \tag{13.10}$$

and with another such that $\Delta f' \approx 0$. It is clear that $|F_{hkl}|$ will be different for the two measurements. Comparison with Fig. 13.1 shows that the effect is the same as that produced by having two isomorphous substances, and the phases can be deduced using the same equations.

The same principle can be applied to the case of a noncentrosymmetric crystal, but, as is true with ordinary isomorphous replacement, an ambiguity remains. Thus, as before, some additional information must be introduced to decide which of the two possible phases is correct.

In practice, this method has several problems. The differences in $\Delta f'$ for different radiations are generally small, and accurate intensity measurements are required to make use of them. The difficulty is accentuated by the fact that those element–radiation combinations that show the largest values of $\Delta f'$ are also those for which the errors in intensity measurements produced by absorption are the most serious. On the other hand, the exact isomorphism available using only a single crystal represents a real advantage over the deviations from ideality that must usually be tolerated when two different compounds are involved.

A fundamentally more serious question arises, however, from the assumption that $\Delta f''$, the imaginary component, can be ignored. Inspection of the tabulated values for $\Delta f'$ and $\Delta f''$ shows that the imaginary term is generally comparable to and frequently much larger than the real term. Thus any accurate analysis of the effects of anomalous scattering should include both.

[21]The notation $\Delta\mathbf{F}'$ represents in vector terms the contribution to the structure factor from the real part of the anomalous scattering factor $\Delta f'$ summed over all the symmetry-related heavy atoms. Similarly, $\Delta\mathbf{F}''$ is the contribution from the imaginary part $\Delta f''$.

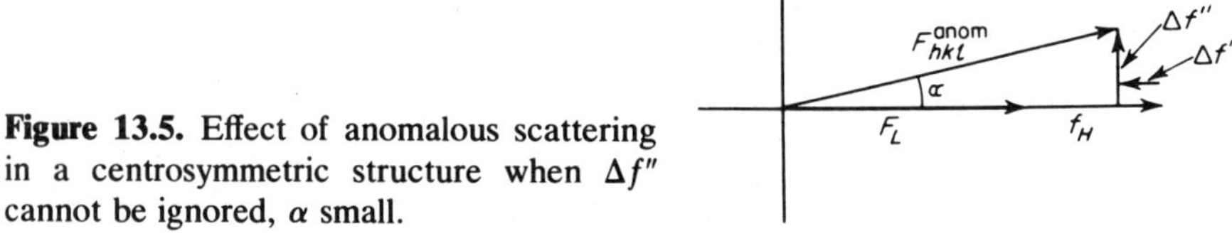

Figure 13.5. Effect of anomalous scattering in a centrosymmetric structure when $\Delta f''$ cannot be ignored, α small.

If both $\Delta f'$ and $\Delta f''$ are appreciable, Fig. 13.4 can be redrawn as shown in Fig. 13.5; that is,

$$\mathbf{F}_{hkl}^{\text{anom}} = \mathbf{F}_{\text{light}} + \mathbf{F}_{\text{heavy}} + \Delta\mathbf{F}' + \Delta\mathbf{F}'' \tag{13.11}$$

In this case the deviation of $\mathbf{F}_{hkl}^{\text{anom}}$ from the x axis is relatively small, and no difficulties in phasing are likely to arise from comparing $|F_{hkl}^{\text{anom}}|$ directly with $|F_{hkl}|$, that is, by ignoring the effects of $\Delta f''$ when comparing two sets of data. Figure 13.6, on the other hand, shows an example in which the deviations between the two vectors is much more pronounced.

A general solution that will handle both cases is obtained in the same way as for the noncentrosymmetric isomorphous replacement. Thus,

$$\mathbf{F}_{hkl}^{\text{anom}} = \mathbf{F}_{hkl} + \Delta\mathbf{F}'_{hkl} + \Delta\mathbf{F}''_{hkl} \tag{13.12}$$

$$\mathbf{F}_{hkl} = \mathbf{F}_{hkl}^{\text{anom}} - \Delta\mathbf{F}'_{hkl} - \Delta\mathbf{F}''_{hkl} \tag{13.13}$$

which implies

$$|F_{hkl}| = |F_{hkl}^{\text{anom}}| - \Delta\mathbf{F}'_{hkl} - \Delta\mathbf{F}''_{hkl} \tag{13.14}$$

Figures 13.7*a* and *b* show the application of Eq. (13.14) to the two examples given above. Note that there is no ambiguity even though the circles for $|F_{hkl}^{\text{anom}}|$ and $|F_{\overline{hkl}}|$ intersect twice, because it is assumed that $\mathbf{F}_{hkl}$ must lie along the x axis.

The problem of the noncentrosymmetric crystal is more complex but

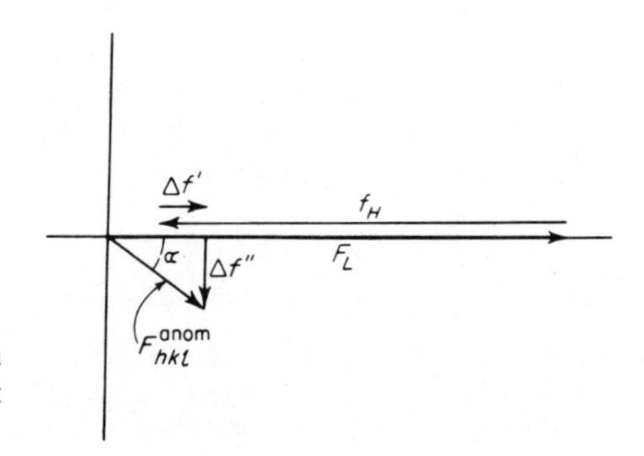

Figure 13.6. Anomalous scattering in a centrosymmetric structure, α not small.

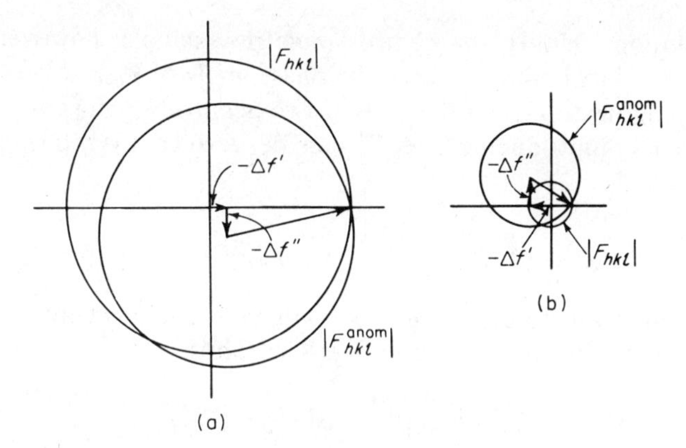

Figure 13.7. Determination of phase from anomalous scattering for centrosymmetric structures. $\mathbf{F}_{hkl}$ without dispersion, $\mathbf{F}_{hkl}^{\text{anom}}$ with dispersion. Compare with Figs. 13.5 and 13.6.

potentially more useful. In principle at least, the measurement of $|F_{hkl}|$, $|F_{hkl}^{\text{anom}}|$, and $|F_{\overline{hkl}}^{\text{anom}}|$ provides an exact solution to the phase problem for those cases to which it is applicable. In real cases it is less effective since its use depends on the precise determination of small intensity differences, but it has nevertheless proved to be a powerful tool.

The analysis of this situation is most easily obtained by considering first the case of one member of the anomalous pair, for example, $|F_{hkl}^{\text{anom}}|$, and the nonanomalous measurement $|F_{hkl}|$. Then Eq. (13.14) can be applied as shown in Fig. 13.8*b* for the example given in Fig. 13.8*a*. Note that in Fig. 13.8*b* the circles for $|F_{hkl}^{\text{anom}}|$ and $|F_{hkl}|$ intersect in two places, indicating two

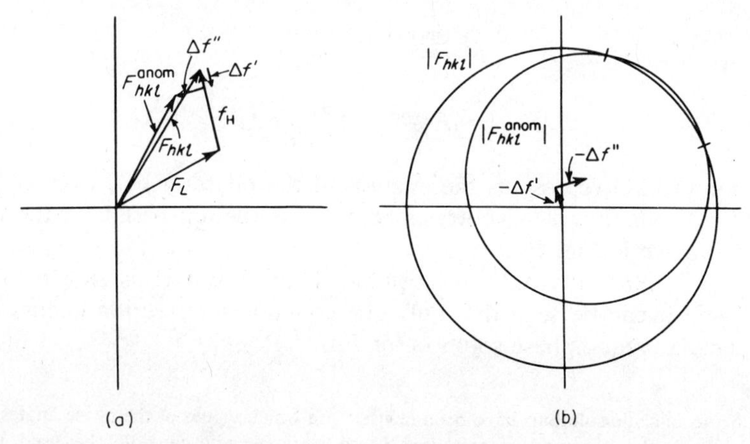

Figure 13.8. Noncentrosymmetric structures. (*a*) Vectors for $\mathbf{F}_{hkl}$ without dispersion and $\mathbf{F}_{hkl}^{\text{anom}}$ with dispersion. (*b*) Ambiguity in phase determination.

possible solutions. Unlike the centrosymmetric example, however, we can make no a priori assumptions about the phase angle of $\mathbf{F}_{hkl}$, so an ambiguity remains analogous to that in the case of two acentric isomorphs.

The effect of introducing $|F_{\overline{hkl}}^{\text{anom}}|$ can be derived by returning to Eq. (13.13). Then

$$\mathbf{F}_{\overline{hkl}} = \mathbf{F}_{\overline{hkl}}^{\text{anom}} - \Delta\mathbf{F}'_{\overline{hkl}} - \Delta\mathbf{F}''_{\overline{hkl}} \tag{13.15}$$

Converting all terms to their complex conjugates is equivalent to reflecting the set of vectors in the x axis (see Fig. 8.7). Thus,

$$\mathbf{F}^*_{\overline{hkl}} = \mathbf{F}_{\overline{hkl}}^{\text{anom}*} - \Delta\mathbf{F}'^*_{\overline{hkl}} - \Delta\mathbf{F}''^*_{\overline{hkl}} \tag{13.16}$$

But by definition (see Section 8.5),

$$\mathbf{F}^*_{\overline{hkl}} = \mathbf{F}_{hkl} \tag{13.17}$$

and

$$\Delta\mathbf{F}'^*_{\overline{hkl}} = \Delta\mathbf{F}'_{hkl} \tag{13.18}$$

Furthermore, it can be shown easily that

$$-\Delta\mathbf{F}''^*_{\overline{hkl}} = \Delta\mathbf{F}''_{hkl} \tag{13.19}$$

(see Fig. 8.15); so

$$\mathbf{F}_{hkl} = \mathbf{F}_{\overline{hkl}}^{\text{anom}*} - \Delta\mathbf{F}'_{hkl} + \Delta\mathbf{F}''_{hkl} \tag{13.20}$$

Since the moduli of a vector and its complex conjugate are always equal (see Section 8.2),

$$|F_{\overline{hkl}}^{\text{anom}*}| = |F_{\overline{hkl}}^{\text{anom}}| \tag{13.21}$$

and so

$$|F_{hkl}| = |F_{\overline{hkl}}^{\text{anom}}| - \Delta\mathbf{F}'_{hkl} + \Delta\mathbf{F}''_{hkl} \tag{13.22}$$

Equation (13.22) represents the analysis of the phase of $\mathbf{F}_{hkl}$ from $|F_{\overline{hkl}}^{\text{anom}}|$, and Fig. 13.9*b* shows the corresponding construction based on the components shown in Fig. 13.9*a*.

If Figs 13.8*b* and 13.9*b* are combined, the result is as shown in Fig. 13.10, and it can be seen that only one common intersection occurs. This represents a unique phase solution for F_{hkl}.[22]

[22] Analogous analytic solutions have been devised[16] in which values of the phase angles α are obtained from the observed magnitudes. These methods will generally be used in any programmed solution of the phase problem following this approach, but they do not give as clear a picture of the technique as does the geometric analysis used here.

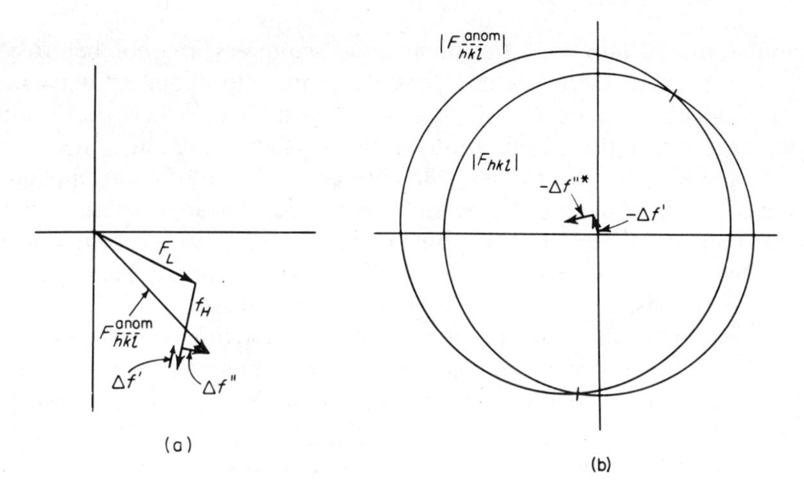

Figure 13.9. (*a*) Vectors for $\mathbf{F}^{\text{anom}}_{\bar{h}\bar{k}\bar{l}}$. (*b*) Use of $|\mathbf{F}^{\text{anom}}_{\bar{h}\bar{k}\bar{l}}|$ to resolve the ambiguity in Fig. 13.8*b*.

The diagrams on which the preceding arguments are based are all derived on the assumption that it is possible to calculate $\mathbf{\Delta F}'$ and $\mathbf{\Delta F}''$. In general this is possible in organic compounds in which the anomalous scatterer is at the same time a heavy atom and can hence be located from the Patterson function by conventional methods.[23] If the structure is centrosymmetric, or if the anomalous scatterers are related by a center of symmetry in an acentric space group (e.g., two heavy atoms in $P2_1$; see Section 12.3), the location is unique and the problem is merely one of

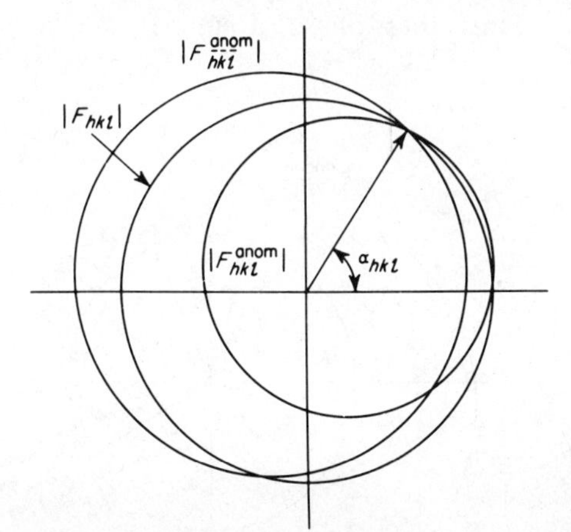

Figure 13.10. Phase circles from Figs. 13.8*b* and 13.9*b*, showing intersection at a unique solution.

[23]See B. W. Matthews, *Acta Cryst.*, **20**, 230 (1966) and G. Kartha and R. Parthasarathy, *Acta Cryst.*, **18**, 745 (1965). For an example, see S. R. Hall and E. N. Maslen, *Acta Cryst.*, **18**, 265 (1965).

computation. If, however, the anomalous scatterers are not centrosymmetrically related, there are two possible enantiomorphous arrangements, either of which will satisfy the Patterson function. One of these corresponds to the true absolute configuration of the crystal, while the other is the mirror image. Only the former will give correct phases when applied as described above, but it is not possible to decide a priori which is the true image. Thus the best approach would appear to be to work out the phases on the basis of each of the two possible models and apply the two sets in Fourier syntheses, expecting one to give better results.

The solution of the phase problem as described involves the collection of three distinct sets of data and the use of two different radiation sources. Most of the experimental determinations that have made use of dispersion have simplified the process by using only $|F^{\text{anom}}_{hkl}|$ and $|F^{\text{anom}}_{\overline{hkl}}|$ and dispensing with the direct measurement of $|F_{hkl}|$. Combining Eqs. (13.13) and (13.20) shows that

$$\mathbf{F}^{\text{anom}}_{hkl} - \Delta\mathbf{F}'_{hkl} - \Delta\mathbf{F}''_{hkl} = \mathbf{F}^{\text{anom}*}_{\overline{hkl}} - \Delta\mathbf{F}'_{hkl} + \Delta\mathbf{F}''_{hkl} \tag{13.23}$$

and thus

$$\mathbf{F}^{\text{anom}}_{hkl} = \mathbf{F}^{\text{anom}*}_{\overline{hkl}} + 2\Delta\mathbf{F}''_{hkl} \tag{13.24}$$

or

$$|F^{\text{anom}}_{hkl}| = |F^{\text{anom}}_{\overline{hkl}}| + 2\Delta\mathbf{F}''_{hkl} \tag{13.25}$$

Figure 13.11 shows Eq. (13.25) applied to the example of Figs. 13.8*b* and 13.9*b*. The ambiguity that remains in this solution has often been removed by utilizing the phase-determining powers of the anomalous scatterer regarded simply as a heavy atom. Thus that phase, found from the

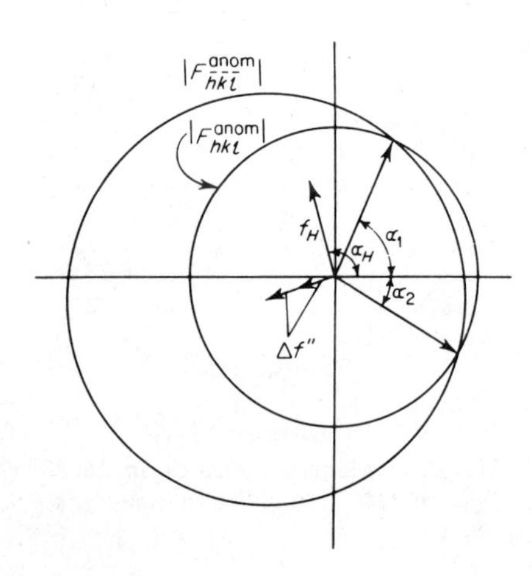

Figure 13.11. Ambiguity in phasing from $|F^{\text{anom}}_{hkl}|$ and $|F^{\text{anom}}_{\overline{hkl}}|$. The correct choice is usually the phase that is closer to α_H, in this case α_1 (cf. Fig. 13.8*a*).

anomalous effects, which lies closer to the phase calculated for the heavy atom alone, is chosen as the more probably correct and is used. This choice is not invariably correct, but it will be so in a majority of the cases, the exact fraction depending on the relative size of the heavy atom.[24]

A more rigorous analysis[25] of the equations relating the magnitudes and phases of Friedel pairs in the presence of dispersion has suggested that the ambiguity described above need not exist and that it is possible to obtain unique, or at least highly probable, phases directly from a single anomalous data set even for the noncentrosymmetric case. Although the method looks promising and practical, it requires further testing in real cases.

Because of the relatively small intensity variations produced by dispersion, accurate intensity measurements are required. If photographic methods are used, it is best if $|F_{hkl}^{\text{anom}}|$ and $|F_{\overline{hkl}}^{\text{anom}}|$ or equivalent reflections (see Section 6.1) can be obtained from a single photograph in order to minimize problems of scaling. Special photographic techniques have been devised for recording two Weissenberg layers simultaneously on one film,[26] for use in cases in which conventional methods do not serve. In general, however, the use of a diffractometer is recommended as providing better accuracy and more secure phasing.

The accuracy required, together with the presence of a heavy atom in the structure, increases the importance of making corrections for absorption. At the very least, efforts should be made to ensure that the absorption effects for the $|F_{hkl}^{\text{anom}}|$ and $|F_{\overline{hkl}}^{\text{anom}}|$ are the same.

BIBLIOGRAPHY

Blundell, T. L., and L. N. Johnson, *Protein Crystallography*, Academic, London, 1976, pp. 151–182, 337–380.

Buerger, M. J., *Crystal–Structure Analysis*, Wiley, New York, 1960, pp. 509–529, 542–547.

Dunitz, J. D., *X-Ray Analysis and the Structure of Organic Molecules*, Cornell University Press, Ithaca, NY, 1979, pp. 117–148.

Glusker, J. P., and K. N. Trueblood, *Crystal Structure Analysis*, 2nd ed., Oxford University Press, New York, 1985, pp. 123–148.

Holmes, K. C., in *Computing Methods in Crystallography*, J. S. Rollett, Ed., Pergamon, Oxford, 1965, pp. 183–203.

Karle, J., in *Computational Crystallography*, D. Sayre, Ed., Clarendon Press, Oxford, 1982, pp. 174–200.

Ladd, M. F. C., and R. H. Palmer, *Structure Determination by X-Ray Crystallography*, Plenum, New York, 1985, pp. 245–263.

[24]An alternative method has been proposed by G. A. Sim, *Acta Cryst.*, **17**, 1072 (1964).
[25]J. Karle, *Acta Cryst.*, **A41**, 387 (1985).
[26]H. P. Stadler, *Acta Cryst.*, **3**, 262 (1950).

Lipson, H., and W. Cochran, *Determination of Crystal Structures*, Cornell University Press, Ithaca, NY, 1966, pp. 170–178, 189–193, 201–233, 389–394.

Luger, P., *Modern X-Ray Analysis on Single Crystals*, de Gruyter, New York, 1980, pp. 216–229.

Ramaseshan, S., and S. C. Abrahams, Eds., *Anomalous Scattering*, Munksgaard, Copenhagen, 1975.

CHAPTER 14

SEARCH METHODS

It would appear reasonable that prior knowledge of the chemical structure of a molecule should simplify the process of determining its crystal structure, especially if the knowledge extends to its actual geometry as well as atomic connectivity. In this chapter we shall consider a number of methods that have been used at various times to incorporate such information into the search for a phasing model.

Most of these methods involve fitting some sort of preconceived model to the observational data. This can be carried out in direct space, Patterson space, or reciprocal space. In the first case, the question is one of packing. Where can the proposed model be physically accommodated in the unit cell without causing unacceptably close approaches between atoms? Although such analyses are rarely capable of solving current problems alone, they do provide useful checks and can often eliminate possible alternatives.[1]

Numerous methods have been proposed for isolating the crystal structure from the information contained in the Patterson map. Many of these have been discussed in theoretical terms or applied only to the reconstruction of very simple known models. The problems thus faced are much less severe than those usually encountered, and therefore the methods cannot be regarded as having been completely reduced to practice. There is also a great tendency for techniques that are fundamentally the same to appear, with slight modifications, under several different names. As in much of crystallography, the number of unique concepts is relatively small. The manner in which these concepts are applied, however, and particularly the

[1]But see S. Perez, C. Vergelati, and V. H. Trant, *Acta Cryst.*, **B41**, 262 (1985) for a complete structure solution from energy and packing analyses.

computational techniques used, often lead to variations that are more apparent than real.

We shall consider two cases for the problem of obtaining a crystal structure from the Patterson map. In the first, at least a part of the chemical structure of the molecules in question is known and is sufficiently rigid that its geometry can be defined with a high degree of probability. In the second, nothing except perhaps the elemental composition is known about the molecules comprising the cell. The first case is a surprisingly common one, including as it does cyclic compounds such as the steroids, many terpenes, sugars, and fused aromatic systems, in all of which the structure and geometry of the basic ring skeleton is known but the position and orientation of substituents remains to be determined. It has always been a point of annoyance to the chemist that the difficulty of solving a crystal structure in which only a minor chemical question was to be answered was almost as great as that of one in which the whole molecular structure was unknown. The use of Patterson methods, however, allows knowledge of the chemical structure to be applied directly to the phase-determining process.

Although these techniques have been largely supplanted by direct phasing for application to small molecules, they have simultaneously seen a rebirth of interest under the name *molecular replacement* as applied to groups of proteins with similar chain folding. In addition, they are proving important as alternatives or adjuncts to direct methods, especially since they are at their most effective for just those structures—for example, regular, rigid planar ones not at all like random sets of atoms—that so often give problems in direct phasing.[2]

Most of the methods to be discussed can be carried out in principle either by sampling the Patterson function, considered as a map of an array of vector points, or by equivalent calculation of special Fourier functions. These functions correspond more or less to performing the search in reciprocal space and fitting the Fourier transform of the search model to the observed intensity data. Although the methods are functionally equivalent, usually one or the other has advantages for computation or logical clarity (unfortunately, one often has one and the alternative the other). We prefer to discuss the Patterson searches in terms of vectors in precalculated maps, but in practice they are often carried out by direct Fourier computation.[3]

[2]For examples and discussion, see E. Egert, *Acta Cryst.*, **A39**, 936 (1983) and W. E. Thiessen and W. R. Busing, *Acta Cryst.*, **A30**, 814 (1974). For an interesting discussion of a program system (PATSEE) combining both direct and Patterson techniques, see E. Egert and G. M. Sheldrick, *Acta Cryst.*, **A41**, 262 (1985).

[3]For references and examples of the combined use of both methods, see J. Ruis and C. Miravittles, *J. Appl. Cryst.*, **20**, 261 (1987).

14.1. TRIAL-AND-ERROR METHODS

Essentially the method of trial-and-error consists of postulating a structure, that is, assuming locations of the atoms in the unique part of the unit cell, calculating structure factors F_c, and comparing their magnitudes with the scaled observed values $|F_o|$. This was for many years the only available method for solving structures, but unfortunately the number of trial structures that must be tested increases enormously with increasing number of parameters.

To appreciate the magnitude of the number of trial structures *if nothing at all is known* about the structure, consider a tiny 10-atom case in which 10 positions are to be tested (e.g., every 0.5 Å over a 5 Å distance) for each of the 30 positional parameters. The total number of possibilities is 10^{30}, a number so large as to preclude the feasibility of testing by any means now in hand.[4] If, however, the 10 atoms are in a known arrangement, the number of parameters is decreased to 6, 3 for position and 3 for orientation. If 10 values for each of the 6 parameters are to be tried, there are 10^6 different trials, still a large number but reduced by a factor of 10^{24}.

In practice, it has been found that phasing attempts based on search models are much simplified if the problem is divided into two parts: rotation searches to define the orientation of the model and translation searches to place the oriented model properly in the cell. Perhaps surprisingly, the former often proves less difficult than the latter, although in part this arises from an accumulation of errors by the time one reaches the translation search. The advantages of this separation are a further reduction in the number of trials needed (for the example above, the problem becomes one of 1000 rotation trials and then only 1000 translations, assuming that the correct orientation can be recognized) and the fact that the two problems can be handled by different methods that can be optimized independently.

[4]On the other hand, the increasing power of computers and the success of methods like RANTAN (Section 11.6), in which random phases refine by careful application of direct methods to a correct set, suggest that the problem may not be quite as impossible as has generally been believed. It is reasonable to suppose that there may exist a practical approach that is essentially the Fourier transform of random direct phasing and that would lead from *some* arbitrary sets of atoms to a final structure. No reliable general procedure of this sort exists at present, however. See A. J. C. Wilson, *Acta Cryst.*, **A33**, 523 (1977). It is also possible that methods will be developed, based on current research in image processing, in which the real space structure will be recovered from the observed magnitudes alone by systematic adjustment (within specified constraints such as reality, nonnegativity, and continuity) of pixels describing the cell contents. In this approach, the idea of atomicity would probably be introduced quite late in the process.

14.2. THE ROTATION SEARCH

The simplest evidence about the orientation of molecules in the cell may come from inspection of the raw intensity data themselves. Highly planar molecules such as extended aromatic systems often lie parallel to low-order lattice planes and at positions that allow most of the electrons of the molecule to scatter effectively. As a result, the reflections corresponding to these planes are extremely intense and stand out conspicuously both in photographs and in lists of F values. Calculation of E's or U's for such reflection will allow an estimate of the degree to which the various atoms are scattering in phase and, combined with a few structure factor graphs, will often give an estimate of the approximate orientation and position of the molecule.

Various more systematic methods have been proposed for deducing the orientation of a general rigid body in a crystal cell from its effects on the observed intensities. Among these are the use of transforms and various analytic approaches. When the detailed shape of a portion of the molecule is known, however, the conceptually simplest approach[5,6] uses the model (e.g., Fig. 14.1*a*) as the source of a set of intramolecular vectors (Fig. 14.1*b*), all translated to a common origin (Fig. 14.1*c*). In practice only one member of each centrosymmetric pair (AB and BA) is taken. The origin of this vector set is placed at the origin of the sharpened Patterson map and rotated through all possible orientations. During this rotation the "fit" between the set of vector points predicted by the model and those actually represented by the Patterson map is tested, and the position or positions giving the best fit are regarded as possible orientations.

The problem of measuring the agreement between a predicted vector distribution and a Patterson function is one we shall meet repeatedly. A number of "functions" have been proposed to describe this agreement.

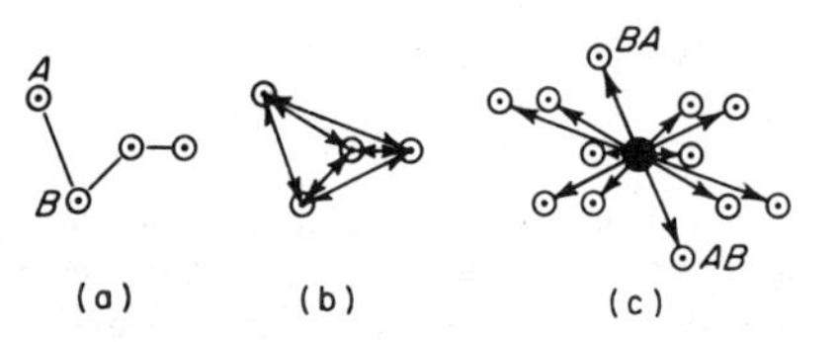

Figure 14.1. (*a*) Atomic model. (*b*) Interatomic vectors. (*c*) Set of vectors translated to a common origin. Model for Patterson rotation search.

[5]C. E. Nordman and K. Nakatsu, *J. Am. Chem. Soc.*, **85**, 353 (1963); C. E. Nordman and S. K. Kumra, *J. Am. Chem. Soc.*, **87**, 2059 (1965).

[6]For a somewhat analogous method using Fourier summation techniques rather than a search in a precalculated Patterson function, see P. Tollin and W. Cochran, *Acta Cryst.*, **17**, 1322 (1964) and D. G. Watson, D. J. Sutor, and P. Tollin, *Acta Cryst.*, **19**, 111 (1965). This approach is now widely used, especially for large molecules, because it has been made very fast by an elegant expansion in terms of spherical harmonic functions rather than standard Fourier coefficients. See R. A. Crowther, in Rossmann, *The Molecular Replacement Method*, pp. 174–178.

Among these are the *sum*, the summation of the values found in the Patterson map at the points predicted for the theoretical distribution; the *product*, their product; and the *minimum*, the smallest value found. Of these the minimum appears to be the most sensitive, since for it to be large, that is, to indicate a good fit, *all* of the predicted vector points must lie in regions of high actual vector density. The sum, on the other hand, can be large when some of the values found are very large and some very small. In addition, the peaks tend to merge increasingly with the background as the number of points sampled becomes large. The product function produces a map with a great deal of local variation, that is, a great deal of "noise." In general, the Min function appears preferable, both in theory and in practice, when working with already calculated Patterson maps. A variant of it, the Min(N) function, which is the sum of the N smallest points found, is sometimes used for large structures.[7] Unfortunately, the minimum function is difficult to incorporate into Fourier calculations and so is restricted to searches in precomputed Patterson maps.

One difficulty with the rotation search is that the number of orientations that have to be examined is surprisingly large. The search pattern can be defined in terms of rotation about three mutually perpendicular axes fixed in the model. These rotations can be performed in various ways, but the most common is probably about A, B, and then A again, in the classic Eulerian fashion.[8] The rotation process can be visualized by imagining a sphere surrounding the axes and following the path of the third rotation axis on it (Fig. 14.2). Thus rotation about A rotates the AC plane, which cuts the sphere in lines corresponding to the lines of longitude on the earth.[9] Rotation about B' carries the A' axis along one of these lines to A''. Thus

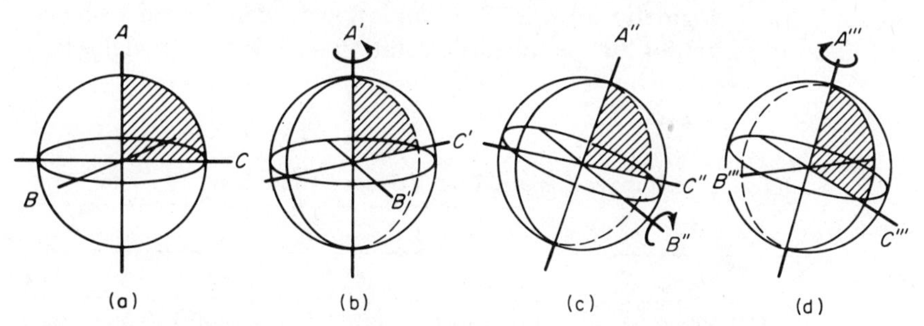

Figure 14.2. (*a*) Sphere centered on origin of coordinate axes. (*b*)–(*d*) Successive rotations indicate search pattern.

[7]D. F. High and J. Kraut, *Acta Cryst.*, **21**, 88 (1966).

[8]H. Margenau and G. M. Murphy, *The Mathematics of Physics and Chemistry*, Van Nostrand, New York, 1943, pp. 272–275.

[9]A small toy globe, 6 in. or so in diameter, on which marks can be made with wax pencil, will be found to be very convenient in visualizing these rotation processes.

by a pair of rotations about A and B', the A'' axis can be located anywhere on the sphere, and a rotation of 360° about A'' provides all possible orientations.

In order to obtain all unique orientations, it is not necessary to perform 360° rotations about all three axes.[10] In particular, symmetry in either the Patterson map or the model reduces the amount of search needed. Table 14.1 gives examples of the rotations required for various symmetries. It should be noted that the "model" actually used is the translated vector set from the molecular model and, like the Patterson function, always has a center of symmetry.

One disadvantage of the search pattern described above is that for steps of a given angular increment in the various rotations, the actual fineness of search is not constant. This arises because of the interaction of the various rotations. Thus for an A, B, A search, the paths followed by the axis make up a lune (Fig. 14.3, cf. Fig. 14.2), with its ends corresponding to $B = 0°$ and 180°. At these positions the rotation of the model is $\Delta A_1 \pm \Delta A_2$ about the initial A axis, so that only a limited number of orientations are actually explored for a large number of angular combinations. When B is nearly, but not exactly, equal to these special values, the various orientations of the A axis form a closely packed set and the search results in the detailed examination of slight modifications of a relatively limited number of orientations. As B approaches 90°, the difference in orientation corresponding to each angular setting increases. The usual solution to this problem is to change the coarseness of the steps used for ΔA as a function of B and thus achieve a more even set of search orientations.

The fineness of search that should be used depends on a large number of factors, among them the size of the model, the sharpness of the Patterson map, and the nature of the program. Actually, too little experience is at hand to allow a dogmatic answer. Structures have been solved using an initial uniform search of 10° on all three rotation axes with a model whose

TABLE 14.1 Rotation Ranges Required for Various Symmetries

	Cell		
Model	$\bar{1}$	$2/m$	mmm
$\bar{1}$	180.360.360	180.180.360	90.180.360
$2/m$	180.180.360	90.180.360	90.180.180 90. 90.360
mmm	90.180.360	90.180.180 90. 90.360	90. 90.180

[10]For theoretical discussions, see P. Tollin, P. Main, and M. G. Rossmann, *Acta Cryst.*, **20**, 404 (1966) and S. N. Rao, J.-H. Jeh, and J. A. Hartsuck, *Acta Cryst.*, **A36**, 878 (1980).

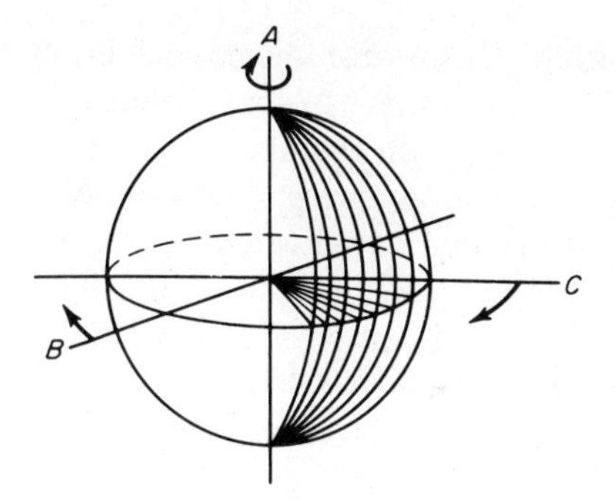

Figure 14.3. Paths followed by the *A* axis for rotation about *A* followed by rotation about *B*.

largest vector was ~7.5 Å. This may have included a significant amount of good luck, however, since the correct orientation was in the region where the interaction of the first and third rotations was producing a finer effective search. To save time, it is probably best to carry out preliminary searches using a small model and a reasonably coarse angular grid, and then extend the model while performing finer searches in those limited areas of interest.

A rotational search of this kind can be carried out with any of the superposition-testing functions described above. If a pure Min function is used, the search is very sensitive, but a very fine angular grid may have to be used in order to hit on an orientation that is recognizably good. The use of a Min(N) function, where N is between one-tenth and one-third of the number of vectors being used, relaxes this requirement considerably and apparently to advantage. Attempts have also been made to use the Min(N) technique to avoid the necessity for interpolation in locating the Patterson values at the ends of the vectors. The results suggest, however, that at least with large models and normal (~0.3 Å grid) sharpened Patterson maps, the time saved at this point is more than lost later through the uncertainties introduced and the extra orientations that must be examined.

14.3. THE TRANSLATION SEARCH

Once a likely orientation (or set of orientations) has been achieved, the next problem is that of locating the molecule properly in the cell. There are a large number of ways in which this can be attempted, no one being obviously superior in all cases.

Computerized Trial and Error[11–13]

The availability of fast computers has made it feasible to test large numbers of trial structures by structure factor calculation. Even so, the process is

[11] A. K. Bhuiya and E. Stanley, *Acta Cryst.*, **17**, 746 (1964).
[12] G. H. Stout, V. F. Stout, and M. J. Welsh, *Tetrahedron*, **19**, 667 (1963).
[13] C. A. Taylor and K. A. Morley, *Acta Cryst.*, **12**, 101 (1959).

slow unless it is done properly. The structure factors for the model in a selected orientation should be calculated first with respect to some arbitrary local origin. Successive sets of structure factors are then calculated for each position of this origin as it is moved about over the unique part of the unit cell and the measure of agreement for each position is indicated by the R factor.

Just how the efficiency of computing is increased can be seen by taking an example of a structure in space group $P\bar{1}$, which has two equivalent positions with atoms in pairs at x, y, z and $\bar{x}$, $\bar{y}$, $\bar{z}$. The structure factor can be written in the form

$$F_{hkl} = \sum_j f_j e^{2\pi i[h(x_j+x_a)+k(y_j+y_a)+l(z_j+z_a)]} + \sum_j f_j e^{2\pi i[h(-x_j-x_a)+k(-y_j-y_a)+l(-z_j-z_a)]} \tag{14.1}$$

where x_a, y_a, z_a are the coordinates of some arbitrary origin with respect to the unit cell origin and the x_j, y_j, z_j are the coordinates of the atoms with respect to this local origin. Factoring out the common terms within each summation gives.

$$F_{hkl} = e^{2\pi i(hx_a+ky_a+lz_a)} \sum_j f_j e^{2\pi i[hx_j+ky_j+lz_j)} + e^{-2\pi i(hx_a+ky_a+lz_a)} \sum_j f_j e^{-2\pi i(hx_j+ky_j+lz_j)} \tag{14.2}$$

Substituting

$$a = \cos 2\pi(hx_a + ky_a + lz_a); \qquad b = \sin 2\pi(hx_a + ky_a + lz_a)$$

$$A = \sum_j f_j \cos 2\pi(hx_j + ky_j + lz_j); \qquad B = \sum_j f_j \sin 2\pi(hx_j + ky_j + lz_j)$$

in Eq. (14.2) gives

$$F_{hkl} = (a + ib)(A + iB) + (a - ib)(A - iB) \tag{14.3}$$

$$F_{hkl} = 2(aA - bB) \tag{14.4}$$

If the model is translated, only x_a, y_a, z_a change; x_j, y_j, z_j do not. Thus only the terms a and b change, and just these two terms need be evaluated for each reflection for each new position.

Those regions where the R factor is relatively low can be searched on a finer grid to locate more exactly the best position for the model. Even better results are obtained if the structure factor program is coupled with a least squares segment that will automatically adjust the position from each of the search grid points to minimize the sum of the squares of the weighted ΔF values, $\sum_w \left||F_o| - |F_c|\right|^2$.

Limited Data

In applying the method of trial and error, it is advantageous to use data within a restricted range of $(\sin\theta)/\lambda$. The computing time is decreased not only because fewer reflections are tested but also because a coarser grid of points can be used to cover the unique volume of the unit cell, that is, fewer positions of the model need be tried.

The courseness of the grid over which the model is to be moved is related to the range in $(\sin\theta)/\lambda$ of the data. This determines the minimum interplanar spacing d_{min} at the limit of the data and hence the resolution.

If the model used in mechanical or other trial-and-error testing is good, the R index is surprisingly sensitive to position.[14] Both theory and experience indicate that the grid spacing at which the model is to be tested can be taken as $d_{min}/8$, one-eighth the limit of resolution of the data used. If, however, the tests are coupled with a least squares adjustment of the position at each point of the test grid, it need not be finer than $d_{min}/4$. In practice the data may be conveniently limited by using only reflections with no index greater than some selected h_{max}. Under these conditions, the limit of resolution, in fractions of a cell edge, is $1/h_{max}$, and the corresponding grids are $1/8h_{max}$ and $1/4h_{max}$.

During the initial stages of testing only the inner core of data need be used, with indices ranging no higher than 3 or 4. Promising regions found in the initial trials will be tested at finer intervals with additional shells of data added, and for the most promising positions sufficient data should be used to attain atomic resolution in an F_o synthesis of the electron density.

The *R* Criterion

In Chapter 9 the residual index R was defined, and its use in assessing a model was discussed. As indicated there, no single value of R can be taken as assuring the correctness of a structure. Values within the range 0.4–0.5 for centrosymmetric structures and 0.3–0.4 for noncentrosymmetric structures can be taken as indicating some degree of significance in the trial structure. In many instances, however, refinement has been successful from considerably higher values of R provided that the major features of the structure were correct.

It should be noted that in cases where heavy atoms dominate the structure (see Chapter 13), considerably lower values of R must be attained before much significance can be attached to them. For example, the

[14]See E. Stanley, *Acta Cryst.*, **17**, 609 (1964).

structure of the $CdCl_2$ salt of α-glycerophosphorylcholine trihydrate[15] in the noncentrosymmetric space group $P2_12_12_1$ is dominated by Cd atoms in an approximately centered arrangement. The R for the one Cd and two Cl atoms alone is 0.35. The other 19 nonhydrogen atoms can be added in quite wrong positions and yet be refined to 0.23. The correct position, however, refines readily to R under 0.10.

Although there is some disagreement on the subject, we believe that the correlation coefficient (Section 9.1) is a superior test of agreement for the sorts of rough fits that are usually obtained by trial-and-error methods. Fortunately, most programs provide both R and the correlation coefficient as part of their output, so one can check both.

If the rotation model is large enough to serve as a reasonably good phasing model, trial-and-error approaches can be used effectively. The model, however, must be significantly larger than that required for simple refinement, because it must yield results that are recognizably better than random even though both the orientation and position may be somewhat in error. Useful results can be achieved under surprising conditions by this method, however. Thus phytolaccagenin isoxazole[16] gave a recognizably best choice for orientation and position with only 34 of 40 atoms, some of which were misplaced (mainly orientation error) by up to 2.2 Å.

Superposition Methods

For smaller models, or if it is desired to continue the determination by using the Patterson function, there are a number of vector methods possible. One of the simplest is to move the oriented model throughout the unique portion of the cell, calculating the intermolecular vectors between the atoms in the model and those in the symmetry-related images, and then to compare the calculated vector points against the Patterson maps. As usual, the correct location should be signaled by a high observed vector density at each of the calculated points.[17] In general, the vectors between an atom and the symmetry images of *other* atoms provide a more sensitive test than those between an atom and its own images. The former are distributed throughout the Patterson map, while the latter are usually restricted to the Harker lines and planes, which are regions of high average vector density and tend to give more accidental agreements.

Another approach to recovering the structure from the Patterson synthesis involves an alternative way of looking at this function. In Chapter 12, the Patterson map was described in terms of the set of vectors between all the constituent atoms of the structure, the vector peaks having weights

[15] M. Sundaralingham and L. H. Jensen, *Science*, **150**, 1035 (1965).
[16] G. H. Stout, B. M. Malofsky, and V. F. Stout, *J. Am. Chem. Soc.*, **86**, 957 (1964).
[17] Also as usual, roughly equivalent Fourier methods can be used as well. See R. A. Crowther and D. M. Blow, *Acta Cryst.*, **23**, 544 (1967).

proportional to the products of the atomic numbers of the atoms they join. An entirely equivalent view regards the Patterson map as a combination of N images of the original crystal structure, each offset by the amount required to bring one of the atoms to the origin. Each image represents the ends of the vectors from some one atom (the one at the origin) to all other atoms in the cell, and as before it is weighted in terms of the weights of the atoms involved. Figure 14.4 shows how the vector map of Fig. 14.1c can be identified with N images of the original point set (Fig. 14.1a). Seen in this way, the problem of completing the structure becomes one of extracting one image from the overlapping set.

It has been shown[18] that the problem is solvable in theory, subject to certain possible ambiguities, for the case of a set of discrete points. The ambiguities can arise from the existence of so-called *homometric* sets,[19] that is, differing arrangements of points that have the same vector sets. Although of theoretical interest, these are exceedingly unlikely to appear in practice and do not pose a real difficulty. More serious, however, are the problems presented by the finite size of the Patterson peaks and their inevitable overlap. It is by no means obvious that what is theoretically true for an ideal point set will also be possible for an ill-resolved vector density map. In fact, very considerable practical difficulties arise because of overlap, but nevertheless structures of 30–50 light atoms have been solved by recovery of the molecule from the Patterson map.

As usual, many methods have been proposed for accomplishing this recovery, but two techniques appear to be the most generally applicable, and we shall limit ourselves to them. Both come under the heading of *superposition methods*, because they involve superimposing several copies of the Patterson map,[20] an operation that was formerly carried out manually

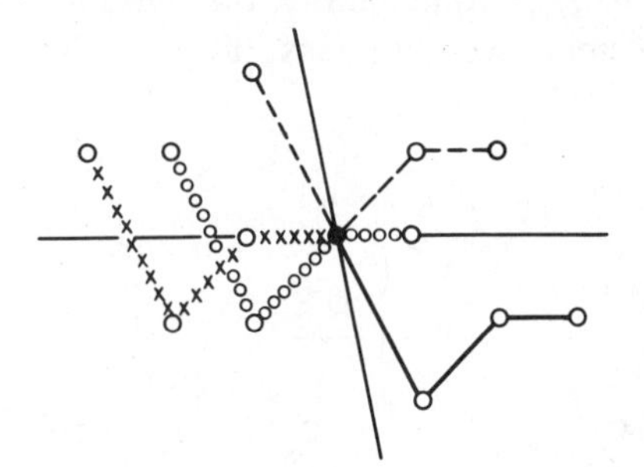

Figure 14.4. The Patterson function for a set of N atoms in the unit cell contains N images of the set.

[18] D. M. Wrinch, *Phil. Mag.*, **27**, 95 (1939). See also the references at the end of the chapter.
[19] A. L. Patterson, *Phys. Rev.*, **65**, 195 (1944).
[20] Although we shall discuss superposition methods in terms of the physical superposition of precalculated Patterson functions, it should be noted that many of the same ends can be achieved by the calculation of Fourier series that are *hybrids* between a normal Fourier series and a Patterson function. For a discussion and examples, see J. Fridricksons and A. McL. Mathieson, *Acta Cryst.*, **15**, 1065 (1962).

(usually in two dimensions) and is now done by computer. We shall refer to one of these techniques as *vector superposition* and the other as *atomic superposition*. Of these, the first is perhaps easier at the start, and the latter easier at the end.

The fundamental principle behind all superposition methods is as follows. Any peak (AB) in a Patterson map represents the end of a vector from some atom A to another atom B. The same Patterson map also contains the images of the original molecular structure with A placed at the origin and with B placed at the origin. Call these $I(A)$ and $I(B)$, respectively. Now if two copies, M and N, of the Patterson map are taken and N is translated so that its origin falls at the point AB on M, the points belonging to $I(A)$ in map M will coincide with those of $I(B)$ in map N. On the other hand, those points that belong to other sets will not coincide except by accident. Thus in principle the structure, or at least a closer approximation to it, can be obtained by selecting only those points that superimpose on the two maps. In actual practice, this superposition is carried out by shifting one copy of the Patterson map (which may be stored in the memory of a computer) and comparing it with an unshifted copy. This comparison is made between corresponding ("overlapping") points on the two maps by using one of the functions described in Section 14.2. The result is a superposition map that contains at every grid point a measure of the extent of coincidence between the two starting maps. The same process can also be applied to the superposition of several Patterson maps at once.

It can be shown[21] that a single superposition of this sort is sufficient to return the structure of a centrosymmetric point set, apart from possible accidental coincidences, provided that the point AB corresponds to the end of only a single vector (Fig. 14.5*c*). Note, however, that the Patterson map of a centrosymmetric model (Figs. 14.5*a* and *b*) of N unique atoms contains $2N(N-1)$ peaks that arise from two vectors, or, equivalently, the overlapping of points in two images, and only $2N$ single-weight peaks. If one of the

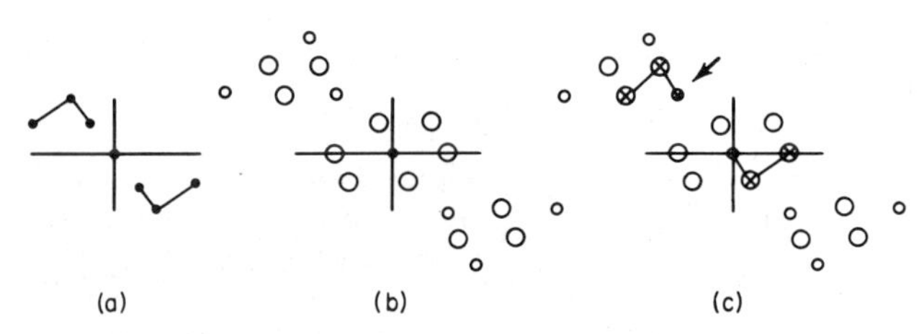

Figure 14.5. (*a*) Fundamental centrosymmetric set. (*b*) Patterson map of the fundamental set. (○) single peaks; (◯) double peaks. (*c*) Patterson map as in (*b*) with the origin of a duplicate superimposed on the single peak indicated by an arrow. Coincidences indicated by × are identical with the fundamental set.

[21]J. Clastre and R. Gay, *Compt. Rend.*, **230**, 1876 (1950).

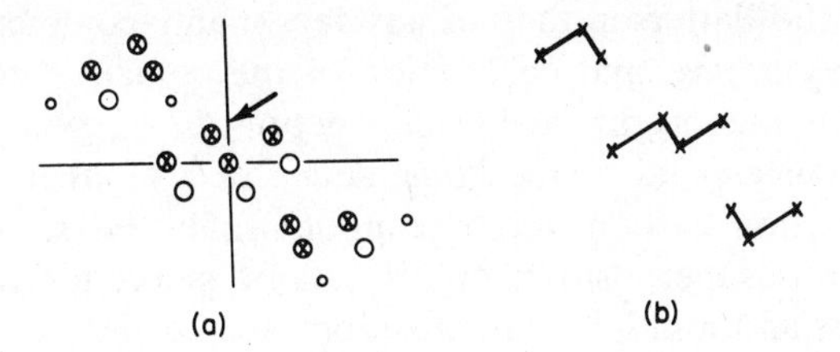

Figure 14.6. (*a*) Patterson map as in Fig. 14.5*b* with the origin of a duplicate superimposed on the *double* peak indicated by arrow. Coincidences indicated by ×. (*b*) Coincidences indicated in (*a*), showing two images of the fundamental set in Fig. 14.5*a*.

double peaks is chosen as *AB*, the superposition will still contain two images and the process will have to be repeated to obtain the true point set (Fig. 14.6). In the some way, although for different reasons, two superpositions are required to recover a noncentrosymmetric set (Fig. 14.7).

Obviously, matters cannot be so simple in practice, or the phase problem would have long since vanished. A number of difficulties appear when an attempt is made to apply the theoretically simple superposition process. Most of these can be traced to the overlap of Patterson peaks. In particular,

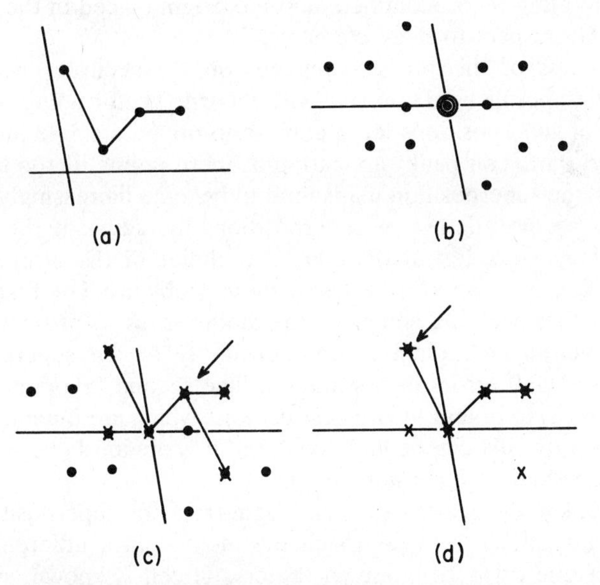

Figure 14.7. (*a*) Fundamental noncentrosymmetric set. (*b*) Patterson map of (*a*). (*c*) Patterson map as in (*b*) with the origin of a duplicate superimposed on the single peak indicated by an arrow. Coincidences indicated by × constitute two images of the fundamental set. (*d*) Coincidences in (*c*) with the origin of a duplicate of (*b*) superimposed on the peak indicated by an arrow. Coincidences indicated by * show only one image of the fundamental set.

examination of the Patterson map of any reasonably complex molecule not containing heavy atoms, and calculation of the actual vector densities by comparison with the origin peak (see Section 12.1), will show that the obvious peaks correspond to the coincidence of a number of vectors and that a single carbon–carbon vector cannot usually be resolved from the background. Thus superposition on the major peaks tends to give back multiple images, and multiple superpositions are needed to obtain the true structure.

If a partial model is available and has been oriented in the cell as described above, it can be used to select the vector peaks that are to be used for superposition. Some one atom of the oriented model is placed at the origin of the Patterson function, and the vectors from it to the remaining $N-1$ atoms serve to select $N-1$ vector peaks corresponding to a single image. The superpositions are then carried out by placing the origins of $N-1$ copies of the Patterson map on the points defined by these vectors. Ideally, the results of this *vector superposition* will be a structure in which one molecule appears around the selected atom, which is located at the origin of the superposition map. This origin is not the origin of the true crystal cell, however, but rather some point within the cell. In order to locate the cell origin it is necessary to extend the size of the superposition cell so that surrounding molecules can be recognized, the symmetry elements generating them identified, and the origin placed in the conventional position with respect to these elements.

The success of this process depends on the accuracy with which the superposition points used coincide with the true vector points and the use of a number of superpositions large enough to obtain a single image from the overlapped Patterson peaks appearing at these points. If the first condition is not met, the superposition maps tend to become increasingly uniform and featureless as the number of superpositions increases; if the second fails, multiple images are left at the end. The choice of the origin atom of the model is affected in opposite ways by these problems. The first suggests the use of an atom near the center of the model so as to provide the shortest possible vectors and minimize the sensitivity of the superposition point locations to small errors in orientation. The second favors atoms near the periphery so as to obtain the longest vectors, which are the most likely to be unique. No firm rule can be laid down, but it is obviously better to have too many images than to have none at all.

The problem of locating the cell origin from the superposition maps has led some investigators to approach the method in a different way. If the location of one or more atoms in the crystal cell is known, superpositions can be performed by placing the origin of the Patterson map in turn at the sites of these atoms and their symmetry-related images (Fig. 14.8). When such an *atomic superposition* is carried out at the positions of atoms A, B, C, and so on, the images $I(A)$, $I(B)$, $I(C)$ superimpose and reinforce. Even if the structure is not revealed by the first superposition maps, probable new

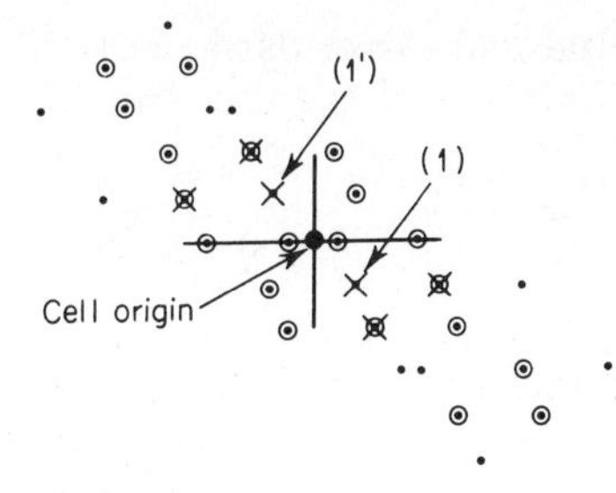

Figure 14.8. Atomic site (1) and centrosymmetrically related (1′) site of the fundamental set in Fig. 14.5*a* indicated by arrows. Origin of Patterson map in Fig. 14.5*b* placed on these two sites and peaks marked. ×'s indicate coincidences between the two Patterson functions and show the fundamental set placed properly with respect to the origin.

atomic sites can often be picked out and added to the original set. The process can then be repeated, using more superpositions and hopefully revealing more of the structure.

The outcome of such a series of superpositions is a structure in which the molecule under study appears *at its correct position in the cell.* Thus it is possible when using this method to abandon it at any time that it appears to have lost its utility and go over directly to Fourier methods, using the atomic positions that have been found. It is this ability to work back and forth easily between superposition and Fourier techniques that is one of the attractive features of atomic as compared with vector superposition.

Atomic superpositions have particular value in two cases. The first is that of crystals containing a heavy atom. These can be placed as described in Chapters 12 and 13, with the various symmetry-related positions then used as locations for superpositions. The situation here is particularly favorable because the images being superimposed are made up of vectors between heavy and light atoms and so are accentuated with respect to the background of light atom–light atom vectors. This method was used in one of the early applications of superposition techniques, the solution of the structure of strychnine hydrobromide.[22] No systematic study has been made of the efficiency of such atomic superpositions as compared with heavy atom Fourier methods, but it is quite possible that they will yield results more quickly if suitable programs are at hand.

The second case is that of an oriented model that has been positioned in the cell by use of intermolecular vectors (see above) but is too small to serve as a phasing model for Fourier refinement. It can, however, be used as a basis for atomic superpositions to expand or complete the structure. Difficulties may be encountered because the accumulated error of the first steps will appear in the atomic positions and may tend to wash out the superpositions. It is possible, however, to use atomic superpositions as a means of refining the preliminary positions, shifting the atoms to the centers of the peaks that come back on the superposition maps. The process is often

[22]C. A. Beevers and J. H. Robertson, *Acta Cryst.*, **3**, 164 (1950). See also J. H. Robertson, *Acta Cryst.*, **4**, 63 (1951) and J. H. Robertson and C. A. Beevers, *Acta Cryst.*, **4**, 270 (1951). For another example, see F. R. Ahmed and W. H. Beevers, *Acta Cryst.*, **16**, 1249 (1963).

a lengthy one, however, and involves much trial and error, particularly if the structure of the missing part is not known.

14.4. THE NO-MODEL PROBLEM[23]

Clearly, if no model exists for a structure, structure factor methods, except possibly those based on random atoms, are not available. The superposition techniques are in principle the same as those described above. The process is much more difficult, however, because of the lack of a good starting point for superpositions and the uncertainty as to what should be looked for in the resulting maps.

Vector superposition can be performed starting with some arbitrary vector point in the Patterson function, but the difficulties involved in selecting a peak corresponding to a single light atom vector from a complex vector map are extreme. If a multiple peak is used, or if the structure is noncentrosymmetric, multiple images arise from the first superposition, and these must be further unraveled. It is not possible to get around this problem by carrying out several superpositions at once on randomly chosen peaks, since the locations used must correspond to a single molecular image for the recovery to succeed. One method to achieve this is to carry out only one superposition at a time, inspecting the results and selecting a likely peak as a point to place the origin in the next stage. Difficulties arise, however, as they do throughout work with superpositions, when peaks appear that are not actually part of the set of retained molecular images but rather result from accidental correspondences of the vector maps in preceding superpositions. When these are used for later work, the structure that one is attempting to follow usually just fades away, and it is necessary to go back to the early stages and begin again.

Alternative approaches have been described that differ in details but involve fundamentally the same methods.[24] It is uncertain how far they can be carried with completely unknown structures, since no really systematic study has been made, but they merit consideration when one is faced with such a problem.

On the other hand, before the general move to direct phasing there was a growing tendency to attack the no-model problem by the use of atomic superpositions. In order to use this method, however, it is necessary to be able to locate an initial atom or atoms in the cell. Once this is done, the techniques are essentially the same as those used in the case of a small known model.

As discussed in Chapter 12, various means are available for finding an initial atomic position with which to begin a series of atomic superpositions.

[23]P. Luger and J. Fuchs, *Acta Cryst.*, **A42**, 380 (1986).
[24]See Buerger, *Vector Space*, pp. 218–309.

For a purely light atom structure the starting point is usually the symmetry minimum function (Section 12.3). This generally eliminates much of the cell but gives back many more possible atomic sites than will actually be occupied. These can be tested systematically by carrying out superpositions at each possible point or at each point and its symmetry-related positions.[25] Because of the large number of points involved, such a process is likely to be extremely time-consuming unless some simplified method is used to evaluate the significance of each set of superpositions. Thus one technique has been to use a modified Patterson map that contains only 0s and 1s, corresponding to values below and above some preselected threshold. The superpositions can then be carried out using the built-in logic of the computer to provide a very rapid Min test, and the quality of the final result evaluated simply by counting the number of 1s that survive the operations.

The result of this testing should be a map showing which of the possible atomic locations give a large number of coincidences when used as sites for superpositions. This process can be helpful in eliminating peaks in the symmetry map that are due to accidental coincidences, but it will not solve the problem of the compression of various parts of the crystal cell into a smaller volume. Nevertheless, it should help in selecting more reliable locations that can be used as a basis for actual superposition maps, which can then be studied in more detail. Since the symmetry minimum function and the maps derived from it all refer to the possible distribution of atoms in the real crystal cell, a true atomic position should appear as a peak in the same position on each. Thus a further elimination of false peaks is possible by comparing the results of the different manipulations.

A further approach to unraveling the symmetry map depends on finding in it those possible atomic sites that are related by a vector in the Patterson function. If a vector of low multiplicity is chosen, there should be relatively few pairs of atoms with a corresponding separation, and only a few locations in the symmetry map should be possible. Since the search now involves two atoms with a specified orientation, it can in certain cases distinguish between two molecular images appearing within the symmetry map and related by the symmetry of the Patterson function. Thus a beginning can be made at unraveling the multiple ambiguities of the symmetry map. The positions that are found can be tested by multiple superpositions involving both of the vector-related atoms and all their symmetry images.

An equivalent approach, one that does not involve the prior calculation of a symmetry map, is based on using as a model the two atoms whose relative positions are defined by a peak in the Patterson function. This model is then translated throughout the unique search volume of the cell, and the vectors between the two atoms and all their symmetry images are calculated and compared with the actual set. The vectors between the atoms and their own images correspond to the calculation of the symmetry

[25] A. D. Mighell and R. A. Jacobson, *Acta Cryst.*, **16**, 443, 1554 (1963).

map, while the cross vectors between one atom and the images of the other test the acceptability of their relative orientation. If many such calculations are to be tried, however, it is more efficient to compute the symmetry map first and use it as a guide for further testing rather than repeat the process for every vector studied.

It should be clear that the effectiveness of methods based on a symmetry map depends on the symmetry elements present in the space group being studied. These must be sufficient to define a unique set of origins for the cell. Thus for $P2_1$, the position of the origin along the y axis is arbitrary and the symmetry map will be the same for any value of y. All planes parallel to the *ac* net will contain peaks for every possible atomic position in the cell and will correspond to a projection of the cell contents down the y axis. This is very likely to be uninterpretable for a cell of any reasonable thickness. Even if the origin is determined by the symmetry elements, the selectivity of the method is enhanced as the number of symmetry-related images increases. It appears to be particularly effective for space groups in which the number of possible origin sites is fewer than eight per cell.

For space groups of lower symmetry, and particularly those for which the origin is not determined, methods involving finding at least two atoms as the starting points are advantageous. On the other hand, as the space group symmetry decreases, the advantages of atomic as compared to vector superpositions likewise decrease. This process reaches a limit in $P1$, where any point can be chosen as the origin and the two methods are the same.

The preceding discussion makes no pretense of providing a complete coverage of superposition methods. To do so would require not a chapter but a book, and many of the answers are not in hand. The use of superposition techniques is often regarded as a last resort after all other methods have failed, and it is certainly true that direct phasing usually provides a simpler and faster solution. Nevertheless, the power and generality of the methods cannot be denied, and the solution of a structure by superposition provides in some ways the final test of one's understanding of the Patterson function and its relationship to direct space.

BIBLIOGRAPHY

Blundell, T. L., and L. N. Johnson, *Protein Crystallography*, Academic, London, 1976, pp. 443–464.

Buerger, M. J., *Vector Space*, Wiley, New York, 1959.

Dunitz, J., *X-Ray Analysis and the Structures of Organic Molecules*, Cornell University Press, Ithaca, NY, 1979, pp. 112–117.

Glusker, J. P., B. K. Patterson, and M. Rossi, Eds., *Patterson and Pattersons. Fifty Years of the Patterson Function*, Oxford University Press, 1987.

Glusker, J. P., and K. N. Trueblood, *Crystal Structure Analysis*, 2nd ed., Oxford University Press, New York, 1985, pp. 113–123.

Lipson, H., and W. Cochran, *The Determination of Crystal Structure*, Cornell University Press, Ithaca, NY, 1966, pp. 178–189.

Nordman, C., and L.-Y. R. Hsu, in *Computational Crystallography*, D. Sayre, Ed., Clarendon Press, Oxford, 1982, pp. 141–149.

Proceedings of the symposium on machine interpretations of Patterson functions and alternative direct approaches, *Trans. Am. Cryst., Assoc.*, **2**, 1–45 (1966).

Rossmann, M. G., Ed., *The Molecular Replacement Method*, Gordon and Breach, New York, 1972.

CHAPTER 15

COMPLETING THE STRUCTURE

Although it is sometimes possible to deduce the location of all atoms in the unit cell during the initial stages of a structure determination—for example, by finding them in an "E-map" or a superposition function—it is not common. Instead the structure is usually developed to the stage of a phasing model by one of the methods described previously. This model is used for the calculation of structure factors (F_c), and these serve as a source of phases for Fourier calculations. In this chapter we shall be concerned with ways in which such calculations can be used to complete the phasing model in preparation for systematic refinement.

A reasonable question at this point is how much of the structure must be in hand before Fourier methods have a satisfactory probability of revealing the remainder. As usual, there is no simple answer, but a structure not containing heavy atoms can generally be completed without excessive difficulty if 50–75% of the electron density is located within an average error of about 0.3 Å. A smaller fraction of the electrons is required if it is concentrated in one heavy atom. To a certain extent the size of the model required is inversely related to the time and thought one is prepared to devote to interpreting the early results and following possible leads. The process is likely to be somewhat easier for centrosymmetric than noncentrosymmetric space groups for comparably good models, but the latter may allow solution and refinement from bad models that would fail with the former.

15.1. F_o SYNTHESIS

The simplest approach to structure completion involves merely the combination of the observed structure amplitudes with the calculated phases and the use of these as the coefficients in a Fourier series. Two questions, however, arise with this process: the first is that of selecting the terms to be used in the synthesis, and the second has to do with the interpretation of the results.

Selection of Data

With regard to the selection of data, it is quite possible to use all of the reflection information at hand as soon as a phasing model is available. Many structures have been solved this way. It may not, however, represent the most rapid or the most efficient method.

As pointed out earlier (Section 9.2), the bulk of the contributions to the Fourier sum come from the stronger reflections. Thus if only F_o syntheses are to be prepared from a cycle of structure factor calculations, weaker reflections (especially unobserveds) can be omitted with very little effect on the resulting map.[1]

More systematic approaches, which are applicable to other Fourier series as well, involve selecting only reflections that lie within a specified distance of the r.l. origin,[2] or those for which $|F_c|$ is at least some specified fraction of $|F_o|$.

The use of reflections limited in $(\sin\theta)/\lambda$ has several real advantages. Since the number of reflections increases with $[(\sin\theta)/\lambda]^3$, it is possible to effect a sizable reduction in the number of terms used at a relatively small expense in resolution. Thus cutting off the data at $(\sin\theta)/\lambda = 0.4$ reduces the number of reflections included to about 25% of those that can be measured within the copper sphere ($[(\sin\theta)/\lambda]_{max} = 0.65$). At the same time the resolution is 1.25 Å, quite adequate to separate carbon atoms in three dimensions.

Another virtue of data limited in this way is that the reflections that remain are relatively less sensitive to small errors in the coordinates of the atoms of the phasing model (see the discussion of Bragg–Lipson charts, Appendix D). Thus while they are not particularly well suited to the final refinement of positional parameters, they are more likely to have calculated phases close to the true ones than are those near the outer limit of the measurable data.[3]

[1]Note that this is not true if ΔF series are to be used (see below). Since these represent a very powerful tool, it is often preferable to calculate the structure factors for all reflections and make any selection before or during the Fourier calculation.

[2]This is usually specified in terms of a maximum value of $(\sin\theta)/\lambda$ or $(\sin\theta)^2/\lambda^2$ or a related minimum spacing (resolution) for the reflecting planes; see Section 2.4.

[3]This is not necessarily true in the case of a phasing model consisting of a heavy atom, since this atom can be located very accurately from the Patterson map and since its fractional contribution to a reflection tends to increase with $(\sin\theta)/\lambda$. On the other hand, the smaller contribution made by the light atoms (because their scattering factor curves fall off more rapidly) means that the high-order reflections will in turn yield relatively little information about the light atom positions when used in a Fourier series.

Imposing such a cutoff on the data affects the resulting synthesis in a way similar to having atoms with abnormally high temperature factors. Their peak densities are lowered and their volumes are increased. This causes them to be somewhat less well resolved from the background but does not usually represent a serious problem. On the other hand, the greater effective size of the atoms means that they will be revealed by a Fourier synthesis calculated with a coarser grid than is generally used. Thus with $[(\sin\theta)/\lambda]_{max} = 0.4$, a 0.5-Å grid is quite adequate.

The question of selecting reflections in terms of the agreement between $|F_c|$ and $|F_o|$ is more complex. Clearly if $|F_c| = 0$, it contributes no useful phase information for application to $|F_o|$. On the other hand, if the synthesis were computed only with those reflections for which $|F_c| \approx |F_o|$, the result would look very much like the initial model[4] and would give little or no additional information. The success of the Fourier approach at this stage depends largely on the fortunate combination of correct phases with $|F_o|$ values that differ significantly from the corresponding $|F_c|$ values. Many structures have been solved by making no attempt to select reflections in terms of the probable correctness of their phases, and this is certainly the simplest course. On the other hand, it can be argued that the signal-to-noise ratio in the resulting synthesis can be improved by omitting at least those coefficients that appear to have nearly random phases. If such a cut is to be made, it is probably better to exclude only the worst cases and take those for which $|F_c| > x|F_o|$, with x between 0.1 and 0.3.

A more sophisticated analysis[5,6] of the partial model (heavy atom) case gives for F_H (= F_c for the heavy atom) and $X = (2|F_o||F_H|)/\sum f_L^2$

$$w = \frac{I_1(X)}{I_0(X)} \tag{15.1}$$

where I_0 and I_1 are zero- and first-order modified Bessel functions.

Interpretation of Results

The property of the F_o Fourier analysis that concerns us at the moment is its ability to reveal probable atomic locations that were not included in the phasing model. These new atoms do not, unfortunately, appear as strongly as do those of the phasing model,[7] so their identification can sometimes be difficult. This is particularly true in noncentrosymmetric structures, for

[4] Note that this is now equivalent to a Fourier synthesis using both calculated phases and calculated magnitudes (F_c Fourier) and will give back just the model used in the structure factor calculations.

[5] G. A. Sim, *Acta Cryst.*, **12**, 813 (1959); **13**, 511 (1960).

[6] M. M. Woolfson, *Acta Cryst.*, **9**, 804 (1956).

[7] V. Luzzati, *Acta Cryst.*, **6**, 142 (1953).

which the errors in the model have a much greater tendency to produce errors in the phase than in centrosymmetric structures. An approximate rule is that in the latter a peak can reasonably be taken as an atomic site if it appears with a peak density about 0.5 that of a corresponding atom already in the model. For the noncentrosymmetric case the factor is about 0.3.

The figures given above are conservative, and it is often possible to recognize atomic peaks with much lower densities if there is other evidence to point to their locations. Thus if no peaks meeting these criteria are available, the most intense ones found should be considered. These may be tested for reasonableness by inquiring whether they are related to the model peaks by a sensible molecular geometry. In addition, true atomic peaks tend to have a more or less spherical distribution of density near the center, although the outer limits may be quite ragged. This rule is not inviolable, but any peak that spreads in a grossly irregular manner should be regarded with suspicion.

15.2. ΔF SYNTHESIS

Introduction and Selection of Data

Another extremely valuable approach to finding atoms lies through the ΔF or difference synthesis.[8] This is a Fourier synthesis that can be expressed in the form

$$\Delta\rho = \frac{1}{V}\sum\sum\sum(|F_o| - |F_c|)e^{i\alpha_c}e^{-2\pi i(hx+ky+lz)} \tag{15.2}$$

where α_c is the phase of F_c (see Section 9.3).

The difference synthesis has two major virtues. First, if the phases α_c equal α_o, it provides a direct measure of the errors between the model used and the true structure implied by the $|F_o|$ values. This property is highly useful for refinement and is considered further in Chapter 16. Second, and more important for our present needs, even when the phases are moderately in error the ΔF synthesis can provide useful information that is difficult or impossible to obtain from F_o syntheses.

This particular advantage of the ΔF synthesis arises from the fact that it permits the selection and use of a partial set of data for which the probabilities are especially high that the assigned phases are close to correct. Furthermore, this partial set proves to be particularly rich in useful information, although it cannot be be extracted by an F_o synthesis. The fundamental arguments behind these assertions may be seen by considering three limiting cases of $|F_o|$ and $|F_c|$ (Fig. 15.1). In the first (Fig. 15.1a),

[8] W. Cochran, *Acta Cryst.*, **4**, 408 (1951).

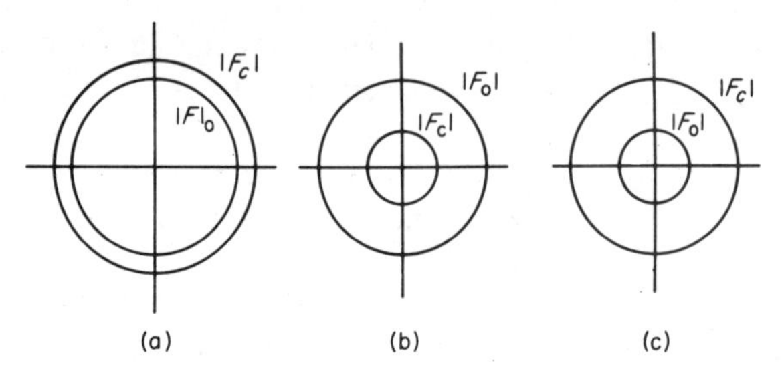

Figure 15.1. (*a*) $|F_o| \approx |F_c|$. (*b*) $|F_o| > |F_c|$. (*c*) $|F_c| > |F_o|$.

$|F_o| \approx |F_c|$. If the phasing model is a suitable one, there is a high probability that α is approximately the value corresponding to the true phase of F_o. This is the assumption on which the F_o synthesis is based. On the other hand, as was pointed out earlier, the reflections for which $|F_o| \approx |F_c|$ tend merely to reproduce the model and add little new information. Furthermore, if α differs from the true value, a large $|F_o|$ can introduce serious errors into the resulting synthesis. Thus the use of these reflections offers at this stage relatively little gain and the possible risk of large distortions. In the difference synthesis, however, $|F_c| - |F_o| \approx 0$, so the effects of these reflections are automatically minimized.

The informative cases, those for which $|F_o| \neq |F_c|$, are considered in Figs. 15.1*b* and *c*. If $|F_o| \gg |F_c|$ (Fig. 15.1*b*), $|F_o|$ will make a significant contribution to the F_o Fourier series and should provide useful information. The probability of correspondence between α_c and α_o, however, is small and diminishes as $|F_c|$ approaches zero. Thus, these reflections are unreliable even though theoretically useful, and it was for this reason that a rejection test based on $|F_c|$ and $|F_o|$ magnitudes was considered above. In the difference synthesis, $|F_o| - |F_c|$ will be large and contribute to the summation, but the phase uncertainty is still present and can again provide grounds for rejection (see below).

The third and most interesting case is $|F_c| \gg |F_o|$ (Fig. 15.1*c*). Although these reflections carry information about the disagreement between the model and the true structure, it cannot be obtained by an F_o synthesis because they are too weak to have much effect on the summation. $||F_o| - |F_c||$, on the other hand, will be large and will be a significant contributor to the difference synthesis. Furthermore, it can be shown that these terms will be more or less correctly phased *regardless* of differences between α_c and α_o.

The truth of the preceding statement can be seen from Fig. 15.2. The coefficients that are actually desired for the difference syntheses are

$$\Delta \mathbf{F} = \mathbf{F}_o - \mathbf{F}_c \tag{15.3}$$

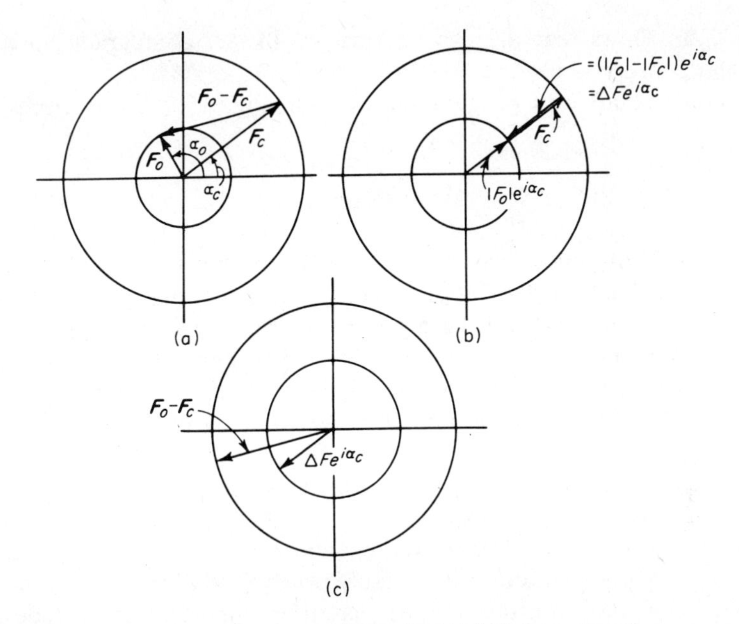

Figure 15.2. (a) Vectors for the case $|F_c| > |F_o|$. (b) Vectors for the same case if α_o is taken as equal to α_c. (c) Comparison of vectors $\mathbf{F}_o - \mathbf{F}_c$ and $\Delta F e^{i\alpha_c}$.

or

$$|\Delta F| e^{i\alpha_\Delta} = |F_o| e^{i\alpha_o} - |F_c| e^{i\alpha_c} \tag{15.4}$$

These expressions correspond to Fig. 15.2a and merely state that the coefficients are the vectors required to correct the current F_c to the true F_o. These vectors are just the structure factors that would result from adding electron density where it is needed and subtracting it where it is in excess. As usual, however, α_o is unknown; so the assumption is made (Fig. 15.2b) that

$$\alpha_o \approx \alpha_c \tag{15.5}$$

$$|\Delta F| e^{i\alpha_\Delta} \approx |F_o| e^{i\alpha_c} - |F_c| e^{i\alpha_c} \tag{15.6}$$

$$|\Delta F| e^{i\alpha_\Delta} \approx (|F_o| - |F_c|) e^{i\alpha_c} \tag{15.7}$$

$$|\Delta F| e^{i\alpha_\Delta} \approx \Delta F e^{i\alpha_c} \tag{15.8}$$

Since we are considering at present only the reflections for which ΔF is always negative, the approximation actually used is

$$|\Delta F| e^{i\alpha_\Delta} \approx -|\Delta F_c| e^{i\alpha_c} \tag{15.9}$$

Figure 15.2*c* shows that the two vectors are in fact comparable in magnitude and phase.

This construction can be generalized as in Fig. 15.3 by rewriting Eq. (15.3) as

$$\Delta\mathbf{F} = -\mathbf{F}_c + \mathbf{F}_o \tag{15.10}$$

Thus the end of $\Delta\mathbf{F}$ must always lie on a circle of radius $|F_o|$ described about the end of $-\mathbf{F}_c$. The smaller this circle, the smaller the range over which α_Δ will vary as α_o changes, and the more nearly Eq. (15.9) will approach an equality. Furthermore, as may be seen from the figure, the approximation used has the most probable phase (for random α_o) and is conservative in its value for $|\Delta F|$. Thus there is no tendency to exaggerate the error and overshoot.

It should be clear from these arguments that the most valuable reflections for use in difference syntheses are those for which $|F_o|$ is very small and $|F_c|$ very large. Unobserved reflections meet the first requirement, and in an incomplete structure it is highly probable that some of them will calculate large. It is for this reason that it is important to include unobserved reflections in the data set.

The selection of other reflections to be used in the synthesis is a more open problem. As $|F_o|$ approaches $|F_c|$, the uncertainty of α_Δ increases for random α_o. On the other hand, the probability also increases that α_c and α_o are correlated while ΔF_c decreases, so no serious errors are expected for a reasonably correct model. When $|F_o|$ becomes larger than $|F_c|$, however, the magnitude of $|\Delta F|$ increases again while the approximation to α_Δ becomes less reliable. For these reasons, the original difference syntheses ("error syntheses")[9] made use of only those reflections for which $|F_c| > |F_o|$ and rejected all the others. Such a scheme limits the number of usable

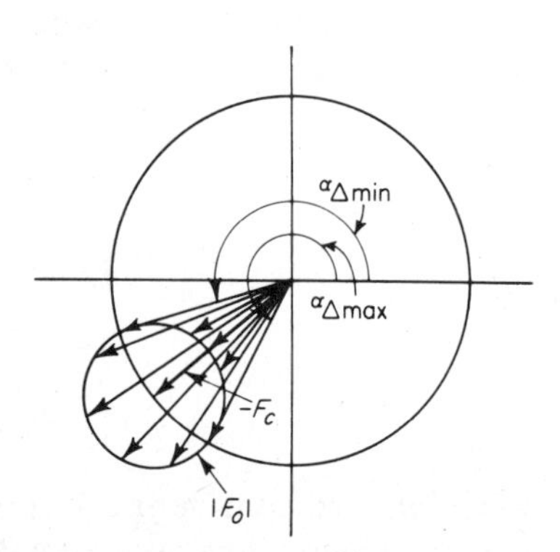

Figure 15.3. Construction showing range of α.

[9] D. Crowfoot, C. W. Bunn, B. W. Rogers-Low, and A. Turner-Jones, *The Chemistry of Penicillin*, Princeton University Press, Princeton, NJ, 1949, p. 310.

reflections very severely, especially for incomplete structures, and in view of the success that has been obtained with F_o syntheses and the usual phasing assumptions appears unnecessarily restrictive.

A generalized alternative would be to use here the same cutoff that was suggested earlier, that is, to reject those reflections for which $|F_c| < x|F_o|$. Again, no systematic study has been made of the effects of varying x, but it is unlikely that the rate of refinement is highly sensitive to the value chosen. Good success has been had with $x = 0.5$, but this represents a quite arbitrary choice.

Another view of this problem is that the extent of contribution of ΔF_c to the synthesis should reflect in some way the probability of the correctness of its associated phase. Various weighting schemes are possible to produce this result, of which the simplest is merely to use as coefficients $w\,\Delta F_c$, where

$$w = \frac{|F_c|}{|F_o|} \tag{15.11}$$

This scheme tends to apply excessive weight to unobserved reflections that calculate too large and to cause gross ripples in the resulting map. Consequently it is often useful to set $w = 1$ for $|F_c| > |F_o|$ and to apply Eq. (15.11) only when $|F_o| > |F_c|$.

The limitation of data by $(\sin\theta)/\lambda$ can be applied to ΔF as well as F_o syntheses. It is not so necessary in one sense because the phases of a ΔF synthesis are more reliable than those of the corresponding F_o Fourier series. On the other hand, the results may be better because the ΔF calculation reduces many of the false maxima that are associated with termination of series errors (see Section 16.2) and that can be particularly noticeable with truncated data.

Interpretation of Results

As mentioned in Chapter 9 and discussed in more detail in Chapter 16, the difference synthesis corresponds to a point-by-point subtraction of an F_c Fourier map from an F_o Fourier map calculated with the same phases. Thus the difference map has a peak everywhere the F_c model fails to provide the electron density implied by the $|F_o|$ data (when taken with the F_c phases) and a hole where it provides too much. As a first approximation, therefore, correctly placed atoms will not appear in the synthesis, incorrectly placed ones will be in holes, and missing ones will appear as peaks. These peaks should correspond to ones coming up in the F_o map, but they are usually much more distinct and obvious. This occurs not only because of better average phasing of the difference map, but also because these new atomic peaks are not overshadowed by much larger ones at the already occupied locations.

If all the data are used for the difference map and if the ΔF_{000} term is included, the peak difference density $\Delta\rho$ for missing atoms should ideally be the same as the corresponding peak heights ρ in the F_o Fourier map. Thus the selection rules suggested in Section 15.1 can be applied in this case as well. If a weighted or truncated set of coefficients is used, the equivalent peak height for a full atom is more difficult to estimate. Consequently it is necessary to select atomic peaks on the basis of their relative sizes, positions, and shapes. Usually, however, they are sufficiently obvious that this poses no difficulties.

The difference synthesis can also be used as a guide to the removal of atoms that are incorrectly placed. Several points must be considered, however. The most important of these is that many such syntheses return perfectly correct atoms in holes that may have minima of 1 e/Å^3 or more. Thus the mere appearance of such holes should not lead automatically to the conclusion that the particular atoms are incorrectly placed, or worse, the whole structure is wrong. These minima appear for various reasons. The simplest, and perhaps the most common, is failure to include a ΔF_{000} term in syntheses for which the model is less than the whole structure. Another is the use of weighting or cutoff functions that reduce the contributions from reflections for which $|F_c| < |F_o|$. The effect of these functions is to make the effective average value of $|F_c|$ too large and the corresponding calculated electron density too high. This excess density is reflected as a hole in the ΔF map.

Physical reasons for the appearance of atoms in holes are the use of temperature factors that are too low, which increases the calculated peak density over what it should be, and the misidentification of atoms. If a carbon atom, for example, is thought to be an oxygen and oxygen scattering factors are used, the two extra electrons will usually cause a significant minimum at the atom site. For this reason it is advisable to call all carbon, nitrogen, or oxygen atoms in a structure carbon during the early stages and change the designation only when the identification becomes unambiguous (see Section 16.4).

For all these reasons, the "zero point" for real atomic positions is best estimated from the average minimum values found for the atoms of the model rather than from the $\Delta\rho = 0$ contour of the synthesis. Values that are much lower than this average are then suspect. Unfortunately, the presence of an atom in the model tends to bias the phasing to favor it, so a false one will rarely appear in a hole equal to its usual peak density. Instead the factors suggested for evaluating positive peaks in an F_o map, 0.5 for centric and 0.3 for acentric, are often applicable. Again, the use of weighted coefficients will affect these values, so comparisons should be made against positive peaks in the same map. Such prominent holes should be expected, however, only for individual wrong atoms in a structure that is essentially correct. If the errors are numerous, and particularly if they are systematic, the average holes will just be deepened, and it will often be impossible to single out a specific atom for rejection.[10]

15.3. OTHER FOURIER SYNTHESES

Various other Fourier syntheses using modified coefficients have been proposed for use with incomplete structures.[11] Of these the one most commonly used, especially by macromolecular crystallographers, has coefficients given by

$$(2|F_o| - |F_c|)e^{i\alpha_c} \tag{15.12}$$

By exactly the same arguments as those used for the ΔF synthesis, this corresponds to a point-by-point summation of an F_o and a ΔF synthesis.[12] It has the advantage of showing the input model as well as the implications of the difference synthesis, so it is easier to visualize how additional structure can be fitted to the structure already present. In addition, it is suggested that since both the F_o and ΔF syntheses return new atoms at some fractional height, the combination will be the sum and thus more easily recognized. This is probably not as effective in practice as is suggested, and it must be balanced against the lesser contrast as compared with a pure ΔF synthesis and the greater phase uncertainty introduced by the F_o synthesis. The choice is likely to be one of personal taste and the details of the problem, but in general we prefer to compute separate F_o and ΔF maps and then plot them together if desired (see below).

Variations on this synthesis such as $3|F_o| - 2|F_c|$ have been proposed[13] in which the contribution of $|F_o|$ is reduced so that the results tend more and more to a pure ΔF synthesis.

15.4. STRUCTURE COMPLETION IN PRACTICE

The principal techniques that are used in this part of a structure analysis have already been discussed. During the process it is often difficult to give any quantitative measure of how well it is going. R should decrease steadily from its starting value, but the change is often slow while any atoms are

[10]It is just this sort of behavior that precludes the attractive but unsuccessful phasing approach of distributing a suitable number of atoms in the cell and then removing, with the aid of difference syntheses, those that do not correspond to the true structure.

[11]See G. N. Ramachandran and R. Srinivasan, *Fourier Methods in Crystallography*, Wiley, New York, 1970, Chapter 7, and P. Main, *Acta Cryst.*, **A35**, 779 (1979).

[12]S. T. Freer, R. A. Alden, C. W. Carter, and J. Kraut, *J. Biol. Chem.*, **250**, 46 (1975).

[13]For discussions, see D. W. Rice, *Acta Cryst.*, **A37**, 491 (1981) and M. Vijayan, *Acta Cryst.*, **A36**, 295 (1980).

missing from the model. Probably the best guide is the continuing appearance of new atoms at chemically reasonable positions and the development of a consistent molecular whole.

A very convenient practical approach to searching for new atoms consists of contouring the F_o and ΔF syntheses simultaneously on the same set of plastic sheets (Section 9.2), using one color for the ρ contours and two others for the plus and minus $\Delta\rho$ contours. The F_o synthesis then provides a map of the current model, while the ΔF reinforces the suggested new atoms. Figure 15.4 shows an example of the comparative results of the two functions.

Once a new atom has been found, the position of its maximum density can be picked off either by inspection or by more accurate curve-fitting methods (Section 16.2). Usually this is more easily done from the ΔF map, just because the peaks are more clearly defined.

Even while the structure is being completed, it is possible to refine the model by using the Fourier techniques discussed in Chapter 16. One case

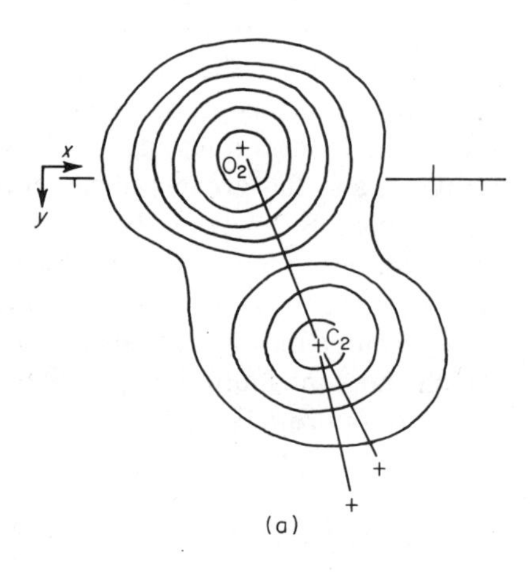

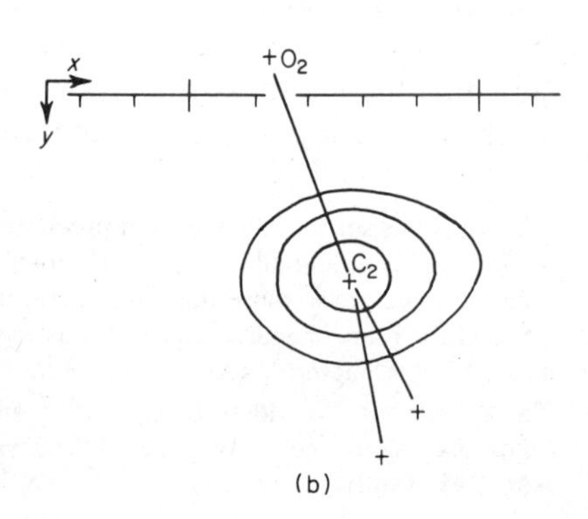

Figure 15.4. Sections of (*a*) F_o and (*b*) ΔF syntheses through an atom (C_2 omitted in the calculation of F_c's). Contours at 1 e/Å^3. Lines show the directions of bonds, the lowest two atoms being out of this section.

that falls between normal refinement and those described above is that of an atom that is misplaced by several tenths of an angstrom. Such an atom is usually in a hole on the difference map but not so badly as to require rejection, and it may be on a gradient that suggests a corrective move in a wrong direction. Nearby, at what will ultimately prove to be the right location, is usually a peak, obviously too close to be another atom. If such a situation persists through several cycles, the best move is to shift the atom to the peak and see if matters are improved. If the situation is sufficiently bad that it is not clear which atom should be moved and where, it is often very effective to remove two or three of the most confusing ones and see where they reappear in the next cycle of calculations. Shifts of an angstrom or so that cannot be seen with the atoms in place can be effected in this way, particularly with data limited in $(\sin\theta)/\lambda$.

After the set of new atoms has been selected and desired changes made for the parameters of the old ones, structure factors are recalculated and the process is repeated. A question that arises at this point is whether to rescale the $|F_o|$'s to match the $|F_c|$'s (see Section 9.1). Although a spurious improvement in R can sometimes be achieved in this way, it is better not to adjust the scale until all the atoms are in place. An exception can be made, however, if the scale is obviously in serious error. A fairly good approximation to the correct rescale value can be obtained for an incomplete model from

$$(\mathrm{LRS})_{\mathrm{incomp}} = \sum |F_c| \Big/ \sum |F_o| \qquad (15.13)$$

$$(\mathrm{LRS})_{\mathrm{true}} = (\mathrm{LRS})_{\mathrm{incomp}}/\sqrt{p} \qquad (15.14)$$

where p is the fraction of total cell electrons in the current model.

Particularly if the structure completion is carried out together with partial refinement, the structure factor calculations immediately following the addition of the last atoms will often show a precipitous drop in R. If it does not appear at this time, it will normally occur after a cycle or two of refinement. It is the sight of this sudden drop, often 0.05–0.10, rather than the attainment of a specified figure, that is the sign to the crystallographer that the refinement is on the right path and that the struggle is nearly over. This phenomenon tends to be more marked with centrosymmetric crystals and does not invariably appear, so its absence need not be interpreted as a sign of failure. Some structures, particularly acentric ones that start from poor phasing models, merely plod steadily downward in the refinement without any particular display of life.

On the other hand, it occasionally happens that a structure initially is well behaved but refuses to refine below an R of 0.20–0.30. At the same time it fails to show any signs of obvious errors such as misplaced atoms.

Such a result often means that the structure is wrong crystallographically in some respect even though the model may be quite correct chemically. The extent of the remaining errors will determine the point at which refinement ceases, and various examples and possible causes are considered in Section 16.10.

BIBLIOGRAPHY

Blundell, T. L., and L. N. Johnson, *Protein Crystallography*, Academic, London, 1976, pp. 381–400, 404–419.

PART III

REFINEMENT AND RESULTS

CHAPTER 16

REFINEMENT OF CRYSTAL STRUCTURES

In the preceding chapters we have dealt with different methods by which a satisfactory phasing model can be obtained and ways in which atoms can be added to it to complete the structure. The refinement stage of a structure analysis will be assumed to begin with a completed trial structure containing all the atoms. Just as no single value of the R factor ensures the correctness of a phasing model, so no one value can be taken as the point below which successful refinement is ensured. Usually, however, a structure can be refined smoothly beginning with R values in the range 0.3–0.4, though much higher values have served as successful starting points. On the other hand, structures with lower R values may possess quite wrong features and fail to refine satisfactorily.[1]

16.1. INTRODUCTION TO WEIGHTING

Much of the art, and probably the science, of proceeding from the first glimmerings of a structure to a detailed final result, fully refined and exhibiting all of the information the native data provide, lies in the choice of proper weights for the data at various stages. At each step the information content of each datum should be judged in terms of its estimated reliability and the degree to which it differs from that predicted by the current model. Those that are both reliable and different provide the information that will point to the next stage of improvement. Successful weighting is the means of favoring just these data.

[1]For an example, see J. Donohue and K. N. Trueblood, *Acta Cryst.*, **9**, 615 (1956).

Weighting is often just an either–or selection process, especially in the early stages of structure development. Thus, for direct methods, only those reflections with large E values are chosen, both because they are strongly different from the average and because they are (one hopes) reliably phased. For the same reasons, in developing the structure, unobserved reflections are chosen for difference syntheses if they calculate large. At the same time, strong reflections that calculate small are often rejected because although they are different the phases are very unreliable.

In the stages of developing the structure and even in the early parts of refinement, the major uncertainty is in the phases. Unfortunately, phase accuracy is difficult to estimate a priori since there is little information about the true phase to measure against. In some cases, such as single isomorphous replacement or phasing from anomalous scattering (see Sections 13.2 and 13.3), the phase probably has one of two values, and a figure of merit derived from the spread between these can be used to weight the contributions to Fourier maps. In most, however, the only evidence of phase accuracy is that judged from the agreement of $|F_o|$ and $|F_c|$. Unfortunately, this is not an entirely reliable guide, and new information tends to zero just as the apparent phase consistency becomes greatest.

Once the structure model has been completed, the presumption is that the phases have largely settled in the vicinity of their final values. There are still likely to be shifts of tens of degrees in some cases, but these are expected to be within the range of normal refinement techniques and no longer to affect the weighting of the data. Now attention turns to getting the best possible fit between the calculated structure factors and the *true* ones for the real crystal. Unfortunately we do not have these ideal measurements, only the observed ones, which contain a collection of measurement errors. Since we must make do with them, the appropriate technique is to weight them individually with our best estimate of their reliability, usually obtained from an analysis of the errors of measurement (see Chapter 7 and Appendix E). Thus large differences are less significant for those measurements whose accuracy we doubt than for those we trust.

16.2. REFINEMENT BY FOURIER SYNTHESES

F_o Syntheses

Although F_o syntheses are not as suitable for refinement as the other methods to be described and cannot be used to refine scale and thermal parameters,[2] nevertheless if atoms are resolved, fairly accurate positional parameters for the atoms can be determined from them. The electron

[2]Scale and thermal parameters are dealt with by comparison between $|F_o|$ and $|F_c|$ as explained in Chapter 9.

density ρ_o is evaluated at the points of an appropriate grid using Eq. (8.47), and the coordinates of the density maxima are determined and used as a basis for a new structure factor calculation. Successive cycles are calculated until there are no significant changes in the parameters or, if the structure is centrosymmetric, until there are no changes in the signs of the F_c's.

The coordinates of the electron density maxima can be located roughly by inspection or graphical interpolation. A more accurate method is based on the observation that the electron density in the vicinity of the maximum is approximately Gaussian.[3] Schoemaker et al.[4] suggested the function

$$\rho = Ae^{-(ax^2+by^2+cz^2+dxy+exz+fyz+g)} \tag{16.1}$$

and used it to fit the electron density at $3 \times 3 \times 3$ grids of points surrounding the peak maxima. Although the calculations are time-consuming if done by hand,[5] they are easily programmed for rapid computation, and peak-locating routines are available in several crystallographic program packages.

Differential Syntheses

Instead of computing ρ_o at a series of fixed grid points, Booth[6] has shown how the *corrections* in atomic positions can be derived from a knowledge of the first derivative or slope of the electron density function at the assumed atomic positions and the second derivative or curvatures of the function at the atomic centers. An approximate relationship between the slope and curvature can be found by expressing the slopes $\partial\rho/\partial x$, $\partial\rho/\partial y$, $\partial\rho/\partial z$ as a Taylor series and neglecting second and higher powers. Thus the slopes at the atomic centers can be written

$$\begin{aligned}
\frac{\partial\rho}{\partial x} &= \frac{\partial\rho}{\partial x_a} + \frac{\partial^2\rho}{\partial x_a^2}(x - x_a) + \frac{\partial^2\rho}{\partial x_a\,\partial y_a}(y - y_a) + \frac{\partial^2\rho}{\partial x_a\,\partial z_a}(z - z_a) \\
\frac{\partial\rho}{\partial y} &= \frac{\partial\rho}{\partial y_a} + \frac{\partial^2\rho}{\partial x_a\,\partial y_a}(x - x_a) + \frac{\partial^2\rho}{\partial y_a^2}(y - y_a) + \frac{\partial^2\rho}{\partial y_a\,\partial z_a}(z - z_a) \qquad (16.2) \\
\frac{\partial\rho}{\partial z} &= \frac{\partial\rho}{\partial z_a} + \frac{\partial^2\rho}{\partial x_a\,\partial z_a}(x - x_a) + \frac{\partial^2\rho}{\partial y_a\,\partial z_a}(y - y_a) + \frac{\partial^2\rho}{\partial z_a^2}(z - z_a)
\end{aligned}$$

[3]W. Costain, Ph. D. thesis, University of Birmingham, 1941, cited by A. D. Booth, *Proc. Roy. Soc.* (*London*), **188A**, 77 (1946).

[4]D. P. Shoemaker, J. Donohue, V. Schomaker, and R. B. Corey, *J. Am. Chem. Soc.*, **72,** 2328 (1950).

[5]See B. Dawson, *Acta Cryst.*, **14**, 999 (1961) for a modification more suitable for hand calculation.

[6]A. D. Booth, *Trans. Faraday Soc.*, **42**, 444 (1946).

where the subscript a refers to values of these quantities at the assumed positions of the atoms. But the slope of the electron density at the true atomic center is zero; so Eqs. (16.2) become

$$\begin{aligned}
\frac{\partial^2 \rho}{\partial x^2}\Delta x + \frac{\partial^2 \rho}{\partial x\,\partial y}\Delta y + \frac{\partial^2 \rho}{\partial x\,\partial z}\Delta z &= -\frac{\partial \rho}{\partial x} \\
\frac{\partial^2 \rho}{\partial x\,\partial y}\Delta x + \frac{\partial^2 \rho}{\partial y^2}\Delta y + \frac{\partial^2 \rho}{\partial y\,\partial z}\Delta z &= -\frac{\partial \rho}{\partial y} \\
\frac{\partial^2 \rho}{\partial x\,\partial z}\Delta x + \frac{\partial^2 \rho}{\partial y\,\partial z}\Delta y + \frac{\partial^2 \rho}{\partial z^2}\Delta z &= -\frac{\partial \rho}{\partial z}
\end{aligned} \tag{16.3}$$

where the subscript a has been dropped because all quantities are to be evaluated at the assumed positions of the atoms and Δx, Δy, and Δz are the corrections to be applied to the assumed positions. Although these equations are not exact, they are good approximations if Δx, Δy, and Δz are no larger than 0.1–0.2 Å. If the partial derivatives in Eqs. (16.3) are evaluated and substituted therein, one such set of equations is obtained for each of the m atoms in the asymmetric unit. Solution of these sets of equations yields the corrections to the assumed atomic positions.

The slopes and curvatures can be obtained from the electron density function written in the form (see Section 8.9)

$$\rho_o = \frac{1}{V}\sum_h \sum_k \sum_l |F_{o,hkl}| \cos 2\pi(hx + ky + lz - \alpha'_{hkl}) \tag{16.4}$$

For the slope at the assumed position of the qth atom,

$$\frac{\partial \rho}{\partial x_q} = \frac{-2\pi}{V}\sum_h \sum_k \sum_l h|F_{o,hkl}| \sin 2\pi(hx_q + ky_q + lz_q - \alpha'_{hkl}) \tag{16.5}$$

and similar expressions for $\partial\rho/\partial y_q$ and $\partial\rho/\partial z_q$. For the curvatures,

$$\begin{aligned}
\frac{\partial^2 \rho}{\partial x_q^2} &= \frac{-4\pi^2}{V}\sum_h \sum_k \sum_l h^2|F_{o,hkl}| \cos 2\pi(hx_q + ky_q + lz_q - \alpha'_{hkl}) \\
\frac{\partial^2 \rho}{\partial x_q\,\partial y_q} &= \frac{-4\pi^2}{V}\sum_h \sum_k \sum_l hk|F_{o,hkl}| \cos 2\pi(hx_q + ky_q + lz_q - \alpha'_{hkl})
\end{aligned} \tag{16.6}$$

and similar expressions for the other second partial derivatives. In each of Eqs. (16.4)–(16.6) it is to be understood that summation is over *all* planes h, k, l within the limiting sphere.

If the electron density is spherically symmetric and the axes orthogonal, the mixed partial derivatives in Eqs. (16.3) are zero. In this case the corrections to the coordinates of the qth atom are simply

$$\Delta x_q = \frac{-\partial\rho/\partial x_q}{\partial^2\rho/\partial x_q^2}; \qquad \Delta y_q = \frac{-\partial\rho/\partial y_q}{\partial^2\rho/\partial y_q^2}; \qquad \Delta z_q = \frac{-\partial\rho/\partial z_q}{\partial^2\rho/\partial z_q^2} \tag{16.7}$$

These equations are approximately correct for atoms in most structures with interaxial angles differing by not more than 10–15° from a right angle.

The computational effort involved in evaluating the partial derivatives in Eqs. (16.5) and (16.6) is not as great as may at first appear since, in general, all trigonometric terms in the derivatives have already been computed in the structure factor calculation. It is merely necessary to sum the terms after multiplying by the appropriate quantities. However, even less computational effort is required by following Booth's suggestion of evaluating the curvatures from the symmetric Gaussian function

$$\rho(r) = Ae^{-pr^2} \tag{16.8}$$

where r is the distance from the atomic center and A and p are adjustable constants. The parameters p may be evaluated by equating the integral over all space of the right-hand side of Eq. (16.8) to the number of orbital electrons, N, equal to the atomic number Z except in the case of ions. Expressing this as an integral we have

$$A\int_0^\infty 4\pi r^2 e^{-pr^2}\, dr = N \tag{16.9}$$

from which

$$p = \pi(A/N)^{2/3} \tag{16.10}$$

Substitution in Eq. (16.10) of the appropriate value of N and the observed peak density A from an F_o synthesis yields the value of p.[7] The curvature is then estimated from the second derivative of Eq. (16.8),

$$\frac{d^2\rho}{dr^2} = -2pAe^{-pr^2}(1-2pr^2) \tag{16.11}$$

[7]The parameter p can also be evaluated graphically from the ρ_o map. By taking the logarithms of both sides of Eq. (16.8) we obtain

$$2.303 \log \rho(r) = 2.303 \log A - pr^2$$

and

$$\log \rho(r) = \log A - (p/2.303)r^2$$

Hence a plot of log $\rho(r)$ versus r^2 will be linear, and p is equal to -2.303 times the slope of the line.

TABLE 16.1 Peak Densities A, p, and Curvatures for C, N, and O Atoms

Atom	A, e/Å^3	p, Å^{-2}	Curvature, e/Å^5
C	7	3.50	−49
N	9	3.72	−67
O	11	3.90	−86

Close to the center of light atoms—that is, for r less than 0.1 Å—$2pr^2$ is small with respect to 1 and can be neglected. Furthermore, the value of the exponential is approximately 1. Hence the curvature is nearly constant and can be taken as

$$\frac{d^2\rho}{dr^2} \approx -2pA \approx -2\pi A\left(\frac{A}{N}\right)^{2/3} \tag{16.12}$$

Curvatures determined from Eqs. (16.11) or (16.12) are usually measured in e/Å^5. On the other hand, the slopes $\partial\rho/\partial x$, $\partial\rho/\partial y$, $\partial\rho/\partial z$ given by Eq. (16.4) are in e/Å^3 per unit cell edge. Consequently, to be dimensionally consistent these latter must be divided by a, b, and c, respectively, before they are used in Eq. (16.3) or (16.7). The resulting shifts are then in angstroms and must be converted to fractional coordinate changes.

Representative curvatures for C, N, and O atoms are listed in Table 16.1 along with A and p. The A values were chosen to approximate atoms with thermal B values in the range 3–5 Å^2. The values for p were then derived from those for A by the use of Eq. (16.10) and curvatures estimated using Eq. (16.12).

It should be noted that the values of A and p for a given element vary widely from structure to structure and even within a given structure. The results in Table 16.1 are therefore to be taken only as representative values.

Series Termination Error[8]

As noted in Chapter 8, the electron density resulting from an F_o synthesis does not exactly match that of the structure. The differences are due not only to experimental errors in the observed amplitudes and to errors in the phases, but also to the omission of significant terms beyond the limit of observation. Atomic positions determined either from ordinary F_o synthesis or by the differential method suffer from such errors. When data are collected with Cu K_α radiation for structures with atoms of approximately equal weight, series termination can lead to errors in position as high as 0.02–0.03 Å and to considerably higher values for light atoms in the

[8]A. D. Booth, *Proc. Roy. Soc. (London)*, **188A**, 77 (1946); **190A**, 482 (1947).

presence of those that are much heavier. Thus in the study of β-isoprene sulfone ($C_5H_8SO_2$), Jeffrey[9] found a maximum series termination error of 0.06 Å, mainly attributable to the single S atom in the molecule.

Booth[8] suggested a method of correcting for series termination error (*back-shift method*). After the last refinement cycle, an F_c synthesis is calculated. Because of the effects of series termination, the peaks in this synthesis will not agree exactly with coordinates used for the structure factor calculation. The differences are the termination errors. Since the $|F_o|$ and $|F_c|$ values are nearly identical at this point, the effects of series termination should be essentially the same regardless of which are used as coefficients in a synthesis. Thus the errors observed in the F_c synthesis can be taken as the unknown errors in the F_o synthesis and applied in the opposite sense to the coordinates from the final F_o synthesis. This process gives results nearly free of series termination errors.

Refinement by ΔF Syntheses

Another Fourier method of refining a structure is by means of difference syntheses (see Chapter 15) in which the coefficients are the differences between the observed and calculated structure factors, $\Delta F = F_o - F_c$. The advantages of a difference synthesis can best be seen by writing it in terms of an F_o synthesis,

$$\rho_o(x, y, z) = \frac{1}{V} \sum_h \sum_k \sum_l F_{o,hkl} e^{-2\pi i(hx+ky+lz)} + R \tag{16.13}$$

and an F_c synthesis representing the assumed electron density

$$\rho_c(x, y, z) = \frac{1}{V} \sum_h \sum_k \sum_l F_{c,hkl} e^{-2\pi i(hx+ky+lz)} + R' \tag{16.14}$$

where R and R' are remainders representing the omitted parts of the series. Subtracting the two term by term gives the series

$$\rho_o(x, y, z) - \rho_c(x, y, z) = \frac{1}{V} \sum_h \sum_k \sum_l (F_o - F_c)_{hkl} e^{-2\pi i(hx+ky+lz)} + R - R' \tag{16.15}$$

Since the termination of series errors R and R' are nearly equal, their difference is very nearly zero and the difference synthesis

$$\rho_o - \rho_c = \frac{1}{V} \sum_h \sum_k \sum_l \Delta F_{hkl} e^{-2\pi i(hx+ky+lz)} \tag{16.16}$$

[9] G. A. Jeffrey, *Acta Cryst.*, **4**, 58 (1951).

represents closely the difference between the actual electron density and that of the model used to calculate F_c. Since R and R' virtually cancel, the parameters from ΔF syntheses are essentially free of series termination error.

Difference syntheses have the additional virtue of showing clearly errors in the structure and can be used as a basis for refinement. The effects of errors *solely in the positional coordinates* are illustrated in Fig. 16.1, where the light line shows the idealized electron density of an atom projected on a line through its center. The dashed line shows the position of the assumed atom used in calculating the F_c's, and the heavy line is $\rho_o - \rho_c$ as would be obtained from a ΔF synthesis. It is evident that an error in a coordinate results in a gradient in the ΔF synthesis and that the correction to an atomic coordinate will be in the direction of the gradient, that is, toward a more positive region.

The correction in the x coordinate is given with sufficient accuracy by the equation[10]

$$\Delta x_q = -\frac{\text{slope}}{\text{curvature}} = -\frac{\partial \Delta \rho / \partial x_q}{\partial^2 \rho / \partial x_q^2} \tag{16.17}$$

with similar equations for Δy_q and Δz_q. The proper values for the curvatures can be derived from an F_o synthesis containing only terms corresponding to those used in the ΔF synthesis. In practice, however, the considerable amount of extra computational effort is unwarranted, since the whole process is iterative and accurate values for the curvatures are not required. Thus those listed in Table 16.1 can be used for the atoms tabulated there and modified to larger or smaller values if there is a pronounced tendency for the corrections to be too large or too small, respectively.

To see how positional shifts are obtained from ΔF syntheses, consider

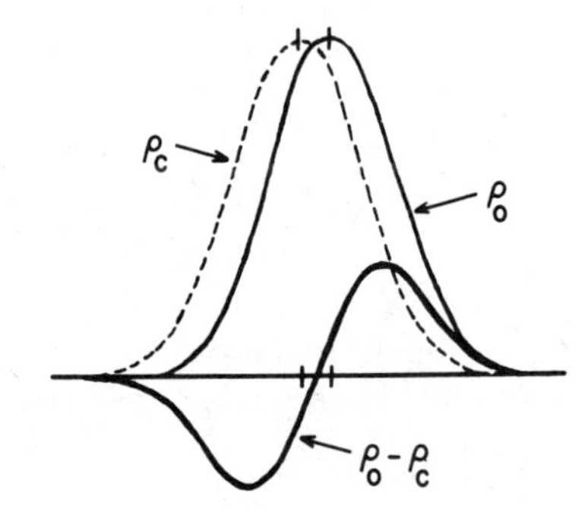

Figure 16.1. Line section through $\rho_o - \rho_c$ synthesis showing effects of an error in position.

[10]Compare Eq. (16.7) and note that with series termination error removed, $\partial \rho_c / \partial x_q = 0$ at the assumed atomic positions, so that $\partial \Delta \rho / \partial x_q = \partial \rho_o / \partial x_q$.

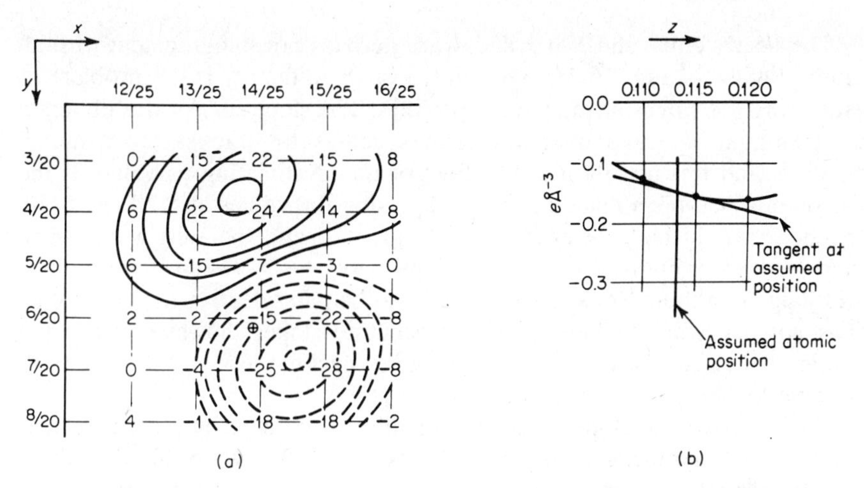

Figure 16.2. (*a*) Part of an *xy* section from a ΔF synthesis for *n*-nonanoic acid hydrazide. The section is at $z = 0.115$ and passes through a carbon atom centered at $z = 0.1137$. Numbers are differences of electron density in e/Å$^3 \times 10^2$. Contours at intervals of 0.05 e/Å^3, zero contour omitted. Positive contours indicated by solid lines, negative by broken lines. Grid spacing along x is 0.300 Å and along y, 0.243 Å. Assumed atomic position indicated by +. (*b*) Plot of values at $x = 14/25$, $y = 6/20$ for three sections along z. Grid spacing along z is 0.294 Å.

the xy section through a carbon atom from such a synthesis (Fig. 16.2*a*). Except on rare occasions, such sections will not go through the centers of the atoms. However, if grid spacings are no more than 0.3 Å, the atomic centers will be within 0.15 Å of a section, and sufficiently good corrections can be obtained from the closest section without interpolation. The approximate slope along x at the assumed atomic site in Fig. 16.2*a* is

$$\text{slope}_x = \frac{(-0.15 - 0.02)\ \text{e/Å}^3}{0.300\ \text{Å}} = \frac{-0.17}{0.300}\ \text{e/Å}^4 = -0.57\ \text{e/Å}^4 \tag{16.18}$$

Therefore the shift Δx from Eq. (16.17) is

$$\Delta x = -\frac{-0.57\ \text{e/Å}^4}{-49\ \text{e/Å}^5} = -0.012\ \text{Å} \tag{16.19}$$

Similarly for y,

$$\text{slope}_y = \frac{(-0.25 - 0.07)/2}{0.243\ \text{Å}}\ \text{e/Å}^3 = \frac{-0.16}{0.243}\ \text{e/Å}^4 = -0.66\ \text{e/Å}^4 \tag{16.20}$$

$$\Delta y = -\frac{-0.66\ \text{e/Å}^4}{-49\ \text{e/Å}^5} = -0.013\ \text{Å} \tag{16.21}$$

The nearly equal shifts in x and y are perhaps not in agreement with an intuitive view of Fig. 16.2*a*. The source of the difficulty is the problem of estimating the slopes at the atomic position. The slope for Δy was obtained by taking an average over two grid spacings to bracket the assumed position, and this usually has the effect of underestimating its value. If the calculation had been made as for Δx, by using the values at 5/20 and 6/20, the slope would have been about 50% greater, and the shift would have increased accordingly. The true value probably lies between these figures, and judgement born of experience is a great help in selecting the best value. It is not necessary to have precisely accurate slopes, however, since the whole process is iterative and the problems become less and less acute as the true location is approached.

The approximate slope along z is readily determined by plotting values from the ΔF synthesis at the grid points $x = 14/25$, $y = 6/20$ from three successive sections (Fig. 16.2*b*). The approximate slope along z is

$$\text{slope}_z = \frac{[-0.18-(-0.13)]/2\ \text{e/Å}^3}{0.294\ \text{Å}} = \frac{-0.025}{0.294}\ \text{e/Å}^4 = -0.085\ \text{e/Å}^4 \quad (16.22)$$

and the shift along z is

$$\Delta z = -\frac{-0.085\ 4\text{e/Å}^4}{-49\ \text{e/Å}^5} = -0.002\ \text{Å} \quad (16.23)$$

The shifts have been given in angstroms so they can be correlated with the appearance of the ΔF synthesis. To be used in calculating F_c, however, the shifts must be converted to fractional coordinates by dividing by the unit-cell translations. In practice the shifts are usually calculated directly in fractional coordinates. Thus the shift Δx is calculated by the equation

$$\Delta x = \frac{-(\Delta\rho_2 - \Delta\rho_1)}{(x_2 - x_1)(\text{curvature})a^2} \quad (16.24)$$

where $\Delta\rho_1$ and $\Delta\rho_2$ are difference electron densities at grid points bracketing the assumed atomic positions, x_1 and x_2 are their x values in fractional coordinates, and a is the length of the cell axis in angstroms.

In Fig. 16.3 the effects of errors solely in the thermal parameters are illustrated.[11] As before, the light line shows the actual electron density, the dashed line shows assumed density, and the heavy line their difference. It is evident that if the thermal parameters have been assumed too large (root-

[11]If the F_o values are not properly scaled, the ΔF synthesis will display an appearance similar to that shown when the thermal parameters are in error. If the scale factor is too small ($\Sigma|F_o|$ smaller than $\Sigma|F_c|$), the appearance is similar to Fig. 16.3*b*; if it is too large, similar to Fig. 16.3*a*. It will be assumed here that the $|F_o|$'s have been scaled to the $|F_c|$'s (see Chapter 9).

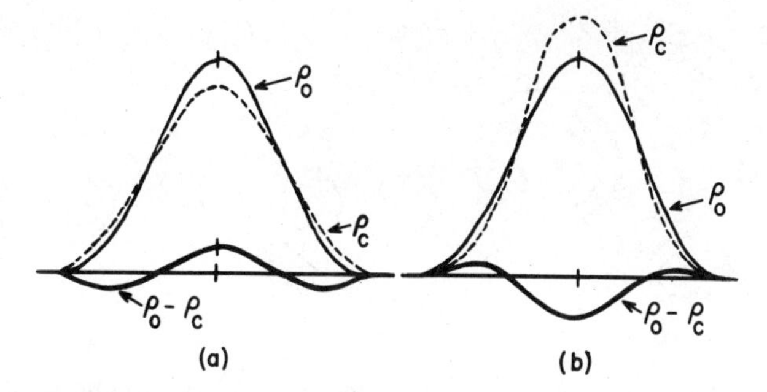

Figure 16.3. Line section through $\rho_o - \rho_c$ synthesis. (*a*) Thermal parameter overestimated. (*b*) Thermal parameter underestimated.

mean-square amplitude of vibration too large, Fig. 16.3*a*), the atomic center in the ΔF map will lie in a positive region; if too small, it will lie in a negative region, Fig. 16.3*b*. It can be shown that the magnitude and sign of the curvature of the maximum or minimum in the ΔF map is a measure of the errors in the parameters. While the correction can be calculated, in practice it requires less computational effort to interpolate (or extrapolate) from the heights of the peaks or depths of the holes at atomic sites in successive ΔF syntheses. The appropriate values by which to shift the parameters are easily determined in this way for the isotropic case.

In general, however, thermal motion of atoms is not spherically symmetrical, that is, it is anisotropic and leads to an ellipsoidal distribution of electron density. This is illustrated in Fig. 16.4*a* by a section through an anisotropically vibrating atom designated ρ_a. Figure 16.4*b* shows a similar section through an identical isotropic atom, ρ_i, with the same average thermal parameter. Figure 16.4*c* shows the difference $\rho_a - \rho_i$ and illustrates the ideal appearance of a section from a ΔF synthesis through an anisotropic atom approximated by an isotropic one. Figure 16.5 is a section from an

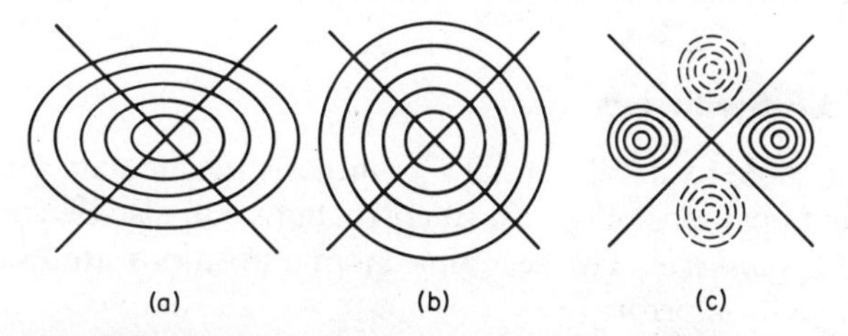

Figure 16.4. (*a*) Section through an anisotropically vibrating atom, idealized F_o synthesis. (*b*) Section through an isotropically vibrating atom, idealized F_c synthesis. (*c*) Difference between (*a*) and (*b*) showing appearance of an anisotropically vibrating atom when it has been assumed to be isotropic in calculating F_c. Positive contours indicated by solid lines, negative by broken lines.

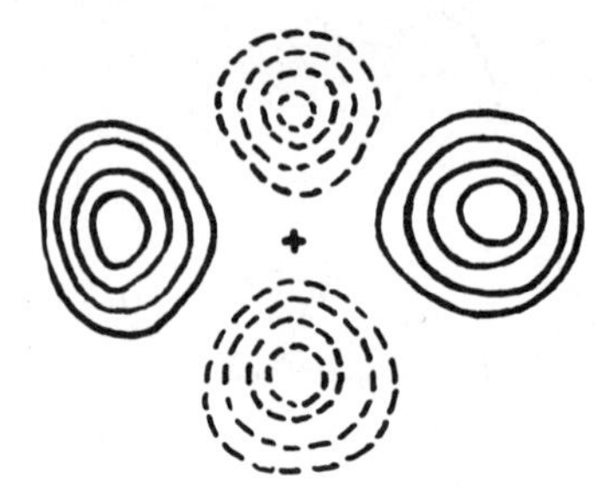

Figure 16.5. Part of an *xz* section from an ΔF synthesis through a carbon atom in *n*-nonanoic acid hydrazide. Contours at intervals of 0.05 e/Å^3. Solid contours represent positive values and begin at +0.1 e/Å^3; broken contours represent negative values and begin at −0.1 e/Å^3.

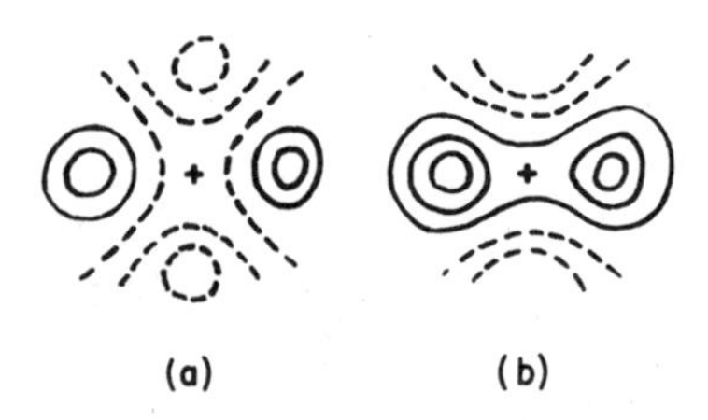

Figure 16.6. ΔF sections showing an atom with anisotropy approximated by (*a*) too small an isotropic thermal parameter, (*b*) too large an isotropic thermal parameter.

actual ΔF synthesis and is characterized by a saddlepoint similar to that in Fig. 16.4*c*. Figure 16.6 again illustrates anisotropic thermal motion, but the anisotropic thermal parameters assumed in calculating the structure factors in Fig. 16.6*a* was appreciably smaller than the true value, while in Fig. 16.6*b* it was appreciably larger.

The general form of the temperature factor expression contains six parameters that determine the magnitudes and orientations of the three principal axes of the general ellipsoidal electron distribution of an atom (see Appendix B). The temperature factor expression can be cast in a form such that the orientation parameters can be directly determined from a ΔF map. The parameters describing the amplitudes of vibration can then be estimated as in the isotropic case.[12] The method is tedious to apply, however, and it is clearly better and easier to determine anisotropic thermal parameters by the method of least squares (see Section 16.5).

Differential ΔF Syntheses

Just as it is possible by a differential F_o synthesis to calculate the corrections to the assumed positions of atoms, so corrections can also be calculated by a *differential* ΔF *synthesis*. The resulting atomic positions are essentially free of series termination error.

In calculating a differential ΔF synthesis, it is the slope of the ΔF map that is calculated by using ΔF's as the coefficients in Eq. (16.5). As before, the curvatures in ρ are necessary, and they may be evaluated as previously

[12] Y. C. Leung, R. E. Marsh, and V. Shomaker, *Acta Cryst.*, **10**, 650 (1957).

described. The corrections are then determined from either the exact or approximate equation, Eq. (16.3) or (16.7), respectively.

Rejection Test

It is often advantageous, particularly in the early stages of a refinement by ΔF or differential ΔF syntheses, to use a rejection test to eliminate terms with questionable phases. A 50% rule, that is, terms with $|F_c| < |F_o|/2$ rejected, has been found useful. If a significant number of terms are eliminated from the synthesis, the curvature used in calculating positional shifts should be modified by the factor n_p/n_t, where n_p is the number of reflections used in the ΔF synthesis and n_t is the total number of reflections. More simply, the calculated shifts can be multiplied by the factor n_t/n_p.

Double Shifts

In applying the methods outlined above to refining structures in noncentrosymmetric space groups, it is found that the errors in the calculated phases due to the incorrect positions of the atoms cause the corrections to be less than their true values. Cruickshank[13] showed that the correction applied in such cases should be increased by a factor of 2. Centrosymmetric projections of noncentrosymmetric space groups, however, are treated in the normal manner, and shifts derived from both centric and acentric data are often modified by an empirical factor between 1 and 2.

16.3. LOCATING HYDROGEN ATOMS

Although data to the limit of the copper sphere provide resolution more than adequate to separate hydrogen atoms from other atoms to which they are bonded, in practice the contrast is not sufficient, and they appear only as bulges on the electron density of the nonhydrogen atoms. A ΔF synthesis, however, provides a means for locating them by subtracting out the electron density of other atoms in the structure.[14] Thus hydrogen atoms *not included* in the F_c's will appear in the ΔF map based on these F_c's. Before attempts are made to locate hydrogen atoms in this way, however, both the positional and thermal parameters of the nonhydrogen atoms should have been refined. Furthermore, good data are required if the locations are to be unequivocal, since hydrogen atoms are easily obscured by errors. The peak positions in a ΔF synthesis showing hydrogen can be located in the ways described for F_o syntheses.

[13]D. W. J. Cruickshank, *Acta Cryst.*, **3**, 10 (1950); see also J. Donohue, *J. Am. Chem. Soc.*, **72**, 949 (1950).

[14]For a good example of the difference in appearance of these syntheses, see I. Ambats and R. E. Marsh, *Acta Cryst.*, **19**, 942 (1965).

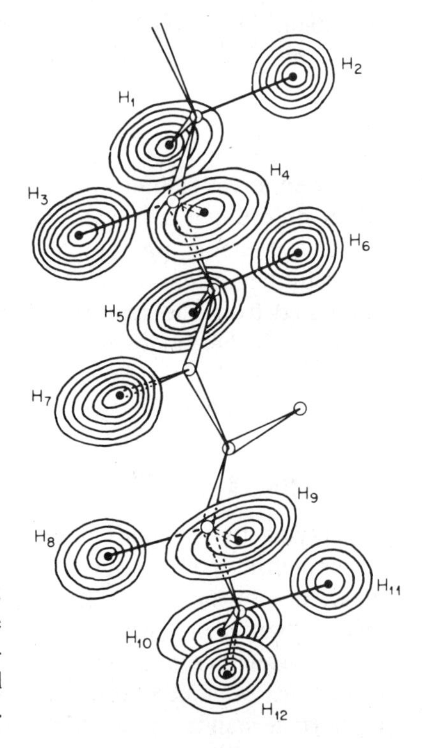

Figure 16.7. ΔF synthesis showing the 12 hydrogen atoms in a half molecule (the asymmetric unit) of *N,N'*-hexamethylenebispropionamide. Contours at integral multiples of 0.05 e/Å^3, beginning at 0.1 e/Å^3.

Figure 16.7 shows the 12 hydrogen atoms in a half molecule (the asymmetric unit) of *N,N'*-hexamethylenebispropionamide.[15] The peak densities range from less than 0.3 e/Å^3 to somewhat more than 0.4 e/Å^3, and values in this range are characteristic of hydrogen with *B*'s in the range 6–10 Å^2. For *B* values less than this, peak densities will, of course, be greater. Values as high as 0.5–0.7 e/Å^3 are observed for data collected at room temperatures, and much higher values for low-temperature data (liquid nitrogen and below). Peak values lower than those in Fig. 16.7 are observed for structures with large thermal motion, and in an extreme case, a rotating methyl group, the hydrogen image appears as a toroid of low electron density.[16]

16.4. IDENTIFICATION OF ATOM TYPES

The identity of a few unknown atoms in a structure can usually be inferred from an F_o synthesis by comparing the peak densities of unknown atoms

[15]L. H. Jensen, *Acta Cryst.*, **15**, 433 (1962).
[16]C. D. Fisher, L. H. Jensen, and W. M. Shubert, *J. Am. Chem. Soc.*, **87**, 33 (1965).

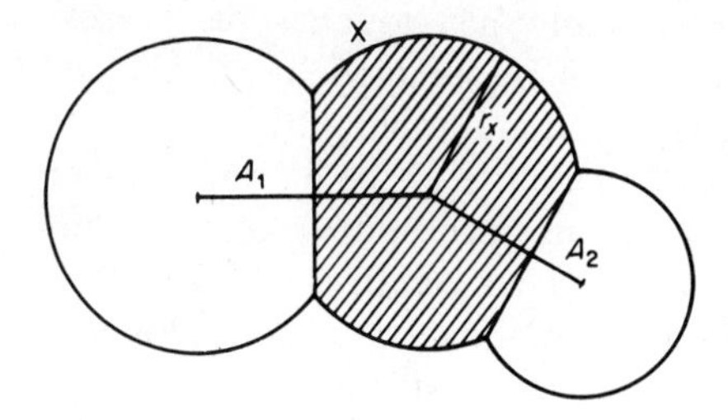

Figure 16.8. Shaded volume is that over which integration is done for an electron count in atom X.

with known ones. If this is not definitive, identification can usually be achieved for atoms with $Z < 20$ by an electron count, that is, by integration of the electron density over an assumed volume for the peak. Although the appropriate volume cannot be known exactly, it is sufficient for a covalently bonded atom X to integrate within the volume of a sphere (Fig. 16.8) determined by a reasonable van der Waals radius, r_x, and limited by planes normal to the bonds at covalent radii R_1 and R_2 from the atomic centers of the atoms to which X is bonded. The appropriate radius r_x is determined by

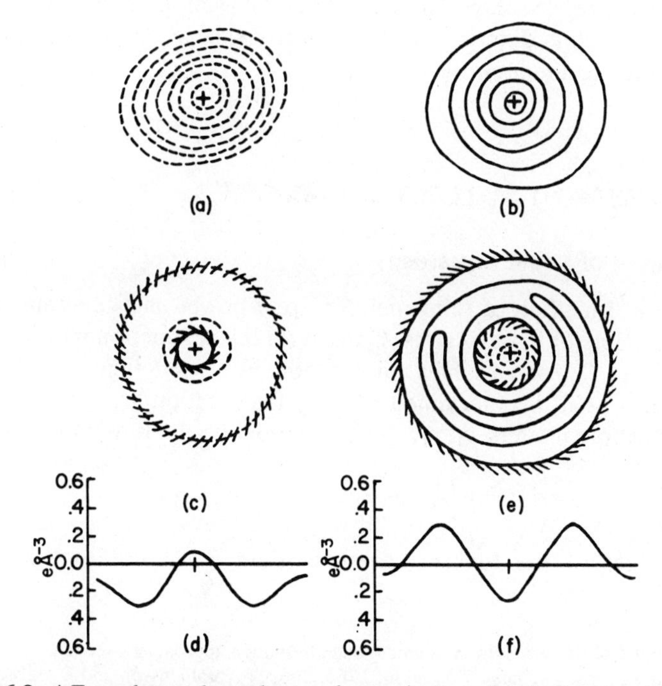

Figure 16.9. ΔF syntheses through an anisotropically refined nitrogen atom, $B_{ave} =$ 3.7 Å^2, contours at 0.2 e/Å^3. (*a*) Oxygen scattering factors used instead of nitrogen. (*b*) Carbon scattering factors used instead of nitrogen. (*c*) Same as (*a*) but after anisotropic refinement of thermal parameters. $B_{ave} = 5.0$. (*d*) Line section through (*c*). (*e*) Same as (*b*) but after anisotropic refinement of thermal parameters. $B_{ave} =$ 1.9. (*f*) Line section through (*c*).

choosing a value such that the unknown atom just touches the closest nonbonded atom. Sufficiently good integration is obtained by simply adding the electron densities at the grid points and multiplying by the volume of the grid element. The value found should be close to the atomic number for covalent compounds or this plus or minus the number of electrons transferred for ionic compounds. The result of an electron count by this method should be checked against counts of one or more known atoms in the structure. The values derived for known atoms will serve to calibrate the method and can be used to correct for possible systematic errors.

In principle, ΔF syntheses can be used to identify atom types. If the correct thermal parameters were used, a nitrogen atom misidentified as oxygen would result in a hole in the ΔF map and one taken as carbon would result in a peak (Figs. 16.9*a* and *b*). In practice, however, thermal parameters are adjusted to flatten the ΔF map, and this will result in an abnormally high value (Figs. 16.9*c* and *d*) to compensate for the hole in Fig. 16.9*a* and an abnormally low value (Figs. 16.9*e* and *f*) to compensate for the peak in Fig. 16.9*b*. Such abnormal values can be recognized by comparing the thermal parameters with those of adjacent atoms of similar *Z*. Least squares (Section 16.5) provides an analogous method in which thermal parameters differing appreciably from those of adjacent atoms can furnish a clue to the atom type.[17]

16.5. REFINEMENT BY LEAST SQUARES[18–21]

The Method of Least Squares

An analytical method of refinement of great power and generality is based on the principle of least squares. Consider a linear function with n variables $x_1, x_2, \ldots, x_n$. These variables can be thought of as defining a space whose value at any point is determined both by the location $x_1, x_2, \ldots, x_n$ and by independent parameters $p_1, p_2, \ldots, p_n$ that define the function. Thus,

$$f = p_1x_1 + p_2x_2 + p_3x_3 + \cdots + p_nx_n \tag{16.25}$$

If the values of the function are measured at m different points with

[17]For examples of this problem in practice, and alternative approaches, see S. C. Nyburg, G. L. Walford, and P. Yates, *Chem. Commun.*, **1965**, 203; S. Abrahamsson and B. Nilsson, *J. Org. Chem.*, **31**, 3631 (1966); and S. M. Rothstein, M. F. Richardson, and W. D. Bell, *Acta Cryst.*, **A34**, 969 (1978).

[18]Whittaker and Robinson, *The Calculus of Observations*, pp. 209–259.

[19]E. W. Hughes, *J. Am. Chem. Soc.*, **63**, 1737 (1941).

[20]Hamilton, *Statistics in Physical Science*, pp. 124–157.

[21]W. E. Deming, *Statistical Adjustment of Data*, Wiley, New York, 1943, pp. 14–58.

$m > n$,[22] the principle of least squares states that the best values for the parameters[23] $p_1, p_2, \ldots, p_n$ are those that minimize the sums of the squares of the properly weighted differences between the observed and calculated values of the function for all the observational points. Thus the quantity to be minimized is given by

$$D = \sum_{r=1}^{m} w_r(f_{o,r} - f_{c,r})^2 \tag{16.26}$$

where w is the weight to be assigned an observation, $f_{o,r}$ is one of the m observed values of the function, and $f_{c,r}$ is the corresponding calculated value. In order to obtain the best fit, it will be necessary to consider the parameters p as variables that can be adjusted to minimize D. This is a straightforward minimization problem that is treated by differentiating the right-hand side of Eq. (16.26) with respect to each of the parameters in turn and setting the derivative equal to zero. This gives

$$\sum_{r=1}^{m} w_r(f_{o,r} - f_{c,r}) \frac{\partial f_{c,r}}{\partial p_j} = 0; \qquad j = 1, 2, \ldots, n \tag{16.27}$$

a set of n equations in n unknowns called the *normal equations*.

In practice, there are m *observational equations* of the form of Eq. (16.25), one for each observation. Since we shall be treating the parameters p_j as quantities to be adjusted, and the x's have different fixed values for each of the m observations, it is customary to reverse their order. Taking the partial derivatives $\partial f_{c,r}/\partial p_j$ for each of the m observational equations and substituting into Eq. (16.27) gives the n normal equations

$$\begin{aligned}
&\sum_{r=1}^{m} w_r(f_{o,r} - x_{r1}p_1 - x_{r2}p_2 - \cdots - x_{rn}p_n)x_{r1} = 0 \\
&\sum_{r=1}^{m} w_r(f_{o,r} - x_{r1}p_1 - x_{r2}p_2 - \cdots - x_{rn}p_n)x_{r2} = 0 \\
&\qquad\qquad\vdots \\
&\sum_{r=1}^{m} w_r(f_{o,r} - x_{r1}p_1 - x_{r2}p_2 - \cdots - x_{rn}p_n)x_{rn} = 0
\end{aligned} \tag{16.28}$$

[22]The various values of $x_1, x_2, \ldots, x_n$ for the points $1, 2, \ldots, m$ can be represented compactly as

point 1: $x_{11}, x_{12}, x_{13}, \ldots, x_{1n}$
point 2: $x_{21}, x_{22}, x_{23}, \ldots, x_{2n}$
⋮
point m: $x_{m1}, x_{m2}, x_{m3}, \ldots, x_{mn}$

[23]There must always be more measurements than quantities to be determined in order that the system be overdetermined. The measurements, however, contain experimental error, and averaging is necessary to have the derived quantities best fit the data.

Rearranging and writing more fully,

$$\begin{aligned}
&\sum_{r=1}^{m} w_r x_{r1}^2 p_1 + \sum_{r=1}^{m} w_r x_{r1} x_{r2} p_2 + \cdots + \sum_{r=1}^{m} w_r x_{r1} x_{rn} p_n = \sum_{r=1}^{m} w_r f_{o,r} x_{r1} \\
&\sum_{r=1}^{m} w_r x_{r2} x_{r1} p_1 + \sum_{r=1}^{m} w_r x_{r2}^2 p_2 + \cdots + \sum_{r=1}^{m} w_r x_{r2} x_{rn} p_n = \sum_{r=1}^{m} w_r f_{o,r} x_{r2} \\
&\qquad\vdots \\
&\sum_{r=1}^{m} w_r x_{rn} x_{r1} p_1 + \sum_{r=1}^{m} w_r x_{rn} x_{r2} p_2 + \cdots + \sum_{r=1}^{m} w_r x_{rn}^2 p_n = \sum_{r=1}^{m} w_r f_{o,r} x_{rn}
\end{aligned} \tag{16.29}$$

Solution of these n equations gives directly the best values of the parameters p_j in the least squares sense.

If the functional form of the observational equations is not linear in the p's, the normal equations are not linear and they become intractable. In such cases, however, they can be made linear by approximating the function as a Taylor series

$$\begin{aligned}
f(p_1, p_2, \ldots, p_n) = f(a_1, a_2, \ldots, a_n) &+ \frac{\partial f(a_1, a_2, \ldots, a_n)}{\partial p_1}(p_1 - a_1) + \cdots \\
&+ \frac{\partial f(a_1, a_2, \ldots, a_n)}{\partial p_n}(p_n - a_n)
\end{aligned} \tag{16.30}$$

$$\begin{aligned}
f(p_1, p_2, \ldots, p_n) = f(a_1, a_2, \ldots, a_n) &+ \frac{\partial f(a_1, a_2, \ldots, a_n)}{\partial p_1}\Delta p_1 + \cdots \\
&+ \frac{f(a_1, a_2, \ldots, a_n)}{\partial p_n}\Delta p_n
\end{aligned} \tag{16.31}$$

where terms in Δp_j of powers higher than the first have been neglected. The a_j are approximate values of p_j, and $f(a_1, a_2, \ldots, a_n)$, $\partial f(a_1, a_2, \ldots, a_n)/\partial p_1, \ldots, \partial f(a_1, a_2, \ldots, a_n)/\partial p_n$ are the function and its derivatives evaluated at these approximate values. If the a_j values are sufficiently good approximations, application of a least squares process to the linear equations (16.31) will give values for the quantities Δp_j such that the a_j's given by

$$a'_j = a_j + \Delta p_j \tag{16.32}$$

are a better approximation to the best values for the parameters p_j than the initial a_j's. But because the series has been truncated by neglecting second and higher powers in the Δp_j, the calculations must be repeated using as approximate values for each repetition the results derived from the preceding calculation. The iterative process is complete when there is no significant change in the parameters between two successive cycles. This is in contrast to the case where the observational equations are truly linear, when no approximations in the parameters are required and the solution of the normal equations gives their values without iteration.

In X-ray diffraction, the functional form of the structure factor is transcendental and so must be approximated by a truncated Taylor series. In this case the quantity most commonly minimized is[24]

$$D = \sum_{hkl} w_{hkl}(|F_o| - |kF_c|)^2 \tag{16.33}$$

over all the observed reflections. Minimization is achieved as before by taking the derivative with respect to each of the parameters and equating it to zero. This leads to the n normal equations [similar to Eq. (16.27)]

$$\sum_{hkl} w_{hkl}(|F_o| - |kF_c(p_1, p_2, \ldots, p_n)|) \frac{\partial |kF_c(p_1, \ldots, p_n)|}{\partial p_j} = 0; \quad j = 1, 2, \ldots, n \tag{16.34}$$

We now express the function $|F_c|$ as a Taylor series and neglect second and higher powers so that

$$|kF_c(p_1, \ldots, p_n)| = |kF_c(a_1, \ldots, a_n)| + \frac{\partial |kF_c|}{\partial p_1} \Delta p_1 + \cdots + \frac{\partial |kF_c|}{\partial p_n} \Delta p_n \tag{16.35}$$

where $p_1, \ldots, p_n$ can be any of the scale, positional, or thermal parameters and $\Delta p_j = p_j - a_j$. Substituting Eq. (16.35) into (16.34) gives

$$\sum_{hkl} w_{hkl} \left(|F_o| - |kF_c(a_1, \ldots, a_n)| - \frac{\partial |kF_c|}{\partial p_1} \Delta p_1 - \cdots - \frac{\partial |kF_c|}{\partial p_n} \Delta p_n \right) \frac{\partial |kF_c|}{\partial p_j} = 0; \quad j = 1, 2, \ldots, n \tag{16.36}$$

or

$$\sum_{hkl} w_{hkl} \left(\Delta F - \frac{\partial |kF_c|}{\partial p_1} \Delta p_1 - \cdots - \frac{\partial |kF_c|}{\partial p_n} \Delta p_n \right) \frac{\partial |kF_c|}{\partial p_j} = 0; \quad j = 1, 2, \ldots, n \tag{16.37}$$

where ΔF now plays the role of the known observational quantity f_o in Eqs.

[24] The scaling parameter k must be applied to F_c rather than to F_o for any single refinement cycle. If k is applied to F_o, least squares will tend to wipe out both F_o and F_c, the first by decreasing k and the second by increasing the thermal parameters, thus arriving at an incorrect solution that nevertheless minimizes $\sum w \, \Delta F^2$. It is legitimate, however, to modify the F_o's after each refinement cycle by the factor $1/k$ and to reset k to 1 for the following cycle. In this way the scaling can be carried out in practice on the F_o's.

(16.25)–(16.29). Expansion and rearrangement of Eq. (16.37) leads to the following set of equations, which can be compared term by term with Eq. (16.29):

$$\begin{aligned}
&\sum_{r=1}^{m} w_r \left(\frac{\partial|kF_{c,r}|}{\partial p_1}\right)^2 \Delta p_1 + \sum_{r=1}^{m} w_r \frac{\partial|kF_{c,r}|}{\partial p_1}\frac{\partial|kF_{c,r}|}{\partial p_2}\Delta p_2 + \cdots \\
&\qquad + \sum_{r=1}^{m} w_r \frac{\partial|kF_{c,r}|}{\partial p_1}\frac{\partial|kF_{c,r}|}{\partial p_n}\Delta p_n = \sum_{r=1}^{m} w_r \Delta F_r \frac{\partial|kF_{c,r}|}{\partial p_1} \\
&\sum_{r=1}^{m} w_r \frac{\partial|kF_{c,r}|}{\partial p_2}\frac{\partial|kF_{c,r}|}{\partial p_1}\Delta p_1 + \sum_{r=1}^{m} \left(\frac{\partial|kF_{c,r}|}{\partial p_2}\right)^2 \Delta p_2 + \cdots \\
&\qquad + \sum_{r=1}^{m} w_r \frac{\partial|kF_{c,r}|}{\partial p_2}\frac{\partial|kF_{c,r}|}{\partial p_n}\Delta p_n = \sum_{r=1}^{m} w_r \Delta F_r \frac{\partial|kF_{c,r}|}{\partial p_2} \qquad (16.38) \\
&\qquad\qquad \vdots \\
&\sum_{r=1}^{m} w_r \frac{\partial|kF_{c,r}|}{\partial p_n}\frac{\partial|kF_{c,r}|}{\partial p_1}\Delta p_1 + \sum_{r=1}^{m} w_r \frac{\partial|kF_{c,r}|}{\partial p_n}\frac{\partial|kF_{c,r}|}{\partial p_2}\Delta p_2 + \cdots \\
&\qquad + \sum_{r=1}^{m} w_r \left(\frac{\partial|kF_{c,r}|}{\partial p_n}\right)^2 \Delta p_n = \sum_{r=1}^{m} w_r \Delta F_r \frac{\partial|kF_{c,r}|}{\partial p_n}
\end{aligned}$$

This is a system of n equations in n unknowns, the Δp_j's. These equations are linear in the Δp_j's and solvable for them. Combination of these with the initial approximation, the a_j's, as in Eq. (16.32) gives better, though still approximate, values for the various parameters. These can be used to repeat the process until convergence is obtained and successive cycles produce no further changes.

The approximations made in constructing the least squares equations constrain this application to the refinement of reasonably accurate initial parameters. Thus it is not possible to solve by this method for the parameters directly from the intensity data without first developing the structure by one of the methods described earlier.

Refinement against $|F_o|^2$

The very weak reflections in an X-ray diffraction data set have relatively large errors (see Appendix E), and those for which the background exceeds the peak will have negative net intensities.[25] They are undefined and cannot, therefore, be used when refining against $|F_o|$. It is customary to eliminate the very weak reflections by use of a threshold; that is, reflections with intensities less than some positive threshold value such as $2\sigma(I)$ are

[25] A negative I is a perfectly valid observation. It simply means that the error is negative and its absolute value is greater than the true value of I.

given zero weight. The use of any threshold, however, introduces a bias in the data and is indefensible in principle; it is much better to refine against $|F_o|^2$ so that all data can be included. Accordingly, instead of Eq. (16.33), the function

$$D' = \sum_{hkl} w'_{hkl}(|F_o|^2 - |kF_c|^2)^2 \qquad (16.39)$$

is minimized. This leads to a set of normal equations similar to Eqs. (16.38) except that w is replaced by w', ΔF by $|F_o|^2 - |kF_c|^2$, and $\partial|kF_{c,r}|/\partial p_j$ by $2|kF_{c,r}|(\partial|kF_{c,r}|/\partial p_j)$.[26]

Weighting Functions[27,28]

The functions minimized in the least squares method, Eq. (16.33) or (16.39), carried a weighting factor for each observation. The weighting should be a measure of the reliability of the observation. If properly chosen, it has the effect of adjusting the contribution of each observation to the normal equations in such a way as to produce the most reliable results. From statistical considerations, it can be shown that the best weight is equal to the reciprocal of the square of the standard deviation of the observation.[27,28] For reasonably good data, however, experience indicates that results from structure refinement are not greatly dependent on a particular weighting scheme,[29] although the use of proper weighting functions can produce a real, though small, improvement in the results from a given set of data. Appendix E should be consulted for ways in which suitable functions can be obtained.

16.6. GOODNESS-OF-FIT PARAMETER

A useful index often output by least squares refinement programs is the so-called *goodness of fit*,

$$\text{GOF} = \left[\frac{\sum w(|F_o| - |F_c|)^2}{n - m}\right]^{1/2} \qquad (16.40)$$

sometimes also termed the *standard deviation of an observation of unit*

[26]For a brief discussion of the differences found when both methods were used in a practical case, see S. T. Freer and J. Kraut, *Acta Cryst.*, **19**, 992 (1965).

[27]D. W. J. Cruickshank, in Rollett, *Computing Methods in Crystallography*, pp. 112–115.

[28]W. Hong and B. E. Robertson, in *Structure and Statistics in Crystallography*, A. J. C. Wilson, Ed., Adenine Press, Guilderland, NY, 1985, pp. 125–136.

[29]See K. Sudaranan and R. A. Young, *Acta Cryst.*, **B25**, 1534 (1969).

weight. It is a measure of the degree to which the found distribution of differences between $|F_o|$ and $|F_c|$ fits the distribution expected from the weights used in the refinement. If these weights are correct, which implies that the errors in the data are strictly random and correctly estimated and if the model properly represents the structure that gives rise to the data, the value of the GOF is 1.0.

In practice, models based on excellent data collected to the limit of the copper sphere with accurately estimated σ's[30] (see Appendix E) do not usually refine to a GOF of 1.0 but rather to a final value in the range of 2–3. It is unfortunately a common practice to force the GOF to 1 by adjusting the weights downwards, usually by using an unrealistically large value (often 3–5%) for the instrumental instability constant. This is incorrect in principle since the common sources of error are not in the measured intensities but in the model, which generally fails to account correctly for lone-pair and bonding electrons as well as for extinction in the crystal specimen. This has been shown clearly by the results of refinements carried out with data beyond the copper sphere.[31] Here the I_o's are small and the electronic model is only the inner-shell electrons, which refine cleanly to a GOF of unity. Poorer data, for which the true random error in the observations is larger, fail to reveal this effect and often refine to lower GOF values than do the best data.

16.7. LEAST SQUARES AND MATRICES

Matrix Equations

The normal equations, Eqs. (16.38), can be written as

$$\begin{aligned} a_{11}x_1 + a_{12}x_2 + \cdots + a_{1n}x_n &= v_1 \\ a_{21}x_1 + a_{22}x_2 + \cdots + a_{2n}x_n &= v_2 \\ &\vdots \\ a_{n1}x_1 + a_{n2}x_2 + \cdots + a_{nn}x_n &= v_n \end{aligned} \tag{16.41}$$

where

$$a_{ij} = \sum_{r=1}^{m} w_r \frac{\partial |F_{c,r}|}{\partial p_i} \frac{\partial |F_{c,r}|}{\partial p_j}; \qquad x_j = \Delta p_j; \qquad v_i = \sum_{r=1}^{m} w_r(\Delta F_r) \frac{\partial |F_{c,r}|}{\partial p_i} \tag{16.42}$$

[30] L. E. Mcandlish, G. H. Stout, and L. C. Andrews, *Acta Cryst.*, **A31**, 245 (1975).
[31] S. Furberg and L. H. Jensen, *Acta Cryst.*, **B26**, 1260 (1970).

This set of equations can be written in matrix form (see Appendix C) as

$$\begin{bmatrix} a_{11} & a_{12} & \dots & a_{1n} \\ a_{21} & a_{22} & \dots & a_{2n} \\ & & \vdots & \\ a_{n1} & a_{n2} & \dots & a_{nn} \end{bmatrix} \begin{bmatrix} x_1 \\ x_2 \\ \vdots \\ x_n \end{bmatrix} = \begin{bmatrix} v_1 \\ v_2 \\ \vdots \\ v_n \end{bmatrix} \tag{16.43}$$

In the matrix of the a_{ij}'s, the number of rows always equals the number of columns, that is, the matrix is *square*. The line from the upper left to the lower right in such a matrix is termed the *principal diagonal* and has the elements a_{ii} lying on it. A check of the coefficients in Eqs. (16.29) and (16.38) will show that $a_{ij} = a_{ji}$, that is, the matrix is *symmetric*. A more compact notation for these equations represents them as a matrix **A** acting on a vector of $\mathbf{x}_i$'s to give another vector of v_i's. Thus

$$\mathbf{Ax} = \mathbf{v} \tag{16.44}$$

It can be shown that if Eqs. (16.43) have a solution, an *inverse matrix*, $\mathbf{A}^{-1}$, exists such that $\mathbf{A}^{-1}\mathbf{A}$ equals the matrix equivalent of 1. Then

$$\mathbf{A}^{-1}\mathbf{Ax} = \mathbf{A}^{-1}\mathbf{v} \tag{16.45}$$

$$\mathbf{x} = \mathbf{A}^{-1}\mathbf{v} \tag{16.46}$$

$\mathbf{A}^{-1}$ is a matrix like **A**, with elements often symbolized by b_{ij}. Although the diagonal elements b_{ii} are not in general reciprocals of a_{ii}, they are related in an inverse fashion, and for the special case where **A** has zeros everywhere except on the principal diagonal (a *diagonal matrix*), $b_{ii} = 1/a_{ii}$.

Correlation Coefficients and the Correlation Matrix

It can be shown that b_{ij}, the ijth element of the inverse matrix, is a measure of the interdependence or correlation of parameters i and j. The *correlation coefficient* is defined in this case by the equation

$$\delta_{ij} = b_{ij}/\sqrt{b_{ii}}\sqrt{b_{jj}} \tag{16.47}$$

In refining crystal structures by the method of least squares, it is often important to know the magnitude of the correlation coefficients between the various parameters. Some programs provide a matrix of these coefficients as a part of the output. If correlation coefficients are large, interdependence among the parameters can seriously impair the refinement, either slowing it or in extreme cases causing the results of successive cycles to oscillate. In such cases it is particularly important to use the full matrix, since off-diagonal terms are important and without them it is not easy to diagnose the difficulty.

Correlation coefficients can range from 0 to ± 1, the latter describing cases in which the two parameters are completely dependent and one can be eliminated. As a rule, correlation coefficients from least squares refinement are less than ~ 0.2 and cause little difficulty in completely refined structures. The principal exceptions are those between scale and temperature parameters, between positional parameters for an atom when interaxial angles differ appreciably ($>10°$) from 90°, and between atoms that are related by centers of symmetry that are not part of the space group symmetry. The latter case can arise, for example, when attempts are made to refine in a noncentrosymmetric space group a structure that properly belongs in a centrosymmetric space group. During refinement, attempts are made to adjust independently atoms that are nearly related by the center, and these show large correlation coefficients.[32]

Modified Matrix Least Squares

The elements on the principal diagonal of the coefficient matrix are sums of squares. The terms contributing to them will always be positive, and as these terms accumulate the sums will become progressively larger. The off-diagonal elements, on the other hand, are sums of products that may be either + or −, and these sums, therefore, might reasonably be expected to be smaller than those on the diagonal.

This suggests a means of greatly reducing the time and computer storage requirements for the very extensive calculations involved in least squares refinement. If all off-diagonal elements are very small relative to the diagonal elements, then with little error they can be neglected, and only the diagonal elements need be calculated. The matrix for the *diagonal least squares* approximation, in which only the diagonal elements are different from zero, is shown in Fig. 16.10.

It is evident, however, from the coefficients of the normal equations that the actual magnitude of the ijth coefficient (ith row, jth column) depends on the joint variation of $|F_c|$ with p_i and p_j. If these parameters are correlated in any way, the contributing terms will *not* cancel in a random

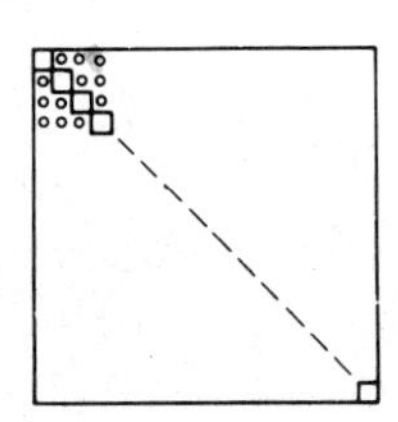

Figure 16.10. Diagonal matrix.

[32]V. Schomaker and R. E. Marsh, *Acta Cryst.*, **B35**, 1933 (1979).

fashion, and this element will tend to have a large positive or negative value. This will be true, for example, for the scale and thermal parameters and will also be true for the parameters of a given atom in a structure where the interaxial angles of the unit cell differ appreciably from 90°. Such terms cannot reasonably be equated to zero.

An alternative to the diagonal approximation is one in which off-diagonal elements involving correlations between scale and thermal parameters and among parameters of the same atom are retained, all others being neglected. This results in a matrix whose nonzero elements are blocks about the diagonal. This is indicated in Fig. 16.11 for a structure with one scale and one thermal parameter (overall isotropic) and three positional parameters per atom. The diagonal thus carries a single 2×2 and a number of 3×3 matrix blocks.

Figure 16.12 shows the case where a single thermal parameter is added for each atom (individual atom isotropic) so that the atom blocks are now 4×4 matrices.

Figure 16.13 shows the general case where six thermal parameters (individual atom anisotropic) are used for each atom, resulting in 9×9 atom blocks.

A useful modification of the block diagonal in the case of anisotropic thermal parameters is one in which the 9×9 matrix for each atom is replaced by a 3×3 and a 6×6 block as shown in Fig. 16.14. This retains the off-diagonal elements involving correlation among positional parameters and among thermal parameters of the same atom but neglects those involving correlation between positional and thermal. Note that although individual atom thermal parameters are included in the latter three cases, the 2×2 block that correlates scale and a single overall thermal parameter is retained. This thermal parameter is not varied during a refinement cycle, and its value corresponds to the average of the individual atom parameters used. While the computational effort and storage requirements for these *block-diagonal least squares* matrices are more than

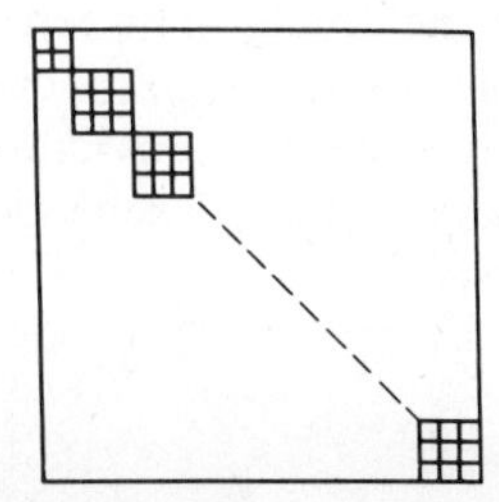

Figure 16.11. Block-diagonal matrix using a 2×2 block for scale and thermal parameters and 3×3 blocks for positional parameters of each atom.

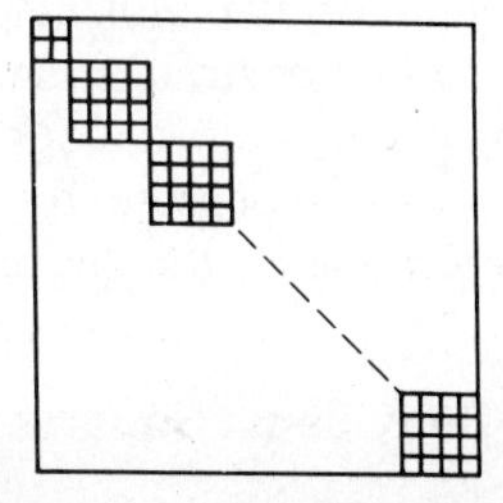

Figure 16.12. Block-diagonal matrix. As in Fig. 16.11 except 4×4 blocks for one thermal and three positional parameters for each atom.

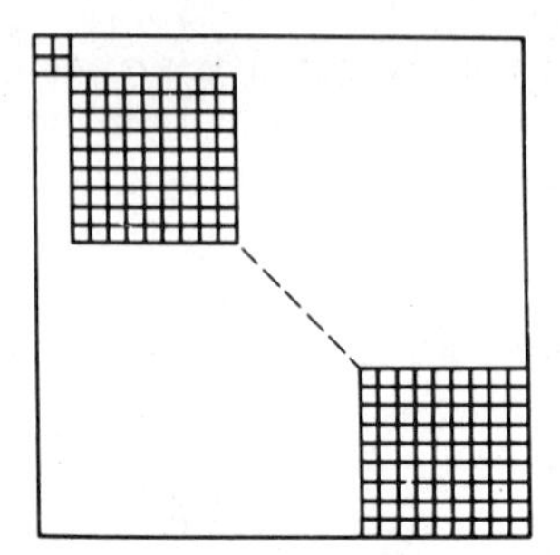

Figure 16.13. Block-diagonal matrix. As in Fig. 16.11 except 9 × 9 blocks for three positional and six thermal parameters for each atom.

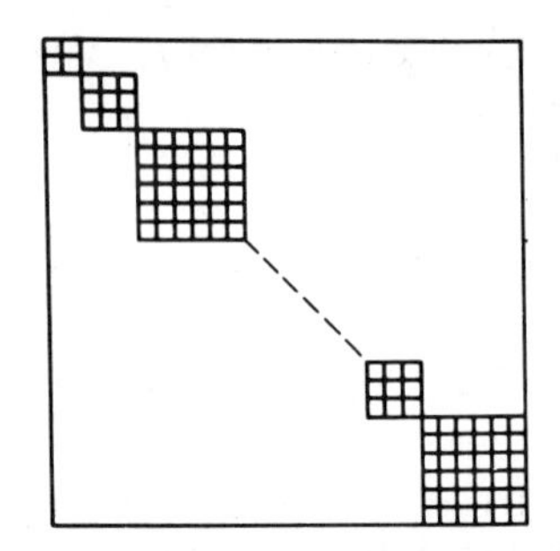

Figure 16.14. Block-diagonal matrix. As in Fig. 16.11 except 3 × 3 blocks for the three positional parameters and 6 × 6 blocks for the six thermal parameters for each atom.

for the diagonal approximation, they are still much less than for a full matrix.[33]

The block-diagonal matrices just noted take into account the off-diagonal terms in the blocks retained in each case, but terms outside the blocks can also be relatively large due to chance correlations between certain atoms, particularly if some atoms in the structure are much heavier than others. Retaining the sizable off-diagonal terms in the matrix, no matter where they occur, and setting all other off-diagonal terms to zero leads to a *sparse* matrix. Such a matrix preserves the most significant coefficients from the normal equations while at the same time requiring considerably less storage.

16.8. MODIFIED LEAST SQUARES REFINEMENTS

While most current refinement is carried out by least squares methods, in many cases it has proved desirable to include additional information about the nature of the molecule or the crystal or the forces between atoms among the constraints acting on the possible solutions. Especially in macromolecular crystallography this has the advantage of providing extra "data" to compensate for the failure to observe enough reflections to overdetermine all the desired atomic parameters.

Rigid-Body Least Squares Refinement

In the case of a rigid molecule of known structure, there are advantages in refining only the position and orientation of the molecule, that is, the

[33]L. I. Hodgson and J. S. Rollett, *Acta Cryst.*, **16**, 329 (1963) and references cited there.

parameters x, y, z of the molecular origin and ϕ_1, ϕ_2, ϕ_3 specifying the molecular orientation. Molecular parameters such as bond lengths and angles are fixed (*constrained*) at their known values with respect to the assumed origin.[34] The number of parameters is reduced, and for a sizable molecule the reduction is dramatic. For example, a 20-atom molecule would require 60 positional parameters if the atoms were refined individually but only the six just noted if the molecule were refined as a rigid group. The quantity minimized is the same as that in Eq. (16.33), but now F_c is a function of only the three positional and three orientation parameters along with any necessary thermal parameters.[35]

Refinements by rigid-body least squares are feasible with far fewer data than are required when individual atom parameters are to be refined. As a result the resolution of the data can be limited by reducing the maximum $(\sin \theta)/\lambda$. Thinking of each X-ray reflection in terms of a three-dimensional Bragg–Lipson "chart" (see Appendix D), we see that the lower-order data with smaller maximum values of h, k, and l have fewer cycles per unit translation along x, y, and z. The effect of limiting the resolution of the data is to broaden the minimum in $\sum w \, \Delta F^2$ space[36] corresponding to the correct structure. This increases the *radius of convergence*, the distance from which the model will refine into the true minimum. As the refinement progresses, the higher-order data should be added to obtain the most precise values of the adjustable parameters.

Few sizable molecules are strictly rigid, but virtually all have rigid groups or regions. Such molecules can be treated as segmented rigid bodies and refined by programs designed to accommodate a collection of varied rigid groups linked by flexible regions in which individual atoms are refined along with the rigid segments.[37,38] Such an approach has been widely applied to the refinement of polynucleotide and polypeptide macromolecules but is not restricted to them.

[34] Such an approach follows directly from the use of a rigid molecule search group in vector and trial-and-error methods of phasing (Chapter 14), and programs for rigid-body refinement are often used as the first stage of refinement for molecules phased in such ways.

[35] R. Riccardo, C. M. Gramaccioli, T. Pilati, and M. Simonetta, *Acta Cryst.*, **A37**, 65 (1981).

[36] It is often convenient to "visualize" problems involving the minimization of a function of a large number of parameters, for example, R or $\sum w \, \Delta F^2$, which depend on all the positional and thermal parameters, in terms of looking for the point of lowest value in an N-dimensional space, where a dimension is allocated to each parameter. Although it is clearly impossible to picture such a space, the analogy with surfaces in 3-space helps one consider questions about local minima in limited regions and the problem of finding the best (*global*) minimum. In this case reducing the resolution tends to smooth the N-dimensional surface and enable the model to find its way to the global minimum with less risk of being tapped in a poorer local minimum.

[37] J. L. Sussman, in Wyckoff et al., *Diffraction Methods for Biological Macromolecules*, pp. 271–303.

[38] J. L. Sussman, S. R. Holbrook, G. M. Church, and S.-H. Kim, *Acta Cryst.*, **A33**, 800 (1977).

Least Squares with Subsidiary Conditions

A more general method of imposing known molecular features, such as bond lengths, bond angles, and the shapes of rigid groups, is to include them as subsidiary conditions (*restraints*) and treat them as additional observational equations.[39] The squares of the residuals from the subsidiary conditions are weighted appropriately and then simply added to the squared residuals from the X-ray data. Thus the function to be minimized is

$$D = \sum_{hkl} w_F(|F_o| - |kF_c|)^2 + \sum_{l} w_l(d_s^2 - d_c^2)^2 + \sum_{\phi} w_\phi(\phi_s^2 - \phi_c^2)^2 + \text{other terms} \tag{16.48}$$

where $d_s, \phi_s, \ldots$ and $d_c, \phi_c, \ldots$ are the standard (ideal) and calculated (current model) values, respectively, of the bond lengths and bond angles. Terms for any other known features of the structure can also be added.[40,41] The weights w_F [$= w_{hkl}$ in Eq. (16.33)] are, as before, those from the experimental observations. The weights $w_l, w_\phi, \ldots$ for the subsidiary conditions should be adjusted so that as the refinement progresses the rms difference between the ideal and current model values approach the estimated reliability of the assumed ideal values.

Refinement with subsidiary conditions may be essential if the observable X-ray data are severely limited, and in any case should be considered if features of a structure such as bond lengths and angles are known a priori with accuracy greater than can be expected by refining solely against the X-ray data. The added information in the subsidiary conditions will result in a more stable refinement and more reliable values of the parameters describing the unknown features of the structure. On the other hand, the more subsidiary conditions one adds, the more vital it is that they be correct. This applies not only to bond lengths and angles, which are now generally known to high precision (although suitable allowance must be made for their natural variability, especially of bond angles), but more particularly to the details of molecular conformation. It is all too easy to believe that since the bond parameters are precise they should be heavily weighted, but even correct bond lengths and angles do not ensure true atomic positions. If some part of the molecule is grossly in error, there may be insufficient power left in the experimental data to correct or even expose the error.

Least Squares with Energy Minimization

This method minimizes both the crystallographic residual and the potential energy of the molecular system. It is similar to refining by least squares with

[39] J. Waser, *Acta Cryst.*, **16**, 1091 (1963).
[40] J. H. Konnert, *Acta Cryst.*, **A32**, 614 (1976).
[41] W. A. Hendrickson, in Wyckoff et al., *Diffraction Methods for Biological Macromolecules*, pp. 252–270.

subsidiary conditions, these being the terms in the complete potential energy function, which includes bond length stretching, bond angle bending, torsional potentials, nonbonding interactions, and electrostatic forces. As programmed by Jack and Levitt,[42] the sum of the potential energy E and the weighted crystallographic residual $k\sum(|F_o|-|F_c|)^2$ is minimized. The weighting factor k is chosen to make the residual comparable in magnitude to E, k being adjusted in successive cycles as E decreases.

Refinement by Simulated Annealing

In recent years computer simulation of atomic motion in biological macromolecules such as proteins has provided a detailed view of the dynamic character of these highly flexible molecules. Each atom is treated as a particle in a potential field moving in accordance with Newton's laws of motion.[43] The process of dynamic simulation begins with atomic positions from a static model. Small random velocities, corresponding to temperatures only a few degrees above absolute zero, are assigned to the atoms, which are then allowed to move through a series of extremely short time steps, typically 100 steps of femtoseconds ($1\ \text{fs} = 10^{-15}$ sec). Progressively greater random velocities are assigned to the atoms in a succession of such time sequences until the mean velocity corresponds to the desired temperature of the simulation.

In refinement by simulated annealing, the basic molecular dynamics equations are modified by adding the crystallographic residual $k\sum(|F_o|-|F_c|)^2$ to the empirical potential energy. The weighting constant k is chosen so that the magnitude of the gradient is comparable to that of the empirical potential energy. In practice the system is raised to a very high temperature (2000° or more) for a few picoseconds ($1\ \text{ps} = 10^{-12}$ sec) with periodic cooling (annealing) followed by minimization.[44] This process increases the convergence range by countering the tendency of pure crystallographic refinements to lodge in local minima. At the same time the crystallographic residual restrains the extreme swings of pure energy refinement.

16.9. RELATIONSHIP BETWEEN FOURIER AND LEAST SQUARES METHODS[45,46]

Cochran[47] showed that the coordinates found for an atom j from the Fourier series

[42] A. Jack and M. Levitt, *Acta Cryst.*, **A34**, 931 (1978).
[43] M. Karplus and J. A. McCammon, *Ann. Rev. Biochem.*, **52**, 263 (1983).
[44] A. T. Brünger, J. Kuriyan, and M. Karplus, *Science*, **235**, 458 (1987). A. T. Brünger, *J. Mol. Biol.*, **203**, 803 (1988); A. T. Brünger, M. Karplus, and G. A. Petsko, *Acta Cryst.*, **A45**, 50 (1989).
[45] See G. A. Jeffrey and R. Shiono, *Acta Cryst.*, **12**, 819 (1959).
[46] D. W. J. Cruickshank, *Acta Cryst.*, **3**, 10 (1950); **5**, 511 (1952).
[47] W. Cochran, *Acta Cryst.*, **1**, 138 (1948); *Nature*, **161**, 765 (1948).

$$\rho(xyz) = \frac{1}{V} \sum_h \sum_k \sum_l |F_{o,hkl}| \cos 2\pi(hx + ky + lz - \alpha) \qquad (16.49)$$

if corrected for series termination error and peak overlap, are the same as those given by minimizing the function

$$\sum \frac{1}{f_j} (|F_o| - |F_c|)^2 \qquad (16.50)$$

Thus if all atoms in a structure are of the same type and if the weighting function in the least squares refinement is taken as $1/f_j$,[48] the resulting coordinates will be the same as those from ΔF or corrected F_o syntheses. For structures with different types of atoms, the results are identical only for those atoms of type j. They are closely similar, however, for other atoms because the scattering curves are similar except for scale.

Although the weights $1/f_j$ are not often used in least squares refinement, they resemble commonly used weighting functions. Thus the results from the two methods of refinement are closely comparable and usually agree within the uncertainties of each.

16.10. PRACTICAL CONSIDERATIONS IN LEAST SQUARES REFINEMENT[49,50]

To illustrate refinement of a model by the method of least squares, consider a small organic structure with 8 hydrogen and 10 nonhydrogen atoms in the asymmetric unit and four asymmetric units in a centrosymmetric cell. If Cu K_α radiation is used, it is reasonable to expect about 1000 independent observations for a structure of this size, and we shall assume the accuracy of the data to be rather high. Assume $R = 0.35$ at the beginning of the refinement.

In the initial stage, computational effort will be reduced if refinement is restricted to the scale factor, a single overall isotropic thermal parameter, and the 30 positional parameters of the nonhydrogen atoms, a total of 32. The structure is overdetermined by a factor of $1000/32 \approx 31$, that is, there are 31 observations for each parameter to be determined. If full matrix least squares is used, two or three cycles are usually sufficient to reach convergence, at this point the limit imposed by the model.[51] The residual R will

[48]This weighting has actually been used in initial least squares refinements in order to obtain rapid convergence. See N. D. Jones, R. E. Marsh, and J. H. Richards, *Acta Cryst.*, **19**, 330 (1965).

[49]See Rollett, *Computing Methods in Crystallography*, pp. 47–56.

[50]R. A. Sparks, in *Computing Methods and the Phase Problem in X-Ray Crystal Analysis*, R. Pepinsky, J. M. Robertson, and J. C. Speakman, Eds., Pergamon, London, 1961, pp. 170–187.

[51]Experience indicates that for acentric structures an extra cycle or two are generally required.

probably fall no lower than a value in the range 0.12–0.20 if there are pronounced differences in the thermal motion of the atoms or if the thermal motion is quite anisotropic.

The next stage is to improve the model by introducing a thermal parameter for each atom. These allow for differences in motion among the atoms, but continue to model the motion as isotropic. The number of parameters is now increased to 41 (one scale factor and 40 atomic parameters), and the overdetermination is reduced to $1000/41 \approx 24$. Two more refinement cycles can be expected to reduce R another 0.03–0.05, although the decrease may be larger or smaller than this depending mainly on how well the model accounts for thermal motion.

If the ultimate in refinement is desired, anisotropic thermal parameters should be introduced and are warranted in this case since the data are assumed to be accurate. Six parameters are required to model the anisotropic motion of each atom (see Appendix B), increasing the total number of parameters to 91 (one scale and 90 atomic parameters) and reducing the overdetermination to $1000/91 \approx 11$. The R index should drop appreciably, often another 0.03–0.05, but how much would depend on the anisotropy of the thermal motion for the particular case.

The final step is the introduction of hydrogen atoms. Their approximate positions can often be calculated from the positions of the nonhydrogen atoms and known bonding geometries, or they can be located from the appropriate $(F_o - F_c)$ synthesis. If three positional and one thermal parameter are included for each hydrogen atom, an additional 32 parameters are added, giving a total of 123, and the overdetermination is reduced to $1000/123 \approx 8$. Two final least squares cycles should be computed, and R can be expected to decrease by 0.02 to 0.03, the final R values probably falling in the range of 0.04 to 0.05. The estimated standard deviations in the positions of the nonhydrogen atoms are expected to be in the range of 0.003 to 0.005 Å, although this will depend on the atom type (C, N, or O) and the magnitude of the thermal parameters. Greater error is associated with atoms having greater thermal motion. A more elegant model would allow for lone-pair and bonding electrons, but such a model is justified only under special conditions and for data of the very highest quality.

For structures of any appreciable complexity, the function $\sum w(|F_o| - |F_c|)^2$, usually minimized by least squares, must have countless local minima. If the trial model is a poor approximation of the real structure and beyond the convergence range of the true structure, sometimes taken to be about $d_{min}/2$ of the data, it will either diverge or converge to a false minimum. The initial R is likely to be disappointingly high and edge downward slowly or haltingly cycle by cycle. This behavior signals major difficulty. It is tempting at this point to attempt to edit the model, adding or deleting atoms as suggested by ΔF maps with or without a rejection test, but it is possible to spend a great deal of time and effort pursuing phantoms in this way. If this course is followed, it is expedient to hold both the scale factor on F_o and the overall temperature factor constant, unless one has the assurance that better values can be substituted. If the trend in R is not unmistakably downward after a very few model-editing cycles, one alternative is to try more brutal means of refinement. Reducing the $(\sin \theta)/\lambda$ cutoff until the atoms are just resolved, or even below that point, may release them so that new and better positions can be found. The use of drastic weighting functions such as $|F_c|/|F_o|$ in difference syntheses and least squares may give sufficient weight to the terms whose phases are reliable that they can overpower the unknown ones that are in error. If this approach fails to work promptly, it is necessary to return to the methods of Chapters 11–15 and redetermine a new model.

The initial dramatic decrease in R, 0.15 to 0.25 in the first few refinement cycles of the hypothetical case just discussed, is typical of models that are within the convergence range of the true minimum and is to be expected as the model settles in. It is also possible, however, for R to experience this rapid drop only to hold up at a lower but unacceptable value. This behavior generally reflects a model that is basically correct but contains some significant systematic error. It may be that a substantial part of the model is within the convergence range of the true minimum but a portion of it is seriously in error. In cases such as this, refinement with subsidiary conditions may be sufficient to pull the offending atoms of the model into convergence range.[39]

Alternatively, the molecular structure may be correct, but the molecule may be incorrectly placed in the cell.[52] We experienced just this behavior in refining the first trial model of rubrofusarin ($C_{15}H_{12}O_5$).[53] The R factor decreased from 0.46 to 0.30 in the first refinement cycle, then only to 0.23 in the next several cycles. The trouble was traced to an error in sign determination (the structure is centrosymmetric, space group $P2_1/c$). A weak relation in the sign-determining process inverted the signs of 20 reflections of the 123 initially determined. Nevertheless, 19 of the 20 nonhydrogen atoms were apparent in the first electron density map, but the molecule was lodged in a false minimum less than 0.25 Å from the true one. Surprisingly, neither ΔF syntheses nor full matrix least squares succeeded in moving the molecule into the correct position.[54] After determining a new trial model that refined satisfactorily, we found that least squares with unconventional weights, F_c/F_o altered from one cycle to the next, moved the molecule into the correct minimum.

If one or more heavy atoms are dominant in a structure and are at or near special positions, caution must be exercised to ensure that the lighter

[52]Such an error is equivalent to fitting correctly all the *intra*molecular vectors in the Patterson function but having errors in the *inter*molecular ones.

[53]G. H. Stout and L. H. Jensen, *Acta Cryst.*, **15**, 45 (1962).

[54]G. H. Stout and L. H. Jensen, *Acta Cryst.*, **15**, 1060 (1962).

atoms are not misplaced. In the example of the $CdCl_2$ salt of α-glycerophosphonylcholine trihydrate ($C_8H_{20}NO_6P \cdot CdCl_2 \cdot 3H_2O$) already cited (Section 14.3), R for the Cd and two Cl atoms alone is 0.35. The trial model, including the other 19 nonhydrogen atoms in quite incorrect positions (but related to the correct ones), converged at $R = 0.23$. With the lighter atoms corrected, refinement converged at R near 0.10. The limitation at that point was not the model, but rather the Cu K_α data, which had not been corrected for the relatively high absorption in the thin, tabular crystals. Refinement against data collected with Mo K_α radiation converged at $R = 0.04$.

When individual atom thermal parameters have been introduced in the refinement, they may provide clues to problems with the model. Expressed in terms of B values, thermal parameters for organic structures often fall in the range 3–5 Å^2. In the case of structures tightly laced together by hydrogen bonds, B values may be as low as 2 Å^2 or even lower. On the other hand, if the molecules are loosely bound together by weak forces, for example, van der Waals forces, B values for some atoms may range upwards to 20 Å^2 and beyond, corresponding to rms atomic excursions in excess of 0.5 Å. For such large oscillations, atomic motion becomes anharmonic, but more important the B values may conceal disorder; that is, one or more atoms of the structure may be distributed over two or more discrete sites. If the temperature factors are expressed in the general anisotropic form, disorder will often be evident in large differences in the B_{ii} (or u_{ii}) parameters and is most easily visualized by plotting the thermal ellipsoids. Large thermal parameters cannot properly model multiple-site atomic positions, and for such models R will hold up at values greater than expected. Thus the refinement of dihydrothymine ($C_5N_2O_2H_8$)[55] converged at $R = 0.073$. The B_{11} parameters for the nonhydrogen atoms fall in the range 5–10 Å^2 except for two of the carbon atoms, which had values of 15 Å^2 and 23 Å^2, suggesting disorder. Accordingly, each of the two sites was modeled as two carbon half-atoms 0.6 Å apart. Refinement of this model converged at $R = 0.043$ with occupancies of the disordered sites in the ratio of 6:4. A statistical test (see Section 17.5) comparing the two results on the basis of R favored the disordered model so strongly as to effectively reject the ordered possibility.

In severe cases of disorder it is sometimes possible to recognize in the Fourier map peaks corresponding to both models (for two orientations) and sometimes to unravel the structure in terms of two overlapping molecules, each weighted by an "occupancy factor" depending on the fraction of each model present in the crystal.[56] The results are rarely entirely satisfactory, however. Such disorder often suggests its presence by causing excessive diffuse scatter, either as halos or streaks, around intense reflections, and

[55] S. Furberg and L. H. Jensen, *J. Am. Chem. Soc.*, **90**, 470 (1968).
[56] See, for example, A. W. Hanson, *Acta Cryst.*, **19**, 19 (1965).

with a little experience this will serve as a warning before the intensity data are collected.

An incorrect space group assignment will lead to a model that is either trivially or seriously in error and one that may fail to refine. Misassignment of the space group (see Section 5.12) is an error that occurs more often than it should.[57] Some space groups, such as *Cc* and $C2/c$, are indistinguishable on the basis of systematic extinctions and may prove troublesome in practice. Although statistical tests or known molecular features can provide additional information, they are not always definitive, especially in those cases in which the real structure has almost but not quite the higher symmetry.

The uncertainty in space group identification is often associated with the question of disorder, since the choice is often between a high-symmetry space group with disorder or a lower-symmetry one without. Can one still determine the space group? It is often assumed that refining the model will provide the answer, but consider the following example. Crystals of the binuclear iron structure shown in Fig. 16.15 are monoclinic, and the extinctions indicate the space group to be either *Cc* or $C2/c$. Statistical tests strongly favor the latter. Direct phase determination positioned the approximate 2-fold axis of the molecule coincident with the 2-fold axis of $C2/c$. Refinement in this space group converged at $R = 0.079$ with apparent disorder in the two phenyl groups and the axial substituent on the middle μ-oxo bridge. Choosing for each of the phenyl rings one of the orientations implied by the ends of the thermal ellipsoids and refining in

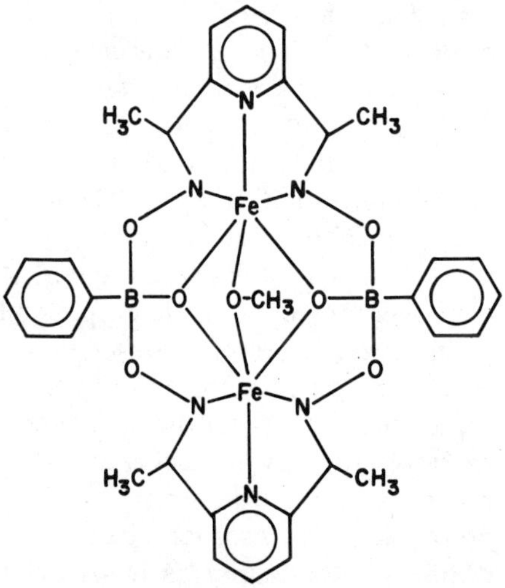

Figure 16.15. Chemical structure of a binuclear iron complex.

[57]For an extreme example of the effects of a wrong space group assignment, see G. H. Stout, S. Turley, L. C. Sieker, and L. H. Jensen, *Proc. Nat. Acad. Sci. USA*, **85**, 1020 (1988).

space group *Cc* reduced R to 0.059 and gave a unique, chemically sensible orientation for the axial substituent. According to the Hamilton R test (see Section 17.5), the lower R for the refinement in *Cc* is highly significant, and this space group is favored as giving the more probably correct description of the crystal.[58] Nevertheless, because most of the structure is essentially centrosymmetric, correlations between many of the atoms must be high.

It is also possible that the data do not really correspond to a good structural model. This may occur because of gross errors in the data-collection process: consistent misindexing, intermittent noise in quantum counters and their associated electronics, very bad level-to-level scaling, mishandling of the data by one of the programs used, or merely very badly measured intensities. Many of these difficulties will be exposed if the data are studied for a systematic pattern of errors, and if the $|F_{rel}|$ values are compared against the original intensity record, films, or strip-chart record.

For large-scale problems, say for structures with hundreds or thousands of parameters, the block-diagonal approximation has much to recommend it. The storage requirements and the computing time involved in building a block-diagonal matrix are approximately proportional to the number of parameters. Therefore, as the magnitude of the problem increases the block diagonal has a progressively greater time advantage over the full matrix. However, the results do not converge as rapidly as those from the full matrix, and there is sometimes a tendency to overcorrect, resulting in oscillation from one cycle to the next. To eliminate such behavior and to improve the rate of convergence, damping factors can be applied, that is, the shifts in parameters can be multiplied by a factor less than 1.0. Because of the slower convergence of block-diagonal refinement, it is questionable whether this method actually provides faster refinement than full matrix methods when the best results are desired. The individual cycles are faster, but more, sometimes many more, are required to ensure reaching the limit. Thus the major advantage of the block diagonal is merely the lessened storage requirements for structures that are too large to be otherwise accommodated. Even these cases are often now treated by being partitioned into a few large blocks, which can be treated individually as full matrix problems. In order to achieve some connection between these blocks, they are usually overlapped by including in each cycle one or more atoms common to all the blocks.[59]

The diagonal least squares approximation requires even less storage and computing time than the block diagonal, but its use cannot be recommended, in particular because the off-diagonal terms correlating the

[58]It is important to note that *Cc* is accepted over *C*2/*c* not only (or not even mainly) because of its lower value on the R test, but also because it gives chemically and structurally sensible results. See W. H. Baur and E. Tillmanns, *Acta Cryst.*, **B42**, 95 (1986).

[59]For an extreme example, see K. D. Watenpaugh, L. C. Sieker, J. R. Herriott, and L. H. Jensen, *Acta Cryst.*, **B29**, 943 (1973).

scale factor and thermal parameters are not small. However, a computation including a 2×2 block involving a scale factor and an overall thermal parameter *may* work satisfactorily for structures in which the angles between the unit cell axes do not differ greatly from 90°. Where this is not true, other off-diagonal terms are not small, and either the block diagonal or full matrix should be used.[60]

BIBLIOGRAPHY

Blundell, T. L., and L. N. Johnson, *Protein Crystallography*, Academic, London, 1976, pp. 420–442.

Booth, A. D., *Fourier Technique in X-Ray Organic Structure Analysis*, Cambridge University Press, Cambridge, 1948.

Buerger, M. J., *Crystal-Structure Analysis*, Wiley, New York, 1960, pp. 585–616.

Dunitz, J., *X-Ray Analysis and the Structures of Organic Molecules*, Cornell University Press, Ithaca, NY, 1979, Chapter 4.

Glusker, J. P., and K. N. Trueblood, *Crystal Structure Analysis*, 2nd ed., Oxford University Press, New York, 1985, Chapter 11.

Hamilton, W. C., *Statistics in Physical Science*, Ronald, New York, 1964, pp. 124–157.

Isaacs, N., in *Computational Crystallography*, D. Sayre, Ed., Clarendon Press, Oxford, 1982, pp. 381–408.

Lipson, H., and W. Cochran, *The Determination of Crystal Structures*, Cornell University Press, Ithaca, NY, 1966, pp. 317–351.

Luger, P., *Modern X-Ray Analysis on Single Crystals*, de Gruyter, Berlin, 1980, Chapter 6.

Rollett, J. S., in *Computing Methods in Crystallography*, Pergamon, Oxford, 1965, pp. 32–37, 47–56.

Schwartzenbach, D., *et al.*, *Acta Cryst.*, **A45**, 63 (1989).

Whittaker, E. T., and G. Robinson, *The Calculus of Observations*, 4th ed., Blackie, London, 1944, Chapter IX.

Wyckoff, H. W., C. H. W. Hirs, and S. N. Timasheff, Eds., *Diffraction Methods for Biological Macromolecules*, Part B, *Methods in Enzymology*, Vol. 115, Academic, New York, 1985, pp. 23–41, 227–337.

[60]For a comparison of the results from several methods of refinement in an actual case, see R. Mason, *Acta Cryst.*, **17**, 547 (1964).

CHAPTER 17

RANDOM AND SYSTEMATIC ERRORS

The true value of an experimentally measured quantity can never be known. The best that can be done is to use in its place the mean value of a number of repeated measurements and attach to this value some measure of the confidence that can be placed in it. This measure can be derived from the scatter of the individual measurements and is usually expressed by attaching an error to the mean value. From this error it is possible to determine the probability that the true value is within any specified limit of the mean value provided the individual measurements are affected only by *random errors*. The first part of the present chapter treats such errors, the standard deviation (σ) used to specify the errors, and how σ values are derived for the results of X-ray structure determination.

There is another class of errors that affect measurements in a systematic way. If such *systematic errors* are present, they will impair the significance of the results and may affect inferences based on the results. The last part of the chapter is concerned with such errors and treats some common examples in X-ray structure analysis.

17.1. RANDOM ERRORS[1–3]

The basis for the prominence of the Gaussian or normal error function in statistical considerations is the central limit theorem of probability theory.

[1]Mendenhall, *Introduction to Statistics*.
[2]E. T. Whittaker and G. Robinson, *The Calculus of Observations*, 4th ed., Blackie and Son, London, 1944, pp. 164–207.
[3]Hamilton, *Statistics in Physical Science*.

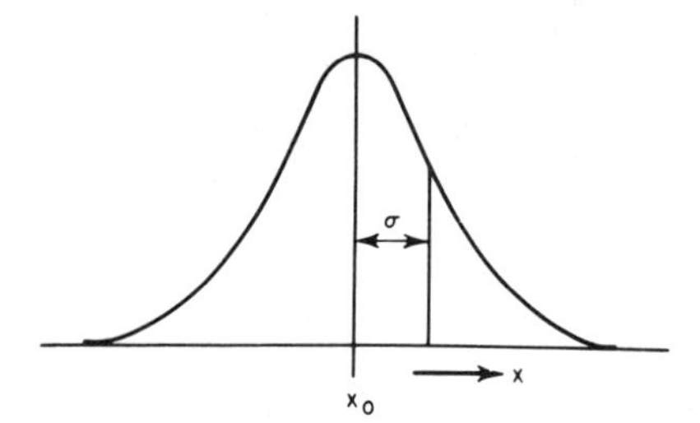

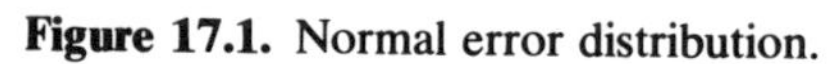
Figure 17.1. Normal error distribution.

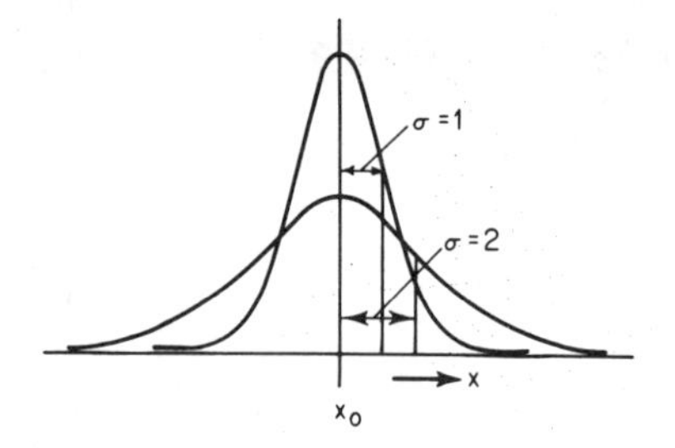

Figure 17.2. Two normal error curves showing effects of changing σ.

This theorem asserts that the sum of n variables tends toward the normal error distribution as n increases, regardless of the distributions of the individual variables. The normal error function is shown in Fig. 17.1 and is expressed by the equation

$$N = \frac{1}{\sigma\sqrt{2\pi}} e^{-(1/2\sigma^2)(x-x_o)^2} \tag{17.1}$$

where N is the relative frequency with which the value x is obtained, x_o is the *true value* of the parameter x, and σ is a constant defining the spread for a given function. Figure 17.2 shows two normal error curves plotted for two different values of σ. If the area under the normal error curve is normalized to unity, as it is in Eq. (17.1), then the area between any two limits of x is just the probability that the value found for any single measurement will fall between these limits.

Equation (17.1) represents a continuous distribution. When measurements of x are made experimentally, however, each observation will result in a discrete value for x, and a set of measurements will give rise to a discrete rather than a continuous function. Such a distribution can be plotted as a histogram. This has been done in Fig. 17.3 for the data in Table 17.1, where the number of measurements of x in each of 15 equal intervals is represented by the length of a vertical bar. For comparison, a normal error curve with $x_o = 20$ and $\sigma = 0.5$ has been superimposed in the histogram. This curve represents the limiting case for the function where the intervals Δx become smaller and smaller while the number of observations increases without bound.

It is clear from Fig. 17.1 that the normal probability function is characterized by a central value x_o, but several measures have been used to describe the spread of the function. The most important of these is the standard deviation σ, but the average error α and the probable error r have also been used.

The standard deviation σ for a set of m measurements is defined as the

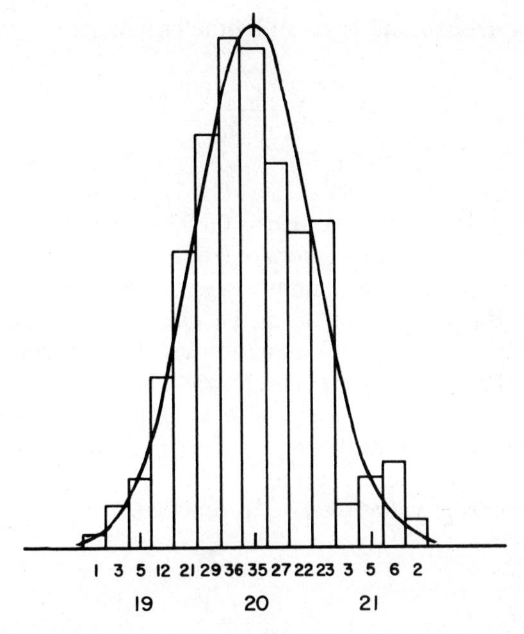

Figure 17.3. Histogram for data from Table 17.1, with superimposed normal error curve.

TABLE 17.1 Hypothetical Data Illustrating a Normal Distribution with Mean Value 20.0 and Standard Deviation 0.5

Range in x	Number of Measurements in Each Range
18.5–18.7	1
18.7–18.9	3
18.9–19.1	5
19.1–19.3	12
19.3–19.5	21
19.5–19.7	29
19.7–19.9	36
19.9–20.1	35
20.1–20.3	27
20.3–20.5	22
20.5–20.7	23
20.7–20.9	3
20.9–21.1	5
21.1–21.3	6
21.3–21.5	2

root-mean-square error and is given by the equation

$$\sigma = \left[\sum_{r=1}^{m} \frac{(x_r - x_o)^2}{m} \right]^{1/2} \tag{17.2}$$

It is this quantity that appears in Eq. (17.1). The area under the normal error curve between the limits $x_o \pm \sigma$ is 0.683 of the total, so that in a series of measurements with random errors one expects on the average two out of every three to fall within these limits.

In practice, Eq. (17.2) can rarely be used to evaluate σ because the errors, $x_r - x_o$, that appear in this equation involve the true value x_o, which, as indicated at the outset, cannot be known. The best that can be done is to use the mean $\langle x \rangle$ as an approximation to x_o. Thus instead of the true errors we have the deviations from the mean, $x_r - \langle x \rangle$. Because $\langle x \rangle$ is derived from a sample of m measurements, its direct substitution for x_o in Eq. (17.2) will lead to overoptimistic values for σ. In practice, therefore, this equation is replaced by

$$s = \left[\sum_{r=1}^{m} \frac{(x_r - \langle x \rangle)^2}{m-1} \right]^{1/2} \tag{17.3}$$

where s, calculated from a sample of m measurements, is an approximation of σ. The value of s calculated in this way is unbiased in the sense that a large number of such determinations will scatter evenly around the true value of σ. However, because of this scatter any single s will be an uncertain estimate of σ in the same sense that $\langle x \rangle$ is an uncertain estimate of x_o. If the sample size is large (>30), this uncertainty is reduced to the point where s and σ can be used interchangeably. With smaller samples, however, the inferences based on s must be regarded with more caution than those based on the true σ. Unfortunately, it is common to denote estimates of a standard deviation by σ with little regard for the size of sample involved and to treat these values as though they were standard deviations. We will follow this practice of using the symbol σ for such estimates since the parameters from structure refinements are usually based on a very large number of observations, but we will refer to them as *estimated standard deviations*.

17.2. NORMAL AND HALF-NORMAL PROBABILITY PLOTS[4]

Probability plots provide a sensitive means of comparing statistically any two independent sets of observed or derived quantities. The observed δ_j

[4]S. C. Abrahams and E. T. Keve, *Acta Cryst.*, **A27**, 157 (1971).

values, defined by the equation

$$\delta_j \equiv \frac{f_{1,j} - f_{2,j}}{(\sigma^2 f_{1,j} + \sigma^2 f_{2,j})^{1/2}} \tag{17.4}$$

are plotted against the expected values of δ_j derived from any assumed distribution. The observed δ_j are then arranged in order of increasing value from negative to positive. If the distribution is assumed to be normal, for example, the expected values of δ_j are obtained from the normal probability function in the form

$$P(x_1, x_2) = \frac{1}{\sqrt{2\pi}} \int_{x_1}^{x_2} e^{-x^2/2}\, dx \tag{17.5}$$

or from tables of the function.

To illustrate the determination of the expected δ_j for a normal distribution, assume an ordered list of 100 observed δ_j values. For the first point, δ_1, the expected value will be the midpoint in the probabilistic sense of the integral

$$\frac{1}{\sqrt{2\pi}} \int_{-\infty}^{x_1} e^{-x^2/2}\, dx = 0.01 \tag{17.6}$$

that is, the point where the integral has the value $0.01/2 = 0.005$. From a

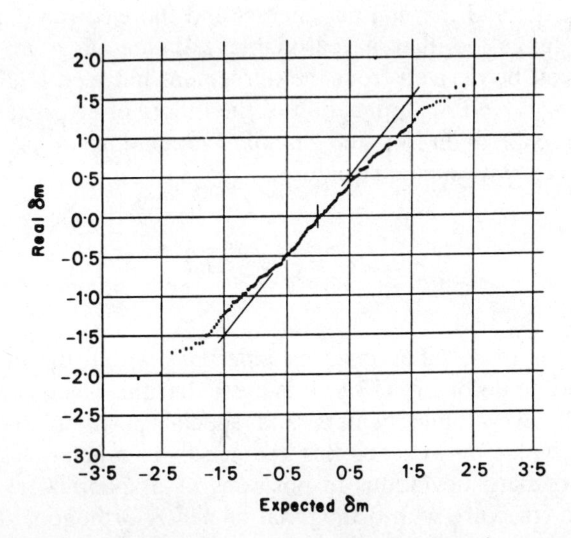

Figure 17.4. Normal probability plot of 254 real δm_j based on measurements of $BaCoF_4$ structure factors using two different crystals. [Modified from S. C. Abrahams and E. T. Keve, *Acta Cryst.*, **A27**, 157 (1971) with permission.]

table of the integral, this point is found to be -2.575. Since the normal error curve is symmetric about the point $x = 0$, the expected value for the last point in the ordered list, δ_{100}, is $+2.575$. Similarly, the expected δ_2 will be the point where the value of the integral is $0.01 + 0.01/2 = 0.015$, found to be -2.17. For δ_{99}, the next to the last point, the expected value is $+2.17$. In a similar way, the expected values for the remaining pairs of δ_j's are evaluated.[4]

In a normal probability analysis plot, the $\delta_{j,\text{observed}}$ values are plotted as ordinates against $\delta_{j,\text{expected}}$ as abscissa. If the distribution matches the one assumed and no systematic errors are present, the plot will be linear with a slope of unity and pass through the origin. Figure 17.4 shows an example.

In cases where only magnitudes are to be compared, the statistic is

$$|\delta_j| = \frac{\big||f_{1,j}| - |f_{2,j}|\big|}{[\sigma^2_{f1,j} + \sigma^2_{f2,j}]^{1/2}} \tag{17.7}$$

and the $|\delta_j|$ are ordered from zero to the maximum. These $|\delta_j|$ values are plotted against the expected δ_j as a half-normal probability plot.

17.3. ESTIMATED STANDARD DEVIATIONS FROM FOURIER REFINEMENT

The immediately derived results from the refinement of a structure are the positional, scale, and thermal parameters and the electron density. Cruickshank[5] has suggested that a reasonable estimate of σ in any of these parameters can be derived from the agreement between $|F_o|$ and $|F_c|$, that is, from $\Delta F = |F_o| - |F_c|$. Application of the theory of errors then leads to an approximate expression for the standard deviation in ρ_o, the electron density in a *centrosymmetric* structure:

$$\sigma_{\rho_o} = \frac{1}{V}\sqrt{\sum \Delta F^2} \tag{17.8}$$

where $\sum \Delta F^2$ is to be taken over *all reflections* within the limiting sphere. From the derivation of Eq. (17.8) it is clear that this is an average σ_{ρ_o} over the unit cell. At atomic centers and special positions such as $0, 0, 0$, however, the value may exceed this average by as much as $\sqrt{2}$.

For the standard deviations in positions of a spherical atom in a centrosymmetric structure with orthogonal or nearly orthogonal axes, Cruick-

[5]D. W. J. Cruickshank, *Acta Cryst.*, (a) **2**, 65 (1949); (b) **3**, 72 (1950); (c) **7**, 519 (1954).

shank[5a,c] has derived the following expressions:

$$\sigma_x = \frac{2\pi}{aV\,(\text{curvature})}\sqrt{\sum h^2\,\Delta F^2}; \qquad \sigma_y = \frac{2\pi}{bV\,(\text{curvature})}\sqrt{\sum k^2\,\Delta F^2};$$

$$\sigma_z = \frac{2\pi}{cV\,(\text{curvature})}\sqrt{\sum l^2\,\Delta F^2} \tag{17.9}$$

where $\sum \Delta F^2$ is again to be taken over *all reflections* within the limiting sphere; a, b, c are the lengths of the unit cell axes in angstroms; and V is the volume of the unit cell in cubic angstroms *if curvatures are expressed in* e/Å^5. While curvatures for C, N, and O atoms could be taken from Table 16.1 it is better to calculate them for each atom from Eq. (16.6)[6] or (16.12). Any terms that may have been omitted from the refinement should also be omitted in calculating curvatures.

For a *noncentrosymmetric* structure, Cruickshank[5b] has further shown that the σ's in Eqs. (17.8) and (17.9) are to be *increased by a factor of* 2. This factor arises because some of the error in parameters is absorbed in shifting the phase angle α from its true (but unknown) value and is not reflected in ΔF.

If reflections of appreciable intensity exist outside the volume studied in reciprocal space, precision will be improved by adding these to the data. Although this increases $\sum h^2\,\Delta F^2$, and so on, the curvature in the denominators of Eqs. (17.9) increases even more rapidly [see Eq. (16.12)] and as a result σ decreases.

A more effective means of improving precision is to collect the data with the crystal at low temperature, thereby decreasing the thermal motion of the atoms. This has the double advantage of increasing the intrinsic curvature of the electron density and at the same time extending the region in reciprocal space where reflections are of measurable magnitude. Both factors increase curvature, and at very low temperatures (liquid nitrogen and below) σ_{position} may be reduced by a factor of 2 or more for light atoms in covalent compounds.[7,8]

Precision can, of course, be improved by improving the data so that $\sum h^2\,\Delta F^2$ and thus the σ values are reduced. In particular, a few large mistakes in $|F_o|$ will result in disproportionately larger standard deviations. This stems in part from the general increase in ΔF due to the incorrect parameters resulting from attempts of the refinement process to adjust to these errors and in part from the fact that a few large ΔF's make a

[6]Note that curvatures calculated from this equation are in electrons per cubic angstrom per cell edge squared and must be converted to e/Å^5 by dividing by the square of the appropriate cell edge in angstroms.

[7]F. L. Hirshfeld and G. M. J. Schmidt, *Acta Cryst.*, **9**, 233 (1956).

[8]R. D. Burbank, *Acta Cryst.*, **6**, 55 (1953).

disproportionately large contribution to $\sum(h^2, k^2, l^2)\,\Delta F^2$. It is very important, therefore, that before the refinement is complete the data be stringently checked to ensure that they are free of mistakes.

17.4. ESTIMATED STANDARD DEVIATIONS FROM LEAST SQUARES[9,10]

In the least squares refinements of a structure, it is standard practice to solve the normal equations by matrix methods and to obtain both the new values of the parameters and their σ's. The general equation for calculating σ for any parameter p_i is

$$\sigma_{p_i} = \left[\frac{b_{ii}(\sum w_r\,\Delta F_r^2)}{m-n}\right]^{1/2} \tag{17.10}$$

where b_{ii} is the ith diagonal element of the inverse matrix, w_r is the relative weight of the rth ΔF, m is the number of observations, and n is the number of parameters.[11] While this equation appears different from Eq. (17.9), it leads to similar results in practice and to similar conclusions concerning ways of improving precision. The effects of increasing the number of observations can be seen by noting that b_{ii} varies inversely as the number of reflections m. Thus $\sigma_{\text{parameters}}$ varies as $\sqrt{1/m}$. Furthermore, if the data are improved, $\sqrt{w}$ will increase proportionately with the increased precision because $\sqrt{w} = 1/\sigma$. Since b_{ii} varies inversely with w (see Section 16.7), it follows that $\sigma_{\text{parameters}}$ is proportional to σ_{F_o}.[12]

In addition to indicating the precision of the final results, the σ's are useful as a guide to the course of the refinement process. Refinement may be considered complete for a given model when the changes in the parameters are small compared with the corresponding σ's. If full matrix least squares is used, the convergence is usually so rapid that a ratio of parameter change to $\sigma_{\text{parameter}}$ of less than 0.3 for each parameter is satisfactory for the last cycle. Block-diagonal methods, however, may creep

[9] D. W. J. Cruickshank, in *Computing Methods in Crystallography*, J. S. Rollett, Ed., Pergamon, Oxford, 1965, pp. 112–114.

[10] *International Tables*, Vol. II, pp. 92–94, 330–331.

[11] If the absolute weights $w = 1/\sigma^2$ are used, and if the model exactly represents the physical basis for the data, the goodness of fit (GOF, see Section 16.6) $[\sum_1^m w_r/\Delta F_r^2(m-n)]^{1/2} = 1$, so that $\sigma_{p_i} = \sqrt{b_{ii}}$. In practice the GOF is often 2 or 3 for very good data, and it is common to use the full Eq. (17.10) as an estimate of the standard deviation. There is no mathematical justification for this practice even though the esd's derived from $\sqrt{b_{ii}}$ are generally too small; it is more a matter of luck than science that the numeric values of the GOF *approximate* the underestimation of σ.

[12] If a constant value is used for $\sqrt{w}$, for example, 1.0, b_{ii} will not change with improved precision, but $\sum w\,\Delta F^2$ will decrease, producing roughly the same effect.

on at this level for a number of cycles, and it has been suggested[13] that refinement should not be regarded as finished until (parameter change)/$\sigma_{\text{parameter}} \approx 0.01$.

A comparison of σ's as calculated from Cruickshank's equations with those from the inverse matrix of least squares can be obtained from the results for *n*-nonanoic acid hydrazide,[14] a structure initially refined by Fourier methods and subsequently also be least squares. The two refinements are not exactly comparable, however, since 16 of the 72 thermal parameters were fixed in the Fourier refinement while all were allowed to vary in least squares. The latter reduced R from 0.073 to 0.052. The average $\sigma_{\text{positional}}$ for a carbon atom from the final least squares inverse matrix was 0.0028 Å, in good agreement with the 0.0037 Å found from Cruickshank's equations applied to the somewhat less refined initial model.

17.5. STANDARD DEVIATIONS AND THE MODEL

It is evident in any of the equations for evaluating standard deviations that significant ΔF values may arise from the limitation of the model in representing accurately the physical situation that yields the F_o values. Thus when refinement ceases, it does not necessarily mean that the limit set by the data has been reached. This should be clear, for example, in the hypothetical example of refinement in Section 16.10. As the model was improved by adding successively individual atom isotropic thermal parameters, then anisotropic, and finally hydrogen atoms, $\sum w\,\Delta F^2$ would decrease, hence the σ's would also. In cases such as that—centrosymmetric organic crystals without heavy atoms—refinement can be expected to cease when R is about 3–5% even for the best data generally available. Careful statistical comparisons of independent sets of data indicate that the precision of the data for a single crystal can be considerably better than this, as low as 1–2%. Therefore, in such cases it is reasonable to assume that the limit of σ is set by the model, and significant and valid improvement can be expected by the introduction of an improved model that is still more sophisticated. Even under conditions where the refinement is limited by the model, however, the σ's obtained from the F's seem to be reasonable measures of the precision, suggesting that errors in the F_c model behave in a random fashion.

The above should not be interpreted, however, as implying that any change in the model that reduces R is significant and meaningful. The introduction of a suitable number of arbitrary parameters can always reduce $\sum w\,\Delta F^2$ to any desired value, and it is for this reason that the term $m-n$

[13] R. A. Sparks, in *Computing Methods and the Phase Problem in X-Ray Crystal Analysis*, R. Papinsky, J. M. Robertson, and J. C. Speakman, Eds., Pergamon, Oxford, 1961, p. 170ff.
[14] L. H. Jensen and E. C. Lingafelter, *Acta Cryst.*, **14**, 507 (1961).

appears in Eq. (17.10). Thus in least squares refinement the observation that a change in the model results in a decrease in σ implies that the improvement is greater than that which would be provided by an equal number of random parameters and is more informative than finding an arbitrarily small drop in R.

Hamilton[15] considered the question of identifying meaningful changes in R produced when the model is altered. Instead of the conventional R, he defined a weighted residual

$$R' = \left[\frac{\sum_j w_j(|F_{o,j}| - |F_{c,j}|)^2}{\sum_j w_j|F_{o,j}|^2}\right]^{1/2} \tag{17.11}$$

The test is based on the ratio of the residuals, $R'(1)/R'(2)$, obtained for two refinements from different models, although Hamilton suggests that in practice the conventional residuals $R(1)$ and $R(2)$ can usually be used since

$$\frac{R'(1)}{R'(2)} \approx \frac{R(1)}{R(2)} \tag{17.12}$$

In applying the method to the present case of two refinements based on models involving different numbers of parameters, we first calculate the ratio $R(1)/R(2)$. If $n(1)$ and $n(2)$ $[n(2) > n(1)]$ are the numbers of parameters used in the models that refine to $R(1)$ and $R(2)$, respectively, the difference $n(2) - n(1)$ is called the dimension of the test. The difference $m - n(2)$, where m is the number of observations, is the degrees of freedom. The ratio is then compared at a given *level of significance*[16] with the tabulated values for the dimensions and degrees of freedom of the particular case.

As an example, apply the test to the hypothetical model used in Chapter 16 to illustrate the course of refinement. Assume the structure with isotropic thermal parameters refined to $R = 0.10$ $[n(1) = 41]$. Introduction of anisotropic parameters $[n(2) = 91]$ and refinement decreased R to 0.07. Is the decrease significant?

The following values are calculated:

$$R(1)/R(2) = 0.10/0.07 = 1.43$$

$$\text{dimensions} = n(2) - n(1) = 50$$

$$\text{degrees of freedom} = 1000 - 91 = 909$$

[15] W. C. Hamilton, *Statistics in Physical Science*, pp. 157–162; *Acta Cryst.*, **18**, 502 (1965).

[16] The level of significance is often designated by α and is given in the heading of the tables in *International Tables*, Vol. IV, pp. 288–292. If the ratio $R(1)/R(2)$ is greater than some appropriate entry in the tables, the probability that it could arise by chance from two equally good models is α. This result is interpreted as implying that model 2 is better than model 1 at the α or $100\alpha\%$ significance level. Significance levels of 0.05 (5%) and 0.01 (1%) are commonly used and entail risks of 1 in 20 or 1 in 100 of accepting as a significant model improvement one that is actually not.

Interpolation in the table for $\alpha = 0.01$ indicates a ratio of about 1.03. Since 1.43 greatly exceeds this, α is much less than 0.01. Thus we can assert with a very high probability of being correct that the decrease is significant.

17.6. SYSTEMATIC ERRORS

The random errors considered thus far result from irregular causes that it is impossible to know or control and that give rise to the spread of the normal error curve. The term *precision* refers to the scatter of the observations and is measured by σ. The lower the value of σ, the higher precision is said to be. If only random errors are present, accuracy can be taken as equal to precision, and the standard deviations calculated from the formulas of Sections 17.3 and 17.4 define the uncertainties in the true values of the parameters.

There is another class of errors, however, that are systematic and produce systematic shifts in the results. They may be inherent in the method of observation or data processing, the apparatus used, or the observer. Such errors may or may not decrease precision. Systematic errors that do not alter the precision are those that do not change the distribution representing the random errors but only displace it. Such errors are especially insidious because there is no evidence in the data themselves to betray their presence. Other systematic errors may alter the distribution of the errors, that is, decrease precision, without shifting the mean value. The most common effect, however, is a combination of these resulting in an unknown change in the accuracy accompanied by a decrease in precision.

Systematic errors can enter in two ways: in the data or through failure of the model. We shall restrict our treatment of such errors to those that are likely to be encountered regardless of the method used to collect the data. The first three systematic errors considered here are most conveniently assigned to operate in the data, and the last two, to operate in the model.

Absorption

In general the effect of absorption is to reduce preferentially the intensity of reflections at low values of $\sin\theta$. If the crystal is regular, absorption will vary regularly with $\sin\theta$, and, in particular, absorption for a spherical crystal varies exponentially with $\sin^2\theta$. If the data from spherical crystals (or equi-inclination Weissenberg data from a cylindrical crystal) are not corrected for absorption, thermal parameters will compensate by adjusting to values less than the true values. This is an example of a systematic error that decreases accuracy but leaves precision unchanged.

For a platelike crystal such as that shown in Fig. 17.5, rays traveling more or less along (1) are more strongly absorbed than those traveling along (2). Thus reflections from planes parallel to (1) are more strongly

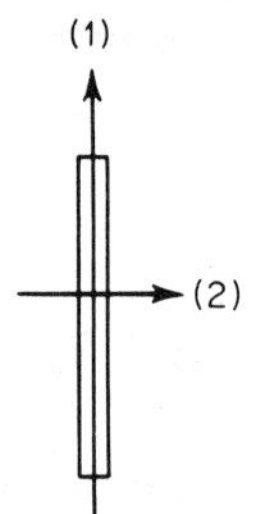

Figure 17.5. Plate crystal showing maximum and minimum directions of absorption.

altered at low sin θ, while those from planes parallel to (2) are most affected at high sin θ. The net result is a decrease in the apparent thermal parameters corresponding to vibration along (2) and an apparent increase in those corresponding to vibration along (1). When anisotropic thermal parameters have been used in the refinement, they will show values that are too small along (2) and too large along (1). In this case the thermal parameters are shifted by varying amounts from their true values but the model cannot compensate accurately for absorption. Thus both accuracy and precision are decreased.

For common crystal shapes, the cross section is approximately centrosymmetric, and the positions of the atoms are not severely affected by absorption although the thermal and scale parameters may shift greatly.[17,18] The reason stems from the centrosymmetry of the transform of the absorption effects. If the crystal shape is not centrosymmetric, however, the positional as well as other parameters will be shifted from their true values.

In order to avoid these errors, measures must be taken to reduce absorption to negligible values as outlined in Chapter 4, or the data must be corrected for it (see Chapter 7). If neither precaution is taken, allowance must be made for the possibility of absorption error in drawing inferences from the results of crystal structure refinement.

Extinction[19,20]

The effect of absorption is to attenuate the X-ray beam as it passes through the crystal. Soon after the discovery of X-ray diffraction, it was recognized that an additional attenuation can occur when the crystal is set at the Bragg angle for a reflection. Darwin,[21] who first treated the effect mathematically, termed it extinction and recognized two different kinds, which he named *primary* and *secondary extinction*.

[17] R. C. Srivastava and E. C. Lingafelter, *Acta Cryst.*, **20**, 918 (1966).
[18] P. Werner, *Acta Chem. Scand.*, **18**, 1851 (1964).
[19] R. W. James, *The Optical Principles of the Diffraction of X-Rays*, Cornell University Press, Ithaca, NY, 1965, pp. 268–299.
[20] M. J. Buerger, *Crystal-Structure Analysis*, Wiley, New York, 1960, pp. 195–204.
[21] C. G. Darwin, *Phil. Mag.*, **43**, 800 (1922).

Primary extinction results from an interference process that occurs within regions of perfectly regular lattice planes, the mosaic blocks illustrated in Fig. 4.4. Although these blocks are not directly visible, they are thought to be related to the dislocations of various kinds that can be observed in crystals. Within a single mosaic block, fixed phase relations exist between the incident and reflected rays so that a block scatters as a unit, that is, coherently, in contrast to rays from separate blocks where no fixed phase relations exist and the blocks scatter individually, that is, incoherently.

The multiple reflections within a mosaic block are portrayed in Fig. 17.6, where a set of planes at the Bragg angle reflects a part of the incident energy. The reflected rays are at the proper angle to reflect a second time from the bottom sides of the planes, and so on for multiple times, the direction of the reflected rays alternating between that of the incident and that of the once-reflected rays. Since a phase change of $\pi/2$ occurs on reflection,[22] any ray multiply reflected n times differs in phase by π from one reflected $n-2$ times. The net effect of the multiple reflections is to reduce the intensity of the reflected ray in the limit to be proportional to $|F|$ rather than $|F|^2$.[19,20]

A crystal in which $I_{hkl} \propto |F_{hkl}|$ is termed *ideally perfect*. While some naturally occurring crystals and some that can be grown under appropriate conditions can be classed as ideally perfect, most crystals embody domain sizes such that

$$I_{hkl} \propto |F_{hkl}|^n \tag{17.13}$$

where $1 < n < 2$ but usually much closer to 2.

Secondary extinction arises for reflections of such intensity that an appreciable fraction of the incident radiation is reflected at a given instant by the first mosaic blocks encountered. The deeper ones thus receive less incident intensity and therefore reflect less power than would otherwise have been the case. The magnitude of secondary extinction will depend critically on the size of the crystal; small crystals will suffer less from the

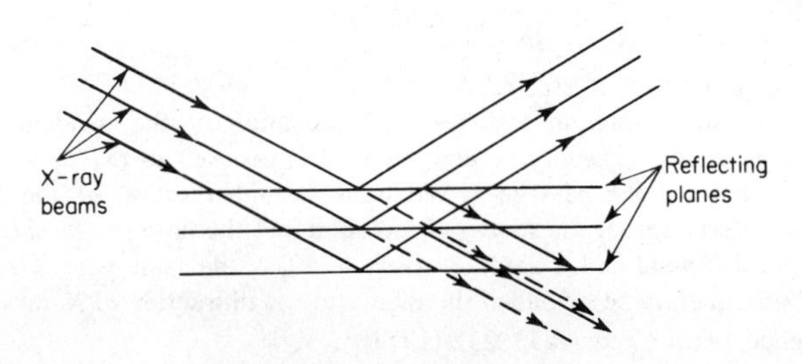

Figure 17.6. Primary extinction.

[22]See James (footnote 19), p. 60.

effect than large ones. Furthermore, crystals with less well aligned blocks will suffer less than those in which the alignment is more nearly parallel, because fewer planes are in position to reflect at a given instant. A mosaic crystal in which secondary extinction is negligible is termed *ideally imperfect*. Real single crystals can be made to approximate this state more closely by submitting them to mechanical shock, for example, by grinding their surfaces, or thermal shock, such as by repeatedly dipping them in liquid air.

The need for accurate intensities in detailed studies of electron density deformations led to renewed interest in extinction and to the development of what was termed a general theory.[23] The theory is based on Darwin's energy transfer equations, which take account of the energy exchange between incident and reflected rays within the crystal. The extinction factor for a reflection is defined by the equation

$$y = I_o^{ext}/I_k \tag{17.14}$$

where I_o^{ext} is the observed integrated intensity on an absolute scale and $I_k = |F_c|^2 Lp$, the subscript k emphasizing the kinematic approximation[24] in the derivation of the structure factor. As defined in Eq. (17.14), y is an overall extinction parameter, but the theory embodies explicitly parameters for both types of extinction, y_p and y_s.

The general theory led to a universal formula for correcting extinction that was said to be valid for the entire range of crystals from the perfect to the ideally mosaic. Application of the theory led to good corrections for secondary extinction if primary extinction was small, but the corrections were inadequate if the latter was severe.[25]

Although it was believed for many years that primary extinction was negligible in most crystals, there is increasing evidence[26–29] that it is a significant contributor in some (many?) cases. In most cases, the functional forms of the corrections are so similar that it is not possible to distinguish the two effects without additional measurements, and in practical cases it is probably best to combine the two as a single extinction correction, which has the effect of an additional absorption correction dependent on intensity.

[23] W. H. Zachariasen, *Acta Cryst.*, **23**, 558 (1967).

[24] The kinematic approximation assumes no attenuation of the incident beam in passing through the crystal, that is, every volume element is exposed to the same incident intensity. The dynamic theory of the passage of radiation through crystals, on the other hand, takes account of the interaction of the scattered radiation and the lattice. The theory was originally published by P. P. Ewald as his doctoral thesis in 1912, the same year X-ray diffraction was discovered. Subsequently he extended the theory to the diffraction of X-rays by crystals.

[25] J. L. Lawrence, *Acta Cryst.*, **A33**, 232 (1977).

[26] W. A. Denne, *Acta Cryst.*, **A28**, 192 (1972).

[27] N. M. Olekhnovich, V. L. Markovich, and A. I. Olekhnovich, *Acta Cryst.*, **A36**, 989 (1980).

[28] P. Seiler and J. D. Dunitz, *Acta Cryst.*, **A34**, 329 (1978).

[29] J. L. Lawrence, *Acta Cryst.*, **A30**, 454 (1974).

Even physical treatments such as thermal shock can affect both primary and secondary extinction and fail to distinguish the two easily.

Becker and Coppens[30] have revised and extended Zachariasen's theory and applied it in studying extinction in a number of structures,[31] but the basic limitations of the energy transfer equations remain.[25,32]

The theory of Zachariasen as revised and extended leads to satisfactory corrections if primary extinction is not severe, and it provides considerable insight concerning the domain structure of real crystals. Nevertheless, if one is interested primarily in molecular parameters, the simple one-parameter correction derived by considering extinction as an additional absorption effect is sufficiently exact. For the case of reflection through a crystal plate, corresponding closely to the normal experimental situation, the relation between the observed and calculated intensities is given by the expression

$$I_o^{\mathrm{ext}} = I_c e^{-2gI_c} \tag{17.15}$$

where g is the extinction coefficient characterizing only a particular crystal for a given radiation.

Rewriting Eq. (17.15) for I_c/I_o, expanding as a series, and neglecting terms in I_c higher than the first, results in the equation[33]

$$I_c/I_o^{\mathrm{ext}} = 1 + 2gI_c \tag{17.16}$$

A plot of I_c/I_o versus I_c should be linear with intercept 1 and slope $2g$. If extinction is present, $g > 0$, and the magnitude of the effect is measured by g, which can be evaluated directly from the plot.[34] Since Eq. (17.16) involves I_c, it cannot be used to evaluate g until refinement is well advanced.

In practice it will usually be more convenient at this stage to work with $|F_o^{\mathrm{ext}}|$ and $|F_c|$. Thus Eq. (17.16) is replaced by

$$|F_c|/|F_o^{\mathrm{ext}}| = 1 + g|F_c|^2 Lp \tag{17.17}$$

where L and p are the Lorentz and polarization factors, respectively. A plot of $|F_c|/|F_o^{\mathrm{ext}}|$ versus $|F_c|^2 Lp$ should again be linear with an intercept of 1 and now with a slope of g. The structure factors are readily corrected according to the equation

$$|F_o^{\mathrm{corrected}}| = |F_o^{\mathrm{ext}}|(1 + g|F|^2 Lp) \tag{17.18}$$

[30]P. Becker and P. Coppens, *Acta Cryst.*, **A30**, 129, 148 (1974).

[31]P. Becker and P. Coppens, *Acta Cryst.*, **A31**, 417 (1975).

[32]For a critical evaluation of the various extinction models and their range of applicability, see the excellent review by P. Becker, *Acta Cryst.*, **A33**, 243 (1977).

[33]This is a good approximation for *most* crystals if no larger than recommended for single-crystal work (see Chapter 4).

[34]I_o and I_c in Eqs. (17.14)–(17.16) are to be taken as $|F_o|^2 Lp$ and $|F_c|^2 Lp$, respectively, where L and p are the Lorentz and polarization factors (see Chapter 7).

Modern least squares refinement programs usually incorporate parameters for both y_p and y_s, but because the two are highly correlated it is usually not feasible to refine both at the same time. Thus in general it is probably preferable to set y_p or $y_s = 1$ and refine the other to absorb the effects of both types of extinction without drawing any conclusions about the physical implications of the result.

Diffuse Scattering

In addition to the usual diffraction pattern that results from the periodicity of the crystal lattice, other diffractional effects arise from the presence of irregularities in the structure. These include stacking faults in layer structures, order–disorder regions common in alloys, certain kinds of structural disorders found in molecular crystals, dislocations and other defects that are present in most crystals, and unusual twinning at the molecular level such as that found in decaborane.[35,36] These and other structural irregularities lead to diffuse diffraction effects that on film may take the form of enlarged, elongated, or variously shaped diffuse reflections, highly diffuse regions, or diffuse lines joining the discrete Bragg reflections. It is desirable to discover the explanation for any such diffuse scattering before proceeding with a structure determination, since the cause could subvert the purpose of the analysis.

Another kind of scattering, *thermal diffuse scattering* (TDS), arises from the correlated displacements caused by *lattice vibrations*, that is, the entire collection of elastic waves that pervade crystals, waves ranging in length from the dimensions of the unit cell to the size of the crystal specimen. The effect of the instantaneous displacements is to add to the general background a highly diffuse component that intensifies in the region of the Bragg reflections and rises to a sharp cusp at the positions of the reciprocal lattice (r.l.) points. This is illustrated in Fig. 17.7, which shows an idealized

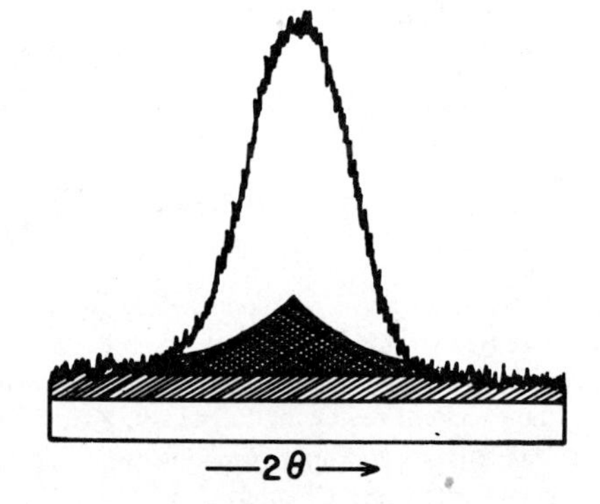

Figure 17.7. TDS contribution under the Bragg peak, represented by shaded areas. Double-shaded areas included in I_o.

[35] H. Jagodinski, *Acta Cryst.*, **A40**, C-1 (1984).
[36] J. S. Kasper, C. M. Lucht, and D. Harker, *Acta Cryst.*, **3**, 436 (1950).

diffractometer scan through a reflection exhibiting TDS. Clearly the net intensity that remains after subtracting the background includes part of the cusp, and it is evident that the longer the scan range (the farther from the reciprocal lattice point that backgrounds are measured), the greater the TDS correction. A more detailed analysis of the effect of TDS on X-ray diffraction data shows that scan mode (e.g., diffractometer ω or $\omega/2\theta$) and geometric features of the experimental arrangement, such as aperture dimensions, are important factors. TDS is usually relatively weak at a distance from r.l. points, although in general it is not spherically distributed about them, but for some crystals it can give rise to substantial intensities that appear as streaks or extra spots on diffraction photographs. In such cases special care must be exercised in collecting data.

At room temperature the magnitude of the TDS inclusion under the Bragg peak can be relatively large, especially for high-order reflections. For example, the 666 reflection of KCl at 20°C (diffractometer ω scan, Mo K_α radiation) has a TDS correction of almost 25% of the Bragg intensity.[37]

Estimates have been made of errors in atomic parameters caused by TDS.[38] The results for dibenzoyl are of particular interest in view of the relatively large TDS expected for the soft crystals with low elasticity. The experimental data were collected on a diffractometer operated in the ω scan mode. At 110 K the maximum error found in C and O positional parameters was less than 0.001 Å, much less than the standard deviations in accurate work. The thermal parameters for the same atoms, on the other hand, are systematically low because TDS increases as $(\sin^2\theta)/\lambda^2$, the errors being 3–9 times the standard deviations, even at 110 K. The decrease of TDS with decreasing temperature provides yet another incentive for collecting data at low temperature.

Errors in Scattering Factors

Scattering factors have been tabulated for all elements and are of varying accuracy depending on the wave functions used, the completeness of the mathematical treatment, and the accuracy of the numerical methods used. These mathematical details do not contribute substantial error to scattering factors as now calculated, but an additional source of error does arise in practice because the scattering contribution from outer electrons either in bonds or in nonbonding orbitals is allowed for only in an average sort of way in the tabulated values. The electron distribution for both the bonding and the nonbonding electrons is nonspherical, so that their scattering depends on the orientation of the reflecting planes.[39] Fortunately, neglect of

[37]B. T. M. Willis, *Acta Cryst.*, **A25**, 277 (1969).

[38]R. B. Helmholdt and A. Vos, *Acta Cryst.*, **A33**, 38 (1977); see also P. A. Kroon and A. Vos, *Acta Cryst.*, **A35**, 675 (1979) and A. Criado, A. Conde, and R. Marquez, *Acta Cryst.*, **A41**, 158 (1985).

[39]R. McWeeny, *Acta Cryst.*, **4**, 513 (1951).

these effects does not result in substantial discrepancy between $|F_o|$ and $|F_c|$ except for low-order reflections.

In one case, that of the hydrogen atom, the difference between the scattering factors of the bonded and nonbounded atoms is relatively large. The error made by using the scattering factor for a free hydrogen atom appears primarily in anomalously low values for hydrogen atom thermal parameters.[40] This follows because the electron in the bonded atom is more localized than that in the free atom. Calculation of the scattering factors for hydrogen in the hydrogen molecule results in values as much as 50% above those for the isolated atom at certain values of sin θ.[41]

An additional effect specific to hydrogen arises because the involvement of its single electron in the C–H or other bond shifts the center of gravity of the electron density toward its partner. As a result, least squares refinement of hydrogen positions from X-ray data give apparent C–H bonds that are about 0.1 Å shorter than those from spectroscopic data or from neutron diffraction, which determines the true nuclear positions.

The effects of anomalous scattering have already been treated (see Chapter 8), and these must be allowed for by the use of proper scattering factors if results of the highest precision are required. The values for f_o^{anom} can be calculated according to Eq. (8.92), where $\Delta f'$ and $\Delta f''$ are chosen for the wavelength used in collecting the data. Note that even for centrosymmetric structures the computer program must be set to calculate in the noncentrosymmetric mode. Methods have been proposed[42] for averaging F_{hkl}^{anom} and $F_{\overline{hkl}}^{\text{anom}}$ to obtain an approximation to $|F_{hkl}|$, but it would appear simpler to calculate F_c^{anom} correctly.

Failure to allow for anomalous scattering can be particularly significant in space groups for which one or more of the positional parameters can be chosen arbitrarily as not being determined by the symmetry elements. This occurs in the (noncentrosymmetric) *polar* space groups derived from the point groups 1, 2, *m*, *mm*2, 4, 4*mm*, 3, 3*m*, 6, and 6*mm* and is known as *polar dispersion error*. Neglect of dispersion can lead to positional errors for the dispersive atoms of 0.02–0.1 Å, many times the calculated σ_{position}.[43,44] It is important to consider both the model and its mirror image as possible structures when correcting for dispersion, since using $\Delta f''$ for the molecule with the wrong hand will lead to *doubling* the errors rather than correcting them. Because of the size of the effect, it is necessary to consider its presence whenever Cu K_α is used with atoms heavier than O or Mo K_α with atoms heavier than S.

[40] F. L. Hirshfeld, S. Sandler, and G. M. J. Schmidt, *J. Chem. Soc.*, **1963**, 2108 (1963); L. H. Jensen and M. Sundaralingam, *Science*, **145**, 1185 (1964).
[41] R. F. Stewart, E. R. Davidson, and W. T. Simpson, *J. Chem. Phys.*, **42**, 3175 (1965).
[42] A. L. Patterson, *Acta Cryst.*, **16**, 1255 (1963).
[43] D. W. Cruickshank and W. S. McDonald, *Acta Cryst.*, **23**, 9 (1967).
[44] H. D. Flack, in Sheldrick et al., *Crystallographic Computing*, 3, pp. 18–28.

In centrosymmetric space groups and those without polar axes, the effect of uncorrected dispersion appears in the thermal parameters rather than in positional parameters. Usually this is of lesser import, but the errors can still be large with respect to the calculated σ's.

Rotary Oscillation

Thermal motion of atoms in crystals can lead to apparent positions that differ in a systematic way from their true values. Thus if a rigid molecule executes rotary oscillation (*libration*) about an axis, the apparent atomic centers will be shifted toward the axis. This can be seen in Fig. 17.8, where the oscillatory motion about the point O causes the atomic center to describe the arc AB. The overlap of electron density when the atom is at different points such as C and D will cause the peak maximum to appear at the midpoint of the *chord CD*, displaced toward the axis about which the atom is oscillating. The combination of all possible atomic positions on the arc leads to a net displacement of the average atomic position toward the axis. To a good degree of approximation, the displacement is given by[45]

$$\delta = \frac{1}{2r}\left(\frac{u^2}{1+2pu^2}\right) \tag{17.19}$$

if the angular oscillation does not exceed about 15°. In this equation r is the distance of the atom from the axis of oscillation, u^2 is the mean square amplitude of oscillation (equal to $B/8\pi^2$), and p is the breadth parameter of the electron distribution of the atom defined by Eq. (16.8). Libration about a second axis at right angles to the first and intersecting it produces an independent shift, and the total displacement of the atom is additive.

For small molecules, root-mean-square librations approaching 10° are not uncommon,[45] so the error in the apparent position of an atom can easily amount to 0.01 Å and more, exceeding the errors from all other sources for structures based on good X-ray data.[46]

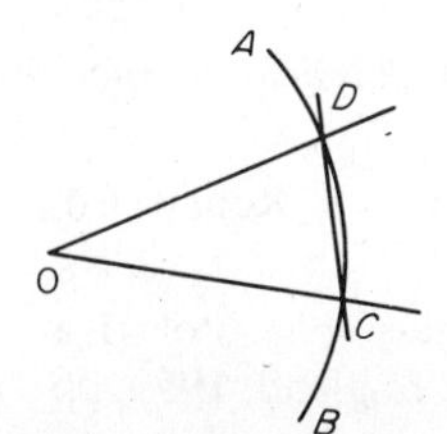

Figure 17.8. Rotary oscillation.

[45]D. W. J. Cruickshank, *Acta Cryst.*, **9**, 757 (1956); **14**, 896 (1961).

[46]C. K. Johnson, in *Crystallographic Computing*, F. R. Ahmed, Ed., Munksgaard, Copenhagen, 1970, p. 220.

The analysis of rigid-body molecular motion is not as straightforward as it may at first appear to be, and the intersection of the libration axes referred to above represents a special case. In a general treatment of the problem, Schomaker and Trueblood[47] show that rigid-body motion may be considered as a composite of three translatory motions and three screw librations about three mutually perpendicular nonintersecting axes. For details of their treatment and the resulting corrections to bond lengths, the original paper should be consulted.

Although the motions of atoms in molecules often have a rigid-body component, in any sizable molecule the actual motions are far more complex, involving all the normal vibratory modes as well as rigid-body oscillations. Busing and Levy[48] have treated the displacements caused by the relative motion of atoms in a general way and have given upper and lower limits to the displacements for certain assumed motions of the atoms. In the case of the O–H bond in $Ca(OH)_2$, for example, the uncorrected value is 0.936 Å, the lower limit is 0.956 Å, and the upper limit is 1.051 Å, with a preferred value of 0.983 Å based on a correction corresponding to what is believed to be the motion of the H atom relative to the O atom.

While the systematic errors treated here are present to a greater or lesser degree in most crystal structure determinations, they are by no means the only ones likely to occur. In particular, the equipment used in collecting the data and the way in which it is used may be subject to important systematic errors. Many of these have been mentioned in previous chapters, but for work of reasonable precision all that is required is to avoid the grossest and most obvious of these. For work of the highest precision, on the other hand, it may be necessary to carry out extensive statistical evaluations in order to determine exactly the sources and limits of error.[26] Allowance must still be made, however, for the possible presence of undetected systematic errors, and different crystallographers regard the estimated σ's with varying amounts of skepticism (see Chapter 19).

BIBLIOGRAPHY

Bevington, P. R., *Data Reduction and Error Analysis for the Physical Sciences*, McGraw-Hill, New York, 1969.

Cruickshank, D. W. J., in *Computing Methods in Crystallography*, J. S. Rollett, Ed., Pergamon, Oxford, 1965, pp. 99–116.

Cruickshank, D. W. J., in *International Tables for X-Ray Crystallography*, Vol. II, J. S. Kasper and K. Lonsdale, Eds., Kynock Press, Birmingham, England, 1959, pp. 84–96.

[47]V. Schomaker and K. Trueblood, *Acta Cryst.*, **B24**, 63 (1968).
[48]W. R. Busing and H. A. Levy, *Acta Cryst.*, **17**, 142 (1964).

Dunitz, J., *X-Ray Analysis and the Structures of Organic Molecules*, Cornell University Press, Ithaca, NY, 1979, pp. 244–265, 290–298.

Hamilton, W. C., *Statistics in Physical Science*, Ronald, New York, 1964.

Mendenhall, W., *Introduction to Statistics*, Wadsworth, Belmont, CA, 1963.

Schwartzenbach, D. *et al.*, *Acta Cryst.*, **A45**, 63 (1989).

Sheldrick, G. M., C. Krüger, and R. Goddard, Eds., *Crystallographic Computing*, 3, *Data Collection*, *Structure Determination*, *Proteins and Data Bases*, Clarendon Press, Oxford, 1985, pp. 3–32.

CHAPTER 18

DERIVED RESULTS

The interest in determining the structure of a molecule often relates to quantities that can be derived from the structural parameters. Preeminent among these are bond lengths and angles. Equations by which these may be evaluated and the accuracy of the derived bond lengths and angles are treated briefly in the first part of this chapter. We then consider least squares planes, the determination of the absolute configuration of molecules, and the interpretation of thermal parameters.

18.1. BOND LENGTHS

For a triclinic lattice, the distance between two points in fractional coordinates (x_1, y_1, z_1) and (x_2, y_2, z_2) is given by the law of cosines in three dimensions,

$$l = \{(\Delta xa)^2 + (\Delta yb)^2 + (\Delta zc)^2 - 2ab\,\Delta x\,\Delta y\cos\gamma - 2ac\,\Delta x\,\Delta z\cos\beta - 2bc\,\Delta y\,\Delta z\cos\alpha\}^{1/2} \tag{18.1}$$

where a, b, c, α, β γ are the unit-cell parameters. Distances in any other system are computed by substituting in Eq. (18.1) the parameters appropriate to the lattice.

18.2. STANDARD DEVIATIONS IN BOND LENGTHS

The general equation for an error in a function f calculated from n *uncorrelated* variables $x_1, x_2, \ldots, x_n$ is[1-3]

$$\sigma_f = \left[\sum_{j=1}^{n} \left(\frac{\partial f}{\partial x_j} \right)^2 \sigma_j^2 \right]^{1/2} \tag{18.2}$$

Applying Eq. (18.2) to the general case of a bond between uncorrelated atoms in a triclinic crystal, we have for the standard deviation in the bond length

$$\begin{aligned}\sigma_l = \Bigg[&(\sigma_{x_1}^2 + \sigma_{x_2}^2) \left(\frac{\Delta x - \Delta y \cos \gamma - \Delta z \cos \beta}{l} \right)^2 \\ &+ (\sigma_{y_1}^2 + \sigma_{y_2}^2) \left(\frac{\Delta y - \Delta x \cos \gamma - \Delta z \cos \alpha}{l} \right)^2 \\ &+ (\sigma_{z_1}^2 + \sigma_{z_2}^2) \left(\frac{\Delta z - \Delta x \cos \beta - \Delta y \cos \alpha}{l} \right)^2 \Bigg]^{1/2}\end{aligned} \tag{18.3}$$

where σ_{x_1} and σ_{x_2} are the standard deviations of atoms 1 and 2 in the x direction, with similar meaning for $\sigma_{y_1}, \sigma_{y_2}, \sigma_{z_1}, \sigma_{z_2}$; Δx is $(x_2 - x_1)$, and so on; and l is the bond length.

If the axes are orthogonal, Eq. (18.3) reduces to

$$\sigma_l = \left[(\sigma_{x_1}^2 + \sigma_{x_2}^2) \left(\frac{\Delta x}{l} \right)^2 + (\sigma_{y_1}^2 + \sigma_{y_2}^2) \left(\frac{\Delta y}{l} \right)^2 + (\sigma_{z_1}^2 + \sigma_{z_2}^2) \left(\frac{\Delta z}{l} \right)^2 \right]^{1/2} \tag{18.4}$$

If the errors are isotropic, that is, if $\sigma_{x_1} = \sigma_{y_1} = \sigma_{z_1}$ and $\sigma_{x_2} = \sigma_{y_2} = \sigma_{z_2}$, without loss of generality we can rotate the orthogonal axes to which Eq. (18.4) applies so that the bond lies along the x axis. For this case, $\Delta y = \Delta z = 0$ and $\Delta x = l$, giving

$$\sigma_l = (\sigma_1^2 + \sigma_2^2)^{1/2} \tag{18.5}$$

where the subscripts x_1 and x_2 have been replaced by 1 and 2 because the σ's are isotropic. It also follows from the isotropic σ's that Eq. (18.5) holds not only in a lattice with orthogonal axes but in any lattice.

[1]Deming, *Statistical Adjustment of Data*, pp. 38–40.
[2]Bevington, *Data Reduction and Error Analysis . . .*, Chapter 4.
[3]Consideration of this general equation leads to the ordinary expression for the standard deviation of a sum or difference; that is, if $y = x_1 \pm x_2$, then $\sigma_y = (\sigma_{x1}^2 + \sigma_{x2}^2)^{1/2}$.

Since Eq. (18.2) applies only when there is no correlation among the x_j, it follows that the positional parameters of the atoms involved in bonds must be uncorrelated for Eqs. (18.3)–(18.5) to hold. If their positions are correlated in any way the expression for σ_l must be modified. Thus, for example, the bond length between two atoms related by a center of symmetry is 2σ, where σ is measured in the direction of a line joining the atomic centers. This follows because any error in bond length caused by an error in position of one atom is just doubled by the accompanying error in the other one.

In general, any two atoms in a structure are neither completely independent nor completely correlated. The *ij*th element of the inverse matrix from a least squares refinement is a measure of the correlation between parameters i and j (see Chapter 16). Although correlations between parameters of different atoms are ordinarily small, special features such as pseudosymmetry or truncated data can result in terms of such magnitude that they cannot properly be neglected as in Eqs. (18.3)–(18.5). Furthermore, errors in unit-cell parameters also contribute[4] to the errors in bond length and must be included in an exact formulation. The completely general expression for the standard deviation in bond lengths is[5]

$$\sigma_l = \left\{\left[\sum_{i=1}^{n}\sum_{j=1}^{n}\left(\frac{\partial l}{\partial p_i}\right)\left(\frac{\partial l}{\partial p_j}\right)V_{ij}\right]+\left[\sum_{i=1}^{6}\sum_{j=1}^{6}\left(\frac{\partial l}{\partial a_i}\right)\left(\frac{\partial l}{\partial a_j}\right)U_{ij}\right]\right\}^{1/2} \quad (18.6)$$

where l is the bond length, p_i, p_j are parameters from the structure refinement (in the case of bond lengths these will be positional parameters), and a_i, a_j are unit-cell parameters. V_{ij} is related to the element b_{ij} of the inverse matrix elements of the refinement by the equation

$$V_{ij} = \left(\sum_{r=1}^{m}\frac{w_r\,\Delta F_r^2}{m-n}\right) b_{ij} \quad (18.7)$$

where w_r is the weight of the rth F_o, $\Delta F_r = |F_{o,r}| - |F_{c,r}|$, m is the number of F_o's measured, and n is the number of parameters. In a similar way the U_{ij}

[4]Errors in unit-cell parameters may not be negligible, since use of photographic methods without special techniques or equipment can lead to lengths that are in error by several tenths of a percent and angles that are in error by several tenths of a degree. An error of only 0.2% in the length of unit-cell edges will result in an error of 0.003 Å in a bond length of 1.50 Å. This equals or exceeds $\sigma_{\text{positional}}$ obtainable from the best data and cannot, therefore, be neglected in calculating $\sigma_{\text{bond length}}$.

Modern cell constants are almost always obtained from the diffractometer orientation matrix found by refinement based on the setting angles measured for a number of reflections. The precision of such constants is generally reported as high (often greater than 1 part in 5000), but the accuracy is considerably more suspect. See R. Taylor and O. Kennard, *Acta Cryst.*, **B42**, 112 (1986).

[5]D. H. Templeton, *Acta Cryst.*, **12**, 771 (1959).

are related to the inverse matrix elements from the least squares refinement of the unit cell parameters; that is,

$$U_{ij} = \left(\sum_{r=1}^{m'} \frac{w'_r \Delta\theta_r^2}{m' - n'} \right) b'_{ij} \tag{18.8}$$

where w'_r is the weight of the rth θ_o, $\Delta\theta_r = \theta_{o,r} - \theta_{c,r}$, m' is the number of measured θ_o's, n' is the number of parameters that define the lattice, and the b'_{ij}'s are the elements of the inverse matrix obtained in the refinement of the unit-cell parameters.[6]

18.3. BOND ANGLES

The angle θ subtended by bonds AB and AC (Fig. 18.1) can be calculated in various ways. If the lengths AB, AC, and BC are known, then the law of cosines provides a direct means of computing the angle:

$$\theta = \cos^{-1}\left[\frac{(AB)^2 + (AC)^2 - (BC)^2}{2(AB)(AC)}\right] \tag{18.9}$$

If the axes of the unit cell are orthogonal, the angle can be calculated from the direction cosines of the line segments AB and AC:

$$\theta = \cos^{-1}(l_1 l_2 + m_1 m_2 + n_1 n_2) \tag{18.10}$$

where l_1, m_1, n_1 are the direction cosines of AB and l_2, m_2, n_2 are the direction cosines of AC.

18.4. STANDARD DEVIATIONS IN BOND ANGLES

If the errors are isotropic and the positions of atoms A, B, and C, Fig. 18.1, are uncorrelated, the standard deviation in a bond angle is given by the

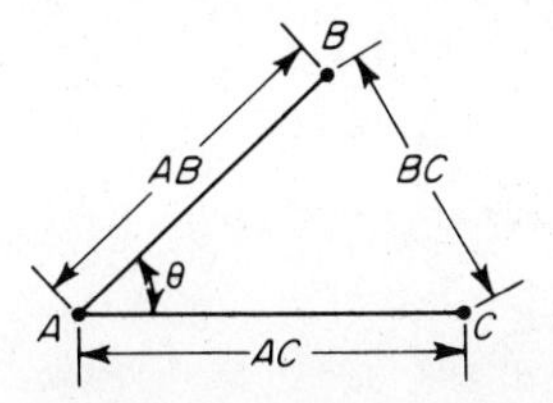

Figure 18.1. Bond angles.

[6]If the unit-cell parameters have not been obtained by a process of least squares (see Section 5.11), U_{ij} with $i \neq j$ can be taken as zero and U_{ii} set equal to the estimated σ_i^2 for the ith parameter of the unit cell.

equation

$$\sigma_\theta = \left[\frac{\sigma_B^2}{(AB)^2} + \frac{\sigma_A^2(BC)^2}{(AB)^2(AC)^2} + \frac{\sigma_C^2}{(AC)^2}\right]^{1/2} \qquad (18.11)$$

where σ_A, σ_B, σ_C are the standard deviations in the positions of the atoms A, B, C.

If the errors are not isotropic, the expression is considerably more complex. It has been given for the case of orthogonal axes by Darlow,[7] and the general case can be treated by an equation similar to (18.6).

18.5. COMPARISON OF BOND LENGTHS AND ANGLES[8]

What is often required from the results of an X-ray diffraction study is a decision whether a bond length or angle in a structure differs significantly from some other bond length or angle in the same or a different structure. Although, as indicated in Chapter 17, we can never know the true value of experimentally derived quantities, we can specify the probability that two measurements of equal quantities will agree within certain limits if only random errors are present.

Table 18.1 tabulates the probabilities[9] p of finding a normally distributed variable more than $\pm\lambda\sigma$ away from its true value. These values will permit us to answer the question: What is the probability that a bond A would be observed to *differ* in length from another bond B when, in fact, they are the

TABLE 18.1 Normal Distribution[a]

p	λ	p	λ
1.00	0.000	0.30	1.04
0.90	0.126	0.20	1.28
0.80	0.253	0.10	1.65
0.70	0.385	0.05	1.96
0.60	0.524	0.01	2.58
0.50	0.674	0.001	3.29
0.40	0.842	0.0001	3.89

[a]The probability that a quantity x differs from its mean value in either direction by more than $\lambda\sigma$ is equal to p.

[7]S. F. Darlow, *Acta Cryst.*, **13**, 683 (1960).

[8]See D. W. J. Cruickshank and A. P. Robertson, *Acta Cryst.*, **6**, 698 (1953). For a more general introduction to the problem of statistical inference, see W. Mendenhall, ***Introduction to Statistics***, Wadsworth, Belmont, CA, 1963, Chapter 8.

[9]This quantity is often designated as α in tables and is the same as the level of significance described in footnote 16 of Chapter 17.

same length? To indicate the significance of any difference, statisticians commonly use the following arbitrary limits:

If $p > 0.05$, the difference *is not* significant.
If $0.05 > p > 0.01$, the difference is possibly significant.
If $0.01 > p$, the difference is significant.

By reference to Table 18.1 these limits can be translated in terms of σ. Thus $p = 0.05$ corresponds to 1.96σ, and $p = 0.01$ to 2.58σ. The choice of these limits to specify significance is arbitrary, and other limits could have been used.

To compare two bond lengths, we first make the assumption that they are actually equal, that is, that the measured values correspond to two measurements of the same quantity. If this is true, the difference observed between two such measurements made repeatedly will be normally distributed around the true or expected value, zero. A test for the reasonableness of the assumed equality can then be constructed in terms of seeing how probable a single observed difference is with respect to this expected value and the uncertainties (σ's) in the individual measurements.

As an example, consider the case of a structure in which the atoms have isotropic $\sigma_{\text{position}} = 0.005$ Å. Two independent bonds are found to differ by 0.03 Å. What is the probability that if these bonds are equal they will differ by this amount? Applying Eq. (18.5) we have

$$\sigma_l = (0.005^2 + 0.005^2)^{1/2} = 0.0071\ \text{Å} \tag{18.12}$$

Since both bonds will have the same error, the standard deviation in their difference is (see footnote 3)

$$\sigma_{\text{difference}} = \sqrt{2}\sigma_l = 0.010\ \text{Å} \tag{18.13}$$

From Table 18.1, the probability is 0.01 that two bonds of the same length will be found to differ by more than 2.58(0.01 Å) = 0.026 Å, a difference less than the 0.03 Å given. Thus on purely statistical grounds, the probability is less than 0.01 that two equal bonds would actually be found to differ by 0.03 Å. The inference, therefore, is that the bonds differ.

In the above example, no allowance has been made for systematic errors. These might shift one or both of the bond lengths so as to invalidate the conclusion. Decreasing the limits suggested above for judging significance would *reduce the risk* of failing to accept two bond lengths that were in fact the same (*Type I error*). But it would also *increase the risk* of accepting as equal two bond lengths that were in fact different (*Type II error*). This emphasizes the importance of identifying and eliminating systematic errors so that statistical considerations provide a valid basis for assessing the results of experimental measurements.

18.6. AVERAGING BOND LENGTHS AND ANGLES[10,11]

It is common and legitimate to average bond lengths of the *same kind* to arrive at a *best value* for the length of a particular type of bond. A great deal of care is required, however, before deciding that bonds of the same chemical type are equivalent. For example, the C–C single bond length is often taken as 1.54 Å, the value found in diamond. The bond length tends to increase, however, with increasing substitution on the carbons. Thus the actual scatter of C–C bond lengths in a complex molecule will be several times larger than that predicted from the σ's of the determination and the assumption of a uniform bond length.

In some cases, inspection of the distribution of bond lengths to be averaged will indicate the lack of validity of the assumption of a single bond length. For example, if 23 values of bonds of supposedly equal length were distributed as in Fig. 18.2, there would be reason to assume that two populations were represented (bimodal), and it would not be legitimate to average the two groups. Differences between groups are usually not as obvious as they have been made to appear in Fig. 18.2, at least not when differences approach the precision of the experimental results and only a few lengths have been measured. On the average, only about one measurement in three will deviate from the mean value by more than σ if the distribution is normal.

In cases where the estimated standard deviations in the lengths of the bonds to be compared do not differ greatly, a test can be made based on the chi square (χ^2) distribution,[12,13] which describes the distribution of sums of squares. This is in common use among statisticians and has been tabulated for various significance levels p and degrees of freedom (Table 18.2).

To answer the question as to whether the bonds in a group are really the same kind, as judged from their length, we wish to compare σ_{sample}, the standard deviation estimated for the group, with $\sigma_{\text{refinement}}$ based on the standard deviations of the atomic positions determined from the refinement. We evaluate first

$$\sigma_{\text{sample}} = \left[\frac{\sum_m (l_m - \langle l \rangle)^2}{m-1}\right]^{1/2} \tag{18.14}$$

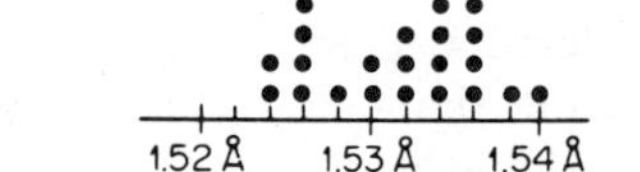

Figure 18.2. Bimodal distribution of bond lengths.

[10] Deming, *Statistical Adjustment of Data*, pp. 17–30.
[11] R. Taylor and O. Kennard, *Acta Cryst.*, **B39**, 517 (1983).
[12] W. Mendenhall, *Introduction to Statistics*, Wadsworth, Belmont, CA, 1963, pp. 197–205.
[13] H. Cramer, *The Elements of Probability Theory*, Wiley, New York, 1955, pp. 122–124, 215–228.

TABLE 18.2 The χ^2 Distribution[a]

Degrees of freedom	p								
	0.99	0.975	0.95	0.90	0.50	0.10	0.05	0.025	0.01
1	0.00	0.00	0.00	0.02	0.45	2.71	3.84	5.02	6.63
2	0.02	0.05	0.10	0.21	1.39	4.61	5.99	7.38	9.21
3	0.11	0.22	0.22	0.35	2.37	6.25	7.81	9.35	11.34
4	0.30	0.48	0.71	1.06	3.36	7.78	9.49	11.14	13.28
5	0.55	0.83	1.15	1.61	4.35	9.24	11.07	12.83	15.09
10	2.56	3.25	3.94	4.87	9.34	15.99	18.31	20.48	23.21
15	5.23	6.26	7.26	8.55	14.34	22.31	25.00	27.49	30.58
20	8.26	9.59	10.85	12.44	19.34	28.41	31.41	34.17	37.57
25	11.52	13.12	14.61	16.47	24.34	34.38	37.65	40.65	44.31
30	14.95	16.79	18.49	20.60	29.34	40.26	43.77	46.98	50.89
40	22.16	24.43	26.51	29.05	39.34	51.80	55.76	59.34	63.69
50	29.71	32.36	34.76	37.69	49.33	63.17	67.50	71.42	76.15
60	37.48	40.48	43.19	46.46	59.33	74.40	79.08	83.30	88.38
70	45.44	48.76	51.74	55.33	69.33	85.53	90.53	95.02	100.42
80	53.54	57.15	60.39	64.28	79.33	96.58	101.88	106.63	112.33
90	61.75	65.65	69.13	73.29	89.33	107.56	113.14	118.14	124.12
100	70.06	74.22	77.93	82.36	99.33	118.50	124.34	129.56	135.81

[a]Values of χ^2 as a function of the number of degrees of freedom (d.f.) and p, where p is the probability that χ^2 assumes a value greater than $\chi^2_{\text{d.f.},p}$. Adapted from the tables prepared by Catherine M. Thompson for *Biometrika*, **32**; reproduced with permission of the editors of *Biometrika*.

where l_m is the length of the mth bond, $\langle l \rangle$ is the mean length, m is the number of bonds, and $m-1$ is the number of degrees of freedom. The ratio

$$\frac{(m-1)\sigma^2_{\text{sample}}}{\sigma^2_{\text{refinement}}} = \frac{\sum_m (l_m - \langle l \rangle)^2}{\sigma^2_{\text{refinement}}} \tag{18.15}$$

follows a χ^2 distribution, and we can determine from the tabulated values the probability that any group of m identical bonds will by chance have a value exceeding this ratio.

As an example, we apply the χ^2 test to the bond lengths plotted in Fig. 18.2. Assume that σ_l from the atomic $\sigma_{\text{refinement}}$ is 0.0035 Å. The mean bond length is 1.532 Å, and the extreme values in the distribution do not differ from this by as much as 2.5σ. The ratio to be tested is

$$\frac{\sum_m (l_m - \langle l \rangle)^2}{\sigma_l^2} = \frac{480}{12.25} \tag{18.16}$$

Interpolation from Table 18.2 for 22 degrees of freedom indicates a probability of only about 0.01 that a random distribution of equal bonds would give a χ^2 value as great as this. One infers, therefore, that the bond lengths probably are not all equal.

When it is appropriate to average a set of bond lengths, the best value is given by the mean

$$\langle l\rangle = \frac{\sum_m l_m}{m} \tag{18.17}$$

or by the weighted mean if the different bond lengths have different σ's.[10] The estimated standard deviation in $\langle l\rangle$ is given by the equation

$$\sigma_{\langle l\rangle} = \left[\frac{\sum_m (l_m - \langle l\rangle)^2}{m(m-1)}\right]^{1/2} \tag{18.18}$$

Comparison with Eq. (17.3) shows $\sigma_{\langle l\rangle}$ to be smaller than σ_l by the factor $1/\sqrt{m}$.

Equation (18.17) applies when the bonds are independent. If bonds have atoms in common, however, they are not independent, and this must be allowed for in calculating $\sigma_{\langle l\rangle}$. For example, consider the case illustrated in Fig. 18.3, where AB and AC are to be averaged. Since atom A is common to the two bonds, their lengths are not independent. Taking this into account but neglecting other correlations and assuming isotropic errors, we have

$$\sigma_l = \left\{\frac{1}{2^2}\left[2\sigma_A^2\cos^2\left(\frac{\theta}{2}\right) + \sigma_B^2\right] + \frac{1}{2^2}\left[2\sigma_A^2\cos^2\left(\frac{\theta}{2}\right) + \sigma_C^2\right]\right\}^{1/2} \tag{18.19}$$

where σ_A, σ_B, σ_C are the standard deviations in positions of atoms A, B, and C and θ is the angle between the bonds. If the errors are not isotropic, the same equation still holds if σ_A is taken as the standard deviation in the direction of the bisector of B and σ_B and σ_C are in the directions of bonds AB and AC.

A similar process can be applied to bond angles, but since the energy necessary to bend bonds is not great, it is to be expected that bond angles of a given type will be quite variable, up to several degrees, depending on packing and other forces in the crystal. Thus it is of questionable value to determine "best values" from solid-state results, although it is useful to tabulate ranges for different types of bond angles.

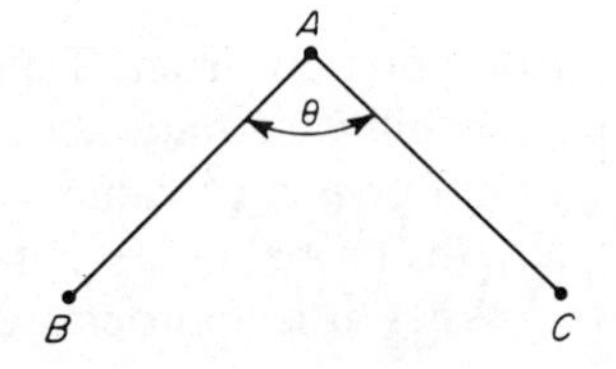

Figure 18.3. Correlation of two bonds having a common atom.

18.7. LEAST SQUARES PLANES

A common question encountered in discussing the results of crystal structure determinations is whether a set of four or more atoms is planar within experimental error. It is usual to discuss planarity in terms of the least squares plane through the set of atoms, that is, the plane that minimizes $\sum_m d_m^2$, where d_m are the perpendicular distances of the m atoms from the plane. The computations involved for the least squares plane have been described in detail and illustrated by Schomaker et al.[14]

Once such a plane has been calculated, the question remains whether the distribution of atoms around it justifies the assumption of planarity.

If the σ's of the atoms in the set do not differ greatly, the χ^2 distribution can be used to provide an answer. As in the case of bond lengths, we first evaluate σ_{plane}, the estimated standard deviation of the atoms from the least squares plane. We find

$$\sigma_{\text{plane}} = \left(\frac{\sum_m d_m^2}{m-3}\right)^{1/2} \tag{18.20}$$

The ratio

$$\frac{(m-3)\sigma_{\text{plane}}^2}{\sigma_{\text{refinement}}^2} = \frac{\sum_m d_m^2}{\sigma_{\text{refinement}}^2} \tag{18.21}$$

follows the χ^2 distribution, and from the tabulated values for $m-3$ degrees of freedom we determine the probability that a planar set of atoms would have a ratio exceeding the one found.

As an example, we test the data in Table 18.3 for planarity. The test

TABLE 18.3 A Set of Atoms for Planarity Test

Atom	$\sigma_{\text{refinement}}$ (Å)	Deviation from l.s. plane[a] (Å)
1	0.007	+0.012
2	0.007	−0.005
3	0.007	+0.001
4	0.008	−0.019
5	0.007	+0.008
6	0.008	+0.002

[a] l.s. = least squares.

[14] V. Schomaker, J. Waser, R. E. Marsh, and G. Gergman, *Acta Cryst.*, **12**, 600 (1959); see also W. C. Hamilton, *Acta Cryst.*, **14**, 185 (1961) and L. C. Andrews and H. J. Bernstein, *Acta Cryst.*, **A32**, 504 (1976).

ratio,

$$\frac{\sum d_m^2}{\sigma_{\text{refinement}}^2} = \frac{599}{54} = 11.1 \qquad (18.22)$$

where the mean-square value has been used for $\sigma^2_{\text{refinement}}$. From Table 18.2, we find this ratio significant at about the 0.01 level. Thus the probability that a truly planar set of atoms will have a χ^2 as large as that found is only 0.01, and we infer that the set of atoms is not rigorously planar within experimental error. In fact, more than three atoms are rarely found to form a true plane within the limits of this test, although they may approach it closely.

18.8. ABSOLUTE CONFIGURATION[15,16]

In 1951 Bijvoet et al.[17] showed that it is possible to determine the absolute configuration of an optically active molecule from the effects of anomalous scattering. Such a substance must crystallize in a noncentrosymmetric space group, and just how the configuration can be determined is evident by considering the Argand diagrams in Fig. 18.4. In Fig. 18.4*a*, F_{hkl} is shown, with $F_{n\,hkl}$ being the contribution from the nonanomalous scatters, and f'_{hkl} and $\Delta f''_{hkl}$ being, respectively, the real and imaginary components of those atoms showing dispersion. Figure 18.4*b* is the same construction for reflection $\overline{hkl}$. In Fig. 18.4*c* the latter has been reflected across the x axis for ease in comparing F_{hkl} and $F_{\overline{hkl}}$. As discussed in Chapter 8, it is evident that Friedel's law no longer holds, and this provides for the determination of absolute configuration. Both F_{hkl} and $F_{\overline{hkl}}$ are measured and compared with computed values. If the effects of dispersion are pronounced, only a few pairs of F_{hkl} and F_{hkl} need be compared in order to establish the configuration, and in principle only one pair would be necessary.[18] If the effects are not much greater than the experimental error of an observation,

[15] P. G. Jones, *Acta Cryst.*, **A40**, 663 (1984).
[16] D. Rogers and F. H. Allen, *Acta Cryst.*, **B35**, 2823 (1979).
[17] J. M. Bijvoet, A. F. Peerdeman, and A. J. Van Bommel, *Nature*, **168**, 271 (1951).
[18] The comparison of the observed and calculated F's is often made by tabulating $F_{o,hkl}/F_{o,\overline{hkl}}$ and $F_{c,hkl}/F_{c,\overline{hkl}}$ and comparing pair by pair. If the ratio of the observed reflections is consistently greater than 1 when the ratio of the calculated reflections is less than 1 and vice versa, the wrong configuration has been chosen. Its enantiomorph is generally obtained by changing the sign of all three coordinates, x to $-x$, y to $-y$, and z to $-z$. In many cases the same effect can be obtained by changing all of the coordinate signs associated with only *one* axis. For this to be successful, however, the axis chosen must be perpendicular to the other two, for example, monoclinic b or any in orthorhombic. Neither of these simple methods holds for enantiomorphic space groups, however.

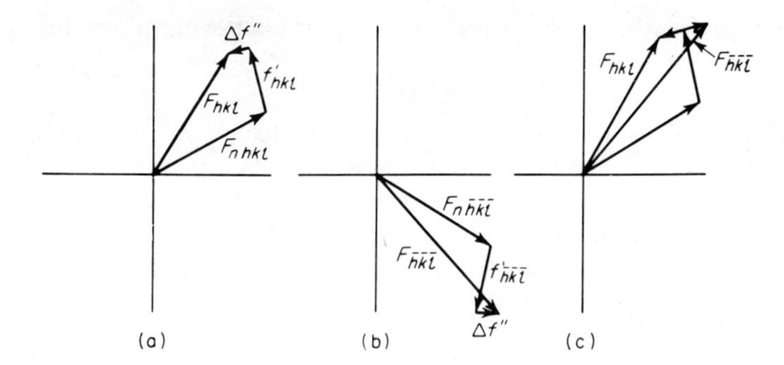

Figure 18.4. Anomalous scattering and absolute configuration. See text for discussion.

however, many pairs of observed and calculated F's might be required before the configuration could be regarded as established.[19]

The distinction is often made in the literature between Friedel pairs (F_{hkl}, $F_{\overline{hkl}}$) and Bijvoet pairs. The latter consist of any two reflections equivalent by Laue symmetry but not by the true point group symmetry of the crystal (see Section 3.6). In principle these serve as well as Friedel pairs to distinguish absolute configuration and are often used, but they are more likely to suffer experimental differences for other reasons, especially absorption.

Other methods have also been used for determining the correct hand, although careful measurement of specific pairs of Friedel-related reflections, selected on the basis of a combination of observed intensity and the magnitude of the anomalous effect, is probably the most reliable in difficult cases. On the other hand, it is often desired to determine absolute configuration from the original unique data set without additional experimental measurements. If the anomalous effect is great enough, this can be done by carrying out the refinement of the structure (without including $\Delta f''$ in the scattering factors) and then comparing the final R factors (with $\Delta f''$) for models in each hand using Hamilton's R-factor test.[20-22]

An alternative and apparently preferable method adds to the least squares refinement one additional parameter, variously called η[22] or x,[23,24]

[19]Although determination of absolute configuration generally involves scattering from relatively heavy atoms, it has been applied even in the case of one oxygen in a structure of 17 carbon atoms; see D. Rabinovich and H. Hope, *Acta Cryst.*, **A36**, 670 (1980). This is an extreme case requiring extraordinary efforts to obtain precise data, but a number of other absolute configurations have been determined for molecules containing a larger fraction of oxygen.

[20]J. Ibers and W. C. Hamilton, *Acta Cryst.*, **17**, 781 (1964).

[21]W. C. Hamilton, *Acta Cryst.*, **18**, 502 (1965).

[22]D. Rogers, *Acta Cryst.*, **A37**, 734 (1981).

[23]H. D. Flack, *Acta Cryst.*, **A39**, 876 (1983).

[24]G. Bernardinelli and H. D. Flack, *Acta Cryst.*, **A41**, 500 (1985).

which converts the expression for the atomic scattering factor to

$$f_o^{\text{anom}} = f_o + \Delta f' + i\eta\, \Delta f'' \tag{18.23}$$

Refinement of η is expected to tend strongly to $+1$, indicating that the model has the correct hand, or -1, indicating that it needs to be reversed. The method appears to be very successful and is being increasingly incorporated into refinement programs.[25] It has the particular advantage of being little affected by the details of experimental conditions, even including uncorrected absorption.[26,27]

18.9. THERMAL MOTION

If proper scattering factors have been used in the calculations and if the data are free of systematic errors, thermal parameters are a measure of the root-mean-square amplitude of the atomic oscillations (Appendix B). If anisotropic thermal parameters have been determined in the structure refinement, both the magnitudes and orientations of the ellipsoids representing the thermal motion can be computed.[28] This information can be presented in a plane figure with the axes and principal sections of the ellipsoids drawn with perspective to indicate the three-dimensional nature.[29] The ultimate, however, is a stereo pair as in Fig. 18.5. If the figure is

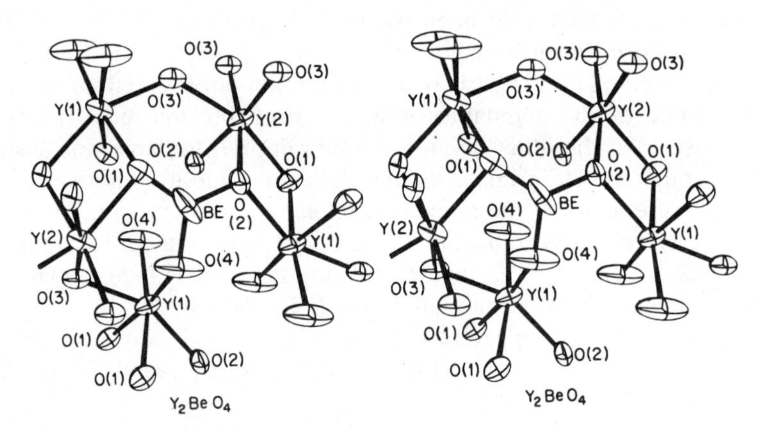

Figure 18.5. Stereoscopic representations of Y_2BeO_4 showing the thermal ellipsoids. (Reproduced by permission from L. A. Harris and H. L. Yakel, *Acta Cryst.*, **22**, 358 (1967).]

[25] P. G. Jones, *Acta Cryst.*, **A40**, 660 (1984).
[26] G. Bernardinelli and H. D. Flack, *Acta Cryst.*, **A43**, 75 (1987).
[27] P. G. Jones and K. Meyer-Bäse, *Acta Cryst.*, **A43**, 79 (1987).
[28] But see K. Lonsdale and J. Milledge, *Acta Cryst.*, **14**, 69 (1961), for a warning on the interpretation of the results.
[29] G. M. Brown and R. E. Marsh, *Acta Cryst.*, **16**, 191 (1963).

observed with a proper viewer, the structure is actually seen in three dimensions. Elegant programs have been written to do the necessary computations for a structure as viewed from any specified direction and distance and to plot the results automatically.[30,31]

Examination of the thermal ellipsoids often shows that the orientation and magnitudes of their principal axes correspond to a reasonable pattern of motion relative to the bonding and packing of the molecules. For example, in long-chain molecules such as *n*-nonanoic acid hydrazide (Fig. 18.6), the thermal motion normal to the chain axis should be greater than that at right angles to it. That this is the case can be seen from the arrows beside each atom, which show the approximate magnitude and direction of motion normal to the viewing direction. Furthermore, since the functional ends of the molecules are linked together by a series of N–H····N and O–H····N hydrogen bonds in a rather rigid net, thermal motion would be expected to be smaller for atoms in this region of the molecule compared with those toward the ends of the chains. Inspection of Fig. 18.6 will again show this to be the case. In general, it is found that terminal atoms such as the O atom in a carbonyl group or the C and H atoms in a methyl group can be expected to have thermal motions greater than the atoms to which they are bonded.

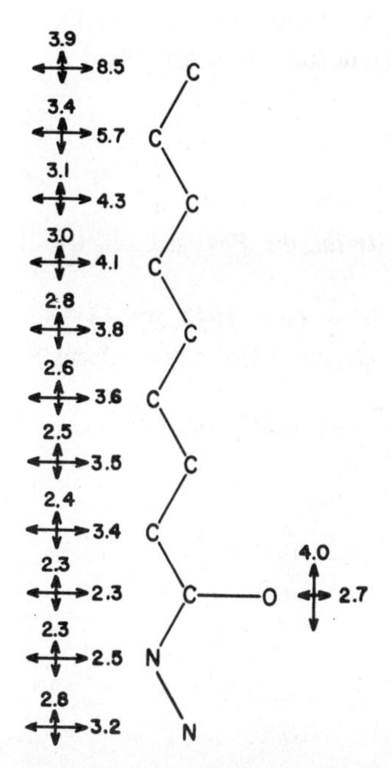

Figure 18.6. Thermal parameters in *n*-nonanoic acid hydrazide; values given are anisotropic *B*'s.

[30]C. K. Johnson, ORTEP: A Fortran Thermal Ellipsoid Plot Program for Crystal Structure Illustrations, Oak Ridge National Laboratory Rept., ORNL-3794, 1965.
[31]T. A. Jones, in *Methods in Enzymology*, Vol. 115, H. W. Wyckoff, C. H. W. Hirs, and S. N. Timasheff, Eds., Academic, New York, 1985, pp. 157–171.

In *n*-nonanoic acid hydrazide, *B* values run from a low of approximately 2.3 Å^2 for the O atom in a direction along the C–O bond to values exceeding 8 Å^2 for motion normal to the chain for the terminal methyl groups. In general, for structures tightly laced together by hydrogen or other bonds, for large cagelike molecules, or for heavy atoms, *B* values can be expected as low as 2 Å^2 or even less. For atoms in structures where packing forces and geometry allow great freedom of motion, *B* values may range up to 15 Å^2 or more. However, for such high values, care must be exercised in the interpretation since it may happen that a structure is disordered and the thermal parameters are adjusting to anomalously high values in order to give the best fit of the assumed model to the actual physical situation.[32]

In this chapter we have been concerned with the most common quantitative deductions from the results of X-ray diffraction and the uncertainties in them. Such results often account for much of the discussion in structure reports, but they should not obscure the fact that features of prime interest in a structure may be its more qualitative aspects. For example, the general features in an organic molecule may provide important information useful to chemists without the quantitative details of bond lengths and angles. Furthermore, in many molecules of biological interest, the relative positions of various atoms or groups of atoms are features of major interest that often provide the primary incentive for the structure determination.

BIBLIOGRAPHY

Bevington, P. R., *Data Reduction and Error Analysis for the Physical Sciences*, McGraw-Hill, New York, 1969.

Deming, W. E., *Statistical Adjustment of Data*, Wiley, New York, 1943, pp. 14–58.

Dunitz, J., *X-Ray Analysis and the Structures of Organic Molecules*, Cornell University Press, Ithaca, NY, 1979, pp. 225–244.

Hamilton, W. E., *Statistics in Physical Science*, Ronald, New York, 1964.

[32]For an example of such disorder, see A. W. Hanson, *Acta Cryst.*, **14**, 365 (1961).

CHAPTER 19

AMBIGUITIES AND UNCERTAINTIES

Now that we have gone through the process of determining a structure by X-ray diffraction, it is worth considering how much the results are to be believed. Aside from the uncertainties that appear to be revealed by statistical error analysis, what other ghosts lurk in the anxiety closet?[1]

In recent years various groups have made efforts to show by independent methods the extent to which crystallographic results, or some subset of them, correspond to reality. We shall consider some of their findings, roughly in the order of the earlier chapters.

19.1. EXPERIMENTAL DATA[2]

The raw materials of crystallographic analysis, experimental measurements of intensities, are generally not measured repeatedly,[3] and their uncertainties are estimated from expected and guessed-at random errors. A number of studies of repeated data measurements have been made, however, corresponding to three proposed models.[4] The American Crystal-

[1]For a delightful review of these problems, see P. G. Jones, *Chem. Soc. Rev.*, **13**, 157 (1984).
[2]S. C. Abrahams, *Acta Cryst.*, **B30**, 261 (1974).
[3]Repeated measurements of the same or symmetry-equivalent reflections are more common in macromolecular data sets, but these suffer from their own characteristic problems to an extent that makes them a poor choice for estimating the accuracy of measurement on crystals of small molecules.
[4]A. McL. Mathieson, *Acta Cryst.*, **A25**, 264 (1969). This paper is highly recommended for its discussion of the problems inherent in trying to extract the true meaning of results from multiple measurements.

lographic Association project[5,6] used a single sphere of CaF_2, hand carried among seven laboratories, in a study designed to examine the replication of measurement of a single specimen. The International Union of Crystallography[7–9] used a single lot of crystals of D(+)-tartaric acid distributed among 16 diffractometers of varying types in an attempt to study the population of average data measurements on organic molecules. Denne[10] undertook a series of measurements of a group of crystals of α-glycine (from one source) using a single very carefully aligned and calibrated diffractometer. The findings of all three studies were disquieting.

Probably the most striking results from both the ACA and IUCr investigations was the sizable fraction of outlier data sets that appeared to contain significant systematic errors, despite the fact that all the data arose from experienced, productive research groups. It is probable that the scatter of results would be smaller today because fewer types of measuring apparatus would be represented, but although there are reasons to believe that the data from the now dominant four-circle diffractometer is generally more reliable than data from the alternatives, loss of variability is not necessarily the same as gain in accuracy.[11]

Even after the outliers were omitted, the conclusions of both the ACA and IUCr projects suggested that the expected variability in F for independent measurements of different crystals was on the order of 5 or 6% at best. Since the reproducibility of a diffractometer can be brought to something less than 1% in I for reflections that are not limited by counting statistics,[10,12] it is clear that either diffractometers or crystals or experimenters differ more than we should like to believe. The evidence at present is that the major source of discrepancies, in the absence of gross experimental error, lies in the variability of individual crystals. Thus to a greater extent than is widely appreciated, a structure determination applies specifically to the particular crystal specimen studied. It has been suggested[9,10] that the principal source of deviation is uncorrected and sample-dependent extinction, but that is certainly only one of the possible ways in which two crystal samples can differ.[13]

Fortunately, most of the errors that lead to the poor agreement between

[5]S. C. Abrahams, L. E. Alexander, T. C. Furnas, W. C. Hamilton, J. Ladell, Y. Okaya, R. A. Young, and A. Zalkin, *Acta Cryst.*, **22**, 1 (1967).

[6]J. K. Mackenzie and V. W. Maslen, *Acta Cryst.*, **A26**, 628 (1968).

[7]S. C. Abrahams, W. C. Hamilton, and A. McL. Mathieson, *Acta Cryst.*, **A26**, 1 (1970).

[8]W. C. Hamilton and S. C. Abrahams, *Acta Cryst.*, **26**, 18 (1970).

[9]J. K. Mackenzie, *Acta Cryst.*, **A30**, 607 (1974).

[10]W. A. Denne, *Acta Cryst.*, **A28**, 192 (1972).

[11]The truth of this statement was, in fact, demonstrated during the IUCr experiment in which the original cell parameters were measured precisely but erroneously, possibly contributing to some of the errors associated with certain measurement techniques.

[12]L. E. McCandlish, G. H. Stout, and L. C. Andrews, *Acta Cryst.*, **A31**, 245 (1975).

[13]H. J. Milledge, *Acta Cryst.*, **A25**, 173 (1969).

individual data sets have comparatively little effect on the final positional parameters derived from them. They do, however, affect the apparent thermal parameters strongly, to the extent that the calculated σ's may be in error by factors of 5–10. This ability of the thermal factors to "soak up" various observational errors[14] is one of the major reasons why data can be refined to R's that are significantly better than the agreement between one data set and another.

19.2. CRYSTAL SYMMETRY

This problem has already been discussed in Chapter 5, but it is well to be reminded of it here. It has been estimated[15,16] that at least 3% of crystals whose structures have been determined are assigned to the wrong space group.

These errors have been classified[16] into three groups: (1) the crystal system is wrong, (2) the crystal system is right but the Laue class is wrong,[17] and (3) both are right but a center of symmetry is missing. The first case should be dealt with by analysis of the reduced cell and the metric symmetry of the lattice, checked by consideration of the intensities of equivalent reflections. Proper application of these methods should prevent errors of this kind before a structure is solved. The second and third cannot be found in this way, but an analysis of the intensities of possibly equivalent reflections should avoid the second and, if dispersion is present, may assist with the third.[18] If the requisite data are not available, it is necessary to analyze the final structure for near or exact symmetry that is not part of the space group.[19,20]

Assignment of the wrong Laue group leads to excess parameters for the final structure, but in general to no particular difficulties in refinement.[21] If a centrosymmetric structure is refined as noncentrosymmetric, however, it produces badly behaved matrices, random and chemically unreasonable distortions in the structure, slow convergence, and high correlation coefficients for centrosymmetrically related atoms. Hence it is important to detect this error if the final structural parameters are to be meaningful.[22]

Errors of space group assignment of other kinds can occur, such as

[14]"Errors" only in a biased sense of disagreeing with the simplistic model being used to describe the real crystal under study.

[15]A. D. Mighell, V. L. Hime, and J. R. Rodgers, *Acta Cryst.*, **A39**, 737 (1983).

[16]W. H. Baur and E. Tillmanns, *Acta Cryst.*, **B42**, 95 (1986).

[17]This is possible only in crystal systems of high symmetry. See Table 3.4.

[18]R. E. Marsh, *Acta Cryst.*, **B43**, 174 (1987).

[19]W. H. Baur, E. Tillmanns, and W. Hofmeister, *Acta Cryst.*, **B39**, 669 (1983).

[20]Y. LePage, *Acta Cryst.*, **B35**, 264 (1987); *J. Appl. Cryst.*, **21**, 983 (1989).

[21]V. Schomaker and R. E. Marsh, *Acta Cryst.*, **B35**, 1933 (1979).

[22]R. E. Marsh, *Acta Cryst.*, **B42**, 193 (1986).

erroneous rejection of a translational symmetry element for the reasons discussed in Chapter 5.[18] One particularly disturbing example occurred in macromolecular crystallography, where a protein was assigned to the space group $P4_32_12$, "solved," and "refined" to an apparently acceptable R value (by the loose standards of the field). After several years, a reinvestigation, incited by a large number of serious internal inconsistencies in the earlier results, showed that the actual space group was the enantiomorphous $P4_12_12$ and that the true structure differed in almost all respects from the one previously proposed.[23]

19.3. STANDARD DEVIATIONS

Given that the data obtained for a crystal pertain more to that specimen than to any other, how well *do* the final results and their σ values relate to the population of other samples of the same material? Fortunately, some evidence on this point has been obtained by a study[24] of the 100 or so compounds whose structures have been determined independently in two different laboratories. Again, the results are less encouraging than we would hope.

Comparison of the observed differences between positional parameters with those predicted on the basis of the σ's derived from the refinement indicated strongly that the uncertainties were underestimated by a factor of about 1.4. These results were in remarkable accord with those of the refined structures based on the data from the IUCr project.[7,8] In addition to this average increase, there were a disproportionate number of larger discrepancies, suggesting the presence of various systematic deviations between data sets.

No comparison was made of the thermal parameters in this study, but considering the large effects that may be ascribed to extinction and TDS (Chapter 17), and the rarity of corrections for these errors, it is likely that the standard deviations in thermal factors are unrealistic by factors of at least 5 to 10 in many cases. In the absence of corrections for absorption, the thermal parameters may at times approach nonsense.

Perhaps most disconcerting was the observation that the estimated standard deviations in cell parameters were seriously underestimated, perhaps by factors of 5 for the cell lengths and 2.5 for the angles. It is clear that the precision with which parameters can be determined on a diffractometer may conceal a serious lack of accuracy. No extended study has been made of the causes of this error, but it is likely that they include variations in crystal purity and errors in the zero settings of the instruments

[23]G. H. Stout, S. Turley, L. C. Sieker, and L. H. Jensen, *Proc. Nat. Acad. Sci. USA*, **85**, 1020 (1988).
[24]R. Taylor and O. Kennard, *Acta Cryst.*, **B42**, 112 (1986).

used, combined with a tendency to obtain orientation matrices based on reflections measured in only one or two octants of reciprocal space (see Section 5.11). Furthermore, the usual approach to obtaining cell constants on a diffractometer spreads the observational errors over *all* the cell parameters even though the space group may require that one or more have *exact* values, forcing greater calculated errors in the others. Although these errors in the cell parameters should not produce the variations observed in the positional parameters, they would contribute to the uncertainties in derived bond lengths (Chapter 18). Cell parameters are often reported with standard deviations of 1 in 5000–10,000, but it is unlikely that they are, in fact, reliable to better than 1 in 1000 or perhaps 2000 unless special techniques have been used.

19.4. FINAL THOUGHTS

The various cautions we have pronounced are intended to warn but should not be allowed to hide the fact that in perhaps 70 to 90% of cases one can mount (and center!) a crystal on a diffractometer, allow the automatic programs to handle the alignment, transfer the data to a multisolution direct methods program, and arrive at an E-map that will give atomic parameters capable of being refined by least squares to an R of a few percent. There are laboratories that hope to maintain a rate of about one structure a week by these procedures, although it is likely that they must practice fairly drastic triage to maintain such a pace.

If the worst that could happen is that a structure would fail to solve in one case in ten, these would be fairly good odds (although the perversity of nature will ensure that the really important problem will be the one that fails). Unfortunately, it is also possible to come close but be wrong, to publish one's result and later to have the error exposed. This is a much more serious fate, unfortunate for the individual and bad for science in general.

In practical terms, it is well to remember that everything leads back to the experimental data. Hence it is well worth the time and effort, no matter how eager one is to have a peek at the final structure, to ensure that the data are correct and accurate. In general it is not very much more difficult to collect good data than bad. Largely it is a matter of being aware of the possible sources of error and attempting to estimate their residual magnitudes after reducing them as much as possible. *Think about what you are doing*!

The intelligent use of an automatic diffractometer reduces the true random errors in a data set, including those of editing and mistranscription. The use of Mo K_α radiation reduces the effects of absorption and extinction sharply, except in a few special cases. A crystal monochromator can improve the signal-to-noise ratio of the data, although it is necessary to be concerned about experimental measurement of the polarization ratio and the effects of inhomogeneities in the incident beam. Maintaining the crystal at or near liquid nitrogen temperatures, although not yet routine, can do a great deal for increasing the average intensity, reducing TDS, probably reducing extinction, and reducing those structural ambiguities associated with atomic motion. More important than any other effect, however, is the ability of deep cooling to increase crystal stability in the X-ray beam, ensuring that the same material is present for the entire data collection.

We hope that the material we have covered will enable the reader to approach X-ray structural problems with greater insight and to understand and deal thoughtfully with questions as they arise. The most necessary tools to avoid mistakes are the ability to recognize uncertainties when they appear, the knowledge to consider possible alternative causes for observations, and the willingness to accept the possibility that one's first opinion may be wrong.

BIBLIOGRAPHY

Schwartzenbach, D. *et al.*, *Acta Cryst.*, **A45**, 63 (1989).

APPENDIX A

BRAGG'S LAW

Although Bragg's law can be justified by showing that its results are the same as those predicted by more rigorous analysis of diffraction from a three-dimensional grating,[1] in the absence of such an analysis students often wonder whether the analogy drawn with reflection is truly legitimate. In particular, doubts arise as to whether diffraction can occur only in the special directions corresponding to reflection from lattice planes.

The correctness of this assumption can easily be demonstrated,[2] however, for the following two-dimensional case of an orthogonal lattice (Fig. A.1), and the arguments can be generalized to three dimensions and a nonorthogonal system. Consider the three noncollinear points A, B, C, each of which scatters one of the parallel incident beams $(1, 2, 3)$ in a direction $(1', 2', 3')$ such that constructive reinforcement occurs. By dropping perpendiculars and considering the requirement that for such reinforcement the difference in path for waves scattered from any two lattice points must be a whole number of wavelengths, it is clear that

$$a - b = h\lambda \tag{A.1}$$

$$c + d = k\lambda \tag{A.2}$$

If ψ and ψ' are the angles of incidence and reflection (not necessarily

[1]For an excellent and simple analysis, see W. L. Bragg, *The Crystalline State*, Vol. I, Macmillan, New York, 1934, pp. 13–20. H. Lipson and W. Cochran, *The Determination of Crystal Structures*, Cornell University Press, Ithaca, NY, 1966, pp. 4–7, give similar results using vectors.

[2]This argument follows closely that of L. R. B. Elton and D. F. Jackson, *Am. J. Phys.*, **34**, 1036 (1966).

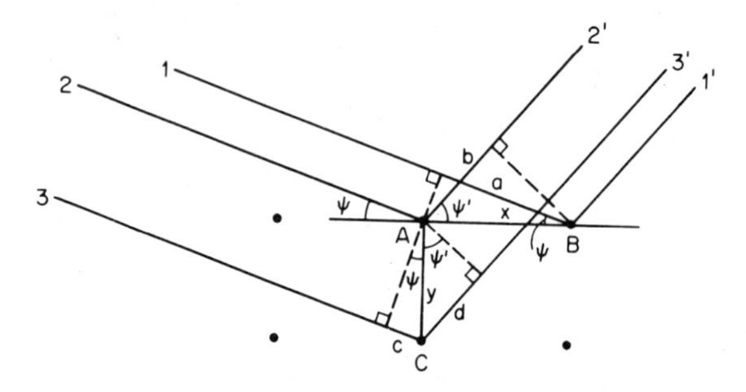

Figure A.1. Parallel scattering from three points in a lattice.

equal) formed with the lattice row AB, Eqs. (A.1) and (A.2) can be rewritten as

$$x \cos \psi - x \cos \psi' = h\lambda \tag{A.3}$$

$$y \sin \psi + y \sin \psi' = k\lambda \tag{A.4}$$

Now consider (Fig. A.2) a plane RS drawn to make equal angles θ with the incident and scattered beams and an angle α with the row AB. Then

$$\theta = \psi + \alpha = \psi' - \alpha \tag{A.5}$$

Rearranging for ψ and ψ' and substituting in (A.3) and (A.4) gives

$$x[\cos(\theta - \alpha) - \cos(\theta + \alpha)] = h\lambda \tag{A.6}$$

$$y[\sin(\theta - \alpha) + \sin(\theta + \alpha)] = k\lambda \tag{A.7}$$

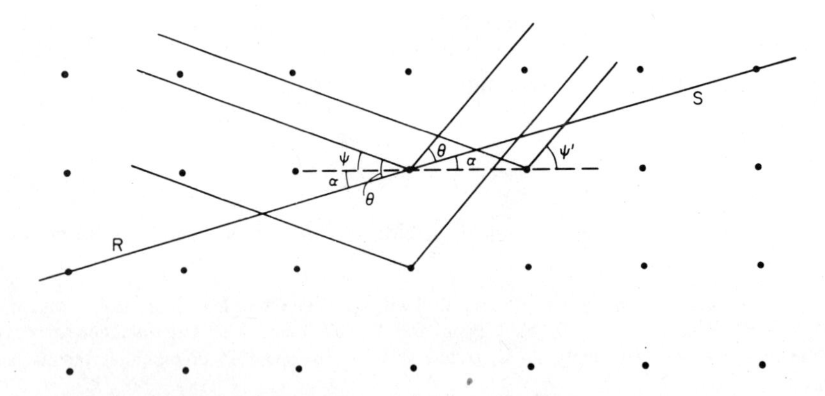

Figure A.2. Equivalent reflection from a lattice plane RS.

Expanding these leads to

$$x(\cos\theta\cos\alpha+\sin\theta\sin\alpha-\cos\theta\cos\alpha+\sin\theta\sin\alpha)=h\lambda \qquad \text{(A.8)}$$

$$y(\sin\theta\cos\alpha-\cos\theta\sin\alpha+\sin\theta\cos\alpha+\cos\theta\sin\alpha)=k\lambda \qquad \text{(A.9)}$$

or

$$2x\sin\theta\sin\alpha=h\lambda \qquad \text{(A.10)}$$

$$2y\sin\theta\cos\alpha=k\lambda \qquad \text{(A.11)}$$

Thus by division

$$\frac{\sin\alpha}{\cos\alpha}=\tan\alpha=\frac{hy}{kx} \qquad \text{(A.12)}$$

The interpretation of Eq. (A.12) is that the plane RS, after leaving A, passes repeatedly through other lattice points successively shifted by h units along y and k units along x. This is equivalent to the definition of a lattice plane (h, k), so the conditions that permit constructive scattering in arbitrary directions are seen to correspond to simple reflection from a suitably oriented lattice plane. Thus the assumption inherent in the usual derivation of the Bragg law is justified.

APPENDIX B

THE GENERAL TEMPERATURE FACTOR EXPRESSION

The temperature factor of an atom for any set of lattice planes (hkl) depends on the interplanar spacing d and on the magnitude of vibration perpendicular to the planes. For the isotropic case, the temperature factor of an atom for a given set of planes (hkl) is

$$\exp\left[-\frac{B}{4}\left(\frac{2\sin\theta_{hkl}}{\lambda}\right)^2\right]=\exp\left[-\frac{B}{4}\left(\frac{1}{d_{hkl}}\right)^2\right] \tag{B.1}$$

where $B = 8\pi^2\langle u^2\rangle$, $\langle u^2\rangle$ is the mean-square amplitude of vibration, and $1/d_{hkl}$ is the reciprocal of the interplanar spacing, that is, the length of the reciprocal lattice vector from the origin to the point hkl. For the general case (any crystal system), the length of this vector is [see Eq. (2.24)]

$$\begin{aligned}\frac{1}{d_{hkl}} = (h^2a^{*2} + k^2b^{*2} + l^2c^{*2} + 2hka^*b^* \cos\gamma^* \\ + 2hla^*c^* \cos\beta^* + 2klb^*c^* \cos\alpha^*)^{1/2}\end{aligned} \tag{B.2}$$

and the temperature factor must have a parameter for every term in this expression since each represents a component perpendicular to the set of planes (hkl). The general temperature factor expression is therefore

$$\begin{aligned}\exp[-\tfrac{1}{4}(B_{11}h^2a^{*2} + B_{22}k^2b^{*2} + B_{33}l^2c^{*2} + 2B_{12}hka^*b^* \cos\gamma^* \\ + 2B_{13}hla^*c^* \cos\beta^* + 2B_{23}klb^*c^* \cos\alpha^*)]\end{aligned} \tag{B.3}$$

where the B_{ij} are the thermal parameters[1] in the same units as the conventional isotropic thermal parameter B.

An equivalent and preferable expression for the general temperature factor is

$$\exp[-2\pi^2(U_{11}h^2a^{*2} + U_{22}k^2b^{*2} + U_{33}l^2c^{*2} + 2U_{12}hka^*b^* \cos \gamma^* + 2U_{13}hla^*c^* \cos \beta^* + 2U_{23}klb^*c^* \cos \alpha^*)] \quad \text{(B.4)}$$

where the U_{ij} are now the thermal parameters expressed in terms of mean-square amplitudes of vibration in angstroms.

Another expression for the anisotropic factor that will be encountered in the literature and in some computer programs is

$$\exp[-(\beta_{11}h^2 + \beta_{22}k^2 + \beta_{33}l^2 + 2\beta_{12}hk + 2\beta_{13}hl + 2\beta_{23}kl)] \quad \text{(B.5)}$$

While this is a relatively compact expression, reporting of thermal parameters as β's is to be discouraged because of the difficulty in comparing β values associated with axes of different lengths.

It should be noted that the factor 2 in the "cross terms" in Eq. (B.5) is sometimes omitted, and this must be taken into account when converting β's into B's or U's. When reporting results in the literature, it is preferable

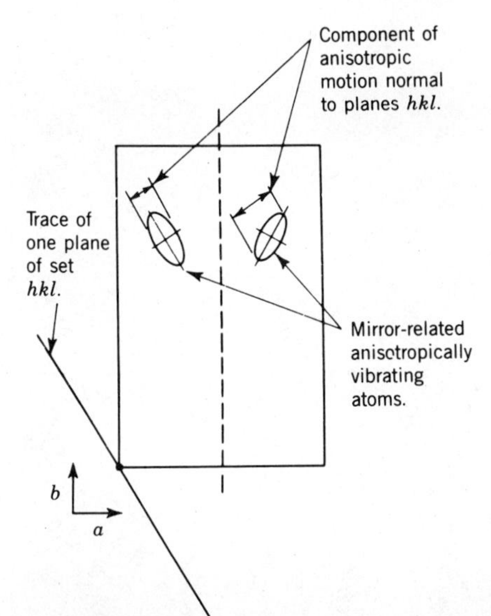

Figure B.1. Mirror-image anisotropic atoms and their relationship to a lattice plane (hkl).

[1]Another justification for the use of six parameters may be had by noting that this is the number necessary to determine a general ellipsoid. Thus the six B_{ij} in the general expression serve to describe the ellipsoidal electron distribution of the anisotropically vibrating atom.

that they appear as U's, since these nominally are physically meaningful. The use of B's is more common, however.

Note that anisotropic thermal parameters cannot, in general, be used with the factored form of the structure factor expression. The reason for this is illustrated in projection in Fig. B.1 for a space group with symmetry m. This shows an anisotropically vibrating atom together with its mirror-related image, and it is evident that the components of vibration perpendicular to the set of planes indicated are quite different for the two atoms. The contribution of each atom to the structure factor must therefore be calculated separately and added.

APPENDIX C

MATRIX OPERATIONS

In Chapter 16 we indicated that a set of n equations in n unknowns,

$$\begin{aligned} a_{11}x_1 + a_{12}x_2 + \cdots + a_{1n}x_n &= v_1 \\ a_{21}x_1 + a_{22}x_2 + \cdots + a_{2n}x_n &= v_2 \\ &\vdots \\ a_{n1}x_1 + a_{n2}x_2 + \cdots + a_{nn}x_n &= v_n \end{aligned} \tag{C.1}$$

could be represented in matrix notation. We shall now explain the meaning of this representation. To do so, we first define matrix equality.[1]

For two matrices to be equal, the numbers of rows and columns of one must equal the numbers of rows and columns of the other, and the elements of one must be equal to the corresponding elements of the other.

It is now evident that Eqs. (C.1) can be written

$$\begin{bmatrix} a_{11}x_1 + a_{12}x_2 + \cdots + a_{1n}x_n \\ a_{21}x_1 + a_{22}x_2 + \cdots + a_{2n}x_n \\ \vdots \\ a_{n1}x_1 + a_{n2}x_2 + \cdots + a_{nn}x_n \end{bmatrix} = \begin{bmatrix} v_1 \\ v_2 \\ \vdots \\ v_n \end{bmatrix} \tag{C.2}$$

[1]For fuller accounts of matrix operations see, for example, V. N. Faddeeva, *Computational Methods of Linear Algebra*, Dover, New York, 1959, pp. 1–23; R. A. Frazer, W. J. Duncan, and R. A. Collar, *Elementary Matrices*, Cambridge University Press, New York, 1960, pp. 1–33; and E. Prince, *Mathematical Techniques in Crystallography and Materials Science*, Springer-Verlag, New York, 1982, Chapters 1 and 4.

where the brackets indicate that the arrays are matrices. Note that the matrix on the right has n rows of one element each and therefore only one column. The matrix on the left also has n rows and one column, but each element is the sum of n terms, $\sum_{j=1}^{n} a_{ij}x_j$, and $i = 1, 2, \ldots, n$ for successive rows.

We next define matrix multiplication. This is defined only if the first matrix in the product has the same number of columns as the second one has rows. Thus the product

$$\mathbf{A}_{mn}\mathbf{B}_{pq} \tag{C.3}$$

where $\mathbf{A}$ is an m-row, n-column matrix and $\mathbf{B}$ is a p-row, q-column matrix, has meaning only when $n = p$.[2] The product matrix has m rows and q columns, so that

$$\mathbf{A}_{mn}\mathbf{B}_{nq} = \mathbf{C}_{mq} \tag{C.4}$$

The ikth element, c_{ik}, of the product matrix $\mathbf{C}_{mq}$ is the sum of the products of corresponding terms in the ith row of the *first* factor and the kth column of the *second* factor; that is,

$$c_{ik} = a_{i1}b_{1k} + a_{i2}b_{2k} + \cdots + a_{in}b_{nk} = \sum_{j=1}^{n} a_{ij}b_{jk} \tag{C.5}$$

where the a's and b's are the elements of $\mathbf{A}$ and $\mathbf{B}$.

In view of the multiplication rule as given by Eq. (C.5), it is readily verified that the left-hand matrix in Eq. (C.2) is the product of an $n \times n$ and an $n \times 1$ matrix:

$$\begin{bmatrix} a_{11} & a_{12} & \cdots & a_{1n} \\ a_{21} & a_{22} & \cdots & a_{2n} \\ & & \vdots & \\ a_{n1} & a_{n2} & \cdots & a_{nn} \end{bmatrix} \begin{bmatrix} x_1 \\ x_2 \\ \vdots \\ x_n \end{bmatrix} \tag{C.6}$$

Thus the matrix equivalent of Eq. (C.1) is

$$\begin{bmatrix} a_{11} & a_{12} & \cdots & a_{1n} \\ a_{21} & a_{22} & \cdots & a_{2n} \\ & & \vdots & \\ a_{n1} & a_{n2} & \cdots & a_{nn} \end{bmatrix} \begin{bmatrix} x_1 \\ x_2 \\ \vdots \\ x_n \end{bmatrix} = \begin{bmatrix} v_1 \\ v_2 \\ \vdots \\ v_n \end{bmatrix} \tag{C.7}$$

[2]The commutative law does not hold for matrix multiplication. In general, therefore, $\mathbf{AB} \neq \mathbf{BA}$. Commuting the terms in Eq. (C.3) to give $\mathbf{B}_{pq}\mathbf{A}_{mn}$ would, in fact, have no meaning unless $m = q$.

This is abbreviated by writing a single letter to represent each matrix as was done in Eqs. (C.3) and (C.4), so that in its simplest form Eq. (C.7) is written

$$\mathbf{Ax} = \mathbf{v} \tag{C.8}$$

as in Eq. (16.44).

When Eqs. (C.1) have a solution, an inverse matrix of $\mathbf{A}$ exists such that

$$\mathbf{A}^{-1}\mathbf{A} = \mathbf{I} \tag{C.9}$$

where $\mathbf{A}^{-1}$ is the inverse matrix with n rows and n columns and $\mathbf{I}$ is the identity matrix, also an $n \times n$ matrix, with 1 for each diagonal element and 0 for all others. In Chapter 16, $\mathbf{I}$ was called the matrix equivalent of 1, since multiplying a matrix by $\mathbf{I}$ leaves it unchanged.

If Eq. (C.8) represents the normal equations obtained in the least squares refinement of a crystal structure, then $\mathbf{A}$ is the matrix of the coefficients of these equations. Multiplying each side of Eq. (C.8) by $\mathbf{A}^{-1}$ gives[3]

$$\mathbf{A}^{-1}\mathbf{Ax} = \mathbf{A}^{-1}\mathbf{v} \tag{C.10}$$

so that

$$\mathbf{x} = \mathbf{A}^{-1}\mathbf{v} \tag{C.11}$$

The full matrix expression for the solution is

$$\begin{bmatrix} x_1 \\ x_2 \\ \vdots \\ x_n \end{bmatrix} = \begin{bmatrix} b_{11} & b_{12} & \cdots & b_{1n} \\ b_{21} & b_{22} & \cdots & b_{2n} \\ & & \vdots & \\ b_{n1} & b_{n2} & \cdots & b_{nn} \end{bmatrix} \begin{bmatrix} v_1 \\ v_2 \\ \vdots \\ v_n \end{bmatrix} \tag{C.12}$$

where the b's are the elements of $\mathbf{A}^{-1}$. Carrying through the indicated multiplication on the right of Eq. (C.12) and noting the term-by-term identity of the product matrix with that on the left, we have

$$\begin{aligned} x_1 &= b_{11}v_1 + b_{12}v_2 + \cdots + b_{1n}v_n \\ x_2 &= b_{21}v_1 + b_{22}v_2 + \cdots + b_{2n}v_n \\ &\vdots \\ x_n &= b_{n1}v_1 + b_{n2}v_2 + \cdots + b_{nn}v_n \end{aligned} \tag{C.13}$$

where the x's are the solutions of the normal equations.

[3]For discussions of the ways in which $\mathbf{A}^{-1}$ can be obtained from the a_{ij}'s, see, for example, V. N. Faddeeva, *Computational Methods of Linear Algebra*, Dover, New York, 1959, pp. 79–145.

APPENDIX D

STRUCTURE FACTOR GRAPHS AND FOURIER TRANSFORMS

The effects of placing atoms in various locations in the cell can be seen most easily by considering plots of the trigonometric part of the structure factor expression for various reflections. Such *structure factor graphs*[1-3] are most simply illustrated for one-dimensional centrosymmetric cells. In Fig. D.1 the trigonometric part of the structure factor expression

$$T = \cos 2\pi hx \tag{D.1}$$

has been plotted for $h = 1, 2, 3$. In this case, the unique part of the graph is the region $0 < x < \frac{1}{2}$, since the other half is related by a center of symmetry. This is the only part of the graph that needs to be plotted, and it is all that needs to be investigated when the graph is used.

To illustrate the use of these graphs, assume that the cell contains two centrosymmetrically related atoms at x and $-x$, a simple one-parameter problem. Suppose the reflection with $h = 1$ is weak. This can be true only if the x coordinate of one atom is near $\frac{1}{4}$ (the centrosymmetrically related atom is near $\frac{3}{4}$). For such a simple, case, this single observation is sufficient to approximate the structure, and it necessarily follows that the reflection with $h = 2$ will be intense. That with $h = 3$ will be a sensitive indicator of how near the parameter actually is to $x = \frac{1}{4}$.

Two-dimensional structure factor graphs, also known as *Bragg–Lipson*

[1]C. W. Bunn, *Chemical Crystallography*, 2nd ed., Oxford, New York, 1961, pp. 279–294.
[2]M. J. Buerger, *Crystal-Structure Analysis*, Wiley, New York, 1960, pp. 283–301.
[3]H. Lipson and W. Cochran, *The Determination of Crystal Structures*, G. Bell and Sons, London, 1957, pp. 65–71, 138–139.

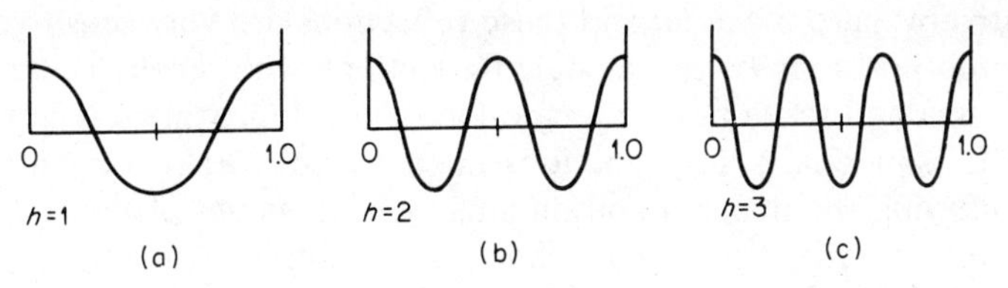

Figure D.1. One-dimensional structure factor graphs, $\cos 2\pi hx$. (*a*) $h=1$; (*b*) $h=2$; (*c*) $h=3$.

charts,[4] have proved particularly useful in practice. For those reflections that have one index zero ($0kl$, $h0l$, $hk0$), the structure factor depends upon only two coordinates, and the graph for any reflection can be plotted on a single sheet of paper. A suitable trial model in two dimensions consists of the full model projected along one the crystal axes. The reflections that can be studied are then those in which the zero index is associated with the projection axis.

The appearance of Bragg–Lipson charts will be illustrated for the $h0l$ and $0kl$ reflections of a structure in space group $P2_1/c$. The trigonometric expression for the observed $h0l$ reflections is

$$T = 4\cos 2\pi(hx + lz) \tag{D.2}$$

For example, the Bragg–Lipson chart for the 102 reflection is shown in Fig. D.2*a*, where it is apparent that there are two cycles along the z axis and one along the x. The maxima and minima of the function are indicated by plus and minus signs, respectively, while a solid line passes through points at which the function is zero. In Fig. D.2*b*, the Bragg–Lipson chart for the 304 reflection is shown, and in Fig. D.2*c* that for reflection $10\bar{2}$.

It is evident from the charts in Fig. D.2 that those for large h and/or l

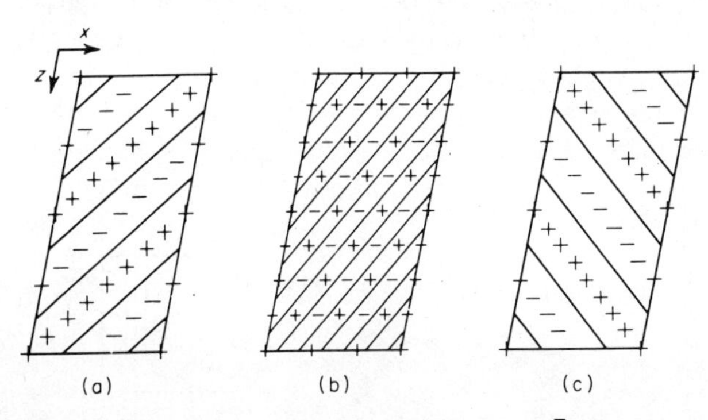

Figure D.2. Bragg–Lipson charts. (*a*) 102; (*b*) 304; (*c*) $10\bar{2}$. Space group $P2_1/c$.

[4] W. L. Bragg and H. Lipson, *Z. Krist.*, **95**, 323 (1936).

have relatively sharp maxima, and these reflections are very sensitive to the exact atomic positions of the model. Reflections with small h and l have broad maxima and serve best to locate the model approximately in the cell.

Bragg–Lipson charts for functions such as $\cos 2\pi(hx + lz)$ are easily made by plotting the maxima and minima at fixed values of lz (or hx), that is, along lines parallel to a (or c). Since the resulting plots are always like those in Fig. D.2, they are most simply made by marking off zero points for h cycles along x and l cycles along z (or $-z$ if the term is $-lz$). Successive points along one axis are then joined by straight lines to successive points along the other.

The trigonometric part of the structure factor expression for the $0kl$ reflections is

$$T = 4 \cos 2\pi ky \cos 2\pi lz \qquad \text{for } k + l \text{ even} \qquad \text{(D.3)}$$

$$T = -4 \sin 2\pi ky \sin 2\pi lz \qquad \text{for } k + l \text{ odd} \qquad \text{(D.4)}$$

The Bragg–Lipson chart for the 011 reflection is shown in Fig. D.3a, and that for 012 in Fig. D.3b. Charts for functions such as $\cos 2\pi ky \cos 2\pi lz$ [or $-\sin 2\pi ky \sin 2\pi lz$] are most easily made by marking off the zeros of the function on the axes and drawing in the zero lines parallel to the cell edges. The maximum and minimum values of ± 1 between successive zero lines should also be indicated. The charts in Figs. D.2 and D.3 are representative of those commonly occurring, although for space groups of higher symmetry they may appear more complex.

Bragg–Lipson charts can also be used for structures in noncentrosymmetric space groups. They are much less convenient, however, since both A and B parts must be plotted for each reflection and $|F|^2$ determined as $A^2 + B^2$.

The most powerful structure factor graphs are those in three dimensions. Unfortunately, they are also the most time-consuming to construct, and no convenient means has been devised whereby a model can be moved through the volume covered by such a graph in the general case. Without

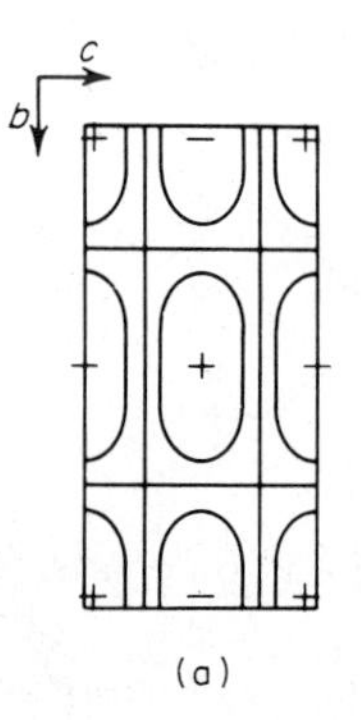

(b)

Figure D.3. Bragg–Lipson charts. (a) 011; (b) 012. Space group $P2_1/c$.

actually constructing such graphs, however, they can be seen in terms of two-dimensional sections, and the visualization is useful as a rough check on the magnitude of general reflections.

In much the same way that atomic contributions to a structure factor calculation involve samples from a continuous trigonometric function in direct space, the observed structure factors can be regarded as samples from a continuous scattering function in reciprocal space. This scattering function represents the X-ray scattering that would be produced by the electrons contained in a *single* unit cell. This is a continuous function, the *Fourier transform*[5,6] of the cell contents, but when many unit cells are combined in a crystal, destructive interference occurs *except* at those places corresponding to r.l. points. Another way of expressing this result is that the Fourier transform of the direct lattice is simply the collection of r.l. points and the observed result is the combination of the two transforms.

The generation of X-ray Fourier transforms and their combination to lead to more or less perfect lattice images can be modeled vividly by using visible light and masks representing various structures. Examination of these *optical transforms*[7,8] is a superb way of becoming familiar with the exactly analogous behavior of X-ray transforms as the sizes, shapes, and numbers of diffracting elements change.

The usual structure factor expression

$$F_{hkl} = \sum_j f_j e^{2\pi i(hx_j + ky_j + lz_j)} \tag{D.5}$$

gives a value to the reciprocal space point h, k, l that is the Fourier transform of the structure with j atoms in the unit cell at x_j, y_j, z_j.[9] Because of the periodic nature of a crystal structure, its observed transform differs from zero only at integral values of h, k, l, and these are the values normally used in calculation. There is no theoretical reason, however, why these indices must be integers, and it is quite possible to calculate the transform at nonintegral points to approximate the continuous distribution

[5] H. Lipson and C. A. Taylor, *Fourier Transforms and X-Ray Diffraction*, G. Bell and Son, London, 1958.

[6] K. J. H. Mackay, *Acta Cryst.*, **15**, 157 (1962).

[7] C. A Taylor and H. Lipson, *Optical Transforms*, G. Bell and Sons, London, 1964.

[8] G. Harburn, C. A. Taylor, and T. R. Welberry, *Atlas of Optical Transforms*, Cornell University Press, Ithaca, NY, 1975.

[9] In exactly the same way, the electron density expressed as a Fourier series

$$\rho(xyz) = \frac{1}{V}\sum_h \sum_k \sum_l F_{hkl} e^{-2\pi i(hx+ky+lz)}$$

is said to be the transform of the structure factors. Thus the electron density in direct space and the structure factors in reciprocal space are transforms of each other.

of the scattering from a single cell. If this transform is plotted in reciprocal space, therefore, its value at the r.l. points (integer indices) will match the observed structure factors.

This process of combining transforms can be extended to the molecules that make up the contents of a unit cell. The Fourier transform of a single molecule, oriented and positioned in space with respect to some axes, can be calculated with Eq. (D.5). The combination of several molecules in a cell under the constraints of a space group corresponds to the similar rotation of the individual transforms, phase shifting to account for molecular translations from the cell origin, and addition to yield the cell transform.

APPENDIX E

WEIGHTING FUNCTIONS

In Fourier methods of refining structures, each observation within the sphere of reflection is given unit weights. To do otherwise would invalidate the physical significance of ΔF or F_o syntheses. The method of least squares has the advantage that weights based on the precision of measurement can be used for each reflection.

The proper weight to be assigned an observation is equal to the reciprocal of the variance of that observation; that is,

$$w_j = \frac{1}{\sigma_j^2} \tag{E.1}$$

where σ_j is the standard deviation. A common method of estimating σ is by making *repeated measurements* of the *same quantity*. The value of σ can then be estimated from the equation (see Section 17.1)[1]

$$\sigma = \left(\frac{\sum_{j=1}^{m} d_j^2}{m-1} \right)^{1/2} \tag{E.2}$$

where d_j is the deviation of the jth measurement from the mean value.

[1]This σ is sometimes referred to as the *standard deviation in a single measurement*. The *standard deviation in the mean* of a multiply measured quantity is given by Eq. (18.18):

$$\sigma_{\text{mean}} = [\textstyle\sum_{j=1}^{m} d_j^2 / m(m-1)]^{1/2}$$

Note that $\sigma_{\text{mean}} = \sigma/\sqrt{m}$; that is, the uncertainty of the mean is smaller by the factor $1/\sqrt{m}$ if the m measurements are assumed to be of equal weight. It follows from Eq. (E.1) that the weight of the mean value of a multiply measured quantity is m times greater than that of a single measurement.

It is not feasible in routine structure determinations involving large numbers of reflections to measure each one a sufficient number of times to obtain reliable estimates of σ. Instead, estimates can be based on multiple measurements of a small number of reflections covering a range of magnitudes at various values of $\sin\theta$ and the results applied to the rest of the reflections assuming σ is a function of the magnitude of the intensity and $\sin\theta$.[2]

Alternatively, a weighting function can be chosen near the end of a refinement by plotting mean $|\Delta F|$ versus mean $|F_o|$ for groups of reflections of similar magnitudes. Although this is not proper in principle because the weights should reflect only the uncertainties in the observations, nevertheless taking the weight for any F as proportional to $1/|\langle\Delta F\rangle|^2$ from such a plot gives a reasonable weighting function that in practice is better than unit weights. An obvious advantage is that such a scheme can be derived without investing additional time in experimental work.

Photographic Observations

In the first application of the method of least squares to crystal structure refinement, Hughes[3] suggested the following weighting scheme:

$$\begin{aligned} \sqrt{w} &= 1 && \text{for } F_o < 4F_{\min} \\ \sqrt{w} &= 4F_{\min}/F_o && \text{for } F_o > 4F_{\min} \end{aligned} \tag{E.3}$$

This scheme gives reasonable weights for structures based on eye-estimated data. For good photographic data, however, it is likely to place too small a weight on intense reflections.[4] The weights given by Eqs. (E.3) are relative and do not provide a check either on the model or on the appropriateness of the weighting scheme itself.

In practice, weights for photographic data can be obtained by plotting mean $|\Delta F|$ versus mean $|F_o|$ as suggested above. If this method is used, the range of magnitude for each group of reflections should be such that at least 50 reflections are in most groups. Figure E.1 is a plot showing an actual case where the ΔF is given approximately by the equation

$$\Delta F = a + bF \tag{E.4}$$

except for the most intense reflections.[4]

[2]See A. de Vries, *Acta Cryst.*, **18**, 1077 (1965) for a further discussion. In fact, the dependence on $\sin\theta$ is small and is usually ignored.

[3]E. W. Hughes, *J. Am. Chem. Soc.*, **63**, 1737 (1941).

[4]While intense reflections for most single crystals of appreciable size suffer from secondary extinction, they should not be given decreased weight by using Eqs. (E.3) or by including additional terms in Eq. (E.4) to fit the increased ΔF's for these reflections, since extinction is a systematic error. Reflections that suffer serious secondary extinction can properly be given zero weight, but it is better to correct the data for the effects (see Chapter 17). Better still, the model should include an additional parameter to be adjusted during refinement.

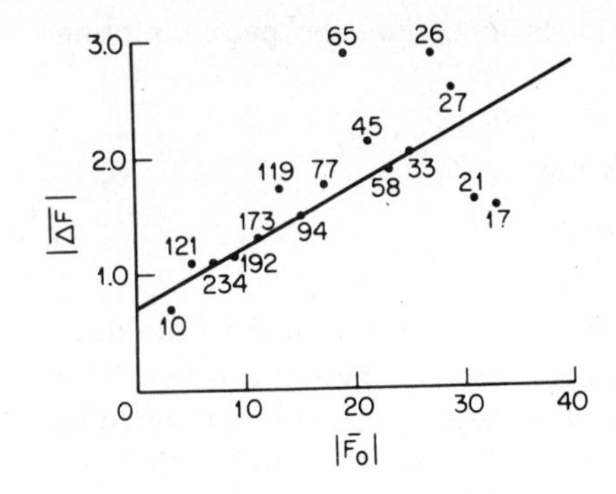

Figure E.1. Plot of $\langle|\Delta F|\rangle$ versus $\langle|F_o|\rangle$ for photographic data. Numbers at points give the number of reflections averaged.

The soundest method of estimating σ is by multiple measurements of reflections as suggested at the outset. Remeasurement of reflections on a given film is unsatisfactory, however, since this involves only the reproducibility of the measuring device. Estimates should be based on measurements made on independent photographic exposures, so that irregularities in development and in the film itself and instability of the X-ray source are included. Estimates of σ obtained in this way should be valid measures of the precision (see Chapter 17) and should lead to reliable weights.

Counter Data

If data have been collected by counter methods, the lower limit to σ is set by counting statistics. It can be shown[5] that the standard deviation in the count due to the random nature of the events is

$$\sigma_{\text{count}} = \sqrt{N} \tag{E.5}$$

where N is the number of counts.

In measuring X-ray reflections, the background count must always be considered. If background has been measured on either side of the reflection for one-half the time used in counting the peak, then the net peak count is

$$N_{\text{pk}} = N_{\text{T}} - N_{\text{bg1}} - N_{\text{bg2}} \tag{E.6}$$

where N_{T} is the total peak count and N_{bg1} and N_{bg2} are the background counts on either side of the peak. The standard deviation in N_{pk} is given by

$$\sigma_{\text{pk}} = (N_{\text{T}} + N_{\text{bg1}} + N_{\text{bg2}})^{1/2} \tag{E.7}$$

[5] E. B. Wilson, Jr., *An Introduction to Scientific Research*, McGraw-Hill, New York, 1952, pp. 191–195.

If the sum of the background times does not equal the peak-counting time, the corresponding equations are

$$N_{pk} = N_T - RN_{bg} \tag{E.8}$$

$$\sigma_{pk} = (N_T + R^2 N_{bg})^{1/2} \tag{E.9}$$

where $R = t_{pk}/(t_{bg1} + t_{bg2})$, the ratio of the peak-counting time to the total background time, and $N_{bg} = N_{bg1} + N_{bg2}$, the total background count. The dependence of σ_{pk} on R^2 can cause great inaccuracy in the measurement of weak reflections if the background is significant and is not counted for enough time to provide a reliable estimate of its value.

Equation (E.7) is the error due only to statistical fluctuations in the counting. An additional error must be included to allow for instrumental instability. This can be determined from the standard reflections, which are measured periodically, and should be in the range 0.005–0.01 for modern instruments. Assuming 0.01 and Eq. (E.7), we have for the standard deviation in I_{rel}

$$\sigma_{I_{rel}} = \sqrt{\sigma_{pk}^2 + (0.01 N_{pk})^2} \tag{E.10}$$

The standard deviation in F can be derived from the equation

$$F = \frac{k}{\sqrt{Lp}} \sqrt{I_{rel}} \tag{E.11}$$

where k is the scale constant and $1/\sqrt{Lp}$ is the Lorentz polarization factor. Differentiating Eq. (E.11) gives

$$dF = \frac{1}{2} \frac{k}{\sqrt{Lp}} \frac{dI_{rel}}{\sqrt{I_{rel}}} \tag{E.12}$$

Dividing both sides by F and substituting from Eq. (E.11) gives

$$\frac{dF}{F} = \frac{1}{2} \frac{dI_{rel}}{I_{rel}} \tag{E.13}$$

Equation (E.13) indicates that, for small errors, the relative error in F is just one-half the relative error in I_{rel}. We now make the approximations $dF = \sigma_F$ and $d_{I_{rel}} = \sigma_{I_{rel}}$. Substituting in Eq. (E.12) results in

$$\sigma_F = \frac{1}{2} \frac{k}{\sqrt{Lp}} \frac{\sigma_{I_{rel}}}{\sqrt{I_{rel}}} \tag{E.14}$$

which in view of Eq. (E.10) can be written

$$\sigma_F = \frac{1}{2} \frac{k}{\sqrt{Lp}} \sqrt{\frac{\sigma_{\text{pk}}^2 + (0.01 N_{\text{pk}})^2}{I_{\text{rel}}}} \tag{E.15}$$

Substituting the value $N_{\text{pk}} \equiv I_{\text{rel}}$ from Eq. (E.8) and σ_{pk} from Eq. (E.9) gives finally

$$\sigma_F = \frac{1}{2} \frac{k}{\sqrt{Lp}} \sqrt{\frac{N_{\text{T}} + R^2 N_{\text{bg}} + (0.01 N_{\text{pk}})^2}{N_{\text{T}} - R N_{\text{bg}}}} \tag{E.16}$$

Figure E.2 shows two plots of $|\langle \Delta F \rangle|$ versus $|\langle F_o \rangle|$ for counter-measured intensities. The first represents data from a small crystal with weak reflections, for which the uncertainties in counting are generally important. The second, on the other hand, shows results from a much larger crystal in

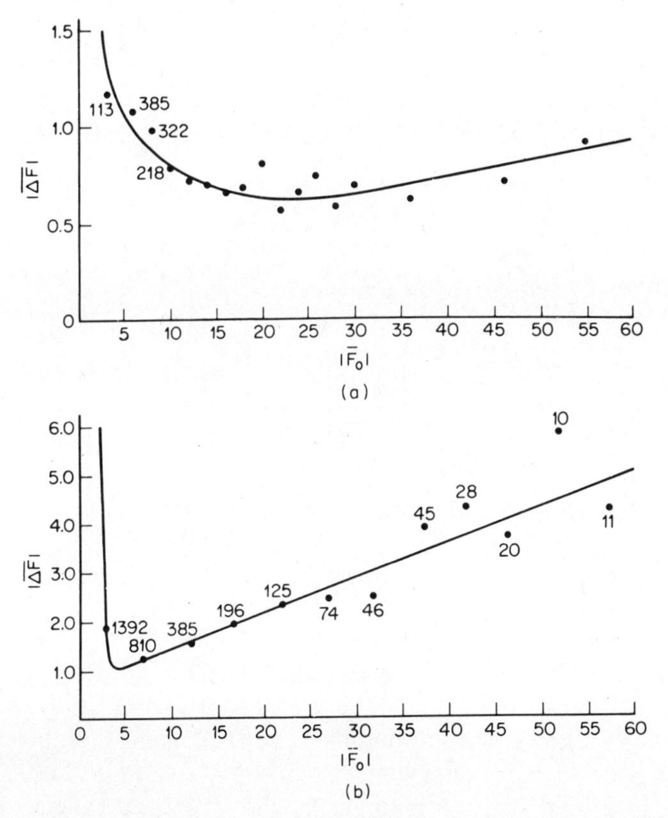

Figure E.2. Plot of $\langle|\Delta F|\rangle$ versus $\langle|F_o|\rangle$ for diffractometer data. Numbers at points give the number of reflections averaged. (*a*) Small crystal; (*b*) large crystal.

which instrumental instabilities predominate. Plots of σ_F versus $|F_o|$ as calculated from Eq. (E.16) for the raw intensity data from these crystals would be very similar. Note that the upturn in $|\Delta F|$ for the weakest reflections, which is due to the unfavorable counting statistics in this region, is absent in the plot from photographic data (Fig. E.1). This occurs because the faintest reflections observed on film result from many times the number of photons counted at ordinary scan rates by the diffractometer. Thus the effects of counting randomness are small compared to other sources of error.

In the above treatment of weights for both photographic and counter data, errors arising from faulty mechanical features of cameras or diffractometers or from misalignment or missetting of the crystal and those arising in the crystal specimen itself have not been explicitly covered. Such errors are likely to be systematic in nature, however (see Chapter 17), and should be eliminated if possible or should be allowed for by correcting the data or including additional parameters in the model. Attempts to account for such errors by selective weighting are often made, but they are unsound in principle.

INDEX

D

E

I

J

K

L

M

Q

R

S

V

W

X

Y

Z